VOLUME SIXTY TWO

PROGRESS IN OPTICS

EDITORIAL ADVISORY BOARD

G.S. Agarwal	*Stillwater, USA*
M.V. Berry	*Bristol, United Kingdom*
C. Brosseau	*Brest, France*
A.T. Friberg	*Stockholm, Sweden*
F. Gori	*Rome, Italy*
D.F.V. James	*Toronto, Canada*
P. Knight	*London, United Kingdom*
G. Leuchs	*Erlangen, Germany*
J.B. Pendry	*London, United Kingdom*
J. Peřina	*Olomouc, Czech Republic*
W. Schleich	*Ulm, Germany*

VOLUME SIXTY TWO

PROGRESS IN OPTICS

Edited by

TACO D. VISSER

VU University Amsterdam, The Netherlands

Contributors
Kasimir Blomstedt, Yangjian Cai, Yahong Chen,
Ari T. Friberg, Lin Liu, Xianlong Liu,
Daniel Malacara-Doblado, Daniel Malacara-Hernández,
Tero Setälä, Tomohiro Shirai, Kurt Bernardo Wolf, Jiayi Yu

Elsevier
Radarweg 29, PO Box 211, 1000 AE Amsterdam, Netherlands
The Boulevard, Langford Lane, Kidlington, Oxford, OX5 1GB, United Kingdom
50 Hampshire Street, 5th Floor, Cambridge, MA 02139, United States

First edition 2017

Copyright © 2017, Elsevier B.V. All rights reserved.

No part of this publication may be reproduced or transmitted in any form or by any means, electronic or mechanical, including photocopying, recording, or any information storage and retrieval system, without permission in writing from the publisher. Details on how to seek permission, further information about the Publisher's permissions policies and our arrangements with organizations such as the Copyright Clearance Center and the Copyright Licensing Agency, can be found at our website: www.elsevier.com/permissions.

This book and the individual contributions contained in it are protected under copyright by the Publisher (other than as may be noted herein).

Notices
Knowledge and best practice in this field are constantly changing. As new research and experience broaden our understanding, changes in research methods, professional practices, or medical treatment may become necessary.

Practitioners and researchers must always rely on their own experience and knowledge in evaluating and using any information, methods, compounds, or experiments described herein. In using such information or methods they should be mindful of their own safety and the safety of others, including parties for whom they have a professional responsibility.

To the fullest extent of the law, neither the Publisher nor the authors, contributors, or editors, assume any liability for any injury and/or damage to persons or property as a matter of products liability, negligence or otherwise, or from any use or operation of any methods, products, instructions, or ideas contained in the material herein.

ISBN: 978-0-12-811999-0
ISSN: 0079-6638

For information on all Elsevier publications
visit our website at https://www.elsevier.com/books-and-journals

Publisher: Zoe Kruze
Acquisition Editor: Poppy Garraway
Editorial Project Manager: Shellie Bryant
Production Project Manager: Vignesh Tamil
Cover Designer: Matthew Limbert

Typeset by SPi Global, India

CONTENTS

Contributors   vii
Preface   ix

1. **Modern Aspects of Intensity Interferometry With Classical Light**   1
   Tomohiro Shirai

   1. Introduction   1
   2. The Hanbury Brown–Twiss Effect With Classical Electromagnetic Beams   3
   3. Ghost Imaging and Diffraction With Classical Light   23
   4. Optical Coherence Tomography Based on Classical Intensity Interferometry   43
   5. Concluding Remarks   62
   Acknowledgments   64
   References   64

2. **Optical Testing and Interferometry**   73
   Daniel Malacara-Hernández and Daniel Malacara-Doblado

   1. Wavefront Representation and Its Characteristics   74
   2. Tests That Measure Wavefront Distortions   89
   3. Tests That Measure Transverse Aberrations   104
   4. Tests That Measure Curvature   120
   5. Interferogram Analysis   128
   6. Phase Shifting Interferometry   132
   7. Testing of Aspherical Surfaces   142
   References   152

3. **Generation of Partially Coherent Beams**   157
   Yangjian Cai, Yahong Chen, Jiayi Yu, Xianlong Liu, and Lin Liu

   1. Introduction   157
   2. Characterization and Generation of Various Partially Coherent Beams   170
   3. Summary   212
   Acknowledgments   213
   References   213

4. Optical Models and Symmetries — 225
Kurt Bernardo Wolf

1. Introduction — 226
2. The Euclidean Group — 228
3. The Fundamental Object of a Model — 230
4. The Geometric Model — 232
5. The Wave and Helmholtz Models — 241
6. Paraxial Models — 245
7. Linear Transformations of Phase Space — 249
8. The Metaxial Regime — 258
9. Discrete Optical Models — 266

Acknowledgments — 282
Appendix A. The $SU(2)$ Wigner Function — 282
References — 288

5. Classical Coherence of Blackbody Radiation — 293
Kasimir Blomstedt, Ari T. Friberg, and Tero Setälä

1. Introduction — 293
2. Blackbody Radiation Inside a Cavity — 298
3. Blackbody Radiation in an Aperture — 317
4. Blackbody Radiation in the Far Zone of an Aperture — 325
5. Summary — 335

Acknowledgments — 336
Appendix A. Blackbody Coherence in Cavities — 336
Appendix B. Derivation of Aperture and Far-Field Mutual Coherence Matrices — 342
References — 344

Author Index for Volume 62 — *347*
Subject Index for Volume 62 — *357*
Cumulative Index-Volumes 1–62 — *365*

CONTRIBUTORS

Kasimir Blomstedt
Institute of Photonics, University of Eastern Finland, Joensuu, Finland

Yangjian Cai
College of Physics, Optoelectronics and Energy & Collaborative Innovation Center of Suzhou Nano Science and Technology; Key Lab of Advanced Optical Manufacturing Technologies of Jiangsu Province & Key Lab of Modern Optical Technologies of Education Ministry of China, Soochow University, Suzhou, China

Yahong Chen
College of Physics, Optoelectronics and Energy & Collaborative Innovation Center of Suzhou Nano Science and Technology; Key Lab of Advanced Optical Manufacturing Technologies of Jiangsu Province & Key Lab of Modern Optical Technologies of Education Ministry of China, Soochow University, Suzhou, China

Ari T. Friberg
Institute of Photonics, University of Eastern Finland, Joensuu, Finland

Lin Liu
College of Physics, Optoelectronics and Energy & Collaborative Innovation Center of Suzhou Nano Science and Technology; Key Lab of Advanced Optical Manufacturing Technologies of Jiangsu Province & Key Lab of Modern Optical Technologies of Education Ministry of China, Soochow University, Suzhou, China

Xianlong Liu
College of Physics, Optoelectronics and Energy & Collaborative Innovation Center of Suzhou Nano Science and Technology; Key Lab of Advanced Optical Manufacturing Technologies of Jiangsu Province & Key Lab of Modern Optical Technologies of Education Ministry of China, Soochow University, Suzhou, China

Daniel Malacara-Doblado
Centro de Investigaciones en Óptica, Colonia Lomas del Campestre, León, Guanajuato, Mexico

Daniel Malacara-Hernández
Centro de Investigaciones en Óptica, Colonia Lomas del Campestre, León, Guanajuato, Mexico

Tero Setälä
Institute of Photonics, University of Eastern Finland, Joensuu, Finland

Tomohiro Shirai
National Institute of Advanced Industrial Science and Technology (AIST), Tsukuba, Japan

Kurt Bernardo Wolf
Instituto de Ciencias Físicas, Universidad Nacional Autónoma de México, Cuernavaca, Morelos, Mexico

Jiayi Yu
College of Physics, Optoelectronics and Energy & Collaborative Innovation Center of Suzhou Nano Science and Technology; Key Lab of Advanced Optical Manufacturing Technologies of Jiangsu Province & Key Lab of Modern Optical Technologies of Education Ministry of China, Soochow University, Suzhou, China

PREFACE

This volume, number 62, contains five chapters describing recent experimental and theoretical progress in modern optics.

The first chapter, written by Tomohiro Shirai, presents an overview of research on classical intensity interferometry, also known as Hanbury Brown-Twiss interferometry. Modern developments such as the generalization from scalar fields to electromagnetic beams, the evolution of correlations on propagation, and classical forms of ghost imaging are discussed.

In Chapter 2, by Daniel Malacara-Hernández and Daniel Malacara-Doblado, a review of optical testing methods is presented. These include geometrical and interferometric techniques to evaluate optical surfaces, wave front deformations, transverse aberrations, local curvatures, and testing of aspheric surfaces.

Chapter 3, by Yangjian Cai, Yahong Chen, Jiayi Yu, Xianlong Liu, and Lin Liu, deals with various kinds of partially coherent beams and their many applications in, for example, particle trapping, and free-space optical communication. Different methods are discussed to generate beams with prescribed properties such as phase, state of polarization, and degree of coherence.

The fourth chapter, contributed by Kurt Bernardo Wolf, is concerned with the recent uses of Lie groups and algebras in optical models. These mathematical tools can be applied successfully in both geometrical and physical optics to describe light in the global, paraxial, and metaxial regimes.

The fifth and final chapter, written by Kasimir Blomstedt, Ari T. Friberg, and Tero Setälä, deals with the coherence properties of blackbody radiation. The classical spectrospatial and spatiotemporal coherence as well as the temporal and spectral polarization of blackbody radiation is analyzed both inside of a cavity and in the far zone.

I hope the reader will find these contributions as informative as I do.

TACO D. VISSER
Amsterdam, February 2017

CHAPTER ONE

Modern Aspects of Intensity Interferometry With Classical Light

Tomohiro Shirai
National Institute of Advanced Industrial Science and Technology (AIST), Tsukuba, Japan

Contents

1. Introduction 1
2. The Hanbury Brown–Twiss Effect With Classical Electromagnetic Beams 3
   2.1 Background 3
   2.2 Formulas for Correlations Between Intensity Fluctuations in Stochastic Electromagnetic Beams 4
   2.3 Properties of the Degree of Cross-Polarization 11
   2.4 Evolution of Correlations Between Intensity Fluctuations on Propagation 13
   2.5 Alternative Approach to the Basic Problem 21
3. Ghost Imaging and Diffraction With Classical Light 23
   3.1 Background 23
   3.2 Methods of Describing Ghost Imaging and Diffraction With Classical Light 25
   3.3 Ghost Imaging of Pure Phase Objects With Classical Light 32
   3.4 Selected Applications of Classical Ghost Imaging and Diffraction 36
4. Optical Coherence Tomography Based on Classical Intensity Interferometry 43
   4.1 Background 43
   4.2 Incorporating Intensity Interferometry Into Optical Coherence Tomography 46
   4.3 Quantum-Mimetic Intensity-Interferometric Optical Coherence Tomography With Dispersion Cancelation 51
5. Concluding Remarks 62
Acknowledgments 64
References 64

1. INTRODUCTION

Intensity interferometry, also referred to as Hanbury Brown–Twiss interferometry after the pioneering work in the 1950s (Hanbury Brown & Twiss, 1954, 1956a), is undoubtedly one of the techniques that revolutionized the development of optics. This significant technique was originally devised in astronomy and provoked controversy immediately afterward with respect to the physics of the phenomenon. Nowadays it is widely known that this technique triggered the emergence of modern quantum optics (Glauber, 2006) and has become a standard tool for measuring the quantum feature of light in various quantum optics experiments. However, it seems to be generally less well appreciated that intensity interferometry still provides new insights into correlations between intensity fluctuations in classical optical fields and finds a wide range of applications within the framework of classical optics. This article is devoted to such recent topics.

As a preliminary, it is worth summarizing very briefly the early attempts to establish intensity interferometry, although they have already been described in a number of publications (Hanbury Brown, 1974; Labeyrie, Lipson, & Nisenson, 2006; Mandel, 1963; Mandel & Wolf, 1965, 1995; Wolf, 2007). The basic idea of intensity interferometry was first conceived to determine the angular diameters of radio stars (Hanbury Brown & Twiss, 1954). The intensity fluctuations of radio waves at two points were measured, and they were correlated by an electronic correlator to obtain the squared modulus of the degree of coherence, from which one could estimate the sizes of radio stars. The principle of intensity interferometry with radio waves was fairly straightforward and commonly accepted. This method was then applied to visible light to determine the angular diameters of visible stars as a successor to Michelson stellar interferometry (Hanbury Brown & Twiss, 1956a, 1956b). However, the principle of the method with light was not quite obvious mainly due to the quantum-mechanical nature of the detection of light, which gave rise to controversy. Hanbury Brown and Twiss demonstrated experimentally the phenomenon now known as the Hanbury Brown–Twiss effect, namely the existence of correlations between the fluctuations in the output photocurrents from two photoelectric detectors (Hanbury Brown & Twiss, 1956a) and succeeded in determining the sizes of visible stars on the basis of such correlation measurements (Hanbury Brown & Twiss, 1956b). On the other hand, some other researchers performed similar experiments and obtained negative results (Ádám, Jánossy, & Varga, 1955; Brannen & Ferguson, 1956). These controversial issues were largely resolved by Purcell (1956) who examined the Hanbury Brown–Twiss effect in terms of a semiclassical theory.

Starting from its use in astronomy, intensity interferometry has recently found useful applications in different branches of physics, such as nuclear physics (Baym, 1998), electron physics (Henny et al., 1999; Kiesel, Renz, & Hasselbach, 2002; Oliver, Kim, Liu, & Yamamoto, 1999), ultracold atom physics (Jeltes et al., 2007; Schellekens et al., 2005; Yasuda & Shimizu, 1996), and atom lasers (Öttl, Ritter, Köhl, & Esslinger, 2005). More recently, the traditional form of intensity interferometry with classical light has received renewed attention in connection with stochastic electromagnetic fields (Hassinen, Tervo, Setälä, & Friberg, 2011; Li, 2014; Shirai & Wolf, 2007; Volkov, James, Shirai, & Wolf, 2008; Wu & Visser, 2014a, 2014b) and, especially, in connection with classical counterparts of quantum imaging (for a review, Gatti, Brambilla, & Lugiato, 2008), namely, ghost imaging and diffraction with classical light (for a review, Erkmen & Shapiro, 2010) and quantum-mimetic optical coherence tomography (for a review, Teich, Saleh, Wong, & Shapiro, 2012). It is interesting to note that there has also been a revival of interest in intensity interferometry in the astronomical community (Borra, 2008; Foellmi, 2009; Jain & Ralston, 2008; Malvimat, Wucknitz, & Saha, 2014; Ofir & Ribak, 2006a, 2006b).

In this article we take an overview of recent findings and up-to-date applications associated with intensity interferometry with classical light. Intensity interferometry as a means of observing the Hanbury Brown–Twiss effect is a rich treasure trove of physics and its application area is still expanding. The topics that this article covers are very limited in this context but are designed to provide new insights into the phenomena associated with classical optics. This article is organized as follows. In Section 2, we discuss recent formulations of the Hanbury Brown–Twiss effect with classical electromagnetic beams as an extension of that with classical scalar beams and manifest some of their consequences. Section 3 focuses on classical counterparts of ghost imaging and diffraction with quantum entangled light, as one of the most striking experiments based on intensity interferometry. Section 4 deals with classical counterparts of quantum optical coherence tomography, as a surely epoch-making application of intensity interferometry. Finally, some concluding remarks are made in Section 5.

2. THE HANBURY BROWN–TWISS EFFECT WITH CLASSICAL ELECTROMAGNETIC BEAMS

2.1 Background

It is known that the phenomenon of correlations between intensity fluctuations at two points in stochastic electromagnetic fields is referred to as the Hanbury Brown–Twiss effect (the HBT effect, for short, below) after Hanbury Brown and Twiss. Most treatments of this effect have been confined to fields which are linearly polarized, and they have been based on a scalar theory of coherence, although limited studies have been devoted to elucidate the effect of partial polarization on the HBT effect (Mandel, 1963; Mandel & Wolf, 1961). In these studies only a few specific expressions were derived for the HBT effect, e.g., with partially coherent unpolarized beams and with fully coherent partially polarized beams. This seems to be mainly because in those days the theory of coherence and that of polarization were developed somewhat independently, and they were not treated in a unified manner, in spite of the fact that these two subjects originate from the same physical phenomenon.

In the early 2000s the unified theory of coherence and polarization of stochastic electromagnetic beams was formulated by Wolf (2003) (see also, Wolf, 2007, Chapter 9), which stimulated us to reexamine the HBT effect with classical electromagnetic beams. In this context, the beam is a special case of general electromagnetic fields in the sense that it propagates close to a specific direction and that the Cartesian components of the field vector along the propagation direction are negligible. This restriction to beams is quite reasonable, since one often encounters various optical beams in practice. With this background we first derive in Section 2.2 a general formula for describing the correlation between intensity fluctuations in classical electromagnetic beams of any state of coherence and polarization (Shirai & Wolf, 2007; Volkov et al., 2008). We then discuss in Section 2.3 some properties of the new quantity, the degree of cross-polarization, introduced in the preceding analysis. Some numerical examples are given in Section 2.4 to illustrate the behavior of the intensity correlation on propagation. We note that there are some other theories formulated in a different way to describe the coherence and polarization of electromagnetic fields (Luis, 2007; Réfrégier & Goudail, 2005; Setälä, Tervo, & Friberg, 2004; Tervo, Setälä, & Friberg, 2003). The related treatment of the problem based on another theory will be described in Section 2.5.

2.2 Formulas for Correlations Between Intensity Fluctuations in Stochastic Electromagnetic Beams

Let us consider a classical stochastic electromagnetic beam which propagates close to the z-direction into the half space $z > 0$, as illustrated in Fig. 1. The fluctuating electric field vector is assumed to be represented by a stationary random process that obeys Gaussian statistics with zero mean. Let $E_x(\boldsymbol{\rho}, z; \omega)$ and $E_y(\boldsymbol{\rho}, z; \omega)$ be the Cartesian components of the electric field vector at frequency ω at a point $(\boldsymbol{\rho}, z)$ in two mutually orthogonal x- and y-directions, perpendicular to the beam axis. The second-order correlation properties of

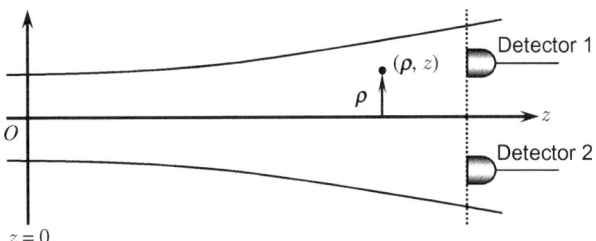

Fig. 1 Illustrating the notation. Two detectors are located in a stochastic electromagnetic beam on a plane perpendicular to the beam axis.

the beam at a pair of points $(\boldsymbol{\rho}_1, z)$ and $(\boldsymbol{\rho}_2, z)$ in a cross-sectional plane $z > 0$ are described by the 2×2 cross-spectral density matrix of the form (Wolf, 2003, 2007, Chapter 9)

$$\mathbf{W}(\boldsymbol{\rho}_1,\boldsymbol{\rho}_2,z;\omega) = \begin{bmatrix} W_{xx}(\boldsymbol{\rho}_1,\boldsymbol{\rho}_2,z;\omega) & W_{xy}(\boldsymbol{\rho}_1,\boldsymbol{\rho}_2,z;\omega) \\ W_{yx}(\boldsymbol{\rho}_1,\boldsymbol{\rho}_2,z;\omega) & W_{yy}(\boldsymbol{\rho}_1,\boldsymbol{\rho}_2,z;\omega) \end{bmatrix}, \quad (1)$$

where

$$W_{ij}(\boldsymbol{\rho}_1,\boldsymbol{\rho}_2,z;\omega) = \langle E_i^*(\boldsymbol{\rho}_1,z;\omega) E_j(\boldsymbol{\rho}_2,z;\omega) \rangle, \quad i,j \in (x,y), \quad (2)$$

the asterisk denotes the complex conjugate, and the angular brackets denote the ensemble average. In this expression, the ensemble is to be understood in the sense of coherence theory in the space–frequency domain (Mandel & Wolf, 1995, Section 4.7; Wolf, 2007, Section 4.1). It is important to note that the 2×2 cross-spectral density matrix defined by Eq. (1) is the space–frequency analog of the 2×2 matrix in the space–time domain, known as the beam coherence-polarization matrix (Gori, 1998; Gori, Santarsiero, Vicalvi, Borghi, & Guattari, 1998).

Using the cross-spectral density matrix given by Eq. (1), the spectral degree of coherence of the electromagnetic beam at two points $(\boldsymbol{\rho}_1, z)$ and $(\boldsymbol{\rho}_2, z)$ is defined by the formula

$$\eta(\boldsymbol{\rho}_1,\boldsymbol{\rho}_2,z;\omega) = \frac{\operatorname{tr} \mathbf{W}(\boldsymbol{\rho}_1,\boldsymbol{\rho}_2,z;\omega)}{\sqrt{\operatorname{tr} \mathbf{W}(\boldsymbol{\rho}_1,\boldsymbol{\rho}_1,z;\omega)}\sqrt{\operatorname{tr} \mathbf{W}(\boldsymbol{\rho}_2,\boldsymbol{\rho}_2,z;\omega)}}, \quad (3)$$

where tr denotes the trace. This definition characterizes quantitatively the visibility of the interference pattern formed by the field at these two points, as a natural generalization of the spectral degree of coherence of the scalar field (Wolf, 2003, 2007, Chapter 9). The absolute value of the spectral degree of coherence is shown to satisfy the relation

$$0 \le |\eta(\boldsymbol{\rho}_1, \boldsymbol{\rho}_2, z; \omega)| \le 1, \tag{4}$$

where the values $|\eta| = 0$ and 1 represent complete spatial incoherence and complete spatial coherence, respectively. The spectral degree of polarization $P(\boldsymbol{\rho}, z; \omega)$, which represents the ratio of the intensity of the polarized portion to the total intensity at frequency ω at a point $(\boldsymbol{\rho}, z)$, is given by the expression

$$P(\boldsymbol{\rho}, z; \omega) = \sqrt{1 - \frac{4 \det \mathbf{W}(\boldsymbol{\rho}, \boldsymbol{\rho}, z; \omega)}{[\operatorname{tr} \mathbf{W}(\boldsymbol{\rho}, \boldsymbol{\rho}, z; \omega)]^2}}, \tag{5}$$

where det denotes the determinant (Wolf, 2007, Chapter 9). One can show that

$$0 \le P(\boldsymbol{\rho}, z; \omega) \le 1, \tag{6}$$

where the value $P = 0$ indicates a completely unpolarized beam, and the value $P = 1$ indicates a completely polarized beam. Note that the matrix $\mathbf{W}(\boldsymbol{\rho}, \boldsymbol{\rho}, z; \omega)$ is the spectral counterpart of the ordinary coherency matrix defined in the space–time domain (Born & Wolf, 1999, Section 10.9).

The fluctuations of the intensity at frequency ω (i.e., spectral density) at a point $(\boldsymbol{\rho}, z)$ are defined by the expression

$$\Delta I(\boldsymbol{\rho}, z; \omega) = I(\boldsymbol{\rho}, z; \omega) - \langle I(\boldsymbol{\rho}, z; \omega) \rangle, \tag{7}$$

where the instantaneous intensity $I(\boldsymbol{\rho}, z; \omega)$ is given by the expression

$$I(\boldsymbol{\rho}, z; \omega) = E_x^*(\boldsymbol{\rho}, z; \omega) E_x(\boldsymbol{\rho}, z; \omega) + E_y^*(\boldsymbol{\rho}, z; \omega) E_y(\boldsymbol{\rho}, z; \omega), \tag{8}$$

and the averaged intensity $\langle I(\boldsymbol{\rho}, z; \omega) \rangle$ is expressible in the form

$$\langle I(\boldsymbol{\rho}, z; \omega) \rangle = \langle E_x^*(\boldsymbol{\rho}, z; \omega) E_x(\boldsymbol{\rho}, z; \omega) \rangle + \langle E_y^*(\boldsymbol{\rho}, z; \omega) E_y(\boldsymbol{\rho}, z; \omega) \rangle$$
$$= \sum_i W_{ii}(\boldsymbol{\rho}, \boldsymbol{\rho}, z; \omega) = \operatorname{tr} \mathbf{W}(\boldsymbol{\rho}, \boldsymbol{\rho}, z; \omega), \quad i \in (x, y). \tag{9}$$

Here, the intensity $I(\boldsymbol{\rho}, z; \omega)$ defined by Eq. (8) is the intensity of a single realization of each field vector component and corresponds to the instantaneous intensity in the space–time domain. In this article we refer to the intensity (in the space–frequency domain) defined in this way as the instantaneous intensity for convenience. It is the correlation between intensity fluctuations at two points $(\boldsymbol{\rho}_1, z)$ and $(\boldsymbol{\rho}_2, z)$ that characterizes the HBT effect. Using the moment theorem for a Gaussian random process, it is given by the expression

$$\langle \Delta I(\boldsymbol{\rho}_1, z; \omega) \Delta I(\boldsymbol{\rho}_2, z; \omega) \rangle = |W_{xx}(\boldsymbol{\rho}_1, \boldsymbol{\rho}_2, z; \omega)|^2 + |W_{xy}(\boldsymbol{\rho}_1, \boldsymbol{\rho}_2, z; \omega)|^2$$
$$+ |W_{yx}(\boldsymbol{\rho}_1, \boldsymbol{\rho}_2, z; \omega)|^2 + |W_{yy}(\boldsymbol{\rho}_1, \boldsymbol{\rho}_2, z; \omega)|^2 \quad (10)$$
$$= \sum_{i,j} |W_{ij}(\boldsymbol{\rho}_1, \boldsymbol{\rho}_2, z; \omega)|^2, \quad i, j \in (x, y),$$

or, equivalently,

$$\langle \Delta I(\boldsymbol{\rho}_1, z; \omega) \Delta I(\boldsymbol{\rho}_2, z; \omega) \rangle = \mathrm{tr}\left[\mathbf{W}^\dagger(\boldsymbol{\rho}_1, \boldsymbol{\rho}_2, z; \omega) \mathbf{W}(\boldsymbol{\rho}_1, \boldsymbol{\rho}_2, z; \omega)\right], \quad (11)$$

where the dagger denotes the Hermitian adjoint (Mandel & Wolf, 1995, Eq. (8.4–18)). Eq. (11) is the fundamental expression for the HBT effect, but it does not provide any insight into the physics of this effect. Hence, we attempt to rearrange this expression to obtain a physically transparent form.

To do this, we first rewrite Eq. (11) in the form

$$\langle \Delta I(\boldsymbol{\rho}_1, z; \omega) \Delta I(\boldsymbol{\rho}_2, z; \omega) \rangle = \left\{ \frac{\mathrm{tr}\left[\mathbf{W}^\dagger(\boldsymbol{\rho}_1, \boldsymbol{\rho}_2, z; \omega) \mathbf{W}(\boldsymbol{\rho}_1, \boldsymbol{\rho}_2, z; \omega)\right]}{|\mathrm{tr}\,\mathbf{W}(\boldsymbol{\rho}_1, \boldsymbol{\rho}_2, z; \omega)|^2} \right\}$$
$$\times \left\{ \frac{|\mathrm{tr}\,\mathbf{W}(\boldsymbol{\rho}_1, \boldsymbol{\rho}_2, z; \omega)|^2}{\mathrm{tr}\,\mathbf{W}(\boldsymbol{\rho}_1, \boldsymbol{\rho}_1, z; \omega)\,\mathrm{tr}\,\mathbf{W}(\boldsymbol{\rho}_2, \boldsymbol{\rho}_2, z; \omega)} \right\}$$
$$\times \langle I(\boldsymbol{\rho}_1, z; \omega) \rangle \langle I(\boldsymbol{\rho}_2, z; \omega) \rangle. \quad (12)$$

It follows from Eq. (3) that the term in the second curly brackets on the right-hand side of Eq. (12) becomes the squared modulus of the spectral degree of coherence, i.e., $|\eta(\boldsymbol{\rho}_1, \boldsymbol{\rho}_2, z; \omega)|^2$. After some rearrangements for the term in the first curly brackets on the right-hand side of Eq. (12), one obtains the expression

$$\langle \Delta I(\boldsymbol{\rho}_1, z; \omega) \Delta I(\boldsymbol{\rho}_2, z; \omega) \rangle = \frac{1}{2}\{1 + [\mathscr{P}(\boldsymbol{\rho}_1, \boldsymbol{\rho}_2, z; \omega)]^2\}$$
$$\times |\eta(\boldsymbol{\rho}_1, \boldsymbol{\rho}_2, z; \omega)|^2 \langle I(\boldsymbol{\rho}_1, z; \omega) \rangle \langle I(\boldsymbol{\rho}_2, z; \omega) \rangle, \quad (13)$$

where the quantity $\mathscr{P}(\boldsymbol{\rho}_1, \boldsymbol{\rho}_2, z; \omega)$ is defined by the formula (Volkov et al., 2008, Eq. (9))

$$\mathscr{P}(\boldsymbol{\rho}_1,\boldsymbol{\rho}_2,z;\omega) = \sqrt{\frac{2\,\mathrm{tr}\left[\mathbf{W}^\dagger(\boldsymbol{\rho}_1,\boldsymbol{\rho}_2,z;\omega)\mathbf{W}(\boldsymbol{\rho}_1,\boldsymbol{\rho}_2,z;\omega)\right]}{|\mathrm{tr}\,\mathbf{W}(\boldsymbol{\rho}_1,\boldsymbol{\rho}_2,z;\omega)|^2} - 1}. \qquad (14)$$

The expression under the square root in Eq. (14) is always real and nonnegative and, thus, the same applies to the quantity $\mathscr{P}(\boldsymbol{\rho}_1,\boldsymbol{\rho}_2,z;\omega)$, as we will show in Section 2.3. When $\boldsymbol{\rho}_1 = \boldsymbol{\rho}_2 = \boldsymbol{\rho}$, the cross-spectral density matrix is evidently Hermitian, i.e., $\mathbf{W}^\dagger(\boldsymbol{\rho},\boldsymbol{\rho},z;\omega) = \mathbf{W}(\boldsymbol{\rho},\boldsymbol{\rho},z;\omega)$, implying that each diagonal element of the matrix is real and that the off-diagonal elements are the complex conjugate of each other. In this case, by making use of the relation $\mathrm{tr}(\mathbf{A}^\dagger\mathbf{A}) = (\mathrm{tr}\mathbf{A})^2 - 2\,\mathrm{det}\mathbf{A}$ which holds for an arbitrary Hermitian matrix $\mathbf{A}$, one readily finds that Eq. (14) reduces exactly to the usual spectral degree of polarization $P(\boldsymbol{\rho},z;\omega)$ given by Eq. (5), i.e.,

$$\mathscr{P}(\boldsymbol{\rho},\boldsymbol{\rho},z;\omega) = P(\boldsymbol{\rho},z;\omega) = \sqrt{1 - \frac{4\,\mathrm{det}\,\mathbf{W}(\boldsymbol{\rho},\boldsymbol{\rho},z;\omega)}{[\mathrm{tr}\,\mathbf{W}(\boldsymbol{\rho},\boldsymbol{\rho},z;\omega)]^2}}. \qquad (15)$$

In general, the cross-spectral density matrix $\mathbf{W}(\boldsymbol{\rho}_1,\boldsymbol{\rho}_2,z;\omega)$ of the beam for a pair of points $(\boldsymbol{\rho}_1, z)$ and $(\boldsymbol{\rho}_2, z)$ is not Hermitian except for the case $\boldsymbol{\rho}_1 = \boldsymbol{\rho}_2$. However, one can show that the matrix is Hermitian, i.e.,

$$\mathbf{W}^\dagger(\boldsymbol{\rho}_1,\boldsymbol{\rho}_2,z;\omega) = \mathbf{W}(\boldsymbol{\rho}_1,\boldsymbol{\rho}_2,z;\omega), \qquad (16)$$

in many situations of practical interests. Specifically, the cross-spectral density matrix in both the near and far zones of an electromagnetic Gaussian Schell-model beam (Li, 2014) or in the far zone of a beam generated by a quasi-homogeneous source (Korotkova, Hoover, Gamiz, & Wolf, 2005) is shown to be symmetric with respect to the interchange of its spatial arguments, i.e.,

$$\mathbf{W}(\boldsymbol{\rho}_2,\boldsymbol{\rho}_1,z;\omega) = \mathbf{W}(\boldsymbol{\rho}_1,\boldsymbol{\rho}_2,z;\omega). \qquad (17)$$

This symmetry condition turns out to be equivalent to the Hermiticity condition given by Eq. (16), if we recall the definition of the cross-spectral density matrix (see Eqs. (1) and (2)). Suppose, for the moment, that the symmetry condition (17) (and thus the Hermiticity condition (16)) is satisfied. Then, in a similar way to the derivation of Eq. (15), we find that the quantity defined by Eq. (14) becomes (Shirai & Wolf, 2007, Eq. (11))

$$\mathscr{P}(\boldsymbol{\rho}_1,\boldsymbol{\rho}_2,z;\omega) = \sqrt{1 - \frac{4\,\mathrm{det}\,\mathbf{W}(\boldsymbol{\rho}_1,\boldsymbol{\rho}_2,z;\omega)}{[\mathrm{tr}\,\mathbf{W}(\boldsymbol{\rho}_1,\boldsymbol{\rho}_2,z;\omega)]^2}}. \qquad (18)$$

Interestingly, Eq. (18) has the same functional form as the usual spectral degree of polarization given by Eq. (5). It is thus easily understood that the quantity given by Eq. (18) reduces to the usual spectral degree of polarization when two points coincide, i.e., $\rho_1 = \rho_2$. It seems, therefore, natural to refer to the quantity $\mathscr{P}(\rho_1,\rho_2,z;\omega)$ defined by Eq. (14) (and Eq. (18)) as the degree of cross-polarization of the beam at two points (ρ_1, z) and (ρ_2, z).

The degree of cross-polarization can be regarded as a two-point generalization of the usual spectral degree of polarization: the (spectral) degree of polarization characterizes correlations between the electric field components at a particular point, while the degree of cross-polarization characterizes correlations in the field components at a pair of points. Further properties of this quantity will be given in Section 2.3.

Eq. (13) is the physically transparent and meaningful formula for the HBT effect. It may be generally taken for granted that properties of stochastic electromagnetic beams are largely described in terms of the degree of coherence and the degree of polarization, together with the averaged intensity, in many practical applications. However, this is not the case for the HBT effect. The formula (13) for the HBT effect depends on the averaged intensity, the degree of coherence, and the degree of cross-polarization in place of the degree of polarization.

To elucidate the implication of the formula (13), let us consider two special cases.

(1) *Linearly polarized beams:* For a beam which is linearly polarized, say, in the x-direction, the cross-spectral density matrix has the form

$$\mathbf{W}(\rho_1,\rho_2,z;\omega) = \begin{bmatrix} W_{xx}(\rho_1,\rho_2,z;\omega) & 0 \\ 0 & 0 \end{bmatrix}. \tag{19}$$

Then, the averaged intensity given by Eq. (9), the spectral degree of coherence given by Eq. (3), and the degree of cross-polarization given by Eq. (14) (and Eq. (18)) become

$$\langle I(\rho_i,z;\omega)\rangle = W_{xx}(\rho_i,\rho_i,z;\omega) \equiv \langle I_x(\rho_i,z;\omega)\rangle, \quad i \in (1,2), \tag{20}$$

$$\eta(\rho_1,\rho_2,z;\omega) = \frac{W_{xx}(\rho_1,\rho_2,z;\omega)}{\sqrt{W_{xx}(\rho_1,\rho_1,z;\omega)}\sqrt{W_{xx}(\rho_2,\rho_2,z;\omega)}} \tag{21}$$

$$\equiv \mu_{xx}(\rho_1,\rho_2,z;\omega)),$$

and

$$\mathscr{P}(\rho_1,\rho_2,z;\omega) = 1. \tag{22}$$

The spectral degree of coherence given by Eq. (21) is now simply the spectral degree of coherence of the scalar beam $E_x(\boldsymbol{\rho}, z; \omega)$. On substituting Eqs. (20)–(22) into the formula (13), we obtain the expression

$$\langle \Delta I(\boldsymbol{\rho}_1, z; \omega) \Delta I(\boldsymbol{\rho}_2, z; \omega) \rangle = \langle I_x(\boldsymbol{\rho}_1, z; \omega) \rangle \langle I_x(\boldsymbol{\rho}_2, z; \omega) \rangle \\ \times |\mu_{xx}(\boldsymbol{\rho}_1, \boldsymbol{\rho}_2, z; \omega)|^2. \qquad (23)$$

This formula is the space–frequency equivalent of the formula derived in the space–time domain by Mandel (1963, Eq. (18)).

(2) *Unpolarized beams:* For an unpolarized beam, the field vector components $E_x(\boldsymbol{\rho}, z; \omega)$ and $E_y(\boldsymbol{\rho}, z; \omega)$ at each point $(\boldsymbol{\rho}, z)$ are uncorrelated and the spectral degree of polarization becomes $P(\boldsymbol{\rho}, z; \omega) = 0$. The cross-spectral density matrix of the beam at coincident points is given by the expression

$$\mathbf{W}(\boldsymbol{\rho}, \boldsymbol{\rho}, z; \omega) = \begin{bmatrix} A(\boldsymbol{\rho}, z; \omega) & 0 \\ 0 & A(\boldsymbol{\rho}, z; \omega) \end{bmatrix}, \qquad (24)$$

when $A(\boldsymbol{\rho}, z; \omega)$ is an arbitrary positive function (Wolf, 2007, Chapter 8). Although the cross-spectral density matrix $\mathbf{W}(\boldsymbol{\rho}_1, \boldsymbol{\rho}_2, z; \omega)$ satisfying Eq. (24) may assume various forms (Visser, Kuebel, Lahiri, Shirai, & Wolf, 2009), one possible form may be

$$\mathbf{W}(\boldsymbol{\rho}_1, \boldsymbol{\rho}_2, z; \omega) = \begin{bmatrix} B(\boldsymbol{\rho}_1, \boldsymbol{\rho}_2, z; \omega) & 0 \\ 0 & B(\boldsymbol{\rho}_1, \boldsymbol{\rho}_2, z; \omega) \end{bmatrix}, \qquad (25)$$

where $B(\boldsymbol{\rho}_1, \boldsymbol{\rho}_2, z; \omega)$ is an arbitrary function satisfying the relation $B(\boldsymbol{\rho}, \boldsymbol{\rho}, z; \omega) = A(\boldsymbol{\rho}, z; \omega)$. Then, the degree of cross-polarization given by Eq. (14) (and Eq. (18)) becomes

$$\mathscr{P}(\boldsymbol{\rho}_1, \boldsymbol{\rho}_2, z; \omega) = 0, \qquad (26)$$

and the spectral degree of coherence given by Eq. (3) is written in the form

$$\eta(\boldsymbol{\rho}_1, \boldsymbol{\rho}_2, z; \omega) = \frac{W_{ii}(\boldsymbol{\rho}_1, \boldsymbol{\rho}_2, z; \omega)}{\sqrt{W_{ii}(\boldsymbol{\rho}_1, \boldsymbol{\rho}_1, z; \omega)} \sqrt{W_{ii}(\boldsymbol{\rho}_2, \boldsymbol{\rho}_2, z; \omega)}} \qquad (27)$$

$$\equiv \mu_{ii}(\boldsymbol{\rho}_1, \boldsymbol{\rho}_2, z; \omega)), \quad i \in (x, y),$$

since $W_{xx}(\boldsymbol{\rho}_1, \boldsymbol{\rho}_2, z; \omega) = W_{yy}(\boldsymbol{\rho}_1, \boldsymbol{\rho}_2, z; \omega)$ according to Eq. (25). This expression again represents the spectral degree of coherence of the scalar beams $E_x(\boldsymbol{\rho}, z; \omega)$ and $E_y(\boldsymbol{\rho}, z; \omega)$. On substituting Eqs. (26) and (27) into the formula (13), we obtain the expression

$$\langle \Delta I(\boldsymbol{\rho}_1,z;\omega)\Delta I(\boldsymbol{\rho}_2,z;\omega)\rangle = \frac{1}{2}\langle I(\boldsymbol{\rho}_1,z;\omega)\rangle\langle I(\boldsymbol{\rho}_2,z;\omega)\rangle \\ \times |\mu_{ii}(\boldsymbol{\rho}_1,\boldsymbol{\rho}_2,z;\omega)|^2, \quad i\in(x,y). \tag{28}$$

This formula is the space–frequency equivalent of the formula derived in the space–time domain by Mandel (1963, Eq. (31b)).

2.3 Properties of the Degree of Cross-Polarization

The degree of cross-polarization was originally introduced to describe the HBT effect with stochastic electromagnetic beams under the symmetry condition given by Eq. (17) (Shirai & Wolf, 2007, Eq. (11)) and later generalized to relax this condition (Volkov et al., 2008, Eq. (9)). This quantity is defined by Eq. (14) (or Eq. (18) under the symmetry condition). It follows from this definition that the degree of cross-polarization can be determined from the knowledge of the cross-spectral density matrix $\mathbf{W}(\boldsymbol{\rho}_1, \boldsymbol{\rho}_2, z; \omega)$. This quantity can also be determined from experiments, since the elements of the cross-spectral density matrix are measurable (Roychowdhury & Wolf, 2003).

In order to find further properties of the degree of cross-polarization, let us express this quantity in terms of the generalized Stokes parameters (Korotkova & Wolf, 2005). These parameters are a two-point generalization of the conventional (spectral) Stokes parameters and are defined by the expressions

$$S_0(\boldsymbol{\rho}_1,\boldsymbol{\rho}_2,z;\omega) = W_{xx}(\boldsymbol{\rho}_1,\boldsymbol{\rho}_2,z;\omega) + W_{yy}(\boldsymbol{\rho}_1,\boldsymbol{\rho}_2,z;\omega) \\ = \operatorname{tr} \mathbf{W}(\boldsymbol{\rho}_1,\boldsymbol{\rho}_2,z;\omega), \tag{29}$$

$$S_1(\boldsymbol{\rho}_1,\boldsymbol{\rho}_2,z;\omega) = W_{xx}(\boldsymbol{\rho}_1,\boldsymbol{\rho}_2,z;\omega) - W_{yy}(\boldsymbol{\rho}_1,\boldsymbol{\rho}_2,z;\omega), \tag{30}$$

$$S_2(\boldsymbol{\rho}_1,\boldsymbol{\rho}_2,z;\omega) = W_{xy}(\boldsymbol{\rho}_1,\boldsymbol{\rho}_2,z;\omega) + W_{yx}(\boldsymbol{\rho}_1,\boldsymbol{\rho}_2,z;\omega), \tag{31}$$

$$S_3(\boldsymbol{\rho}_1,\boldsymbol{\rho}_2,z;\omega) = i[W_{yx}(\boldsymbol{\rho}_1,\boldsymbol{\rho}_2,z;\omega) - W_{xy}(\boldsymbol{\rho}_1,\boldsymbol{\rho}_2,z;\omega)]. \tag{32}$$

Using these parameters, the cross-spectral density matrix may be written in the form (see, e.g., Mandel & Wolf, 1995, Eq. (6.2–43b))

$$\mathbf{W}(\boldsymbol{\rho}_1,\boldsymbol{\rho}_2,z;\omega) = \frac{1}{2}\sum_{n=0}^{3} S_n(\boldsymbol{\rho}_1,\boldsymbol{\rho}_2,z;\omega)\boldsymbol{\sigma}_n, \tag{33}$$

where $\boldsymbol{\sigma}_0$ is the 2×2 unit matrix and $\boldsymbol{\sigma}_n$, $n \in (1, 2, 3)$, are the Pauli spin matrices, viz.,

$$\boldsymbol{\sigma}_1 = \begin{bmatrix} 1 & 0 \\ 0 & -1 \end{bmatrix}, \quad \boldsymbol{\sigma}_2 = \begin{bmatrix} 0 & 1 \\ 1 & 0 \end{bmatrix}, \quad \boldsymbol{\sigma}_3 = \begin{bmatrix} 0 & i \\ -i & 0 \end{bmatrix}. \tag{34}$$

These matrices satisfy the relation

$$\text{tr}(\boldsymbol{\sigma}_m \boldsymbol{\sigma}_n) = 2\delta_{mn}, \quad m, n \in (0, 1, 2, 3), \tag{35}$$

where δ_{mn} denotes the Kronecker symbol. Using the relations (33) and (35), Eq. (11) becomes

$$\text{tr}\left[\mathbf{W}^\dagger(\boldsymbol{\rho}_1, \boldsymbol{\rho}_2, z; \omega) \mathbf{W}(\boldsymbol{\rho}_1, \boldsymbol{\rho}_2, z; \omega)\right] = \frac{1}{2} \sum_{n=0}^{3} |S_n(\boldsymbol{\rho}_1, \boldsymbol{\rho}_2, z; \omega)|^2. \tag{36}$$

On substituting Eqs. (29) and (36) into Eq. (14), we obtain for the degree of cross-polarization the expression

$$\mathscr{P}(\boldsymbol{\rho}_1, \boldsymbol{\rho}_2, z; \omega) = \frac{\sqrt{|S_1(\boldsymbol{\rho}_1, \boldsymbol{\rho}_2, z; \omega)|^2 + |S_2(\boldsymbol{\rho}_1, \boldsymbol{\rho}_2, z; \omega)|^2 + |S_3(\boldsymbol{\rho}_1, \boldsymbol{\rho}_2, z; \omega)|^2}}{|S_0(\boldsymbol{\rho}_1, \boldsymbol{\rho}_2, z; \omega)|}. \tag{37}$$

It immediately follows from Eq. (37) that the degree of cross-polarization is real and nonnegative. It is to be noted that the quantity $\mathscr{P}(\boldsymbol{\rho}_1, \boldsymbol{\rho}_2, z; \omega)$ can take on any nonnegative value, i.e., $0 \leq \mathscr{P}(\boldsymbol{\rho}_1, \boldsymbol{\rho}_2, z; \omega) < \infty$. To see this, we consider two extreme cases.

On substituting Eqs. (30)–(32) into the numerator on the right-hand side of Eq. (37), we find that $\mathscr{P}(\boldsymbol{\rho}_1, \boldsymbol{\rho}_2, z; \omega) = 0$ if, and only if,

$$W_{xx}(\boldsymbol{\rho}_1, \boldsymbol{\rho}_2, z; \omega) = W_{yy}(\boldsymbol{\rho}_1, \boldsymbol{\rho}_2, z; \omega) \tag{38}$$

and

$$W_{xy}(\boldsymbol{\rho}_1, \boldsymbol{\rho}_2, z; \omega) = W_{yx}(\boldsymbol{\rho}_1, \boldsymbol{\rho}_2, z; \omega) = 0. \tag{39}$$

This condition for the components of the cross-spectral density matrix has already been discussed in Section 2.2 to describe unpolarized beams (see Eq. (25)). On the other hand, it is expected that $\mathscr{P}(\boldsymbol{\rho}_1, \boldsymbol{\rho}_2, z; \omega) \to \infty$ when the denominator on the right-hand side of Eq. (37) is zero, i.e., $|S_0(\boldsymbol{\rho}_1, \boldsymbol{\rho}_2, z; \omega)| = 0$. Now, according to Eqs. (3) and (29), the spectral degree of coherence of the stochastic electromagnetic beam can be expressed, in terms of one of the generalized Stokes parameters, by the expression

$$\eta(\boldsymbol{\rho}_1,\boldsymbol{\rho}_2,z;\omega) = \frac{S_0(\boldsymbol{\rho}_1,\boldsymbol{\rho}_2,z;\omega)}{\sqrt{S_0(\boldsymbol{\rho}_1,\boldsymbol{\rho}_1,z;\omega)}\sqrt{S_0(\boldsymbol{\rho}_2,\boldsymbol{\rho}_2,z;\omega)}}. \qquad (40)$$

We thus find that the condition $|S_0(\boldsymbol{\rho}_1,\boldsymbol{\rho}_2,z;\omega)|=0$ is essentially equivalent to $|\eta(\boldsymbol{\rho}_1,\boldsymbol{\rho}_2,z;\omega)|=0$, implying that the beam is completely incoherent. Therefore, it turns out that the degree of cross-polarization at the points $(\boldsymbol{\rho}_1,z)$ and $(\boldsymbol{\rho}_2,z)$ is infinite if, and only if, the beam is completely incoherent at these two points. Note that if the numerator on the right-hand side of Eq. (37) is zero in addition to the condition $|S_0(\boldsymbol{\rho}_1,\boldsymbol{\rho}_2,z;\omega)|=0$ (i.e., the denominator on the right-hand side of Eq. (37) is also zero), all the components of the cross-spectral density matrix are zero. Obviously, the cross-spectral density matrix of this kind does not characterize any beams in reality.

When $\boldsymbol{\rho}_1=\boldsymbol{\rho}_2=\boldsymbol{\rho}$, Eq. (37) becomes the familiar formula for the conventional spectral degree of polarization $P(\boldsymbol{\rho},z;\omega)\equiv\mathscr{P}(\boldsymbol{\rho},\boldsymbol{\rho},z;\omega)$ in terms of the spectral Stokes parameters $s_n(\boldsymbol{\rho},z;\omega)\equiv S_n(\boldsymbol{\rho},\boldsymbol{\rho},z;\omega)$, $n \in (0,1,2,3)$, namely,

$$P(\boldsymbol{\rho},z;\omega) = \frac{\sqrt{|s_1(\boldsymbol{\rho},z;\omega)|^2+|s_2(\boldsymbol{\rho},z;\omega)|^2+|s_3(\boldsymbol{\rho},z;\omega)|^2}}{|s_0(\boldsymbol{\rho},z;\omega)|}. \qquad (41)$$

We have shown in Section 2.2 that the degree of cross-polarization has the same functional forms as the usual spectral degree of polarization under the symmetry condition (see Eq. (18)). However, these two formulas (37) and (41) clearly reveal that there is the remarkable similarity between the degree of cross-polarization and the degree of polarization even under the general condition, if these quantities are expressed in terms of the generalized and conventional Stokes parameters, respectively.

Some other properties justifying the argument that the degree of cross-polarization is a natural generalization of the degree of polarization from a one-point quantity to a two-point quantity are derived by Kuebel (2009), who examined the degree of cross-polarization in the space–time domain using the concept of the statistical similarity (Roychowdhury & Wolf, 2005).

2.4 Evolution of Correlations Between Intensity Fluctuations on Propagation

The propagation of stochastic beams, whether they are scalar or vector (electromagnetic), has been the subject of great interest over many years (for a review, Gbur & Visser, 2010, Section 4). So far, the spectrum, the (spectral) degree of coherence, the (spectral) degree of polarization, and the state of polarization of the beams have been extensively explored, and it is now widely known that these quantities may change on propagation, even in free space. However, it is only quite recently that the evolution of the correlation between intensity fluctuations, which characterizes the HBT effect, in the stochastic electromagnetic beams on propagation has been examined in some detail (Al-Qasimi, Lahiri, Kuebel, James, & Wolf, 2010; Li, 2014; Wu & Visser, 2014a, 2014b). In this subsection, we outline some of the important results with illustrative examples, mainly on the basis of the analyses by Wu and Visser (2014b) and Al-Qasimi et al. (2010).

Let us consider again a stochastic, statistically stationary, electromagnetic beam propagating close to the z-direction from the plane $z = 0$ into the half space $z > 0$. Suppose that the beam is generated by an electromagnetic Gaussian Schell-model (GSM) source located in the plane $z = 0$. The beam of this kind, which is referred to as the electromagnetic GSM beam (Gori et al., 2001; Korotkova, Salem, & Wolf, 2004), is mathematically the most tractable and has played an important role in a wide range of studies relating to the statistical properties of electromagnetic beams (Gori, Ramírez-Sánchez, Santarsiero, & Shirai, 2009; Liu, Huang, Chen, Guo, & Cai, 2015; Liu, Wang, Zhang, & Cai, 2015; Shirai, 2005; Shirai & Wolf, 2004; Zhang & Zhao, 2015).

For the electromagnetic GSM source, the elements of the cross-spectral density matrix have the form

$$W_{ij}(\boldsymbol{\rho}_1,\boldsymbol{\rho}_2,0;\omega) = \sqrt{\langle I_i(\boldsymbol{\rho}_1,0;\omega)\rangle}\sqrt{\langle I_j(\boldsymbol{\rho}_2,0;\omega)\rangle}\,\mu_{ij}(\boldsymbol{\rho}_2-\boldsymbol{\rho}_1,0;\omega), \quad (42)$$

where

$$\langle I_i(\boldsymbol{\rho},0;\omega)\rangle \equiv W_{ii}(\boldsymbol{\rho},\boldsymbol{\rho},0;\omega) = A_i^2 \exp\left(-\frac{\rho^2}{2\sigma_i^2}\right), \quad (43)$$

is the averaged intensity of the component $E_i(\boldsymbol{\rho},\ 0;\ \omega)$ of the field at frequency ω in the source plane $z = 0$ and

$$\mu_{ij}(\boldsymbol{\rho}_2-\boldsymbol{\rho}_1,0;\omega) = B_{ij}\exp\left[-\frac{(\boldsymbol{\rho}_2-\boldsymbol{\rho}_1)^2}{2\delta_{ij}^2}\right], \quad i,j \in (x,y), \quad (44)$$

is the spectral degree of *correlation* between the components $E_i(\boldsymbol{\rho}_1,\ 0;\ \omega)$ and $E_j(\boldsymbol{\rho}_2,\ 0;\ \omega)$. The quantities σ_i and δ_{ij} represent the *rms* width of the averaged intensity $\langle I_i(\boldsymbol{\rho}, 0; \omega)\rangle$ and that of the correlation $\mu_{ij}(\boldsymbol{\rho}_2 - \boldsymbol{\rho}_1, 0; \omega)$, respectively. Parameters A_i, B_{ij}, σ_i, and δ_{ij} are independent of position, but in general depend on frequency. We note that all the parameters but A_i are not independent of each other. From the definition of the cross-spectral density matrix, it follows that these parameters must satisfy the relations (Wolf, 2007, Section 9.4.2)

$$\begin{cases} B_{ij} = 1 & \text{when } i = j, \\ |B_{ij}| \leq 1 & \text{when } i \neq j, \\ B_{ij} = B_{ji}^*, \\ \delta_{ij} = \delta_{ji}, \quad i, j \in (x, y). \end{cases} \quad (45)$$

Moreover, the nonnegative definiteness of the cross-spectral density matrix and the beam condition in the sense that the source generates a beam yield, respectively, the constraints

$$\max\{\delta_{xx}, \delta_{yy}\} \leq \delta_{xy} \leq \min\left\{\frac{\delta_{xx}}{\sqrt{|B_{xy}|}}, \frac{\delta_{yy}}{\sqrt{|B_{xy}|}}\right\} \quad (46)$$

(Gori, Santarsiero, Borghi, & Ramírez-Sánchez, 2008; Roychowdhury & Korotkova, 2005) and

$$\frac{1}{4\sigma_x^2} + \frac{1}{\delta_{xx}^2} \ll \frac{2\pi^2}{\lambda^2}, \quad \frac{1}{4\sigma_y^2} + \frac{1}{\delta_{yy}^2} \ll \frac{2\pi^2}{\lambda^2}, \quad (47)$$

where $\lambda = 2\pi c/\omega$ is the wavelength with c being the speed of light in vacuum (Korotkova et al., 2004; Shirai, Korotkova, & Wolf, 2005). From now on, we assume that the *rms* widths of the averaged intensities $\langle I_x(\boldsymbol{\rho}, 0; \omega)\rangle$ and $\langle I_y(\boldsymbol{\rho}, 0; \omega)\rangle$ of the x and y components of the field in the source plane $z = 0$ are equivalent, i.e., $\sigma_x = \sigma_y \equiv \sigma$.

The elements of the cross-spectral density matrix in a cross-sectional plane $z > 0$ are determined, in terms of those in the source plane $z = 0$, by the paraxial propagation formula (Wolf, 2007, Section 9.4.1)

$$W_{ij}(\boldsymbol{\rho}_1, \boldsymbol{\rho}_2, z; \omega) = \left(\frac{k}{2\pi z}\right)^2 \iint W_{ij}(\boldsymbol{\rho}_1', \boldsymbol{\rho}_2', 0; \omega)$$

$$\times \exp\left\{-i\frac{k}{2z}\left[(\boldsymbol{\rho}_1 - \boldsymbol{\rho}_1')^2 - (\boldsymbol{\rho}_2 - \boldsymbol{\rho}_2')^2\right]\right\} d^2\rho_1' d^2\rho_2', \quad (48)$$

where $k = \omega/c$ is the wave number. On substituting Eq. (42) (along with Eqs. (43) and (44)) into Eq. (48), we obtain the expressions

$$W_{ij}(\boldsymbol{\rho}_1,\boldsymbol{\rho}_2,z;\omega) = \frac{A_i A_j B_{ij}}{\Delta_{ij}^2(z)} \exp\left[-\frac{(\boldsymbol{\rho}_1+\boldsymbol{\rho}_2)^2}{8\sigma^2 \Delta_{ij}^2(z)}\right] \\ \times \exp\left[-\frac{(\boldsymbol{\rho}_2-\boldsymbol{\rho}_1)^2}{2\Omega_{ij}^2 \Delta_{ij}^2(z)} + \frac{ik(\rho_2^2-\rho_1^2)}{2R_{ij}(z)}\right], \quad (49)$$

where

$$\Delta_{ij}(z) = \sqrt{1 + \left(\frac{z}{k\sigma\Omega_{ij}}\right)^2}, \quad (50)$$

$$\frac{1}{\Omega_{ij}^2} = \frac{1}{4\sigma^2} + \frac{1}{\delta_{ij}^2}, \quad (51)$$

and

$$R_{ij}(z) = \left[1 + \left(\frac{k\sigma\Omega_{ij}}{z}\right)^2\right]z, \quad i,j \in (x,y). \quad (52)$$

By making use of the formula (49), we can examine the statistical properties of the electromagnetic GSM beam on propagation, including the evolution of the correlation between intensity fluctuations in the beam. In the rest of this subsection, we focus on two specific problems.

As the first problem, we compare the evolution of the correlation between intensity fluctuations with that of the spectral degree of coherence (Wu & Visser, 2014b). To do this, we introduce the normalized quantity for the correlation between intensity fluctuations as follows:

$$C_N(\boldsymbol{\rho}_1,\boldsymbol{\rho}_2,z;\omega) = \frac{\langle \Delta I(\boldsymbol{\rho}_1,z;\omega) \Delta I(\boldsymbol{\rho}_2,z;\omega) \rangle}{\langle I(\boldsymbol{\rho}_1,z;\omega) \rangle \langle I(\boldsymbol{\rho}_2,z;\omega) \rangle}, \quad (53)$$

where $\langle I(\boldsymbol{\rho}, z;\omega) \rangle$ is the averaged intensity given by Eq. (9). It can be shown that the value of $C_N(\boldsymbol{\rho}_1, \boldsymbol{\rho}_2, z; \omega)$ is bounded by zero and unity (Hassinen et al., 2011). We have derived some different forms for the correlation between intensity fluctuations in Section 2.2. In the present problem, it is convenient to employ the expression (10), since all the elements of the cross-spectral density matrix are given explicitly by the formula (49). Substituting Eqs. (9) and (10) into Eq. (53), we find that

$$C_N(\boldsymbol{\rho}_1,\boldsymbol{\rho}_2,z;\omega) = \frac{\sum_{i,j}|W_{ij}(\boldsymbol{\rho}_1,\boldsymbol{\rho}_2,z;\omega)|^2}{\sum_{i}W_{ii}(\boldsymbol{\rho}_1,\boldsymbol{\rho}_1,z;\omega)\sum_{i}W_{ii}(\boldsymbol{\rho}_2,\boldsymbol{\rho}_2,z;\omega)}, \quad (54)$$

$i,j \in (x,y).$

On the other hand, using Eqs. (1) and (3), the spectral degree of coherence of the beam becomes

$$\eta(\boldsymbol{\rho}_1,\boldsymbol{\rho}_2,z;\omega) = \frac{\sum_{i}W_{ii}(\boldsymbol{\rho}_1,\boldsymbol{\rho}_2,z;\omega)}{\sqrt{\sum_{i}W_{ii}(\boldsymbol{\rho}_1,\boldsymbol{\rho}_1,z;\omega)}\sqrt{\sum_{i}W_{ii}(\boldsymbol{\rho}_2,\boldsymbol{\rho}_2,z;\omega)}}, \quad (55)$$

$i,j \in (x,y).$

On the basis of the expressions (54) and (55), together with the formula (49), we are now in a position to examine the evolution properties of the second-order (i.e., the spectral degree of coherence) and the fourth-order (i.e., the correlation between intensity fluctuations) correlations of the electromagnetic GSM beam on propagation in free space.

We begin by considering the asymptotic behavior of these quantities after a long propagation. It is not difficult to show that, in this limit, the normalized correlation between intensity fluctuations and the spectral degree of coherence are given, respectively, by the expressions[1]

$$\lim_{z \to \infty} C_N(\boldsymbol{\rho}_1,\boldsymbol{\rho}_2,z;\omega) = \frac{\sum_{i,j}A_i^2 A_j^2 |B_{ij}|^2 \Omega_{ij}^4}{(A_x^2 \Omega_{xx}^2 + A_y^2 \Omega_{yy}^2)^2}, \quad i,j \in (x,y), \quad (56)$$

and

$$\lim_{z \to \infty} \eta(\boldsymbol{\rho}_1,\boldsymbol{\rho}_2,z;\omega) = 1. \quad (57)$$

These expressions show that the spectral degree of coherence is unity in this limit, as expected, whereas the normalized correlation between intensity fluctuations is not necessarily unity and may take on various values (but, between zero and unity, as mentioned before), depending on the parameters of the beam. However, Eq. (56) gives the maximum value of unity, if the beam is linearly polarized. These asymptotic values are independent of the points $\boldsymbol{\rho}_1$ and $\boldsymbol{\rho}_2$.

---

[1] These expressions were originally derived for all pairs of points **0** and $\boldsymbol{\rho}_2$ by Wu and Visser (2014b), but they hold for all pairs of arbitrary points $\boldsymbol{\rho}_1$ and $\boldsymbol{\rho}_2$.

In order to show some numerical examples, we assume $\boldsymbol{\rho}_1 = \mathbf{0}$ in accordance with the treatment adopted by Wu and Visser (2014b). Under this assumption, we examine the correlation properties of the electromagnetic GSM beams at two points $(\mathbf{0}, z)$ and $(\boldsymbol{\rho}_2, z)$ in a cross-sectional plane $z > 0$. Fig. 2 shows the contour maps of the normalized correlation between intensity fluctuations, $C_N(\mathbf{0}, \boldsymbol{\rho}_2, z; \omega)$, and the modulus of the spectral degree of coherence, $|\eta(\mathbf{0}, \boldsymbol{\rho}_2, z; \omega)|$, plotted against the propagation distance z and the position $\rho_2 = |\boldsymbol{\rho}_2|$. Comparison between Figs. 2A and B reveals that the modulus of the spectral degree of coherence increases monotonically on propagation for each value of the position ρ_2, while the normalized correlation between intensity fluctuations varies in more complicated way. These behaviors are clearly elucidated by the curves plotted as a function of the propagation distance z for some selected values of the position ρ_2, as illustrated in Fig. 3.

The second problem is to demonstrate that two electromagnetic GSM sources with the same intensities at frequency ω, the same spectral degrees of coherence, and the same spectral degrees of polarization may have different degrees of cross-polarization and, furthermore, that such sources may generate beams with the different correlation between intensity fluctuations at a pair of points in a cross-sectional plane $z > 0$ (Al-Qasimi et al., 2010).

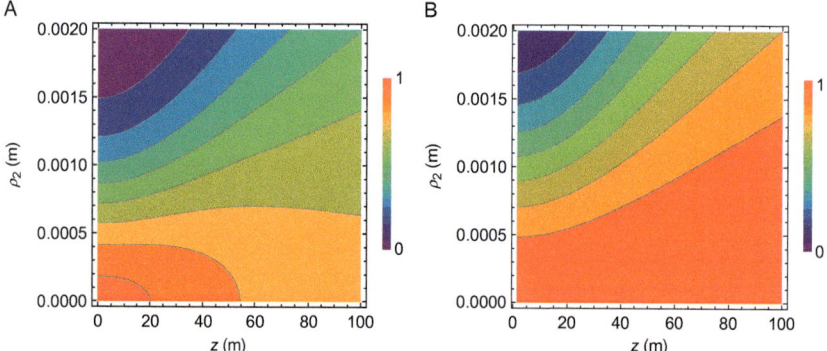

Fig. 2 Contours of (A) the normalized correlation between intensity fluctuations $C_N(\mathbf{0}, \rho_2, z; \omega)$ and (B) the modulus of the spectral degree of coherence $|\eta(\mathbf{0}, \rho_2, z; \omega)|$ plotted against the propagation distance z and the position $\rho_2 = |\boldsymbol{\rho}_2|$ under the conditions that $\lambda = 0.6328$ μm, $\sigma = 4$ mm, $A_x = 1$, $A_y = 3$, $|B_{xy}| = 0.2$, $\delta_{xy} = 3$ mm, $\delta_{xx} = 2.7$ mm, $\delta_{yy} = 1$ mm. *Adapted with permission from Wu, G., & Visser, T. D. (2014). Hanbury Brown–Twiss effect with partially coherent electromagnetic beams. Optics Letters, 39, 2561–2564. https://doi.org/10.1364/OL.39.002561. ©2014 Optical Society of America.*

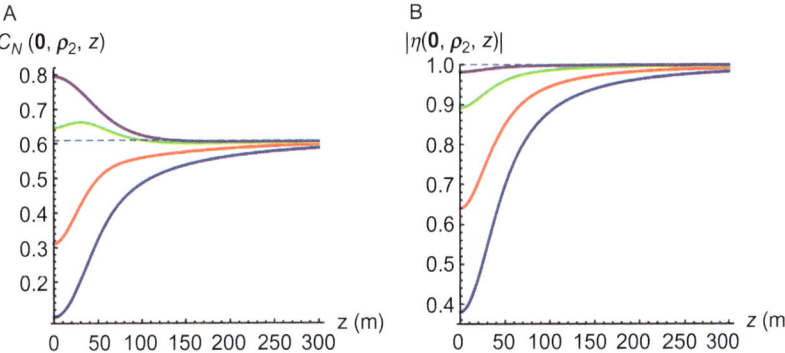

Fig. 3 Evolution of (A) the normalized correlation between intensity fluctuations $C_N(\mathbf{0},\boldsymbol{\rho}_2,z;\omega)$ and (B) the modulus of the spectral degree of coherence $|\eta(\mathbf{0},\boldsymbol{\rho}_2,z;\omega)|$ plotted against the propagation distance z for some selected values of the position $\rho_2 = |\boldsymbol{\rho}_2|$ under the same condition as in Fig. 2. From *bottom* to *top*: $\rho_2 = 1.5$ mm (*blue*), 1 mm (*red*), 0.5 mm (*green*), and 0.2 mm (*purple*). The *dashed lines* are the asymptotic values given by Eqs. (56) and (57). Reproduced with permission from Wu, G., & Visser, T. D. (2014). Hanbury Brown–Twiss effect with partially coherent electromagnetic beams. Optics Letters, 39, 2561–2564. https://doi.org/10.1364/OL.39.002561. ©2014 Optical Society of America.

Let us consider two electromagnetic GSM sources, labeled by α and β. The source α is assumed to be characterized by parameters $A_x = A_y = 1$, $B_{xy} = B_{yx} = 9/16$, $\delta_{xx} = \delta_{yy} = \delta$, and $\delta_{xy} = \delta_{yx} = (4/3)\delta$. The cross-spectral density matrix of the source α at the plane $z = 0$ is then given by the expression

$$\mathbf{W}^{(\alpha)}(\boldsymbol{\rho}_1,\boldsymbol{\rho}_2,0;\omega) = \exp\left(-\frac{\rho_1^2+\rho_2^2}{4\sigma^2}\right)\begin{bmatrix} \mathscr{A} & \mathscr{B} \\ \mathscr{B} & \mathscr{A} \end{bmatrix}, \tag{58}$$

where

$$\mathscr{A} = \exp\left[-\frac{(\boldsymbol{\rho}_2-\boldsymbol{\rho}_1)^2}{2\delta^2}\right] \tag{59}$$

and

$$\mathscr{B} = \frac{9}{16}\exp\left[-\frac{9(\boldsymbol{\rho}_2-\boldsymbol{\rho}_1)^2}{32\delta^2}\right]. \tag{60}$$

For the source β, we assume that $A_x = A_y = 1$, $B_{xy} = B_{yx} = 9/16$, and $\delta_{xx} = \delta_{yy} = \delta_{xy} = \delta_{yx} = \delta$. Then, the cross-spectral density matrix of the source β at the plane $z = 0$ has the form

$$\mathbf{W}^{(\beta)}(\boldsymbol{\rho}_1,\boldsymbol{\rho}_2,0;\omega)=\exp\left(-\frac{\rho_1^2+\rho_2^2}{4\sigma^2}\right)\begin{bmatrix}\mathscr{A}&\mathscr{C}\\\mathscr{C}&\mathscr{A}\end{bmatrix}, \tag{61}$$

where $\mathscr{A}$ is given by Eq. (59) and

$$\mathscr{C}=\frac{9}{16}\exp\left[-\frac{(\boldsymbol{\rho}_2-\boldsymbol{\rho}_1)^2}{2\delta^2}\right]. \tag{62}$$

Using the definitions of each quantity (Eqs. (3), (5), and (9)) and these cross-spectral density matrices (Eqs. (58) and (61)), we can readily show that both sources have the same averaged intensities at frequency ω, viz.,

$$\langle I(\boldsymbol{\rho},0;\omega)\rangle^{(\alpha)}=\langle I(\boldsymbol{\rho},0;\omega)\rangle^{(\beta)}=2\exp\left(-\frac{\rho^2}{2\sigma^2}\right), \tag{63}$$

the same spectral degrees of coherence, viz.,

$$\eta^{(\alpha)}(\boldsymbol{\rho}_1,\boldsymbol{\rho}_2,0;\omega)=\eta^{(\beta)}(\boldsymbol{\rho}_1,\boldsymbol{\rho}_2,0;\omega)=\exp\left[-\frac{(\boldsymbol{\rho}_2-\boldsymbol{\rho}_1)^2}{2\delta^2}\right], \tag{64}$$

and the same spectral degree of polarization, viz.,

$$P^{(\alpha)}(\boldsymbol{\rho},0;\omega)=P^{(\beta)}(\boldsymbol{\rho},0;\omega)=\frac{9}{16}. \tag{65}$$

On the other hand, the degree of cross-polarization defined by Eq. (14) becomes

$$\mathscr{P}^{(\alpha)}(\boldsymbol{\rho}_1,\boldsymbol{\rho}_2,0;\omega)=\frac{|\mathscr{B}|}{|\mathscr{A}|}=\frac{9}{16}\exp\left[\frac{7(\boldsymbol{\rho}_2-\boldsymbol{\rho}_1)^2}{32\delta^2}\right] \tag{66}$$

for the source α and

$$\mathscr{P}^{(\beta)}(\boldsymbol{\rho}_1,\boldsymbol{\rho}_2,0;\omega)=\frac{|\mathscr{C}|}{|\mathscr{A}|}=\frac{9}{16} \tag{67}$$

for the source β. These two sources have obviously different degrees of cross-polarization, i.e., $\mathscr{P}^{(\alpha)}(\boldsymbol{\rho}_1,\boldsymbol{\rho}_2,0;\omega)\neq\mathscr{P}^{(\beta)}(\boldsymbol{\rho}_1,\boldsymbol{\rho}_2,0;\omega)$, unless $\boldsymbol{\rho}_1=\boldsymbol{\rho}_2$.

Now, we examine the correlation between intensity fluctuations in two beams generated by these two sources. For simplicity, we assume $\boldsymbol{\rho}_1=\boldsymbol{\rho}$, $\boldsymbol{\rho}_2=-\boldsymbol{\rho}$ in accordance with Al-Qasimi et al. (2010), namely, that the correlation is evaluated at a pair of diametrically opposite points in a cross-sectional plane $z > 0$. From the formulas (10) and (49), it follows that the correlations between intensity fluctuations are given by the expressions

$$\langle \Delta I(\boldsymbol{\rho}, z; \omega) \Delta I(-\boldsymbol{\rho}, z; \omega) \rangle^{(\alpha)} = \frac{2}{\Delta^4(z)} \exp\left[-\frac{4\rho^2}{\Omega^2 \Delta^2(z)}\right] + \frac{81}{128 \Delta'^4(z)} \exp\left[-\frac{4\rho^2}{\Omega'^2 \Delta'^2(z)}\right] \quad (68)$$

for the source α and

$$\langle \Delta I(\boldsymbol{\rho}, z; \omega) \Delta I(-\boldsymbol{\rho}, z; \omega) \rangle^{(\beta)} = \frac{337}{128 \Delta^4(z)} \exp\left[-\frac{4\rho^2}{\Omega^2 \Delta^2(z)}\right] \quad (69)$$

for the source β, where

$$\Delta(z) = \sqrt{1 + \left(\frac{z}{k\sigma\Omega}\right)^2}, \quad \frac{1}{\Omega^2} = \frac{1}{4\sigma^2} + \frac{1}{\delta^2} \quad (70)$$

and

$$\Delta'(z) = \sqrt{1 + \left(\frac{z}{k\sigma\Omega'}\right)^2}, \quad \frac{1}{\Omega'^2} = \frac{1}{4\sigma^2} + \frac{9}{16\delta^2}. \quad (71)$$

From these expressions, one finds that two beams generated by the sources α and β have the different correlations between intensity fluctuations. These results imply that the intensity at frequency ω, the spectral degree of coherence, and the spectral degree of polarization of the source are not sufficient to determine the correlation between intensity fluctuations in the beam that the source generates. The degree of cross-polarization of a field affects, in general, the correlation between intensity fluctuations in an electromagnetic beam.

2.5 Alternative Approach to the Basic Problem

Let us go back to the basic problem dealt with in Section 2.2. We have derived several different expressions for the correlation between intensity fluctuations in a stochastic electromagnetic beam. One of them is the formula (13) expressed in terms of the spectral degree of coherence, the degree of cross-polarization, and the averaged intensity at frequency ω. In this formula, the spectral degree of coherence of the electromagnetic beam, defined by Eq. (3), quantifies the sharpness of the interference pattern formed in Young's interference experiment, as well as the definition of the spectral degree of coherence of the scalar field (Wolf, 2003, 2007, Chapter 9).

Note that the formula (13) has been derived on the basis of the spectral degree of coherence of this kind and, as a natural consequence of this approach, the degree of cross-polarization has been introduced in this formula.

As we mentioned earlier, on the other hand, some other theories of coherence and polarization of electromagnetic fields have also been put forward in the 2000s, where different definitions of the degree of coherence of the electromagnetic fields have been proposed (Luis, 2007; Réfrégier & Goudail, 2005; Setälä et al., 2004; Tervo et al., 2003). It is thus inferred that other formulations may be possible in the same problem if one relies on those different theoretical frameworks. In what follows we show briefly that this is indeed possible (Hassinen et al., 2011).

As an alternative approach to the same problem as in Section 2.2, we employ the theoretical framework established by Tervo et al. (2003) (see also, for a recent review, Friberg & Setälä, 2016). The degree of coherence of the electromagnetic beam was first proposed in the space–time domain (Tervo et al., 2003) and extended to the space–frequency domain shortly afterward (Setälä et al., 2004). Their definition of the spectral degree of coherence of the electromagnetic beam is given by the expression

$$\eta_A^2(\rho_1,\rho_2,z;\omega) = \frac{\mathrm{tr}[\mathbf{W}(\rho_1,\rho_2,z;\omega)\mathbf{W}(\rho_2,\rho_1,z;\omega)]}{\mathrm{tr}\,\mathbf{W}(\rho_1,\rho_1,z;\omega)\,\mathrm{tr}\,\mathbf{W}(\rho_2,\rho_2,z;\omega)}. \qquad (72)$$

It is shown elsewhere that this formula satisfies the relation $0 \leq \eta_A(\rho_1, \rho_2, z; \omega) \leq 1$. The definition given by Eq. (72) is interpreted as a measure of the strength of electric field correlations at a pair of points, as opposed to the definition given by Eq. (3). Therefore, this formula takes on the limiting values zero when no correlation exists between any of the field components and unity when all the field components are fully correlated. To put it more physically, the spectral degree of coherence defined by Eq. (72) characterizes both the visibility of the interference fringe pattern and the modulation due to the state of polarization in a two-pinhole interference experiment (Friberg & Setälä, 2016, Section 5).

Using the relation $\mathbf{W}(\rho_1, \rho_2, z; \omega) = \mathbf{W}^\dagger(\rho_2, \rho_1, z; \omega)$ which is a direct consequence of the definition of the cross-spectral density matrix and $\mathrm{tr}(AB) = \mathrm{tr}(BA)$ which holds for arbitrary square matrices A and B, one finds that the numerator on the right-hand side of Eq (72) becomes $\mathrm{tr}[\mathbf{W}^\dagger(\rho_1,\rho_2,z;\omega)\mathbf{W}(\rho_1,\rho_2,z;\omega)]$. This expression is equivalent to the correlation between intensity fluctuations given by Eq. (11). It thus follows from Eqs. (9), (11), and (72) that

$$\langle \Delta I(\boldsymbol{\rho}_1, z; \omega) \Delta I(\boldsymbol{\rho}_2, z; \omega) \rangle = \mathrm{tr}\left[\mathbf{W}^\dagger(\boldsymbol{\rho}_1, \boldsymbol{\rho}_2, z; \omega) \mathbf{W}(\boldsymbol{\rho}_1, \boldsymbol{\rho}_2, z; \omega)\right]$$
$$= \langle I(\boldsymbol{\rho}_1, z; \omega) \rangle \langle I(\boldsymbol{\rho}_2, z; \omega) \rangle \, \eta_A^2(\boldsymbol{\rho}_1, \boldsymbol{\rho}_2, z; \omega). \tag{73}$$

The correlation between intensity fluctuations in the electromagnetic beam is now expressed in terms of the spectral degree of coherence defined by Eq. (72) and the averaged intensity at frequency ω, exactly as in the scalar theory (e.g., Wolf, 2007, Chapter 7). This is in contrast to the formula (13) established in Section 2.2. The difference in the formulas for the correlation between intensity fluctuations is, of course, due to the different definition of the spectral degree of coherence of the electromagnetic beam.

3. GHOST IMAGING AND DIFFRACTION WITH CLASSICAL LIGHT

3.1 Background

Since the advent of intensity interferometry, the phenomenon of the intensity correlation which is interpreted as photon coincidence counting in the low-flux limit has been employed in various quantum optics experiments. One of the most striking experiments among them is ghost imaging and diffraction demonstrated experimentally for the first time in 1995 (Pittman, Shih, Strekalov, & Sergienko, 1995; Strekalov, Sergienko, Klyshko, & Shih, 1995).

Ghost imaging is a novel technique for obtaining an image of an object by measuring the correlation between two beams (for a review, Erkmen & Shapiro, 2010). These two beams propagate along two different paths, called the test arm and the reference arm. The object is located in the test arm. The beam after passing through the object is detected by a bucket detector with no spatial resolution, while the other beam in the reference arm is simply detected by a point-like detector. In this setup, neither detector alone can produce an image of the object. Nevertheless, the intensity correlation measurements (the photon coincidence counting in the low-flux limit) based on these two detectors produce the image of the object by scanning transversely the point-like detector located in the reference arm. Ghost diffraction is a similar technique to obtain the diffraction pattern of the object in much the same way. The term "ghost" stems from the unconventional feature that the spatial information associated with the object is retrieved even though the beam detected by the scanning point-like detector in the reference arm never interacts with the object.

The first experimental demonstrations of ghost imaging and diffraction utilized entangled photon pairs generated by spontaneous parametric down-conversion, and hence the phenomenon was naturally ascribed to the quantum entanglement of photons (see also Abouraddy, Saleh, Sergienko, & Teich, 2001). However, it was shown experimentally that ghost imaging can be realized with classical, mutually correlated beams (Bennink, Bentley, & Boyd, 2002). Soon after this experimental work, the possibility of ghost imaging with classical thermal light was examined theoretically (Gatti, Brambilla, Bache, & Lugiato, 2004a, 2004b), and it was verified experimentally by use of pseudothermal light generated by a combination of a rotating ground glass plate and a laser beam (Ferri et al., 2005; Gatti et al., 2006). It was also shown that the phenomenon of ghost imaging and diffraction can be described within the framework of classical coherence theory (Cai & Zhu, 2004, 2005; Cheng & Han, 2004). Note that, although those experimental studies demonstrated that quantum entanglement was not necessary for ghost imaging, there was a similar experimental study leading to a different conclusion (Valencia, Scarcelli, D'Angelo, & Shih, 2005). Furthermore, Scarcelli, Berardi, and Shih (2006) claimed in a subsequent study that ghost imaging together with the HBT effect is a two-photon correlation phenomenon that has to be described quantum mechanically, regardless of whether the source is classical or quantum, giving rise to a lively argument concerning the true physics of ghost imaging and the HBT effect (Gatti, Bondani, Lugiato, Paris, & Fabre, 2007; Meyers, Deacon, & Shih, 2008; Scarcelli, Berardi, & Shih, 2007).

This hot dispute seemed to be settled by the analysis based on the Gaussian states of the quantized electromagnetic field (Erkmen & Shapiro, 2008) and also by the unified theoretical treatment applied to ghost imaging and the HBT effect (Wang, Qamar, Zhu, & Zubairy, 2009). In particular, the proposal of computational ghost imaging played a decisive role in demonstrating that ghost imaging with pseudothermal light can be fully described within the framework of classical optics (Shapiro, 2008), where a rotating ground glass plate for generating pseudothermal light was replaced with a computer-controlled spatial light modulator and the intensity correlation was measured between the object beam captured by a detector and the reference beam calculated in a computer. In fact, it turns out that computational ghost imaging is essentially equivalent to the single-pixel imaging conceived in the field of signal processing (Duarte et al., 2008; Sun et al., 2013). The single-pixel imaging is nothing but a kind of signal processing and evidently independent of any quantum phenomenon. It is now widely appreciated that ghost imaging with pseudothermal light cannot be regarded as a quantum effect, whereas ghost imaging with entangled photons generated by spontaneous parametric downconversion is a manifestation of the quantum effect (for a review, Shapiro & Boyd, 2012).

Ghost imaging and diffraction with classical light have distinct advantages over their quantum counterparts mainly because cost-effective, bright light sources are readily available. This allows us to employ the techniques of ghost imaging and diffraction in some practical applications. In this section, we first provide theoretical methods of describing ghost imaging and diffraction with classical light on the basis of classical coherence theory in the space–frequency domain. We then discuss, as one of the important practical issues in imaging, how one can obtain ghost images of pure phase objects with classical light. Some selected applications of classical ghost imaging and diffraction are also addressed.

3.2 Methods of Describing Ghost Imaging and Diffraction With Classical Light

Let us consider a basic setup for lensless ghost imaging and diffraction with classical thermal light (Basano & Ottonello, 2006; Chen, Liu, Luo, & Wu, 2009; Scarcelli et al., 2006), as illustrated in Fig. 4. To simplify the analysis, we ignore the vector nature of light and treat the field as a scalar quantity in this section. Scalar theory provides enough information about the phenomenon if the object to be imaged is insensitive to polarization. A beam of light from a classical thermal source is split by a beam splitter (BS) into two beams. The beam in the reference arm is detected, after free space propagation, by a scanning point-like single-pixel detector (say, detector 1), located at a point specified by a two-dimensional vector $\boldsymbol{\rho}_1$. On the other hand, the beam in the test arm is incident on the object and the transmitted light is detected by a fixed single-pixel detector (say, detector 2), located at a point specified by a two-dimensional vector $\boldsymbol{\rho}_2$. We assume that detector 2 is also point-like for the moment. The effect of the extended active area of detector 2 on the resultant ghost image and diffraction pattern will be examined later on.

In the first experimental demonstration, the photon coincidence counting was performed to produce a ghost image and diffraction pattern (Pittman et al., 1995; Strekalov et al., 1995). In the present configuration with bright classical light, the intensity correlation (precisely, the correlation

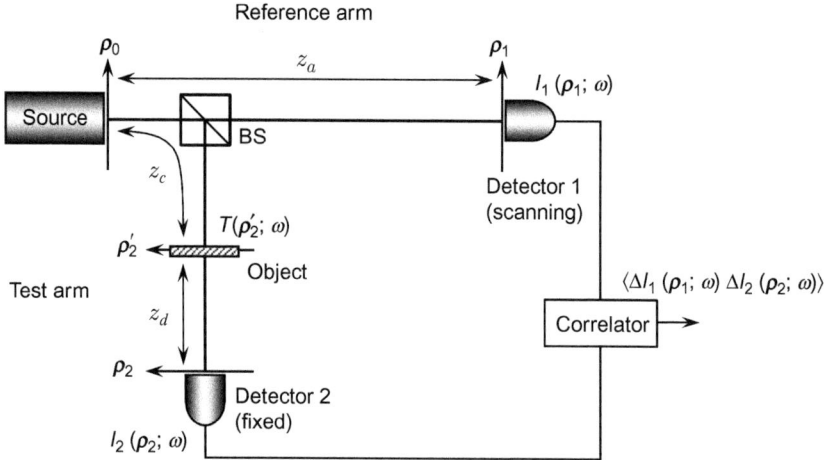

Fig. 4 Basic setup for lensless ghost imaging and diffraction with classical thermal light. Spatially incoherent light from the source is split into two arms, called the test and the reference arms. An object with amplitude transmittance T is located in the test arm. The instantaneous intensities are measured by two point-like single-pixel detectors located in each arm and the correlation between intensity fluctuations at these two detectors is examined. BS, beam splitter. *Adapted with permission from Shirai, T., Setälä, T., & Friberg, A. T. (2011). Ghost imaging of phase objects with classical incoherent light. Physical Review A, 84, 041801(R). https://doi.org/10.1103/PhysRevA.84.041801. ©2011 American Physical Society.*

between photocurrents generated by each detector) should be employed, instead of the photon coincidence counting. Using intensity fluctuations $\Delta I_i(\boldsymbol{\rho}_i; \omega) = I_i(\boldsymbol{\rho}_i; \omega) - \langle I_i(\boldsymbol{\rho}_i; \omega) \rangle$, $i \in (1, 2)$, the intensity correlation is expressible in the form

$$\langle I_1(\boldsymbol{\rho}_1; \omega) I_2(\boldsymbol{\rho}_2; \omega) \rangle = \langle I_1(\boldsymbol{\rho}_1; \omega) \rangle \langle I_2(\boldsymbol{\rho}_2; \omega) \rangle + \langle \Delta I_1(\boldsymbol{\rho}_1; \omega) \Delta I_2(\boldsymbol{\rho}_2; \omega) \rangle, \quad (74)$$

where $I_1(\boldsymbol{\rho}_1;\omega)$ and $I_2(\boldsymbol{\rho}_2;\omega)$ are the instantaneous intensities at a point $\boldsymbol{\rho}_1$ in the reference arm and at a point $\boldsymbol{\rho}_2$ in the test arm, respectively. The angular brackets denote the ensemble average, as before. The first term on the right-hand side of Eq. (74) is a product of two averaged intensities and represents a featureless background that reduces the visibility of the resultant ghost image and diffraction pattern, while the second term is responsible for the formation of the ghost image and diffraction pattern. If we assume that the field fluctuation obeys Gaussian statistics with zero mean as is often the case, the second term on the right-hand side of Eq. (74), i.e., the correlation between intensity fluctuations at points $\boldsymbol{\rho}_1$ and $\boldsymbol{\rho}_2$, is given by the expression (see Eq. (10) in Section 2.2)

$$\langle \Delta I_1(\boldsymbol{\rho}_1;\omega)\Delta I_2(\boldsymbol{\rho}_2;\omega)\rangle = |W_{12}(\boldsymbol{\rho}_1,\boldsymbol{\rho}_2;\omega)|^2, \tag{75}$$

where

$$W_{12}(\boldsymbol{\rho}_1,\boldsymbol{\rho}_2;\omega) = \langle U_1^*(\boldsymbol{\rho}_1;\omega)U_2(\boldsymbol{\rho}_2;\omega)\rangle \tag{76}$$

is the cross-spectral density function between the fields $U_1(\boldsymbol{\rho}_1;\omega)$ at a point $\boldsymbol{\rho}_1$ (at detector 1) in the reference arm and $U_2(\boldsymbol{\rho}_2;\omega)$ at a point $\boldsymbol{\rho}_2$ (at detector 2) in the test arm.

In terms of the cross-spectral density function $W_s(\boldsymbol{\rho}_0,\boldsymbol{\rho}_0';\omega)$ of the source, the cross-spectral density function $W_{12}(\boldsymbol{\rho}_1,\boldsymbol{\rho}_2;\omega)$ between the two fields at detectors 1 and 2 is determined by the propagation formula

$$\begin{aligned}W_{12}(\boldsymbol{\rho}_1,\boldsymbol{\rho}_2;\omega) = \iint & W_s(\boldsymbol{\rho}_0,\boldsymbol{\rho}_0';\omega) \\ & \times K_1^*(\boldsymbol{\rho}_1,\boldsymbol{\rho}_0;\omega)K_2(\boldsymbol{\rho}_2,\boldsymbol{\rho}_0';\omega)\,\mathrm{d}^2\rho_0\mathrm{d}^2\rho_0',\end{aligned} \tag{77}$$

where $K_1(\boldsymbol{\rho}_1,\boldsymbol{\rho}_0;\omega)$ and $K_2(\boldsymbol{\rho}_2,\boldsymbol{\rho}_0';\omega)$ are the impulse response functions for the reference arm and the test arm, respectively. Here and hereafter, the range of integrations extends over the whole two-dimensional space unless otherwise specified. The classical thermal source is essentially spatially incoherent and may be characterized by the cross-spectral density function of the form

$$W_s(\boldsymbol{\rho}_0,\boldsymbol{\rho}_0';\omega) = I_0\delta^{(2)}(\boldsymbol{\rho}_0'-\boldsymbol{\rho}_0), \tag{78}$$

where I_0 and $\delta^{(2)}$ denote a positive constant and the two-dimensional Dirac delta function, respectively. On substituting Eq. (78) into Eq. (77), we obtain the expression

$$W_{12}(\boldsymbol{\rho}_1,\boldsymbol{\rho}_2;\omega) = I_0\int K_1^*(\boldsymbol{\rho}_1,\boldsymbol{\rho}_0;\omega)K_2(\boldsymbol{\rho}_2,\boldsymbol{\rho}_0;\omega)\,\mathrm{d}^2\rho_0. \tag{79}$$

Eq. (75), along with Eq. (79), is the basic formula for analyzing ghost imaging and diffraction with classical thermal (incoherent) light.

The impulse response functions $K_1(\boldsymbol{\rho}_1,\boldsymbol{\rho}_0;\omega)$ for the reference arm and $K_2(\boldsymbol{\rho}_2,\boldsymbol{\rho}_0;\omega)$ for the test arm are given, according to the paraxial propagation law, by the expressions

$$K_1(\boldsymbol{\rho}_1,\boldsymbol{\rho}_0;\omega) = -\frac{ik}{2\pi z_a}\exp\left[i\frac{k}{2z_a}(\boldsymbol{\rho}_1-\boldsymbol{\rho}_0)^2\right] \tag{80}$$

and

$$K_2(\boldsymbol{\rho}_2,\boldsymbol{\rho}_0;\omega) = -\frac{k^2}{(2\pi)^2 z_c z_d} \int T(\boldsymbol{\rho}_2';\omega)$$
$$\times \exp\left\{i\frac{k}{2}\left[\frac{(\boldsymbol{\rho}_2'-\boldsymbol{\rho}_0)^2}{z_c} + \frac{(\boldsymbol{\rho}_2-\boldsymbol{\rho}_2')^2}{z_d}\right]\right\} d^2\rho_2', \tag{81}$$

where $k = \omega/c$ is the wave number, $T(\boldsymbol{\rho}_2';\omega)$ is amplitude transmittance of the object, and z_a, z_c, and z_d are positive constants indicating the propagation distances, as specified in Fig. 4. In these expressions, multiplicative phase factors have been omitted since they do not affect the final results. Substitution of Eqs. (80) and (81) into Eq. (79) yields

$$W_{12}(\boldsymbol{\rho}_1,\boldsymbol{\rho}_2;\omega) = -\frac{iI_0 k^3}{(2\pi)^3 z_a z_c z_d} \iint T(\boldsymbol{\rho}_2';\omega) \exp\left[-i\frac{k}{2z_a}(\boldsymbol{\rho}_1-\boldsymbol{\rho}_0)^2\right]$$
$$\times \exp\left\{i\frac{k}{2}\left[\frac{(\boldsymbol{\rho}_2'-\boldsymbol{\rho}_0)^2}{z_c} + \frac{(\boldsymbol{\rho}_2-\boldsymbol{\rho}_2')^2}{z_d}\right]\right\} d^2\rho_0 d^2\rho_2'. \tag{82}$$

Suppose first the condition that $z_a = z_c$. Then, the integration with respect to ρ_0 on the right-hand side of Eq. (82), viz.,

$$F(\boldsymbol{\rho}_1,\boldsymbol{\rho}_2';\omega) = \int \exp\left\{-i\frac{k}{2}\left[\frac{(\boldsymbol{\rho}_1-\boldsymbol{\rho}_0)^2}{z_a} - \frac{(\boldsymbol{\rho}_2'-\boldsymbol{\rho}_0)^2}{z_c}\right]\right\} d^2\rho_0, \tag{83}$$

becomes

$$F(\boldsymbol{\rho}_1,\boldsymbol{\rho}_2';\omega) = \left(\frac{2\pi z_a}{k}\right)^2 \delta^{(2)}(\boldsymbol{\rho}_1-\boldsymbol{\rho}_2') \exp\left[-i\frac{k}{2z_a}(\rho_1^2-\rho_2'^2)\right]. \tag{84}$$

Combining Eqs. (82)–(84) with Eq. (75) and performing the integration with respect to ρ_2', we obtain for the correlation between intensity fluctuations the expression

$$\langle \Delta I_1(\boldsymbol{\rho}_1;\omega)\Delta I_2(\boldsymbol{\rho}_2;\omega)\rangle = \left(\frac{I_0 k}{2\pi z_d}\right)^2 |T(\boldsymbol{\rho}_1;\omega)|^2. \tag{85}$$

This expression indicates that, when the relation $z_a = z_c$ is satisfied, the correlation between intensity fluctuations at points $\boldsymbol{\rho}_1$ and $\boldsymbol{\rho}_2$ generates, by the process of transversely scanning detector 1, the perfect image of the object whose amplitude transmittance is given by T. The quality (resolution) of the image is independent of the distance z_d between the object and detector 2.

The relation $z_a = z_c$ is the ghost imaging condition for the setup depicted in Fig. 4.

Next suppose the condition that $z_a = z_c + z_d$. The integration given by Eq. (83) then becomes

$$F(\boldsymbol{\rho}_1, \boldsymbol{\rho}_2'; \omega) = \frac{2\pi i z_a z_c}{k z_d} \exp\left[-i\frac{k}{2z_d}(\boldsymbol{\rho}_1 - \boldsymbol{\rho}_2')^2\right]. \qquad (86)$$

In a similar way to the derivation of Eq. (85), we obtain the expression

$$\langle \Delta I_1(\boldsymbol{\rho}_1; \omega) \Delta I_2(\boldsymbol{\rho}_2; \omega) \rangle = I_0^2 \left(\frac{k}{2\pi z_d}\right)^4$$

$$\times \left| \int T(\boldsymbol{\rho}_2'; \omega) \exp\left[-i\frac{k}{z_d}(\boldsymbol{\rho}_2 - \boldsymbol{\rho}_1) \cdot \boldsymbol{\rho}_2'\right] d^2\rho_2' \right|^2. \qquad (87)$$

This expression indicates that, when the relation $z_a = z_c + z_d$ is satisfied, the correlation between intensity fluctuations at points $\boldsymbol{\rho}_1$ and $\boldsymbol{\rho}_2$ generates, by the process of transversely scanning one of the two detectors, the squared modulus of the Fourier transform (i.e., the diffraction pattern) of the object whose amplitude transmittance is given by T. The relation $z_a = z_c + z_d$ is the ghost diffraction condition for the setup depicted in Fig. 4.

Let us now consider the case that the point-like detector located in the test arm (i.e., detector 2) is replaced by a bucket detector. The bucket detector works essentially as a single-pixel detector whose active area is not point-like, but extended. Most ghost imaging setups adopted in the previous studies utilized a bucket detector in the test arm and a scanning point-like detector in the reference arm, whereas all ghost diffraction setups proposed hitherto utilized point-like detectors in both arms (one is fixed and the other is scanning). We now examine briefly the effect of the bucket detector located in the test arm on the resultant ghost image and diffraction pattern.

The intensity fluctuations detected by the bucket detector may be written in the form

$$\Delta I_2'(\omega) = \int_D \Delta I_2(\boldsymbol{\rho}_2; \omega) \, d^2\rho_2, \qquad (88)$$

where the range of integration extends over the active area D of the detector. Consequently, the correlation between intensity fluctuations in this modified configuration becomes

$$\langle \Delta I_1(\boldsymbol{\rho}_1;\omega)\Delta I_2'(\omega)\rangle = \left\langle \Delta I_1(\boldsymbol{\rho}_1;\omega)\int_D \Delta I_2(\boldsymbol{\rho}_2;\omega)\,\mathrm{d}^2\rho_2\right\rangle \qquad (89)$$
$$= \int_D \langle \Delta I_1(\boldsymbol{\rho}_1;\omega)\Delta I_2(\boldsymbol{\rho}_2;\omega)\rangle\,\mathrm{d}^2\rho_2.$$

Under the ghost imaging condition, Eq. (89) is written, using Eq. (85), in the form

$$\langle \Delta I_1(\boldsymbol{\rho}_1;\omega)\Delta I_2'(\omega)\rangle = D\left(\frac{I_0 k}{2\pi z_d}\right)^2 |T(\boldsymbol{\rho}_1;\omega)|^2, \qquad (90)$$

where D is the active area of the bucket detector located in the test arm. Eq. (90) clearly shows that the image of the object is obtained even when the bucket detector is employed in place of the point-like detector and that the quality of the image remains unchanged, irrespective of the type of the detector located in the test arm. Furthermore, the distance z_d between the object and detector 2 has no effect on the quality of the resultant ghost image, whether detector 2 is bucket-like or point-like. On the other hand, under the ghost diffraction condition, one readily finds that Eq. (89) with Eq. (87) no longer represents correctly the diffraction pattern (the squared modulus of the Fourier transform) of the object. The diffraction pattern is blurred by the effect of the bucket detector. Ghost diffraction is not achievable exactly when detector 2 is bucket-like, in contrast to ghost imaging.

Up to now, we have focused on the image-bearing term, namely, the correlation between intensity fluctuations given by Eq. (75). According to Eq. (74), the intensity correlation contains necessarily the product of two averaged intensities acting as a background. In order to examine the effect of this background on the resultant ghost image, we define the visibility as

$$V = \frac{[\langle \Delta I_1(\boldsymbol{\rho}_1;\omega)\Delta I_2(\boldsymbol{\rho}_2;\omega)\rangle]_{\max}}{[\langle I_1(\boldsymbol{\rho}_1;\omega)I_2(\boldsymbol{\rho}_2;\omega)\rangle]_{\max}}, \qquad (91)$$

where the symbol $[\cdots]_{\max}$ indicates the maximum of the function within the brackets. This definition represents the maximum of the ratio of ghost image intensity to the total intensity including a background (Gatti et al., 2006, 2004a). The visibility defined in this way is shown to be an important measure, since it is closely related to the signal-to-noise ratio (Erkmen & Shapiro, 2008; Gatti et al., 2006). Using Eqs. (74) and (75), Eq. (91) may be rewritten in the form

$$V = \frac{[|W_{12}(\boldsymbol{\rho}_1,\boldsymbol{\rho}_2;\omega)|^2]_{\max}}{\langle I_1(\boldsymbol{\rho}_1;\omega)\rangle\langle I_2(\boldsymbol{\rho}_2;\omega)\rangle + [|W_{12}(\boldsymbol{\rho}_1,\boldsymbol{\rho}_2;\omega)|^2]_{\max}}$$

$$= \frac{[|W_{12}(\boldsymbol{\rho}_1,\boldsymbol{\rho}_2;\omega)|^2]_{\max}}{W_{11}(\boldsymbol{\rho}_1,\boldsymbol{\rho}_1;\omega)W_{22}(\boldsymbol{\rho}_2,\boldsymbol{\rho}_2;\omega) + [|W_{12}(\boldsymbol{\rho}_1,\boldsymbol{\rho}_2;\omega)|^2]_{\max}}, \quad (92)$$

where

$$W_{ii}(\boldsymbol{\rho}_i,\boldsymbol{\rho}_i;\omega) = \langle U_i^*(\boldsymbol{\rho}_i;\omega)U_i(\boldsymbol{\rho}_i;\omega)\rangle = \langle I_i(\boldsymbol{\rho}_i;\omega)\rangle, \quad i \in (1,2), \quad (93)$$

is the averaged intensity described in terms of the cross-spectral density function. By use of the Schwarz inequality, we obtain the relation

$$|W_{12}(\boldsymbol{\rho}_1,\boldsymbol{\rho}_2;\omega)|^2 \leq W_{11}(\boldsymbol{\rho}_1,\boldsymbol{\rho}_1;\omega)W_{22}(\boldsymbol{\rho}_2,\boldsymbol{\rho}_2;\omega) \quad (94)$$

and, therefore, $[|W_{12}(\boldsymbol{\rho}_1,\boldsymbol{\rho}_2;\omega)|^2]_{\max} = W_{11}(\boldsymbol{\rho}_1,\boldsymbol{\rho}_1;\omega)W_{22}(\boldsymbol{\rho}_2,\boldsymbol{\rho}_2;\omega)$. As a consequence, it turns out that

$$V = \frac{1}{2}. \quad (95)$$

This indicates that the visibility in ghost imaging with classical light does not exceed the value 1/2. Although a polarized beam is shown to be better than an unpolarized one to produce a high-visibility ghost image if one considers the vector nature of light, the maximum attainable value of visibility defined by Eq. (91) again does not exceed 1/2 (Shirai, Kellock, Setälä, & Friberg, 2011). In other words, the background term is unavoidable in ghost imaging with classical light as far as the intensity correlation (i.e., the correlation between total intensities detected by each detector) is examined. Of course, this problem can be circumvented if the background intensity is subtracted from the total intensity to examine the correlation between intensity fluctuations. In contrast, the visibility can approach unity in ghost imaging with quantum entangled light, even when the photon coincidence counting is performed without background subtraction (Gatti et al., 2004a).

Finally, let us make a brief remark on the case when the source is not spatially incoherent, but spatially partially coherent. In this case, the resultant ghost image is generally degraded by the effect of coherence. However, it is shown theoretically using a Gaussian Schell-model source that the quality of the ghost image improves, at the cost of the visibility, with increasing the source size or with decreasing the source coherence (Cai & Zhu, 2004, 2005). Striking a balance between the source size and the source coherence is essential in ghost imaging with classical partially coherent light, since the quality and the visibility are never compatible.

3.3 Ghost Imaging of Pure Phase Objects With Classical Light

Optical imaging is an indispensable tool for modern biomedical research. Biological specimens examined with microscopes are mostly transparent and they must be regarded as phase objects. Optical imaging of phase objects has thus been the subject of great importance over many years (for a recent review, Mir, Bhaduri, Wang, Zhu, & Popescu, 2012).

It is known that, in general, pure phase objects cannot be seen with an ordinary imaging technique. Special techniques are required to visualize them, such as the phase contrast method developed by Zernike (Born & Wolf, 1999, Section 8.6.3). The same applies to ghost imaging. Pure phase objects have not been observed until recently with a framework of classical ghost imaging. Instead, either ghost diffraction (Bache et al., 2006; Borghi, Gori, & Santarsiero, 2006; Zhang et al., 2007) or phase retrieval (Gong & Han, 2010) has been employed to obtain phase information about the object rather indirectly.

Ghost imaging is basically coherent imaging even though the object is illuminated by incoherent thermal light (Gatti et al., 2006). It is thus expected that pure phase objects can be visualized directly if Fourier-plane filtering could be performed somehow with a framework of classical ghost imaging (Saleh & Teich, 2007, Section 4.4B). In this subsection, we show that this is achievable in accordance with a recent study by Shirai, Setälä, and Friberg (2011).

We first consider the conventional setup for lensless ghost imaging and diffraction, as illustrated in Fig. 4. According to the analysis in Section 3.2, the ghost image obtained in this setup is given by Eq. (85) (or, Eq. (90) for the bucket detection). If the object is a pure phase object of amplitude transmittance $T(\boldsymbol{\rho};\omega) = \exp[i\varphi(\boldsymbol{\rho};\omega)]$, one finds that the image of the object does not emerge in this setup since Eq. (85) (or, Eq. (90)) becomes a constant. To solve this problem, we next consider a modified setup for lensless ghost imaging, as illustrated in Fig. 5. This setup differs from the conventional one (depicted in Fig. 4) in that a spatial filter of amplitude transmittance T_1 is located in the reference arm and that a lens of focal length f is placed right behind the object. The object is now characterized by amplitude transmittance T_2, instead of T employed in Section 3.2.

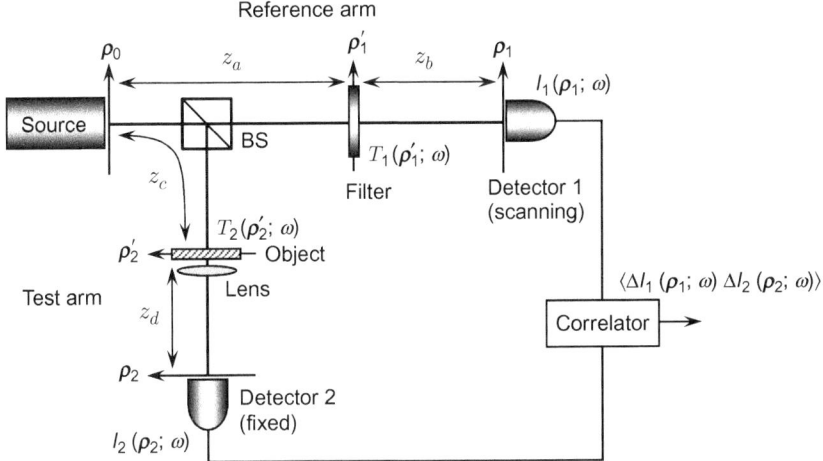

Fig. 5 Setup for lensless ghost imaging of pure phase objects with classical thermal light. This setup is constructed by modifying the basic setup for lensless ghost imaging and diffraction illustrated in Fig. 4 in such a way that a spatial filter (amplitude transmittance T_1) acting as a Fourier-plane filter is located in the reference arm and a lens (focal length f) is placed right behind the object (amplitude transmittance T_2) to be imaged. BS, beam splitter. *Adapted with permission from Shirai, T., Setälä, T., & Friberg, A. T. (2011). Ghost imaging of phase objects with classical incoherent light. Physical Review A, 84, 041801(R). https://doi.org/10.1103/PhysRevA.84.041801. ©2011 American Physical Society.*

The impulse response functions $K_1(\boldsymbol{\rho}_1, \boldsymbol{\rho}_0; \omega)$ for the reference arm and $K_2(\boldsymbol{\rho}_2, \boldsymbol{\rho}_0; \omega)$ for the test arm are determined by means of the paraxial propagation law for the cross-spectral density function. They are given by the expressions

$$K_1(\boldsymbol{\rho}_1, \boldsymbol{\rho}_0; \omega) = -\frac{k^2}{(2\pi)^2 z_a z_b} \int T_1(\boldsymbol{\rho}_1'; \omega)$$
$$\times \exp\left\{i\frac{k}{2}\left[\frac{(\boldsymbol{\rho}_1' - \boldsymbol{\rho}_0)^2}{z_a} + \frac{(\boldsymbol{\rho}_1 - \boldsymbol{\rho}_1')^2}{z_b}\right]\right\} d^2\rho_1', \quad (96)$$

and

$$K_2(\boldsymbol{\rho}_2, \boldsymbol{\rho}_0; \omega) = -\frac{k^2}{(2\pi)^2 z_c z_d} \int T_2(\boldsymbol{\rho}_2'; \omega) \exp\left(-i\frac{k}{2f}\rho_2'^2\right)$$
$$\times \exp\left\{i\frac{k}{2}\left[\frac{(\boldsymbol{\rho}_2' - \boldsymbol{\rho}_0)^2}{z_c} + \frac{(\boldsymbol{\rho}_2 - \boldsymbol{\rho}_2')^2}{z_d}\right]\right\} d^2\rho_2', \quad (97)$$

where z_a, z_b, z_c, and z_d are positive constants indicating the propagation distances, as shown in Fig. 5.

To proceed, we first assume (i) the condition $z_a + z_b = z_c$ which is the ghost imaging condition for the conventional lensless ghost imaging setup depicted in Fig. 4. The formula (79), along with Eqs. (96) and (97), then becomes

$$W_{12}(\boldsymbol{\rho}_1,\boldsymbol{\rho}_2;\omega) = -\frac{iI_0 k^3}{(2\pi)^3 z_b^2 z_d} \iint T_1^*(\boldsymbol{\rho}_1';\omega) T_2(\boldsymbol{\rho}_2';\omega)$$

$$\times \exp\left(-i\frac{k}{2f}\boldsymbol{\rho}_2'^2\right) \exp\left[-i\frac{k}{2z_b}(\boldsymbol{\rho}_1-\boldsymbol{\rho}_1')^2\right] \quad (98)$$

$$\times \exp\left\{i\frac{k}{2}\left[\frac{(\boldsymbol{\rho}_1'-\boldsymbol{\rho}_2')^2}{z_b} + \frac{(\boldsymbol{\rho}_2-\boldsymbol{\rho}_2')^2}{z_d}\right]\right\} d^2\boldsymbol{\rho}_1' d^2\boldsymbol{\rho}_2'.$$

We next assume that (ii) the focal length f of the lens located right behind the object satisfies the relation

$$\frac{1}{f} = \frac{1}{z_b} + \frac{1}{z_d}, \quad (99)$$

and that (iii) the location of detector 2 is fixed at $\boldsymbol{\rho}_2 = \boldsymbol{0}$. If these three conditions (i)–(iii) are all satisfied, the basic formula (75), along with Eq. (79), for ghost imaging and diffraction with classical thermal light is expressible in the form

$$\langle \Delta I_1(\boldsymbol{\rho}_1;\omega) \Delta I_2(\boldsymbol{\rho}_2;\omega) \rangle = A \left| \int T_2(\boldsymbol{\rho}_2';\omega) \mathcal{T}_1^*(\boldsymbol{\rho}_1-\boldsymbol{\rho}_2';\omega) d^2\boldsymbol{\rho}_2' \right|^2, \quad (100)$$

where $A = (k/2\pi)^6 (I_0/z_b^2 z_d)^2$ is a positive constant and

$$\mathcal{T}_1(\boldsymbol{\rho};\omega) = \int T_1(\boldsymbol{\rho}';\omega) \exp\left(-i\frac{k}{z_b}\boldsymbol{\rho}\cdot\boldsymbol{\rho}'\right) d^2\boldsymbol{\rho}' \quad (101)$$

is the Fourier transform of amplitude transmittance T_1 of the spatial filter. Note that the right-hand side of Eq. (100) is the squared modulus of the convolution of two functions T_2 and $\mathcal{T}_1^*$.

Interestingly, Eq. (100) has the same functional form as the expression representing the intensity of the image produced in an imaging system with a Fourier plane. To see this more clearly, let us consider a 4-f coherent imaging system illustrated in Fig. 6. The object of amplitude transmittance $g(\boldsymbol{\rho}_0)$ is

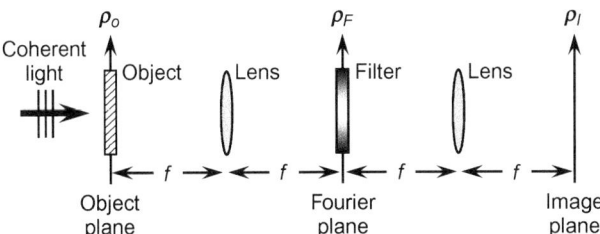

Fig. 6 The 4-f coherent imaging system with a spatial filter in the Fourier plane. The focal length of each lens is f. Reproduced with permission from Shirai, T., Setälä, T., & Friberg, A. T. (2011). Ghost imaging of phase objects with classical incoherent light. Physical Review A, 84, 041801(R). https://doi.org/10.1103/PhysRevA.84.041801. ©2011 American Physical Society.

located in the object plane and illuminated by spatially coherent light. The amplitude distribution in the Fourier plane is then given, aside from unimportant multiplicative factors, by the Fourier transform of the amplitude transmittance of the object, i.e., $G(\boldsymbol{\rho}_F) = \text{FT}[g(\boldsymbol{\rho}_o)]$, where $\text{FT}[\cdot]$ denotes the Fourier transform. Suppose that a spatial filter of amplitude transmittance $H(\boldsymbol{\rho}_F)$ is located in the Fourier plane. Then, the intensity distribution $I_{\text{im}}(\boldsymbol{\rho}_I)$ in the image plane is given by the squared modulus of the Fourier transform of the product $G(\boldsymbol{\rho}_F)H(\boldsymbol{\rho}_F)$, i.e.,

$$I_{\text{im}}(\boldsymbol{\rho}_I) = |\text{FT}[G(\boldsymbol{\rho}_F)H(\boldsymbol{\rho}_F)]|^2. \tag{102}$$

Using the convolution theorem, Eq. (102) is expressible in the form

$$I_{\text{im}}(\boldsymbol{\rho}_I) = |g(-\boldsymbol{\rho}_I) \otimes h(-\boldsymbol{\rho}_I)|^2, \tag{103}$$

where $H(\boldsymbol{\rho}_F) = \text{FT}[h(\boldsymbol{\rho}_o)]$ and the symbol $\otimes$ denotes the convolution operation. Comparison between Eqs. (100) and (103) reveals that these two expressions are essentially equivalent to each other if one regards g as T_2 and h as $\mathcal{T}_1^*$ (equivalently, H as T_1^*).

It follows from the above discussion that the modified setup for lensless ghost imaging illustrated in Fig. 5 has exactly the same function as the 4-f coherent imaging system illustrated in Fig. 6. The spatial filter located in the reference arm in the ghost imaging setup under consideration works as the Fourier-plane filter in the 4-f system. As a consequence, it is possible to control the properties of ghost imaging by an appropriate choice of the spatial filter. In particular, on the basis of this principle, one can observe a pure phase object with the framework of classical ghost imaging.

To prove this, we consider a spatial filter of amplitude transmittance

$$T_1(\boldsymbol{\rho}_1'; \omega) = 1 - [1 - \exp(-ip)]\mathrm{circ}\left(\frac{|\boldsymbol{\rho}_1'|}{\epsilon}\right), \quad (104)$$

where $p = \pi/2$ or $3\pi/2$, ϵ is a small distance from the optical axis, and $\mathrm{circ}(\cdot)$ is the circle function. The spatial filter of this kind yields a phase change of $-p$ over a small circular area close to the axis. A phase object is assumed to be very weak and characterized by amplitude transmittance of the form

$$T_2(\boldsymbol{\rho}_2'; \omega) = \exp\{i[\varphi_0 + \Delta\varphi(\boldsymbol{\rho}_2'; \omega)]\}, \quad (105)$$

where φ_0 is a constant and $|\Delta\varphi(\boldsymbol{\rho}_2'; \omega)| \ll 1$ represents a small phase variation. It is not difficult to show that Eq. (100), together with Eqs. (104) and (105), can be approximated by the expression

$$\langle \Delta I_1(\boldsymbol{\rho}_1; \omega)\Delta I_2(\boldsymbol{\rho}_2; \omega)\rangle \approx \left(\frac{I_0 k}{2\pi z_d}\right)^2 [1 \pm 2\Delta\varphi(\boldsymbol{\rho}_1; \omega)], \quad (106)$$

where the plus ($p = \pi/2$) and minus ($p = 3\pi/2$) signs corresponds to positive and negative contrasts, respectively. This expression clearly shows that the correlation between intensity fluctuations at the output of the ghost imaging setup illustrated in Fig. 5 provides information about the pure phase object. This is a ghost imagining analog of the Zernike phase contrast microscope (Goodman, 2005, Section 8.1.3) and may be referred to as phase contrast ghost imaging (Shirai, Setälä, & Friberg, 2011).

3.4 Selected Applications of Classical Ghost Imaging and Diffraction

3.4.1 Classical Ghost Imaging Through Turbulence

In ghost imaging, whether it is quantum or classical, the incident light that interacts with an object to be imaged is detected by a fixed bucket or point-like detector without spatial resolution. The configuration based on the bucket detector would be useful when an ordinary imager cannot be exploited to capture an image from the very weak light after interacting with the object. The example may be standoff sensing for surveillance as an alternative to a laser radar (for a review, Shapiro & Boyd, 2012, Section 5.1). Since the first demonstration of ghost imaging in the reflective geometry (Meyers et al., 2008), standoff sensing for surveillance appears to be widely recognized as one of the most promising applications of ghost imaging. In this application the bright source for illuminating the object is essential, so that the classical treatment would be especially preferable to the quantum one.

Standoff sensing is expected to be implemented mostly in the open air and, therefore, necessarily undergo atmospheric turbulence. The earliest attempt to analyze the effect of atmospheric turbulence on the resultant ghost image was made by Cheng (2009) on the basis of classical coherence theory. With the help of the extended Huygens–Fresnel principle (Andrews & Phillips, 1998, Section 12.2; Fante, 1985, Section 2.1), one can incorporate the effect of atmospheric turbulence into the impulse response functions for both the object and test arms. In accordance with this method, Cheng showed theoretically that the quality of classical ghost imaging is significantly affected by atmospheric turbulence. Specifically, the resultant ghost image is degraded further with an increase of the turbulence strength, an increase of the propagation distance, and a decrease of the source size. All these trends were confirmed in a similar, but somewhat extended theoretical study by Li, Wang, Pu, Zhu, and Rao (2010). A more detailed analysis was conducted by Hardy and Shapiro (2011), who showed on the basis of the Gaussian-state framework that the effects of turbulence on the quality of the resultant ghost image in the reflective geometry are similar to those derived by Cheng (2009) in the transmissive geometry.

Meanwhile, Meyers, Deacon, and Shih (2011, 2012) reported rather surprising experimental results that the quality of the ghost image was not degraded even though heating elements inducing atmospheric turbulence were inserted in both arms. A pseudothermal source produced by a combination of a rotating ground glass plate and a laser beam was adopted in their experiments. The source was operated at sufficiently low flux and the correlation measurement was performed by use of photon coincidence counting. Meyers et al. explained their experimental results in terms of the quantum two-photon interference. These results were in stark contrast to the preceding theoretical predictions based on classical coherence theory and the Gaussian-state treatment. The effect of turbulence on ghost imaging with entangled photon pairs generated by spontaneous parametric down-conversion was also examined theoretically and experimentally (Chan et al., 2011; Dixon et al., 2011). As a result, it was found that quantum ghost imaging is not essentially immune to turbulence, as opposed to the preceding results by Meyers et al., but the degradation of the resultant ghost image can be minimized by means of appropriate arrangements of the source, the detectors, and turbulence. Further clarifications are necessary on these conflicting findings.

Aside from these controversial aspects of ghost imaging, following an entirely different approach, it was shown theoretically that turbulence-free imaging is possible by means of a modified configuration for conventional ghost diffraction with classical light (Shirai, Kellock, Setälä, & Friberg, 2012).

3.4.2 Classical Ghost Imaging and Diffraction With X Rays

Another example in which the bucket detection in the test arm is effective may be X-ray imaging for medicine and biology. In these applications, radiation exposure must be as small as possible to reduce the damage of the subject and the sample. However, this effort often restricts the achievable resolution limit in such imaging. Ghost imaging with X rays is expected to provide a high-resolution image of the object even when the radiation exposure is very small.

All ghost imaging and diffraction experiments so far performed have utilized visible and infrared light. Very recently, ghost imaging with X rays has been demonstrated experimentally by use of a synchrotron hard X-ray beam that can be regarded as a natural thermal source (Pelliccia, Rack, Scheel, Cantelli, & Paganin, 2016).[2] In the experiment, a thin silicon crystal was employed as a beam splitter working in Laue diffraction. An ultrafast imaging camera was employed to capture both the test and reference beams simultaneously. The bucket detection in the test arm was performed computationally by integrating the intensity of the test beam over a certain area. As a consequence, the ghost image of a copper wire was successfully obtained on the basis of the true intensity correlation extracted from noisy experimental data.

On the other hand, ghost diffraction with X rays has also been demonstrated experimentally (Yu et al., 2016). In contrast to ghost imaging with X rays, it may be difficult to mitigate the radiation damage in this geometry, since the bucket detection is not available there (see Section 3.2). However, ghost diffraction with X rays has some different advantages, as discussed in the early theoretical study (Cheng & Han, 2004).

In recent years, a good deal of attention has been given to phase imaging using a coherent X-ray beam (for recent reviews, Liu, Nelson, Holzner, Andrews, & Pianetta, 2013; Momose, 2005). However, it is still not so easy

---

[2] As another example of ghost imaging *without* (visible and infrared) light, ghost imaging with atoms has also been demonstrated experimentally though it utilizes quantum correlations between atoms (Khakimov et al., 2016).

to utilize a coherent X-ray beam in laboratory experiments. Ghost diffraction is a method of obtaining a diffraction pattern of an object by use of an incoherent beam instead of a coherent beam. Since the phase and the amplitude of the object can be fully retrieved, in principle, from the modulus of its Fourier transform (Fienup, 1982, 1987; McBride, O'Leary, & Allen, 2004; Miao, Sayre, & Chapman, 1998), the principle of ghost diffraction is applicable to phase imaging. Accordingly, ghost diffraction with X rays is expected to work as an alternative to coherent X-ray phase imaging. Furthermore, the whole information about the diffraction pattern can be acquired in this principle, whereas the low-frequency components of the diffraction pattern are always missing due to a beam stop in traditional X-ray diffraction experiments.

In the experimental demonstration by Yu et al. (2016), a pseudothermal X-ray source consisting of a synchrotron hard X-ray beam and a moving porous gold film were employed. The setup had only one arm, as opposed to the conventional ghost diffraction setup. Time-series measurements were performed to imitate the function of the conventional dual-arm setup in this single-arm one. Specifically, two intensity measurements were performed when the sample was inserted into the beam and when the sample was removed from the beam within a duration in which the X-ray speckle pattern from the pseudothermal source did not change, and then their product was calculated. These processes were repeated to take an average over these products. As a consequence, the intensity correlation obtained in this way clearly showed a diffraction pattern of a slit array object. The amplitude and the phase of this object were also successfully retrieved from this ghost diffraction pattern.

3.4.3 Classical Ghost Imaging and Diffraction in the Time Domain

We have seen in this section that ghost imaging and diffraction with classical light are exclusively applied to tangible spatial objects. However, their temporal counterparts have also been examined theoretically (Chen, Li, Li, Shi, & Zeng, 2013; Setälä, Shirai, & Friberg, 2010; Shirai, Setälä, & Friberg, 2010; Torres-Company, Lajunen, Lancis, & Friberg, 2008) and demonstrated experimentally (Devaux, Moreau, Denis, & Lantz, 2016; Ryczkowski, Barbier, Friberg, Dudley, & Genty, 2016) in recent years, by taking into account the space–time analogy in optics (Kolner, 1994). To give a brief account of this temporal ghost imaging and diffraction with classical light, we start with the relationship between spatial and temporal optics.

It is shown elsewhere (Diels & Rudolf, 2006, Chapter 1; Kolner, 1994; Siegman, 1986, Section 9.2) that the equation describing a beam propagation through free space under the paraxial approximation is mathematically identical to that describing a pulse propagation in a dispersive medium under the assumptions that what is called the slowly varying envelope approximation is satisfied and that the dielectric constant of the medium changes slowly within the pulse spectrum. This fact clearly indicates the analogy between the diffraction and the dispersion problem. One can therefore translate at once the phenomenon of the spatial spreading of a beam propagating through free space (due to diffraction) into that of the temporal spreading of a pulse propagating through a dispersive medium (due to dispersion). In this manner we reformulate briefly the treatment given in Section 3.2.

Fig. 7 shows a basic setup for lensless temporal ghost imaging and diffraction with classical light, as a temporal counterpart of the setup illustrated in Fig. 4, where free space through which the light propagates is replaced by a dispersive medium, such as a single-mode optical fiber. The optical fiber is characterized by the group delay dispersion (GDD) parameters $\Phi_l \equiv \beta_{2l} z_l$, $l \in (a, c, d)$, with β_{2l} and z_l being the group velocity dispersion (GVD) coefficients and the lengths of the optical fibers, respectively. The spatial object is, of course, replaced by a temporal object, such as a temporal modulator providing a temporal pattern $m(t)$.

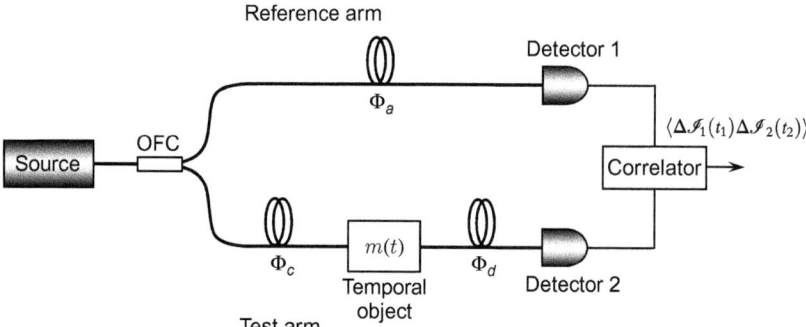

Fig. 7 Basic setup for lensless temporal ghost imaging and diffraction with classical light. The test arm consists of a temporal object and two dispersive media, while the reference arm consists simply of a dispersive medium. The typical example of the dispersive medium is a single-mode optical fiber characterized by the group delay dispersion (GDD) parameter. The instantaneous intensities are measured by the two detectors located at the end of both arms and the correlation between intensity fluctuations at these two detectors is examined. OFC, optical fiber coupler; Φ_a, Φ_c, Φ_d, GDD parameter.

Similarly to usual spatial ghost imaging and diffraction, the correlation between intensity fluctuations $\Delta\mathcal{I}_1(t_1)$ at detector 1 and $\Delta\mathcal{I}_2(t_2)$ at detector 2 is responsible for the formation of the temporal ghost image and diffraction pattern. It is given by the expression

$$\langle\Delta\mathcal{I}_1(t_1)\Delta\mathcal{I}_2(t_2)\rangle = |\Gamma(t_1,t_2)|^2, \quad (107)$$

where

$$\Gamma(t_1,t_2) = \langle E_1^*(t_1)E_2(t_2)\rangle \quad (108)$$

is the correlation between the field envelopes $E_1(t_1)$ at time t_1 in the reference arm and $E_2(t_2)$ at time t_2 in the test arm. If we assume that the source emits temporally incoherent stationary light, the correlation function given by Eq. (108) becomes

$$\Gamma(t_1,t_2) = I_0 \int_{-\infty}^{\infty} \mathcal{K}_1^*(t_1,t_0)\mathcal{K}_2(t_2,t_0)\,dt_0 \quad (109)$$

where I_0 is a positive constant. Here, the temporal impulse response functions $\mathcal{K}_1(t_1,t_0)$ for the reference arm and $\mathcal{K}_2(t_2,t_0)$ for the test arm are given, respectively, by the expressions

$$\mathcal{K}_1(t_1,t_0) = \sqrt{\frac{i}{2\pi\Phi_a}}\exp\left[-i\frac{(t_1-t_0)^2}{2\Phi_a}\right] \quad (110)$$

and

$$\mathcal{K}_2(t_2,t_0) = \frac{1}{2\pi}\sqrt{\frac{i}{\Phi_c}}\sqrt{\frac{i}{\Phi_d}}\int_{-\infty}^{\infty} m(t_2')$$
$$\times \exp\left\{-\frac{i}{2}\left[\frac{(t_2'-t_0)^2}{\Phi_c}+\frac{(t_2-t_2')^2}{\Phi_d}\right]\right\}dt_2'. \quad (111)$$

Eq. (107) along with Eqs. (109)–(111) is, in general, shown to be given as a fractional Fourier transform of the object (Setälä et al., 2010). In a special case when $\Phi_a = \Phi_c$, this equation reduces to the expression

$$\langle\Delta\mathcal{I}_1(t_1)\Delta\mathcal{I}_2(t_2)\rangle = \frac{I_0^2}{2\pi|\Phi_d|}|m(t_1)|^2, \quad (112)$$

showing the perfect image of the temporal object as a function of the time t_1 (Shirai et al., 2010). The condition $\Phi_a = \Phi_c$ is thus the temporal ghost imaging condition for the setup illustrated in Fig. 7. On the other hand, in another special case when $\Phi_a = \Phi_c + \Phi_d$, we obtain the expression

$$\langle \Delta \mathcal{I}_1(t_1) \Delta \mathcal{I}_2(t_2) \rangle = \left(\frac{I_0}{2\pi \Phi_d} \right)^2 \times \left| \int_{-\infty}^{\infty} m(t_2') \exp\left[-i\left(\frac{t_1 - t_2}{\Phi_d} \right) t_2' \right] dt_2' \right|^2, \quad (113)$$

showing the squared modulus of the Fourier transform (i.e., the diffraction pattern) of the temporal object (Torres-Company et al., 2008). The condition $\Phi_a = \Phi_c + \Phi_d$ is thus the temporal ghost diffraction condition for the preceding setup.

In the above treatment, each detector is implicitly assumed to be a fast detector which can resolve the rapid temporal fluctuation of the incident light. The detector of this kind is a temporal counterpart of a point-like detector employed in spatial ghost imaging and diffraction. To mimic the bucket detection employed in the test arm of spatial ghost imaging, the fast detector located in the test arm in Fig. 7 must be replaced by a slow detector whose response is so slow that it cannot resolve even the temporal object. The correlation between intensity fluctuations at *fast* detector 1 and at *slow* detector 2 is expressible in the form

$$\langle \Delta \mathcal{I}_1(t_1) \Delta \mathcal{I}_2' \rangle = \int_{-\tau/2}^{\tau/2} \langle \Delta \mathcal{I}_1(t_1) \Delta \mathcal{I}_2(t_2) \rangle \, dt_2$$
$$= \frac{\tau I_0^2}{2\pi |\Phi_d|} |m(t_1)|^2, \quad (114)$$

where τ is a positive constant characterizing the response of detector 2. It follows from the above expression that the perfect image of the temporal object is obtained again even when the slow detector is employed in the test arm. It is also found that the dispersion Φ_d between the temporal object and detector 2 does not affect the resultant temporal ghost image, irrespective of the type of the detector located in the test arm. Very recently these features of temporal ghost imaging have been verified experimentally by Ryczkowski, Barbier, et al. (2016).

If the incident light is temporally incoherent nonstationary pulsed light, the resultant temporal ghost image is generally distorted by the effect of the incident pulse (Shirai et al., 2010). However, the effect of the distortion is expected to be negligible when the effective temporal width of the incident pulse is much longer than the temporal object.

4. OPTICAL COHERENCE TOMOGRAPHY BASED ON CLASSICAL INTENSITY INTERFEROMETRY

4.1 Background

Stellar intensity interferometry was conceived to achieve a performance better than Michelson stellar interferometry in determining angular diameters of stars (see Section 1). This is the first and successful application of intensity interferometry as an alternative to conventional amplitude interferometry. Another such application, which is modern and surely epoch-making, may be optical coherence tomography (OCT).

OCT is a powerful technique for high-resolution cross-sectional imaging, in particular, of biological samples (for a review, Fercher, Drexler, Hitzenberger, & Lasser, 2003). Since its first application in medicine (Huang et al., 1991), OCT has become an indispensable tool for various clinical diagnoses, especially in ophthalmology. The principle of OCT is based on white-light amplitude interferometry. The typical setup is illustrated in Fig. 8A. The broadband light is split by a beam splitter into two beams and incident on the sample and the reference mirror. The light reflected from the sample and that from the reference mirror are combined

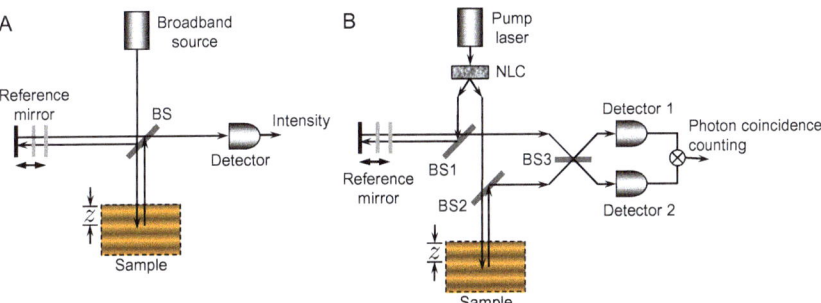

Fig. 8 (A) Setup for conventional time-domain optical coherence tomography (TD-OCT). The broadband light is split into two beams and incident on the sample and the reference mirror. The interference fringe intensity is recorded as a function of the mirror displacement. (B) Setup for quantum optical coherence tomography (Q-OCT). The frequency-entangled photon pairs are generated in a nonlinear crystal (NLC) by the process of spontaneous parametric downconversion. The photon coincidence counting is recorded as a function of the mirror displacement. *BS*, beam splitter.

by the same beam splitter to produce an interference fringe. By scanning the reference mirror, one can obtain the axial profile of the sample directly from the interference fringe pattern produced as a function of the time delay between the reference beam and the sample beam. The axial resolution is generally improved with broadening the source spectrum or, equivalently, shortening the temporal coherence length of the source. However, for broadband light, the effect of the GVD associated with the sample and the optical setup can no longer be ignored and this reduces the axial resolution. The GVD control is a key issue for high-resolution OCT.

The GVD effect may, in principle, be compensated for by inserting a dispersive plate in the reference arm to balance GVD in both arms of the interferometer. However, implementation of this physical method is not straightforward when the GVD associated with the sample is complicated or unknown. A similar difficulty is encountered also with numerical methods (Fercher et al., 2001; Wojtkowski et al., 2004), since intensive computations are required to eliminate the complicated GVD, limiting the acquisition speed of the cross-sectional images and thereby degrading the overall performance of the system.

A novel idea for OCT has paved the way for the solution to this problem. By exploiting the frequency-entangled two-photon light and performing the photon coincidence counting in a specific setup illustrated in Fig. 8B, the novel OCT, termed quantum OCT (Q-OCT), acquires both a factor-of-two improvement in axial resolution and an immunity to all even-order dispersions, including GVD (which is the second-order dispersion), associated with the sample and the optical setup (Abouraddy, Nasr, Saleh, Sergienko, & Teich, 2002; Nasr, Saleh, Sergienko, & Teich, 2003, 2004). However, its use in clinical diagnoses or even in basic biomedical research is expected to be quite difficult due to the weak and fragile nature of the quantum entangled light.

To circumvent this difficulty, a number of attempts have so far been made to emulate main features of Q-OCT with classical light, rather than cumbersome quantum light. Consequently, one of the two important features (i.e., the axial resolution improvement) or both of them (i.e., plus the immunity to GVD) have successfully been achieved with these classical counterparts of Q-OCT. They include phase-conjugate OCT (Erkmen & Shapiro, 2006; Gouët, Venkatraman, Wong, & Shapiro, 2010), chirped-pulse OCT (Kaltenbaek, Lavoie, Biggerstaff, & Resch, 2008; Lavoie, Kaltenbaek, & Resch, 2009; Mazurek, Schreiter, Prevedel, Kaltenbaek, & Resch, 2013), and intensity-interferometric OCT in which only the method of detecting the output signal is altered (Lajunen, Torres-Company, Lancis, & Friberg, 2009; Pe'er, Bromberg, Dayan, Silberberg, & Friesem, 2007; Shirai & Friberg, 2013; Zerom, Piredda, Boyd, & Shapiro, 2009) or some other modifications are made to mimic closely Q-OCT (Resch, Puvanathasan, Lundeen, Mitchell, & Bizheva, 2007; Ryczkowski, Turunen, Friberg, & Genty, 2016; Shirai, 2015, 2016a; Shirai & Friberg, 2014). All these attempts except for phase-conjugate OCT are based on, or closely related to, the principle of intensity interferometry. In this section, we outline these classical counterparts of Q-OCT established on the basis of classical intensity interferometry.

Before going into specifics, we make a brief comment on the two types of OCT: time-domain OCT (TD-OCT) and spectral-domain OCT (SD-OCT). TD-OCT developed originally performs axial measurements of the sample by scanning the reference mirror, as illustrated in Fig. 8A. On the other hand, in SD-OCT developed subsequently (Fercher, Hitzenberger, Kamp, & Elzaiat, 1995; Häusler & Lindner, 1998), the reference mirror is fixed and the (point-like) detector is replaced by a spectrometer usually consisting of a diffraction grating, a focusing lens, and a one-dimensional detector. By performing the Fourier transform of the (properly rescaled and resampled) spectral interference fringe pattern captured by the spectrometer, one can obtain the same information about the axial profile of the sample without mechanical scanning. It is shown that SD-OCT is generally superior to TD-OCT, and it is currently widely employed in various applications due to its high-speed and high-sensitivity measurements in comparison with TD-OCT (Choma, Sarunic, Yang, & Izatt, 2003; Leitgeb, Hitzenberger, & Fercher, 2003; Yaqoob, Wu, & Yang, 2005). We note finally that the principle of Q-OCT proposed originally is based on TD-OCT, so that mechanical scanning of the reference mirror is required to obtain the axial profile of the sample.

In what follows, we first show that the incorporation of the principle of intensity interferometry into conventional OCT yields the axial resolution improvement by a factor of $\sqrt{2}$ (rather than two) in comparison with conventional OCT. We then describe an immediate method of acquiring the feature of the insensitivity to GVD, in addition to the axial resolution improvement. It should be kept in mind that chirped-pulse OCT is also promising as a different method of acquiring these two features. A brief overview of this technique is given in a recent review by Teich et al. (2012).

4.2 Incorporating Intensity Interferometry Into Optical Coherence Tomography

Let us first consider the basic features of conventional TD-OCT, illustrated in Fig. 8A. To keep the analysis as simple as possible, we consider a lossless partially reflecting mirror of (real-valued) amplitude reflectance r as the sample and ignore the effects of diffraction and polarization. However, we take into account the effect of dispersion to examine whether the OCT setup to be discussed is immune to GVD. To do so, a lossless dispersive plate characterized by the wave number $\beta(\omega)$ at frequency ω is intentionally located in front of the sample. The thickness of the dispersive plate is assumed to be $L/2$. Expanding $\beta(\omega)$ about ω_0 up to second order, we obtain the expression (Diels & Rudolf, 2006, Section 1.2)

$$\beta(\omega) \approx \beta(\omega_0) + \left.\frac{d\beta}{d\omega}\right|_{\omega_0} (\omega - \omega_0) + \frac{1}{2}\left.\frac{d^2\beta}{d\omega^2}\right|_{\omega_0} (\omega - \omega_0)^2 \\ \equiv \beta_0 + \beta_1(\omega - \omega_0) + \frac{1}{2}\beta_2(\omega - \omega_0)^2, \quad (115)$$

where β_1 and β_2 denote the inverse of the group velocity at ω_0 and the GVD coefficient at ω_0, respectively, with ω_0 being the center frequency of the source.

Let $V_0(t)$ be the analytic signal (Born & Wolf, 1999, Section 10.2; Mandel & Wolf, 1995, Section 3.1; Wolf, 2007, Section 2.3), representing the statistically stationary broadband field at time t, generated by the source, i.e.,

$$V_0(t) = \int_0^\infty v_0(\omega) \exp(-i\omega t)\, d\omega. \quad (116)$$

Using the function $v_0(\omega)$ in Eq. (116), the spectral density (or, the spectrum) $S(\omega)$ of the source is given by the expression

$$\langle v_0^*(\omega)v_0(\omega')\rangle = S(\omega)\delta(\omega - \omega'), \quad (117)$$

where the angular brackets denote time or ensemble average and δ denotes the Dirac delta function. The analytic signals representing the fields arrived at the detector via the reference mirror and the sample are given, respectively, by the expressions

$$V_R(t) = \int_0^\infty v_0(\omega) \exp(i\omega t_1) \exp(-i\omega t) \, d\omega \qquad (118)$$
$$= V_0(t - t_1)$$

and

$$V_S(t) = r \int_0^\infty v_0(\omega) \exp\{i[\omega t_2 + \beta(\omega)L]\} \exp(-i\omega t) \, d\omega, \qquad (119)$$

where t_1 and t_2 denote the times needed for light to travel from the source to the detector via the reference mirror and the sample, respectively. To be precise, the propagation time t_2 excludes the time for light to pass through the dispersive plate twice, so that the path length from the source to the detector via the sample equals to $L + ct_2$, with c being the speed of light in vacuum. Substitution of Eq. (115) into Eq. (119) yields

$$V_S(t) = r \exp(i\phi) V_0'(t - t_2 - \beta_1 L + \beta_2 L \omega_0), \qquad (120)$$

where $\phi = \beta_0 L - \beta_1 L \omega_0 + \beta_2 L \omega_0^2 / 2$ and

$$V_0'(t) = \int_0^\infty v_0(\omega) \exp\left(\frac{1}{2} i \beta_2 L \omega^2\right) \exp(-i\omega t) \, d\omega. \qquad (121)$$

The interference fringe intensity $I_D = \langle |V_R(t) + V_S(t)|^2 \rangle$ at the detector is given by the expression

$$I_D = \langle |V_R(t)|^2 \rangle + \langle |V_S(t)|^2 \rangle + 2 \operatorname{Re}\left[\langle V_R^*(t) V_S(t) \rangle\right], \qquad (122)$$

where Re denotes the real part. On substituting Eqs. (118) and (120) into Eq. (122) and using the stationarity of the field, we obtain the expression

$$I_D = I_0 + r^2 I_0' + 2r \exp(i\phi) \operatorname{Re}\left[\Gamma(\tau')\right], \qquad (123)$$

where $I_0 = \langle |V_0(t)|^2 \rangle$, $I_0' = \langle |V_0'(t)|^2 \rangle$, $\tau' = t_1 - t_2 - \beta_1 L + \beta_2 L \omega_0$, and

$$\Gamma(\tau) = \langle V_0^*(t) V_0'(t + \tau) \rangle \qquad (124)$$

is the correlation function between the fields from the reference mirror and the sample. The formula (124) is known as the self-coherence function when the dispersive plate is removed from the setup, so that $V_0'(t)$ defined by Eq. (121) becomes $V_0(t)$ defined by Eq. (116). Using Eqs. (116), (117), and (121), the correlation function $\Gamma(\tau)$ defined by Eq. (124) is expressible in the form

$$\Gamma(\tau) = \int_0^\infty S(\omega) \exp\left(\frac{1}{2} i\beta_2 L\omega^2\right) \exp(-i\omega\tau) \, d\omega. \quad (125)$$

The last term on the right-hand side of Eq. (123) is the OCT signal providing information about the sample. Specifically, the height and the width of the envelope of the signal determine the amplitude transmittance of the sample and the axial resolution of this TD-OCT setup, respectively. It follows from Eq. (125) that the axial resolution is generally improved (i.e., the width of the envelope of the function $\text{Re}[\Gamma(\tau')]$ decreases) with broadening the source spectrum $S(\omega)$ and that the GVD characterized by the coefficient β_2 induces a degradation of the axial resolution, i.e., an increase in the width of the envelope of the function $\text{Re}[\Gamma(\tau')]$.

Let us next modify the conventional TD-OCT setup illustrated in Fig. 8A so as to incorporate the principle of intensity interferometry. This is done by doubling the optical path to the detector in such a way that the intensity of light reflected from the reference mirror and that from the sample can be detected independently (Lajunen et al., 2009), as illustrated in Fig. 9. We then perform necessary calculations based on these two intensities to obtain an intensity interference fringe, rather than a conventional amplitude interference fringe. The attempt to combine conventional TD-OCT with intensity interferometry is stimulated by the advent of Q-OCT, in which the photon coincidence counting (essentially equivalent to the intensity correlation measurement) is performed using two detectors.

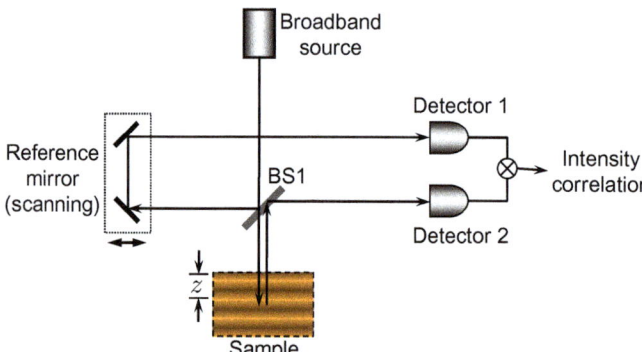

Fig. 9 Setup for intensity-interferometric time-domain optical coherence tomography (I-TD-OCT). The broadband light is split by a beam splitter (BS1) into two beams that are incident on the sample and the reference mirror, respectively. The instantaneous intensity of light reflected from the sample and that from the reference mirror are detected by two detectors. The correlation between these two intensities is examined as a function of the mirror displacement.

Suppose that the two detectors in Fig. 9 detect the instantaneous intensity of light reflected from the reference mirror and that from the sample denoted, respectively, by $|V_R(t)|^2$ ($= |V_0(t-t_1)|^2$, according to Eq. (118)) and $|V_S(t)|^2$ ($= r^2|V_0'(t-t_2-\beta_1 L+\beta_2 L\omega_0)|^2$, according to Eq. (120)). If we assume that the field fluctuation obeys Gaussian statistics with zero mean as is often the case, the correlation between these two intensities is given by the expression

$$\mathscr{C}_T(\tau') = r^2 \langle |V_0(t-t_1)|^2 |V_0'(t-t_2-\beta_1 L+\beta_2 L\omega_0)|^2 \rangle$$
$$= r^2 [I_0 I_0' + |\Gamma(\tau')|^2], \tag{126}$$

where, as before, $I_0 = \langle |V_0(t)|^2 \rangle$, $I_0' = \langle |V_0'(t)|^2 \rangle$, $\tau' = t_1 - t_2 - \beta_1 L + \beta_2 L \omega_0$, and $\Gamma(\tau')$ is defined by Eq. (124). Eq. (126), derived by making use of the Gaussian moment theorem and the stationarity of the field, describes the intensity interference fringe as a function of the time delay τ' caused by scanning the reference mirror. The three quantities I_0, I_0', and $\Gamma(\tau')$ in this equation are exactly the same as those presented in Eq. (123) describing the amplitude interference fringe. The last term on the right-hand side of Eq. (126) expresses the signal of the modified TD-OCT setup, illustrated in Fig. 9, to which we refer as intensity-interferometric TD-OCT (I-TD-OCT) based on two detectors. One thus finds that the axial resolution is now determined by the width of the envelope of the function $|\Gamma(\tau')|^2$, rather than that of the function $\text{Re}[\Gamma(\tau')]$ in conventional TD-OCT. In order to compare the axial resolution in both OCT setups, we assume that the function $\Gamma(\tau')$ takes a Gaussian form. Then, one finds at once that the width (of the envelope) of the function $|\Gamma(\tau')|^2$ is a factor-of-$\sqrt{2}$ smaller than that of the function $\text{Re}[\Gamma(\tau')]$. Therefore, the axial resolution in I-TD-OCT is improved by a factor of $\sqrt{2}$ in comparison with that in conventional TD-OCT. However, the effect of GVD still remains in I-TD-OCT as well as in conventional TD-OCT, since the axial resolutions in both cases are evaluated using the same correlation function $\Gamma(\tau')$ given by Eq. (125).

For I-TD-OCT based on two detectors to work properly, the two detectors need to track the rapid intensity fluctuation. More quantitatively, they are required to be fast enough to resolve the time duration over which the intensity fluctuation remains stable. The time duration given in this way and its corresponding length are known as the coherence time τ_c and the coherence length $l_c = c\tau_c$ of light, respectively. The axial resolution of typical OCT setups for medical applications is around 10 μm ($= l_c/2$, due to the reflection geometry) and hence the coherence time τ_c of light employed in such setups is roughly 70 fs. On the other hand, ultrafast photodiodes currently readily available can resolve up to roughly 10 ps time duration, which is two orders of magnitude longer than the coherence time τ_c of light employed in typical OCT setups. Therefore, in reality, it is difficult to realize I-TD-OCT based on two detectors in its original form, especially for medical applications.

This difficulty can be overcome by performing the ultrafast intensity correlation optically, rather than electronically on the basis of the output electronic signals from the two detectors. One possible method to achieve this is the use of sum-frequency generation (SFG) in a nonlinear crystal (Pe'er et al., 2007). In this method, the two fast detectors in Fig. 9 are removed and, instead, a nonlinear crystal and a slow detector commonly employed in conventional TD-OCT are placed. Moreover, the stationary broadband source is replaced by a source of ultrashort pulses. Then, the beam of pulses reflected from the reference mirror and that from the sample are mixed in the nonlinear crystal. As a consequence, various fields arising from second-harmonic generation, SFG, difference-frequency generation, and optical rectification are generally excited (Boyd, 2008, Section 1.2). However, one can readily extract only the SFG field from the other fields by an appropriate spectral or spatial filtering. The SFG field is proportional to the product of the reference and sample fields. Therefore, the intensity of the SFG field captured by the slow detector provides information about the intensity correlation given by Eq. (126). However, the intensities of the two beams to be mixed in the nonlinear crystal must be strong enough to excite the SFG field, which may limit the practical use of this I-TD-OCT based on SFG, especially in ophthalmology.

Another possible and rather practical method to perform the ultrafast intensity correlation is the use of two-photon absorption (TPA) in a semiconductor (Zerom et al., 2009). A semiconductor photodetector working on the principle of TPA, under the condition that photons with energies inducing single-photon absorption are completely blocked, generates a photocurrent only when two photons with prescribed energies are absorbed simultaneously or, actually, within a time interval smaller than a few femtoseconds in the visible and near infrared range (for a review, Rumi & Perry, 2010). Therefore, a detector of this kind can be utilized to measure the photon coincidence count or, equivalently, the intensity correlation (Boitier et al., 2011; Boitier, Godard, Rosencher, & Fabre, 2009). Furthermore, importantly, TPA can be excited in a GaAs photomultiplier tube (PMT) by a low-power beam of photons on the order of 1 μW (Roth, Murphy, & Xu, 2002), contrary to the general belief that a high-power beam of photons is necessary to excite TPA.

The OCT setup based on TPA is very similar to the conventional OCT setup illustrated in Fig. 8A, rather than the setup illustrated in Fig. 9. Some important features in this OCT, aptly termed intensity-interferometric TD-OCT (I-TD-OCT) based on TPA, are that the GaAs PMT is employed as the detector, a long-wave-pass filter is placed in front of the detector to eliminate single-photon absorption, and a stationary broadband source with a center wavelength around 1.5 μm is employed to excite TPA in this detector. The output signal from the GaAs PMT provides directly the intensity interference fringe as a function of the time delay due to the scanning of the reference mirror. Accordingly, I-TD-OCT based on TPA is very simple and promising for its practical realization, in comparison with that based on two detectors and that based on SFG. Of course, as well as the preceding two setups, I-TD-OCT based on TPA is not immune to GVD, but achieves a factor-of-$\sqrt{2}$ axial resolution improvement in comparison with conventional TD-OCT. Note, however, that the axial resolution improvement in these I-TD-OCT setups is not as good as that in Q-OCT which achieves a factor-of-two improvement.

4.3 Quantum-Mimetic Intensity-Interferometric Optical Coherence Tomography With Dispersion Cancelation

It has been shown in Section 4.2 that combining conventional TD-OCT with intensity interferometry enhances the axial resolution in cross-sectional imaging. However, this treatment failed to yield the feature of dispersion cancelation that is absolutely crucial for high-resolution imaging with broadband light. The present subsection is devoted to a method of acquiring simultaneously the immunity to GVD and the enhancement in axial resolution. As one means to this end, further modifications to the setup for I-TD-OCT based on two detectors are made and a new method of signal processing is adopted with a view to mimicking Q-OCT more closely (Resch et al., 2007; Shirai & Friberg, 2014). In the following we first outline the analysis in the spectral domain, in accordance with the recent study by Shirai and Friberg (2014), to establish what we call quantum-mimetic intensity-interferometric SD-OCT (quantum-mimetic I-SD-OCT)[3] that enables axially scanless cross-sectional imaging with an immunity to GVD and a factor-of-$\sqrt{2}$ improvement in axial resolution. As an extension of this analysis, we then show that quantum-mimetic I-SD-OCT can be realized in a very simple manner by means of a slightly modified conventional SD-OCT setup (Shirai, 2015). The corresponding analysis in the time domain is also addressed in brief to establish quantum-mimetic intensity-interferometric TD-OCT (quantum-mimetic I-TD-OCT) with a GVD immunity and a factor-of-$\sqrt{2}$ resolution improvement.

4.3.1 Theory in the Spectral Domain

Fig. 10 shows the setup for quantum-mimetic I-SD-OCT (Shirai & Friberg, 2014). To mimic the Q-OCT setup closely, a lossless beam splitter BS2 is inserted at the output of the spectral counterpart of the I-TD-OCT setup illustrated in Fig. 9 to form a Hong-Ou-Mandel (HOM) interferometer

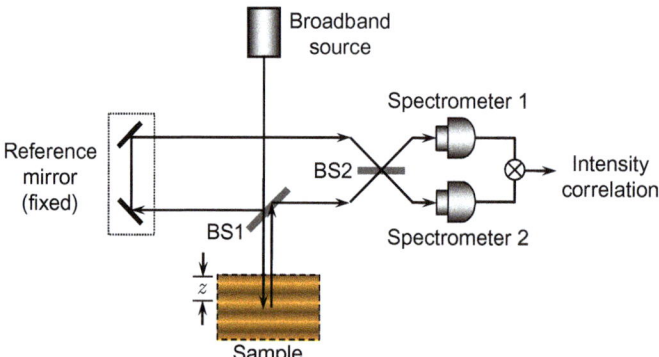

Fig. 10 Setup for quantum-mimetic intensity-interferometric spectral-domain optical coherence tomography (quantum-mimetic I-SD-OCT). This setup was devised by combining the concepts of Q-OCT and conventional SD-OCT. In this setup, the broadband light is split into two beams and incident on the sample and the reference mirror. The averaged spectral intensities are detected by two spectrometers. The Fourier transform of the product of these two (mutually shifted) spectral intensities is performed to obtain information about the sample without scanning the reference mirror. *BS*, beam splitter.

---

[3] Quantum-mimetic I-SD-OCT to be discussed in this subsection was simply termed I-SD-OCT in some recent publications (Shirai, 2015, 2016a; Shirai & Friberg, 2014). However, the prefix "quantum-mimetic" is added in this article to distinguish it from a similar terminology that appeared in Section 4.2.

(Hong, Ou, & Mandel, 1987). The two spectrometers capture the spectral intensities of the fields at the output ports of the HOM interferometer. The reference mirror is fixed, as opposed to the I-TD-OCT setup discussed previously. The source is assumed to emit classical broadband light whose fluctuation obeys Gaussian statistics with zero mean. Following the treatment adopted in Section 4.2, we consider again that a lossless partially reflecting mirror of amplitude reflectance r is employed as the sample and that a lossless dispersive plate with a thickness of $L/2$ characterized by Eq. (115) is inserted in front of the sample. In contrast to the previous treatment, the analysis in this subsection is based on classical coherence theory in the space–frequency domain, formulated in terms of a statistical ensemble of monochromatic realizations (Mandel & Wolf, 1995, Section 4.7; Wolf, 2007, Section 4.1).

If we assume that BS2 is symmetric in the sense that the reflected fields of light (from the upper and lower surfaces of BS2) suffer a $\pi/2$ phase shift relative to each transmitted field of light (Loudon, 2000, Section 3.2), the fields $U_1(\omega)$ and $U_2(\omega)$ arriving respectively at spectrometer 1 and 2 are given by the expressions

$$\begin{cases} U_1(\omega) = iU_R(\omega) + U_S(\omega), \\ U_2(\omega) = U_R(\omega) + iU_S(\omega), \end{cases} \quad (127)$$

with

$$\begin{cases} U_R(\omega) = U_0(\omega)\exp(i\omega t_1), \\ U_S(\omega) = rU_0(\omega)\exp\{i[\omega t_2 + \beta(\omega)L]\}, \end{cases} \quad (128)$$

where $U_0(\omega)$ denotes the field across the source. The propagation times t_1 and t_2 are essentially the same as those defined in Section 4.2 and hence the path lengths from the source to both spectrometers (under the assumption that the path lengths from BS2 to both spectrometers are equivalent) via the reference mirror and the sample become $z_1 = ct_1$ and $z_2 = L + ct_2$, respectively. The source spectrum is defined by the expression

$$S(\omega - \omega_0) = \langle |U_0(\omega)|^2 \rangle \quad (129)$$

and assumed to satisfy the symmetric property $S(\omega') = S(-\omega')$, in analogy with the theoretical treatment in Q-OCT (Abouraddy, Nasr, et al., 2002).

To mimic the frequency-entangled two-photon light employed in Q-OCT, we consider the correlation between different spectral intensities given by the expression

$$C_S(\omega') = \langle |U_1(\omega_0 + \omega')|^2 |U_2(\omega_0 - \omega')|^2 \rangle, \tag{130}$$

where $|U_1(\omega_0 + \omega')|^2$ and $|U_2(\omega_0 - \omega')|^2$ are the instantaneous spectral intensities captured by spectrometer 1 and 2, respectively. Eq. (130) is not a complete replica of the quantum entangled light under consideration, but could be regarded as its classical analog. Using the Gaussian moment theorem and Eqs. (127) and (128), the intensity correlation (130) is expressible in the form

$$C_S(\omega') = \langle |U_1(\omega_0 + \omega')|^2 \rangle \langle |U_2(\omega_0 - \omega')|^2 \rangle \\ \times \left[1 + |\mu(\omega_0 + \omega', \omega_0 - \omega')|^2 \right], \tag{131}$$

where

$$\mu(\omega_1, \omega_2) = \frac{\langle U_0^*(\omega_1) U_0(\omega_2) \rangle}{\sqrt{\langle |U_0(\omega_1)|^2 \rangle} \sqrt{\langle |U_0(\omega_2)|^2 \rangle}} \tag{132}$$

is the complex degree of spectral coherence of the source, which satisfies the inequality $0 \leq |\mu(\omega_1, \omega_2)| \leq 1$.

The value of $\mu(\omega_0 + \omega', \omega_0 - \omega')$ depends on the properties of the source. For statistically strictly stationary sources, the modulus $|\mu(\omega_0 + \omega', \omega_0 - \omega')|$ vanishes unless $\omega' = 0$ since different spectral components of stationary fields are mutually uncorrelated (cf., Eq. (117)). For fully coherent nonstationary pulses, the modulus becomes unity. However, this situation must be considered as the limit of vanishing field fluctuations. For partially coherent nonstationary pulses, the modulus depends generally on the frequency ω'. Suppose now, as a first approximation, that the modulus is effectively constant (i.e., $|\mu(\omega_0 + \omega', \omega_0 - \omega')| = \mu_A$ is a positive constant) over the spectral range of the source. Then, Eq. (131) becomes

$$C_S(\omega') = A \langle |U_1(\omega_0 + \omega')|^2 \rangle \langle |U_2(\omega_0 - \omega')|^2 \rangle, \tag{133}$$

where $A = 1 + \mu_A^2$ is a positive constant characterizing the coherence property of the source. Eq. (133) reveals that the correlation function

(130) reduces to a product of two averaged spectral intensities $\langle |U_1(\omega_0+\omega')|^2\rangle$ captured by spectrometer 1 and $\langle |U_2(\omega_0-\omega')|^2\rangle$ captured by spectrometer 2. In other words, one needs only to measure these two averaged spectral intensities and calculate their product to obtain the correlation function $C_S(\omega')$. Therefore, fast detectors (spectrometers) are not necessary in this technique, in distinct contrast to I-TD-OCT discussed in Section 4.2.

By analogy with conventional SD-OCT, we next perform the Fourier transform of the function $C_S(\omega')$ in accordance with the formula

$$\hat{C}_S(x) = \int_{-\infty}^{\infty} C_S(\omega')\exp(-ix\omega')\,d\omega', \qquad (134)$$

where the explicit expression for $C_S(\omega')$ can be obtained by combining Eqs. (115), (127), and (128) with Eq. (133). After some straightforward calculations, we find that the ensuring expression for $\hat{C}_S(x)$ when $x \geq 0$ is given by a summation of three terms, viz.,

$$\hat{C}_S(x) = A[\hat{c}_0(x) + \hat{c}_1(x) + \hat{c}_2(x)], \qquad (135)$$

where

$$\hat{c}_0(x) = (r^2+1)^2 G(x) + 2r^2 \operatorname{Re}[\exp(-2i\alpha)G_{d2}(x)], \qquad (136)$$

$$\hat{c}_1(x) = -2ir(r^2+1)\operatorname{Re}[\exp(-i\alpha)G_{d1}(x-\tau')], \qquad (137)$$

$$\hat{c}_2(x) = -r^2 G(x-2\tau'), \qquad (138)$$

with $\alpha = \omega_0\tau + \beta_0 L$, $\tau = t_2 - t_1$, and $\tau' = \tau + \beta_1 L$. In these expressions, the function $G(x)$ is the Fourier transform of the square of the source spectrum, i.e.,

$$G(x) = \int_{-\infty}^{\infty} S^2(\omega')\exp(-ix\omega')\,d\omega', \qquad (139)$$

and the functions $G_{d1}(x)$ and $G_{d2}(x)$ are the corresponding Fresnel transforms defined by

$$G_{d1}(x) = \int_{-\infty}^{\infty} S^2(\omega')\exp\left(-\frac{1}{2}i\beta_2 L\omega'^2\right)\exp(-ix\omega')\,d\omega' \qquad (140)$$

and

$$G_{d2}(x) = \int_{-\infty}^{\infty} S^2(\omega') \exp\left(-i\beta_2 L\omega'^2\right) \exp\left(-ix\omega'\right) d\omega'. \tag{141}$$

The three terms $\hat{c}_0(x)$, $\hat{c}_1(x)$, and $\hat{c}_2(x)$ are well localized around $x = 0$, τ', and $2\tau'$, respectively, if the source spectrum is sufficiently broad, as is often the case with OCT. The last term $\hat{c}_2(x)$ provides complete information about the sample, and hence it is referred to as the I-SD-OCT signal. Specifically, similar to the conventional SD-OCT signal, the magnitude and the location of the I-SD-OCT signal determine the reflectance and the location of the sample, respectively. The width of the signal determines the axial resolution of this quantum-mimetic I-SD-OCT system. It is important to note that the I-SD-OCT signal $\hat{c}_2(x)$ given by Eq. (138) is independent of the GVD coefficient β_2, although the dispersive plate is intentionally located in front of the sample. Accordingly, on the basis of the I-SD-OCT signal $\hat{c}_2(x)$, one can perform axially scanless cross-sectional imaging that is immune to GVD. Incidentally, the remaining two terms $\hat{c}_0(x)$ and $\hat{c}_1(x)$ given, respectively, by Eqs. (136) and (137) are broadened due to GVD. The I-SD-OCT signal $\hat{c}_2(x)$ can be readily extracted from the other terms since $\hat{c}_2(x)$ is real, but the adjacent term $\hat{c}_1(x)$ is purely imaginary.

The axial resolution in quantum-mimetic I-SD-OCT is determined by the width of the signal $\hat{c}_2(x)$ or, equivalently, that of the function $G(x - 2\tau')$ representing the Fourier transform of the square of the source spectrum $S^2(\omega')$. On the other hand, it is shown elsewhere that the axial resolution in conventional OCT is determined by the width of the function representing the Fourier transform of the source spectrum $S(\omega)$ (cf., Eq. (125) when $\beta_2 = 0$). As a consequence, if the source spectrum is assumed to take a Gaussian form, one readily finds that the width of the I-SD-OCT signal is a factor-of-$\sqrt{2}$ larger than that of the conventional OCT signal. Therefore, at first glance, the axial resolution in quantum-mimetic I-SD-OCT is deteriorated by a factor of $\sqrt{2}$, in comparison with that in conventional OCT. However, care must be taken of the location of the signal. The location of the I-SD-OCT signal on the x-axis is $2\tau'$, whereas that of the conventional SD-OCT signal is τ' according to the conventional SD-OCT theory (e.g., Fercher et al., 2003, Section 4.2). This difference in the location of the signal causes a factor-of-two resolution improvement in quantum-mimetic I-SD-OCT, as compared with conventional OCT. Combining these two effects, we find that the overall enhancement of the axial resolution in quantum-mimetic I-SD-OCT is a factor of $\sqrt{2}$. This is exactly the same as the axial resolution enhancement in I-TD-OCT discussed in Section 4.2.

It is to be emphasized that the immunity to GVD is successfully acquired by making use of the specific signal processing based on the intensity correlation (130) which mimics the frequency-entangled two-photon light employed in Q-OCT. If we use the correlation $\langle |U_1(\omega)|^2 |U_2(\omega)|^2 \rangle$ between the same spectral intensities in the present analysis, the axial resolution is improved by a factor of $\sqrt{2}$, but it is deteriorated by the effect of GVD (Shirai & Friberg, 2013). The immunity to GVD is not acquired with this correlation function which has no connection with such quantum entangled light.

To illustrate the preceding analysis by numerical examples, let us consider a sample consisting of two partially reflecting surfaces (amplitude transmittance: $r_1 = 0.1$ and $r_2 = 0.15$) separated by 150 μm of air, as depicted at the top of Fig. 11. Broadband light with FWHM bandwidth $\Delta\lambda = 100$ nm centered at $\lambda_0 = 800$ nm is incident on the sample. A dispersive BK7 glass plate with a thickness $L/2 = 20$ mm and a GVD coefficient $\beta_2 = 4.5 \times 10^{-26}$ s^2/m is located in front of the sample.

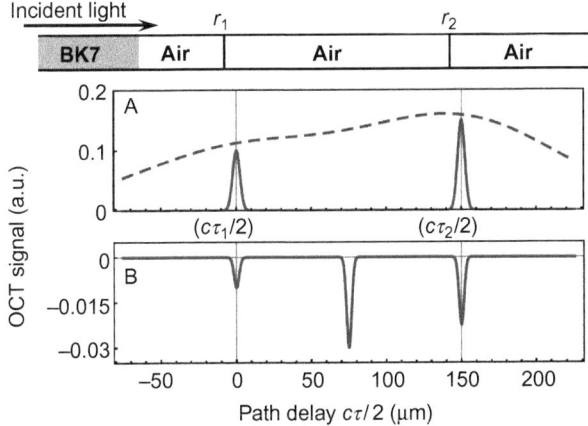

Fig. 11 OCT signals obtained by (A) conventional SD-OCT and (B) quantum-mimetic I-SD-OCT. Two peaks and two dips representing the locations of the two reflecting surfaces of the sample are clearly observed when the BK7 glass plate is removed, as indicated by the *solid curves* in (A) and (B). Another dip between the two dips in (B) is an unwanted artifact. When the BK7 glass plate is set in place, the conventional SD-OCT signal is severely broadened by the effect of GVD, as indicated by the *dashed curve* in (A), while the I-SD-OCT signal remains unchanged. *Reproduced with permission from Shirai, T., & Friberg, A. T. (2014). Intensity-interferometric spectral-domain optical coherence tomography with dispersion cancellation. Journal of the Optical Society of America A, 31, 258–263. https://doi.org/10.1364/JOSAA.31.000258. ©2014 Optical Society of America.*

We first remove the BK7 glass plate and calculate the OCT signals obtained by conventional SD-OCT and quantum-mimetic I-SD-OCT. The locations of two peaks in the conventional SD-OCT signal represent those of the two reflecting surfaces of the sample, as indicated by the solid curve in Fig. 11A. Two dips occurring at the same locations in Fig. 11B are the true I-SD-OCT signal which provides information about the sample. The I-SD-OCT signal consists of dips, rather than peaks, as also deduced from Eq. (138) for a simple model sample with a single reflecting surface. This is reminiscent of the well-known HOM dip observed in two-photon interference experiment using an HOM interferometer (Hong et al., 1987). Another dip existing between these two dips is an artifact that we will touch on later. The width of these two dips in Fig. 11B is narrower than that of the two peaks in Fig. 11A by a factor of $\sqrt{2}$. This proves that the axial resolution in quantum-mimetic I-SD-OCT is enhanced by a factor of $\sqrt{2}$.

When the BK7 glass plate is set in place, the conventional SD-OCT signal is severely broadened by the effect of GVD, as indicated by the dashed curve in Fig. 11A and thus the locations of the two reflecting surfaces of the sample are no longer estimated from this signal. However, the I-SD-OCT signal remains unchanged even in this case and it is exactly the same as that obtained when the BK7 glass plate is removed, as shown in Fig. 11B. This proves obviously that quantum-mimetic I-SD-OCT is immune to GVD. These theoretical findings have recently been verified by experiments using a setup constructed in accordance with the original proposal illustrated in Fig. 10 (Ryczkowski, Turunen, et al., 2016).

4.3.2 A Practical Method of Realization
The quantum-mimetic I-SD-OCT setup illustrated in Fig. 10 has a great resemblance to the Q-OCT setup illustrated in Fig. 8B, except that a cumbersome entangled two-photon source is replaced by a very common classical broadband source. Hence, the practical realization of quantum-mimetic I-SD-OCT in its original form is much simpler than that of Q-OCT. However, the original quantum-mimetic I-SD-OCT setup is still complicated in comparison with a conventional SD-OCT setup, since two spectrometers and an additional beam splitter are required in quantum-mimetic I-SD-OCT, whereas conventional SD-OCT works properly with only one spectrometer and no additional beam splitter. We thus explore a method of realizing quantum-mimetic I-SD-OCT more simply.

The spectral interference fringe patterns detected by each spectrometer in quantum-mimetic I-SD-OCT are basically the same, but they are the reverse of each other in intensity. This kind of reversal in spectral intensity is caused by the π phase difference, owing to the effect of BS2, between the field arriving at spectrometer 1 via the reference mirror and that arriving at spectrometer 2 via the sample. Consequently, it turns out that one can emulate the function of quantum-mimetic I-SD-OCT using a slightly modified conventional SD-OCT setup, where the reference mirror is located on a translation stage (Shirai, 2015). Specifically, we first capture a spectral intensity by the spectrometer and store it in a computer for signal processing. We then generate the π phase shift by translating the stage along the optical path by half the center wavelength of the incident light.[4] Subsequently, we capture another spectral intensity after the π phase shift by the same spectrometer and store it again in the same computer. As a next step, according to the quantum-mimetic I-SD-OCT theory described before, we calculate the product of these two spectral intensities and perform the Fourier transform of this product to obtain the I-SD-OCT signal.

We have shown in Fig. 11 that the artifact emerges at the center of two dips corresponding to the two reflecting surfaces of the sample. It is known that such an artifact emerges also in Q-OCT and chirped-pulse OCT. It is not difficult to show that its magnitude is proportional to the factor $\cos[2(\omega_0/c)\Delta z]$ with $\Delta z = c(t_2 - t_1)/2$ being the optical path length between the two reflecting surfaces. Thus one readily finds that the artifact can be reduced, in principle, by an adjustment for the center frequency ω_0 of the incident light due to the oscillatory nature of the cosine function, as also mentioned in the study of Q-OCT. However, this method is not so straightforward in practice. As an alternative to this method, we have demonstrated theoretically and experimentally that the unwanted artifact that deteriorates the quality of the resultant image can be reduced by means of either a mechanical displacement of the detector (inside the spectrometer) for capturing spectral intensity patterns in the modified SD-OCT setup that we have just discussed or, essentially equivalently, an analogous numerical displacement of the spectral intensity patterns stored in a computer (Shirai, 2016a). This method is very simple and easily implemented in practice, as

---

[4] The π phase shift produced in this way is not exact for all the wavelengths of the broadband light, especially when the bandwidth is large. To produce the precise π phase shift for all the wavelengths, one has to employ an achromatic phase shifter, such as a method operating on the geometric phase (Hariharan & Ciddor, 1994; Helen, Kothiyal, & Sirohi, 1998).

compared with the method of controlling the source spectrum. A schematic diagram of the experimental setup for verifying the artifact reduction method is illustrated in Fig. 12. This setup, which emulates the function of quantum-mimetic I-SD-OCT, is exactly the same as a conventional SD-OCT setup except that the reference mirror is located on a translation stage. The typical experimental results obtained with this setup are shown in Fig. 13. In this experiment, a microscope cover glass with a thickness of approximately 70 μm was employed as a sample having two reflecting surfaces. Absolute values of the I-SD-OCT signals are plotted in Fig. 13, so that the signal appears in the form of peaks, rather than dips. Fig. 13 clearly shows that the artifact located between the two peaks corresponding to the two reflecting surfaces (i.e., the true I-SD-OCT signal) is successfully reduced by slightly displacing the detector, while the true I-SD-OCT signal remains unchanged. The amount of the displacement was 80 μm, which was less than 0.3% of the active length of the line scan CCD camera employed as the detector.

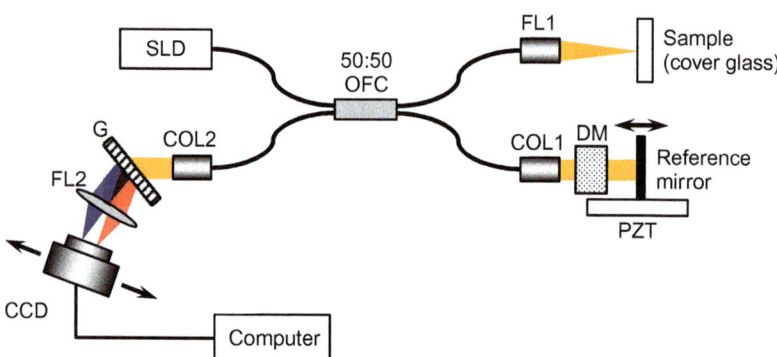

Fig. 12 Experimental setup for verifying the artifact reduction methods. This setup, which emulates the function of quantum-mimetic I-SD-OCT, is exactly the same as a conventional SD-OCT setup except that the reference mirror is located on a translation stage. A combination of a diffraction grating (G), a focusing lens (FL2), and a line scan CCD camera (CCD) forms a spectrometer. The unwanted artifact can be reduced by means of either a mechanical displacement of the detector (CCD) for capturing spectral intensity patterns or an analogous numerical displacement of the spectral intensity patterns stored in a computer. *COL*, collimator lens; *DM*, dispersive medium; *FL*, focusing lens; *OFC*, optical fiber coupler; *SLD*, superluminescent diode; *PZT*, piezoelectric translator. *Reproduced with permission from Shirai, T. (2016). Improving image quality in intensity-interferometric spectral-domain optical coherence tomography. Journal of Optics, 18, 075601. https://doi.org/10.1088/2040-8978/18/7/075601. ©2016 IOP Publishing.*

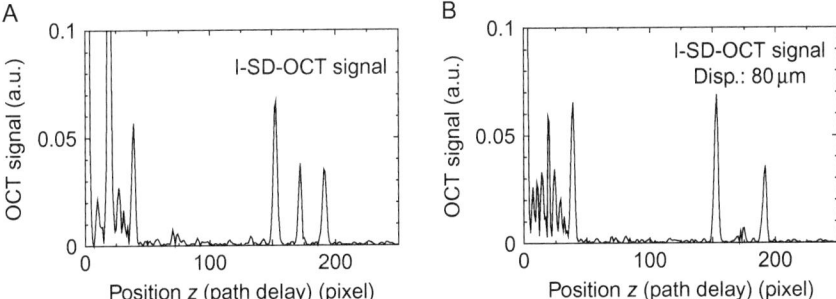

Fig. 13 Experimental results for the I-SD-OCT signals (A) before and (B) after the displacement of the detector. Absolute values of the signals are plotted. These results were obtained with the setup illustrated in Fig. 12. The amount of the displacement was 80 μm. One pixel along the horizontal axis is equivalent to approximately 2.6 μm. Reproduced with permission from Shirai, T. (2016). Improving image quality in intensity-interferometric spectral-domain optical coherence tomography. *Journal of Optics, 18*, 075601. https://doi.org/10.1088/2040-8978/18/7/075601. ©2016 IOP Publishing.

4.3.3 Theory in the Time Domain

Up to now we have dealt with the analysis in the spectral domain to establish quantum-mimetic I-SD-OCT. Finally, let us take a brief look at the corresponding analysis in the time domain to establish what we call quantum-mimetic I-TD-OCT. As we have shown before, the correlation between different spectral intensities defined by Eq. (130) plays a key role in acquiring the immunity to GVD. Although this correlation function reduces to the expression (133) representing a product of two averaged spectral intensities, the spectral manipulation in each spectral intensity is still necessary. This kind of spectral manipulation would be crucial even in the temporal counterpart of quantum-mimetic I-SD-OCT. As a consequence, the quantum-mimetic I-TD-OCT setup may be equivalent to a slightly modified quantum-mimetic I-SD-OCT setup, where the modification is made in such a way that the reference mirror is located on a translation stage to perform axial scanning. In order to obtain the intensity correlation signal in the time domain with this setup, we need to integrate all the spectral components after the necessary spectral manipulation as follows:

$$C_T(\tau) = \int_{-\infty}^{\infty} C_S(\omega') \, d\omega'$$
$$= A \int_{-\infty}^{\infty} \langle |U_1(\omega_0 + \omega')|^2 \rangle \langle |U_2(\omega_0 - \omega')|^2 \rangle \, d\omega'. \tag{142}$$

Eq. (142) is essentially equivalent to the basic expression given by Resch et al. (2007, Eq. (6) with obvious changes in notation) to establish a classical counterpart of Q-OCT. The intensity correlation signal in the time domain obtained in this way is shown to be a mixture of the true I-TD-OCT signal and rapidly oscillating unwanted signals. These unwanted signals can be removed by means of a low-pass Fourier filter. Therefore, one can obtain only the true I-TD-OCT signal which provides information about the sample. On the basis of the true I-TD-OCT signal, one can perform cross-sectional imaging, in principle, with an immunity to GVD and a factor-of-$\sqrt{2}$ improvement in axial resolution. One of these important features (i.e., the immunity to GVD) was verified experimentally by Resch et al. (2007). Note that, in contrast to quantum-mimetic I-SD-OCT discussed above, the axial scanning of the reference mirror is necessary in this time-domain technique.

5. CONCLUDING REMARKS

In this article we have reviewed some recent topics involved with classical intensity interferometry. Special emphasis has been placed on modern applications of intensity interferometry with classical light. All the applications dealt with in this article are classical counterparts of techniques based on quantum optics. These what we call quantum-mimetic technologies are very promising in the sense that almost the same performance, which is usually novel and excellent, as in the corresponding quantum techniques is readily achievable with cost-effective bright classical sources. This is evident from the discussions on ghost imaging and diffraction given in Section 3 and those on OCT given in Section 4.

Quantum techniques inducing their classical counterparts as quantum-mimetic technologies are not limited to the topics that we have discussed in this article, but there are some others. One of the most important examples is quantum optical lithography aiming at increasing the resolution of the lithographic process in interferometric lithography beyond the classical Rayleigh diffraction limit (Boto et al., 2000; D'Angelo, Chekhova, & Shih, 2001). Its classical counterpart has been realized using second-harmonic generation in a nonlinear crystal by Bentley and Boyd (2004). In this connection, it is to be noted that phase difference amplification for enhancing sensitivity and accuracy in interferometric measurements has recently been demonstrated on the basis of classical intensity interferometry, although this technique does not beat the classical limit (Shirai, 2016b).

Another important example may be entangled-photon ellipsometry which makes it possible to characterize thin films without source and detector calibration and moreover without a reference sample (Abouraddy, Toussaint, Sergienko, Saleh, & Teich, 2001, 2002). Its classical counterpart has only recently been proposed by Hannonen, Friberg, and Setälä (2016).

Since the advent of intensity interferometry in the 1950s, the heart of the apparatus (i.e., the intensity interferometer) has been a combination of two detectors and an electronic correlator. However, these components are less effective when the intensity fluctuation of light is so rapid that the detectors cannot trace it precisely. A semiconductor detector working on the principle of two-photon absorption becomes a powerful alternative in such a case, as we have seen in Section 4.2. However, this device is applicable only when two beams to be correlated are overlapped, like an OCT configuration. Nevertheless, interestingly, this device has been employed in a wide variety of applications based on intensity interferometry, such as measurements of the ultrashort pulse shape (Panasenko & Fainman, 2002), those of the surface profile (Tanaka, Sako, Kurokawa, Tsuda, & Takeda, 2003), and those of the temporal degree of coherence (Shevchenko, Roussey, Friberg, & Setälä, 2015).

Finally it is worth noting that there is a classical intensity-interferometric technique which is long-standing and well-established, but still plays a major role, especially in modern biomedical research. The technique is known as fluorescence correlation spectroscopy (FCS) pioneered by Magde, Elson, and Webb (1972), which is a sort of photon correlation spectroscopy employed in dynamic light scattering experiments (Berne & Pecora, 2000; Chu, 1991). FCS measures intensity fluctuations of emission from fluorescent particles to determine some basic quantities of the particles, such as their concentration and their diffusion coefficient. The intensity fluctuations observed in this technique are due to the motion of the particles, rather than the optical field itself. By applying this technique to biomolecules tagged with fluorescent dyes, one can obtain significant information about biological systems, such as the dynamics of conformational fluctuations of DNA molecules (for reviews, Elson, 2011; Krichevsky & Bonnet, 2002).

Intensity interferometry in its early days prompted the birth of modern quantum optics, as we have mentioned in Section 1. After many years intensity interferometry with classical light serves to establish various quantum-mimetic technologies. Hereafter intensity-interferometric techniques, be they quantum or classical, are expected to maintain their position as not only fundamental but also practical tools for opening up new fields of science and technology.

ACKNOWLEDGMENTS

I wish to thank Ari T. Friberg for fruitful discussions and helpful comments on the article. I am also grateful to Taco D. Visser for providing me with original data for Figs. 2 and 3. This work was supported, in part, by JSPS KAKENHI Grant Number 16K04990.

REFERENCES

Abouraddy, A. F., Nasr, M. B., Saleh, B. E. A., Sergienko, A. V., & Teich, M. C. (2002). Quantum-optical coherence tomography with dispersion cancellation. *Physical Review A*, *65*(5), 053817.
Abouraddy, A. F., Saleh, B. E. A., Sergienko, A. V., & Teich, M. C. (2001). Role of entanglement in two-photon imaging. *Physical Review Letters*, *87*(12), 123602.
Abouraddy, A. F., Toussaint, K. C., Sergienko, A. V., Saleh, B. E. A., & Teich, M. C. (2001). Ellipsometric measurements by use of photon pairs generated by spontaneous parametric downconversion. *Optics Letters*, *26*(21), 1717–1719.
Abouraddy, A. F., Toussaint, K. C., Sergienko, A. V., Saleh, B. E. A., & Teich, M. C. (2002). Entangled-photon ellipsometry. *Journal of the Optical Society of America B*, *19*(4), 656–662.
Ádám, A., Jánossy, L., & Varga, P. (1955). Beobachtungen mit dem elektronenvervielfacher an kohärenten lichtstrahlen. *Annalen der Physik*, *451*(5–8), 408–413.
Al-Qasimi, A., Lahiri, M., Kuebel, D., James, D. F. V., & Wolf, E. (2010). The influence of the degree of cross-polarization on the Hanbury Brown–Twiss effect. *Optics Express*, *18*(16), 17124–17129.
Andrews, L. C., & Phillips, R. L. (1998). *Laser beam propagation through random media*. Bellingham, WA: SPIE Press.
Bache, M., Magatti, D., Ferri, F., Gatti, A., Brambilla, E., & Lugiato, L. A. (2006). Coherent imaging of a pure phase object with classical incoherent light. *Physical Review A*, *73*(5), 053802.
Basano, L., & Ottonello, P. (2006). Experiment in lensless ghost imaging with thermal light. *Applied Physics Letters*, *89*(9), 091109.
Baym, G. (1998). The physics of Hanbury Brown–Twiss intensity interferometry: From stars to nuclear collisions. *Acta Physica Polonica B*, *29*(7), 1839–1884.
Bennink, R. S., Bentley, S. J., & Boyd, R. W. (2002). "Two-photon" coincidence imaging with a classical source. *Physical Review Letters*, *89*(11), 113601.
Bentley, S. J., & Boyd, R. W. (2004). Nonlinear optical lithography with ultra-high sub-Rayleigh resolution. *Optics Express*, *12*(23), 5735–5740.
Berne, B. J., & Pecora, R. (2000). *Dynamic light scattering: With applications to chemistry, biology, and physics*. Mineola, NY: Dover Publications.
Boitier, F., Godard, A., Dubreuil, N., Delaye, P., Fabre, C., & Rosencher, E. (2011). Photon extrabunching in ultrabright twin beams measured by two-photon counting in a semiconductor. *Nature Communications*, *2*, 425.
Boitier, F., Godard, A., Rosencher, E., & Fabre, C. (2009). Measuring photon bunching at ultrashort timescale by two-photon absorption in semiconductors. *Nature Physics*, *5*(4), 267–270.
Borghi, R., Gori, F., & Santarsiero, M. (2006). Phase and amplitude retrieval in ghost diffraction from field-correlation measurements. *Physical Review Letters*, *96*(18), 183901.
Born, M., & Wolf, E. (1999). *Principles of optics* (7th ed.). Cambridge: Cambridge University Press.

Borra, E. F. (2008). Observations of time delays in gravitational lenses from intensity fluctuations: The coherence function. *Monthly Notices of the Royal Astronomical Society, 389*(1), 364–370.

Boto, A. N., Kok, P., Abrams, D. S., Braunstein, S. L., Williams, C. P., & Dowling, J. P. (2000). Quantum interferometric optical lithography: Exploiting entanglement to beat the diffraction limit. *Physical Review Letters, 85*(13), 2733–2736.

Boyd, R. W. (2008). *Nonlinear optics* (3rd ed.). Burlington, MA: Academic Press.

Brannen, E., & Ferguson, H. I. S. (1956). Question of correlation between photons in coherent light rays. *Nature, 178*(4531), 481–482.

Cai, Y., & Zhu, S. Y. (2004). Ghost interference with partially coherent radiation. *Optics Letters, 29*(23), 2716–2718.

Cai, Y., & Zhu, S. Y. (2005). Ghost imaging with incoherent and partially coherent light radiation. *Physical Review E, 71*(5), 056607.

Chan, K. W. C., Simon, D. S., Sergienko, A. V., Hardy, N. D., Shapiro, J. H., Dixon, P. B., … Boyd, R. W. (2011). Theoretical analysis of quantum ghost imaging through turbulence. *Physical Review A, 84*(4), 043807.

Chen, X. H., Liu, Q., Luo, K. H., & Wu, L. A. (2009). Lensless ghost imaging with true thermal light. *Optics Letters, 34*(5), 695–697.

Chen, Z., Li, H., Li, Y., Shi, J., & Zeng, G. (2013). Temporal ghost imaging with a chaotic laser. *Optical Engineering, 52*(7), 076103.

Cheng, J. (2009). Ghost imaging through turbulent atmosphere. *Optics Express, 17*(10), 7916–7921.

Cheng, J., & Han, S. (2004). Incoherent coincidence imaging and its applicability in X-ray diffraction. *Physical Review Letters, 92*(9), 093903.

Choma, M. A., Sarunic, M. V., Yang, C., & Izatt, J. A. (2003). Sensitivity advantage of swept source and Fourier domain optical coherence tomography. *Optics Express, 11*(18), 2183–2189.

Chu, B. (1991). *Laser light scattering: Basic principles and practice* (2nd ed.). San Diego, CA: Academic Press.

D'Angelo, M., Chekhova, M. V., & Shih, Y. (2001). Two-photon diffraction and quantum lithography. *Physical Review Letters, 87*(1), 013602.

Devaux, F., Moreau, P. A., Denis, S., & Lantz, E. (2016). Computational temporal ghost imaging. *Optica, 3*(7), 698–701.

Diels, J. C., & Rudolf, W. (2006). *Ultrashort laser pulse phenomena* (2nd ed.). Burlington, MA: Academic Press.

Dixon, P. B., Howland, G. A., Chan, K. W. C., O'Sullivan-Hale, C., Rodenburg, B., Hardy, N. D., … Howell, J. C. (2011). Quantum ghost imaging through turbulence. *Physical Review A, 83*(5), 051803(R).

Duarte, M. F., Davenport, M. A., Takhar, D., Laska, J. N., Sun, T., Kelly, K. F., & Baraniuk, R. G. (2008). Single-pixel imaging via compressive sampling. *IEEE Signal Processing Magazine, 25*(2), 83–91.

Elson, E. L. (2011). Fluorescence correlation spectroscopy: Past, present, future. *Biophysical Journal, 101*(12), 2855–2870.

Erkmen, B. I., & Shapiro, J. H. (2006). Phase-conjugate optical coherence tomography. *Physical Review A, 74*(4), 041601(R).

Erkmen, B. I., & Shapiro, J. H. (2008). Unified theory of ghost imaging with Gaussian-state light. *Physical Review A, 77*(4), 043809.

Erkmen, B. I., & Shapiro, J. H. (2010). Ghost imaging: From quantum to classical to computational. *Advances in Optics and Photonics, 2*(4), 405–450.

Fante, R. L. (1985). Wave propagation in random media: A systems approach. In E. Wolf (Ed.), *Progress in optics* (Vol. 22, pp. 341–398). Amsterdam: Elsevier.

Fercher, A. F., Drexler, W., Hitzenberger, C. K., & Lasser, T. (2003). Optical coherence tomography—Principles and applications. *Reports on Progress in Physics, 66*(2), 239–303.

Fercher, A. F., Hitzenberger, C. K., Kamp, G., & Elzaiat, S. Y. (1995). Measurement of intraocular distances by backscattering spectral interferometry. *Optics Communications, 117*(1–2), 43–48.

Fercher, A. F., Hitzenberger, C. K., Sticker, M., Zawadzki, R., Karamata, B., & Lasser, T. (2001). Numerical dispersion compensation for partial coherence interferometry and optical coherence tomography. *Optics Express, 9*(12), 610–615.

Ferri, F., Magatti, D., Gatti, A., Bache, M., Brambilla, E., & Lugiato, L. A. (2005). High-resolution ghost image and ghost diffraction experiments with thermal light. *Physical Review Letters, 94*(18), 183602.

Fienup, J. R. (1982). Phase retrieval algorithms: A comparison. *Applied Optics, 21*(15), 2758–2769.

Fienup, J. R. (1987). Reconstruction of a complex-valued object from the modulus of its Fourier transform using a support constraint. *Journal of the Optical Society of America A, 4*(1), 118–123.

Foellmi, C. (2009). Intensity interferometry and the second-order correlation function $g^{(2)}$ in astrophysics. *Astronomy & Astrophysics, 507*(3), 1719–1727.

Friberg, A. T., & Setälä, T. (2016). Electromagnetic theory of optical coherence. *Journal of the Optical Society of America A, 33*(12), 2431–2442.

Gatti, A., Bache, M., Magatti, D., Brambilla, E., Ferri, F., & Lugiato, L. A. (2006). Coherent imaging with pseudo-thermal incoherent light. *Journal of Modern Optics, 53*(5–6), 739–760.

Gatti, A., Bondani, M., Lugiato, L. A., Paris, M. G. A., & Fabre, C. (2007). Comment on "Can two-photon correlation of chaotic light be considered as correlation of intensity fluctuations?" *Physical Review Letters, 98*(3), 039301.

Gatti, A., Brambilla, E., Bache, M., & Lugiato, L. A. (2004a). Correlated imaging, quantum and classical. *Physical Review A, 70*(1), 013802.

Gatti, A., Brambilla, E., Bache, M., & Lugiato, L. A. (2004b). Ghost imaging with thermal light: Comparing entanglement and classical correlation. *Physical Review Letters, 93*(9), 093602.

Gatti, A., Brambilla, E., & Lugiato, L. (2008). Quantum imaging. In E. Wolf (Ed.), *Progress in optics* (Vol. 51, pp. 251–348). Amsterdam: Elsevier.

Gbur, G., & Visser, T. D. (2010). The structure of partially coherent fields. In E. Wolf (Ed.), *Progress in optics* (Vol. 55, pp. 285–341). Amsterdam: Elsevier.

Glauber, R. J. (2006). Nobel lecture: One hundred years of light quanta. *Reviews of Modern Physics, 78*(4), 1267–1278.

Gong, W., & Han, S. (2010). Phase-retrieval ghost imaging of complex-valued objects. *Physical Review A, 82*(2), 023828.

Goodman, J. W. (2005). *Introduction to Fourier optics* (3rd ed.). Englewood, CO: Roberts & Company.

Gori, F. (1998). Matrix treatment for partially polarized, partially coherent beams. *Optics Letters, 23*(4), 241–243.

Gori, F., Ramírez-Sánchez, V., Santarsiero, M., & Shirai, T. (2009). On genuine cross-spectral density matrices. *Journal of Optics A: Pure and Applied Optics, 11*(8), 085706.

Gori, F., Santarsiero, M., Borghi, R., & Ramírez-Sánchez, V. (2008). Realizability condition for electromagnetic Schell-model sources. *Journal of the Optical Society of America A, 25*(5), 1016–1021.

Gori, F., Santarsiero, M., Piquero, G., Borghi, R., Mondello, A., & Simon, R. (2001). Partially polarized Gaussian Schell-model beams. *Journal of Optics A: Pure and Applied Optics, 3*(1), 1–9.

Gori, F., Santarsiero, M., Vicalvi, S., Borghi, R., & Guattari, G. (1998). Beam coherence-polarization matrix. *Pure and Applied Optics: Journal of the European Optical Society Part A*, 7(5), 941–951.

Gouët, J. L., Venkatraman, D., Wong, F. N. C., & Shapiro, J. H. (2010). Experimental realization of phase-conjugate optical coherence tomography. *Optics Letters*, 35(7), 1001–1003.

Hanbury Brown, R. (1974). *The intensity interferometer: Its application to astronomy*. London: Taylor & Francis.

Hanbury Brown, R., & Twiss, R. Q. (1954). A new type of interferometer for use in radio astronomy. *Philosophical Magazine*, 45(366), 663–682.

Hanbury Brown, R., & Twiss, R. Q. (1956a). Correlation between photons in two coherent beams of light. *Nature*, 177, 27–29.

Hanbury Brown, R., & Twiss, R. Q. (1956b). Test of a new type of stellar interferometer on Sirius. *Nature*, 178(4541), 1046–1048.

Hannonen, A., Friberg, A. T., & Setälä, T. (2016). Classical spectral ghost ellipsometry. *Optics Letters*, 41(21), 4943–4946.

Hardy, N. D., & Shapiro, J. H. (2011). Reflective ghost imaging through turbulence. *Physical Review A*, 84(6), 063824.

Hariharan, P., & Ciddor, P. E. (1994). An achromatic phase-shifter operating on the geometric phase. *Optics Communications*, 110(1–2), 13–17.

Hassinen, T., Tervo, J., Setälä, T., & Friberg, A. T. (2011). Hanbury Brown–Twiss effect with electromagnetic waves. *Optics Express*, 19(16), 15188–15195.

Häusler, G., & Lindner, M. W. (1998). "Coherence radar" and "spectral radar"—New tools for dermatological diagnosis. *Journal of Biomedical Optics*, 3(1), 21–31.

Helen, S. S., Kothiyal, M. P., & Sirohi, R. S. (1998). Achromatic phase shifting by a rotating polarizer. *Optics Communications*, 154(5–6), 249–254.

Henny, M., Oberholzer, S., Strunk, C., Heinzel, T., Ensslin, K., Holland, M., & Schönenberger, C. (1999). The fermionic Hanbury Brown and Twiss experiment. *Science*, 284(5412), 296–298.

Hong, C. K., Ou, Z. Y., & Mandel, L. (1987). Measurement of subpicosecond time intervals between two photons by interference. *Physical Review Letters*, 59(18), 2044–2046.

Huang, D., Swanson, E. A., Lin, C. P., Schuman, J. S., Stinson, W. G., Chang, W., ... Fujimoto, J. G. (1991). Optical coherence tomography. *Science*, 254(5035), 1178–1181.

Jain, P., & Ralston, J. P. (2008). Direct determination of astronomical distances and proper motions by interferometric parallax. *Astronomy & Astrophysics*, 484(3), 887–895.

Jeltes, T., McNamara, J. M., Hogervorst, W., Vassen, W., Krachmalnicoff, V., Schellekens, M., ... Westbrook, C. I. (2007). Comparison of the Hanbury Brown–Twiss effect for bosons and fermions. *Nature*, 445(7126), 402–405.

Kaltenbaek, R., Lavoie, J., Biggerstaff, D. N., & Resch, K. J. (2008). Quantum-inspired interferometry with chirped laser pulses. *Nature Physics*, 4(11), 864–868.

Khakimov, R. I., Henson, B. M., Shin, D. K., Hodgman, S. S., Dall, R. G., Baldwin, K. G. H., & Truscott, A. G. (2016). Ghost imaging with atoms. *Nature*, 540(7631), 100–103.

Kiesel, H., Renz, A., & Hasselbach, F. (2002). Observation of Hanbury Brown–Twiss anticorrelations for free electrons. *Nature*, 418(6896), 392–394.

Kolner, B. H. (1994). Space-time duality and the theory of temporal imaging. *IEEE Journal of Quantum Electronics*, 30(8), 1951–1963.

Korotkova, O., Hoover, B. G., Gamiz, V. L., & Wolf, E. (2005). Coherence and polarization properties of far fields generated by quasi-homogeneous planar electromagnetic sources. *Journal of the Optical Society of America A*, 22(11), 2547–2556.

Korotkova, O., Salem, M., & Wolf, E. (2004). Beam conditions for radiation generated by an electromagnetic Gaussian Schell-model source. *Optics Letters*, 29(11), 1173–1175.

Korotkova, O., & Wolf, E. (2005). Generalized Stokes parameters of random electromagnetic beams. *Optics Letters*, *30*(2), 198–200.

Krichevsky, O., & Bonnet, G. (2002). Fluorescence correlation spectroscopy: The technique and its applications. *Reports on Progress in Physics*, *65*(2), 251–297.

Kuebel, D. (2009). Properties of the degree of cross-polarization in the space-time domain. *Optics Communications*, *282*(17), 3397–3401.

Labeyrie, A., Lipson, S. G., & Nisenson, P. (2006). *An introduction to optical stellar interferometry*. Cambridge: Cambridge University Press.

Lajunen, H., Torres-Company, V., Lancis, J., & Friberg, A. T. (2009). Resolution-enhanced optical coherence tomography based on classical intensity interferometry. *Journal of the Optical Society of America A*, *26*(4), 1049–1054.

Lavoie, J., Kaltenbaek, R., & Resch, K. J. (2009). Quantum-optical coherence tomography with classical light. *Optics Express*, *17*(5), 3818–3825.

Leitgeb, R., Hitzenberger, C. K., & Fercher, A. F. (2003). Performance of Fourier domain vs. time-domain optical coherence tomography. *Optics Express*, *11*(8), 889–894.

Li, C., Wang, T., Pu, J., Zhu, W., & Rao, R. (2010). Ghost imaging with partially coherent light radiation through turbulent atmosphere. *Applied Physics B*, *99*(3), 599–604.

Li, Y. (2014). Correlations between intensity fluctuations in stochastic electromagnetic Gaussian Schell-model beams. *Optics Communications*, *316*, 67–73.

Liu, L., Huang, Y., Chen, Y., Guo, L., & Cai, Y. (2015). Orbital angular moment of an electromagnetic Gaussian Schell-model beam with a twist phase. *Optics Express*, *23*(23), 30283–30296.

Liu, X., Wang, F., Zhang, M., & Cai, Y. (2015). Experimental demonstration of ghost imaging with an electromagnetic Gaussian Schell-model beam. *Journal of the Optical Society of America A*, *32*(5), 910–920.

Liu, Y., Nelson, J., Holzner, C., Andrews, J. C., & Pianetta, P. (2013). Recent advances in synchrotron-based hard X-ray phase contrast imaging. *Journal of Physics D*, *46*(49), 494001.

Loudon, R. (2000). *The quantum theory of light* (3rd ed.). Oxford: Oxford University Press.

Luis, A. (2007). Degree of coherence for vectorial electromagnetic fields as the distance between correlation matrices. *Journal of the Optical Society of America A*, *24*(4), 1063–1068.

Magde, D., Elson, E., & Webb, W. W. (1972). Thermodynamic fluctuations in a reacting system—Measurement by fluorescence correlation spectroscopy. *Physical Review Letters*, *29*(11), 705–708.

Malvimat, V., Wucknitz, O., & Saha, P. (2014). Intensity interferometry with more than two detectors? *Monthly Notices of the Royal Astronomical Society*, *437*(1), 798–803.

Mandel, L. (1963). Fluctuations of light beams. In E. Wolf (Ed.), *Progress in optics* (Vol. 2, pp. 181–248). Amsterdam: North-Holland.

Mandel, L., & Wolf, E. (1961). Correlation in the fluctuating outputs from two square-law detectors illuminated by light of any state of coherence and polarization. *Physical Review*, *124*(6), 1696–1702.

Mandel, L., & Wolf, E. (1965). Coherence properties of optical fields. *Reviews of Modern Physics*, *37*(2), 231–287.

Mandel, L., & Wolf, E. (1995). *Optical coherence and quantum optics*. Cambridge: Cambridge University Press.

Mazurek, M. D., Schreiter, K. M., Prevedel, R., Kaltenbaek, R., & Resch, K. J. (2013). Dispersion-cancelled biological imaging with quantum-inspired interferometry. *Scientific Reports*, *3*, 01582.

McBride, W., O'Leary, N. L., & Allen, L. J. (2004). Retrieval of a complex-valued object from its diffraction pattern. *Physical Review Letters*, *93*(23), 233902.

Meyers, R., Deacon, K. S., & Shih, Y. (2008). Ghost-imaging experiment by measuring reflected photons. *Physical Review A*, *77*(4), 041801(R).

Meyers, R. E., Deacon, K. S., & Shih, Y. (2011). Turbulence-free ghost imaging. *Applied Physics Letters, 98*(11), 111115.

Meyers, R. E., Deacon, K. S., & Shih, Y. (2012). Positive-negative turbulence-free ghost imaging. *Applied Physics Letters, 100*(13), 131114.

Miao, J., Sayre, D., & Chapman, H. N. (1998). Phase retrieval from the magnitude of the Fourier transforms of nonperiodic objects. *Journal of the Optical Society of America A, 15*(6), 1662–1669.

Mir, M., Bhaduri, B., Wang, R., Zhu, R., & Popescu, G. (2012). Quantitative phase imaging. In E. Wolf (Ed.), *Progress in optics* (Vol. 57, pp. 133–217). Amsterdam: Elsevier.

Momose, A. (2005). Recent advances in X-ray phase imaging. *Japanese Journal of Applied Physics, 44*(Pt. 1, Number 9A), 6355–6367.

Nasr, M. B., Saleh, B. E. A., Sergienko, A. V., & Teich, M. C. (2003). Demonstration of dispersion-canceled quantum-optical coherence tomography. *Physical Review Letters, 91*(8), 083601.

Nasr, M. B., Saleh, B. E. A., Sergienko, A. V., & Teich, M. C. (2004). Dispersion-cancelled and dispersion-sensitive quantum optical coherence tomography. *Optics Express, 12*(7), 1353–1362.

Ofir, A., & Ribak, E. N. (2006a). Offline, multidetector intensity interferometers—I. Theory. *Monthly Notices of the Royal Astronomical Society, 368*(4), 1646–1651.

Ofir, A., & Ribak, E. N. (2006b). Offline, multidetector intensity interferometers—II. Implications and applications. *Monthly Notices of the Royal Astronomical Society, 368*(4), 1652–1656.

Oliver, W. D., Kim, J., Liu, R. C., & Yamamoto, Y. (1999). Hanbury Brown and Twiss-type experiment with electrons. *Science, 284*(5412), 299–301.

Öttl, A., Ritter, S., Köhl, M., & Esslinger, T. (2005). Correlations and counting statistics of an atom laser. *Physical Review Letters, 95*(9), 090404.

Panasenko, D., & Fainman, Y. (2002). Single-shot sonogram generation for femtosecond laser pulse diagnostics by use of two-photon absorption in a silicon CCD camera. *Optics Letters, 27*(16), 1475–1477.

Pe'er, A., Bromberg, Y., Dayan, B., Silberberg, Y., & Friesem, A. A. (2007). Broadband sum-frequency generation as an efficient two-photon detector for optical tomography. *Optics Express, 15*(14), 8760–8769.

Pelliccia, D., Rack, A., Scheel, M., Cantelli, V., & Paganin, D. M. (2016). Experimental X-ray ghost imaging. *Physical Review Letters, 117*(11), 113902.

Pittman, T. B., Shih, Y. H., Strekalov, D. V., & Sergienko, A. V. (1995). Optical imaging by means of two-photon quantum entanglement. *Physical Review A, 52*(5), R3429–R3432.

Purcell, E. M. (1956). Question of correlation between photons in coherent light rays. *Nature, 178*(4548), 1449–1450.

Réfrégier, P., & Goudail, F. (2005). Invariant degrees of coherence of partially polarized light. *Optics Express, 13*(16), 6051–6060.

Resch, K. J., Puvanathasan, P., Lundeen, J. S., Mitchell, M. W., & Bizheva, K. (2007). Classical dispersion-cancellation interferometry. *Optics Express, 15*(14), 8797–8804.

Roth, J. M., Murphy, T. E., & Xu, C. (2002). Ultrasensitive and high-dynamic-range two-photon absorption in a GaAs photomultiplier tube. *Optics Letters, 27*(23), 2076–2078.

Roychowdhury, H., & Korotkova, O. (2005). Realizability conditions for electromagnetic Gaussian Schell-model sources. *Optics Communications, 249*(4–6), 379–385.

Roychowdhury, H., & Wolf, E. (2003). Determination of the electric cross-spectral density matrix of a random electromagnetic beam. *Optics Communications, 226*(1–6), 57–60.

Roychowdhury, H., & Wolf, E. (2005). Statistical similarity and the physical significance of complete spatial coherence and complete polarization of random electromagnetic beams. *Optics Communications, 248*(4–6), 327–332.

Rumi, M., & Perry, J. W. (2010). Two-photon absorption: An overview of measurements and principles. *Advances in Optics and Photonics, 2*(4), 451–518.

Ryczkowski, P., Barbier, M., Friberg, A. T., Dudley, J. M., & Genty, G. (2016). Ghost imaging in the time-domain. *Nature Photonics, 10*(3), 167–170.

Ryczkowski, P., Turunen, J., Friberg, A. T., & Genty, G. (2016). Experimental demonstration of spectral intensity optical coherence tomography. *Scientific Reports, 6*, 22126.

Saleh, B. E. A., & Teich, M. C. (2007). *Fundamentals of photonics* (2nd ed.). Hoboken, NJ: John Wiley & Sons, Inc.

Scarcelli, G., Berardi, V., & Shih, Y. (2006). Can two-photon correlation of chaotic light be considered as correlation of intensity fluctuations? *Physical Review Letters, 96*(6), 063602.

Scarcelli, G., Berardi, V., & Shih, Y. H. (2007). Scarcelli, Berardi, and Shih reply. *Physical Review Letters, 98*(3), 039302.

Schellekens, M., Hoppeler, R., Perrin, A., Gomes, J. V., Boiron, D., Aspect, A., & Westbrook, C. I. (2005). Hanbury Brown Twiss effect for ultracold quantum gases. *Science, 310*(5748), 648–651.

Setälä, T., Shirai, T., & Friberg, A. T. (2010). Fractional Fourier transform in temporal ghost imaging with classical light. *Physical Review A, 82*(4), 043813.

Setälä, T., Tervo, J., & Friberg, A. T. (2004). Complete electromagnetic coherence in the space-frequency domain. *Optics Letters, 29*(4), 328–330.

Shapiro, J. H. (2008). Computational ghost imaging. *Physical Review A, 78*(6), 061802(R).

Shapiro, J. H., & Boyd, R. W. (2012). The physics of ghost imaging. *Quantum Information Processing, 11*(4), 949–993.

Shevchenko, A., Roussey, M., Friberg, A. T., & Setälä, T. (2015). Ultrashort coherence times in partially polarized nonstationary optical beams measured by two-photon absorption. *Optics Express, 23*(24), 31274–31285.

Shirai, T. (2005). Polarization properties of a class of electromagnetic Gaussian Schell-model beams which have the same far-zone intensity distribution as a fully coherent laser beam. *Optics Communications, 256*(4–6), 197–209.

Shirai, T. (2015). Modifications of intensity-interferometric spectral-domain optical coherence tomography with dispersion cancellation. *Journal of Optics, 17*(4), 045605.

Shirai, T. (2016a). Improving image quality in intensity-interferometric spectral-domain optical coherence tomography. *Journal of Optics, 18*(7), 075601.

Shirai, T. (2016b). Phase difference enhancement with classical intensity interferometry. *Optics Communications, 380*, 239–244.

Shirai, T., & Friberg, A. T. (2013). Resolution improvement in spectral-domain optical coherence tomography based on classical intensity correlations. *Optics Letters, 38*(2), 115–117.

Shirai, T., & Friberg, A. T. (2014). Intensity-interferometric spectral-domain optical coherence tomography with dispersion cancellation. *Journal of the Optical Society of America A, 31*(2), 258–263.

Shirai, T., Kellock, H., Setälä, T., & Friberg, A. T. (2011). Visibility in ghost imaging with classical partially polarized electromagnetic beams. *Optics Letters, 36*(15), 2880–2882.

Shirai, T., Kellock, H., Setälä, T., & Friberg, A. T. (2012). Imaging through an aberrating medium with classical ghost diffraction. *Journal of the Optical Society of America A, 29*(7), 1288–1292.

Shirai, T., Korotkova, O., & Wolf, E. (2005). A method of generating electromagnetic Gaussian Schell-model beams. *Journal of Optics A: Pure and Applied Optics, 7*(5), 232–237.

Shirai, T., Setälä, T., & Friberg, A. T. (2010). Temporal ghost imaging with classical nonstationary pulsed light. *Journal of the Optical Society of America B, 27*(12), 2549–2555.

Shirai, T., Setälä, T., & Friberg, A. T. (2011). Ghost imaging of phase objects with classical incoherent light. *Physical Review A*, *84*(4), 041801(R).

Shirai, T., & Wolf, E. (2004). Coherence and polarization of electromagnetic beams modulated by random phase screens and their changes on propagation in free space. *Journal of the Optical Society of America A*, *21*(10), 1907–1916.

Shirai, T., & Wolf, E. (2007). Correlations between intensity fluctuations in stochastic electromagnetic beams of any state of coherence and polarization. *Optics Communications*, *272*(2), 289–292.

Siegman, A. E. (1986). *Lasers*. Sausalito, CA: University Science Books.

Strekalov, D. V., Sergienko, A. V., Klyshko, D. N., & Shih, Y. H. (1995). Observation of two-photon ghost interference and diffraction. *Physical Review Letters*, *74*(18), 3600–3603.

Sun, B., Edgar, M. P., Bowman, R., Vittert, L. E., Welsh, S., Bowman, A., & Padgett, M. J. (2013). 3D computational imaging with single-pixel detectors. *Science*, *340*(6134), 844–847.

Tanaka, Y., Sako, N., Kurokawa, T., Tsuda, H., & Takeda, M. (2003). Profilometry based on two-photon absorption in a silicon avalanche photodiode. *Optics Letters*, *28*(6), 402–404.

Teich, M. C., Saleh, B. E. A., Wong, F. N. C., & Shapiro, J. H. (2012). Variations on the theme of quantum optical coherence tomography: A review. *Quantum Information Processing*, *11*(4), 903–923.

Tervo, J., Setälä, T., & Friberg, A. T. (2003). Degree of coherence for electromagnetic fields. *Optics Express*, *11*(10), 1137–1143.

Torres-Company, V., Lajunen, H., Lancis, J., & Friberg, A. T. (2008). Ghost interference with classical partially coherent light pulses. *Physical Review A*, *77*(4), 043811.

Valencia, A., Scarcelli, G., D'Angelo, M., & Shih, Y. (2005). Two-photon imaging with thermal light. *Physical Review Letters*, *94*(6), 063601.

Visser, T. D., Kuebel, D., Lahiri, M., Shirai, T., & Wolf, E. (2009). Unpolarized light beams with different coherence properties. *Journal of Modern Optics*, *56*(12), 1369–1374.

Volkov, S. N., James, D. F. V., Shirai, T., & Wolf, E. (2008). Intensity fluctuations and the degree of cross-polarization in stochastic electromagnetic beams. *Journal of Optics A: Pure and Applied Optics*, *10*(5), 055001.

Wang, L. G., Qamar, S., Zhu, S. Y., & Zubairy, M. S. (2009). Hanbury Brown–Twiss effect and thermal light ghost imaging: A unified approach. *Physical Review A*, *79*(3), 033835.

Wojtkowski, M., Srinivasan, V. J., Ko, T. H., Fujimoto, J. G., Kowalczyk, A., & Duker, J. S. (2004). Ultrahigh-resolution, high-speed, Fourier domain optical coherence tomography and methods for dispersion compensation. *Optics Express*, *12*(11), 2404–2422.

Wolf, E. (2003). Unified theory of coherence and polarization of random electromagnetic beams. *Physics Letters A*, *312*(5-6), 263–267.

Wolf, E. (2007). *Introduction to the theory of coherence and polarization of light*. Cambridge: Cambridge University Press.

Wu, G., & Visser, T. D. (2014a). Correlation of intensity fluctuations in beams generated by quasi-homogeneous sources. *Journal of the Optical Society of America A*, *31*(10), 2152–2159.

Wu, G., & Visser, T. D. (2014b). Hanbury Brown–Twiss effect with partially coherent electromagnetic beams. *Optics Letters*, *39*(9), 2561–2564.

Yaqoob, Z., Wu, J., & Yang, C. (2005). Spectral-domain optical coherence tomography: A better OCT imaging strategy. *BioTechniques*, *39*(Suppl. 6), S6–S13.

Yasuda, M., & Shimizu, F. (1996). Observation of two-atom correlation of an ultracold neon atomic beam. *Physical Review Letters*, *77*(15), 3090–3093.

Yu, H., Lu, R., Han, S., Xie, H., Du, G., Xiao, T., & Zhu, D. (2016). Fourier-transform ghost imaging with hard X rays. *Physical Review Letters*, *117*(11), 113901.

Zerom, P., Piredda, G., Boyd, R. W., & Shapiro, J. H. (2009). Optical coherence tomography based on intensity correlations of quasi-thermal light. In *2009 conference on lasers and electro-optics (CLEO)*. (JWA48.pdf).

Zhang, M., Wei, Q., Shen, X., Liu, Y., Liu, H., Cheng, J., & Han, S. (2007). Lensless Fourier-transform ghost imaging with classical incoherent light. *Physical Review A, 75*(2), 021803(R).

Zhang, Y., & Zhao, D. (2015). Hanbury Brown–Twiss effect with partially coherent electromagnetic beams scattered by a random medium. *Optics Communications, 350*, 1–5.

CHAPTER TWO

Optical Testing and Interferometry

Daniel Malacara-Hernández, Daniel Malacara-Doblado
Centro de Investigaciones en Óptica, Colonia Lomas del Campestre, León, Guanajuato, Mexico

Contents

1. Wavefront Representation and Its Characteristics 74
   1.1 Mathematical Wavefront Representations 74
   1.2 Transverse Aberrations 78
   1.3 Least Squares Fitting 79
   1.4 Gram–Schmidt Orthogonalization 81
   1.5 Zernike Polynomials 82
   1.6 Aspheric Optical Surface Representation 85
2. Tests That Measure Wavefront Distortions 89
   2.1 Newton Interferometer 90
   2.2 Fizeau Interferometer 93
   2.3 Twyman–Green Interferometer 96
   2.4 Common Path Interferometers 101
3. Tests That Measure Transverse Aberrations 104
   3.1 Foucault or Knife-Edge Test 104
   3.2 Ronchi Test 107
   3.3 Hartmann Test 111
   3.4 Shack–Hartmann and Other Modified Hartmann Tests 112
   3.5 Lateral Shearing Interferometers 116
4. Tests That Measure Curvature 120
   4.1 Hartmann Test With Four Sampling Points 120
   4.2 Irradiance Transport Equation 121
5. Interferogram Analysis 128
   5.1 Sparse Sampling of the Fringes 128
   5.2 Digital Interferometry 129
   5.3 Single Interferogram Analysis With a Spatial Carrier 130
6. Phase Shifting Interferometry 132
   6.1 Instrumentation 132
   6.2 Algorithms 135
7. Testing of Aspherical Surfaces 142
   7.1 Autocompensating Configurations 143
   7.2 Compensators to Test Aspherical Surfaces 146
   7.3 Wavefront Stitching 150
   7.4 Two Wavelengths Measurements 151
References 152

Progress in Optics, Volume 62
ISSN 0079-6638
http://dx.doi.org/10.1016/bs.po.2016.12.001

© 2017 Elsevier B.V.
All rights reserved.

1. WAVEFRONT REPRESENTATION AND ITS CHARACTERISTICS

The components in modern optical systems require a high quality, which increases day-by-day. The reason is that new technologies in many areas like astronomical telescopes and photolithographic techniques used in electronics are demanding ever better optical instruments. During the manufacturing steps, the optical surfaces of the optical components must be accurately polished and configured to obtain the correct shape within a strict tolerance. But we must remember that a surface can be made only as good as it can be tested.

The newer optical designs have optical surfaces with more complicated and accurate shapes than a few years ago. Even off-axis and asymmetrical surfaces are often required. The optical quality should be such that the surface departures from the ideal theoretical shape are quite small, frequently a small fraction of the wavelength of the visible light.

Testing methods to measure the shape of optical surfaces have been developed for many decades. Recently, new methods have been discovered due to the advent of newer tools, like computers, lasers, light detectors, etc. In this chapter, the most common methods to test optical surfaces will be described. Most modern methods can be considered as modifications of the older classical methods, where the new tools make them more automatic, accurate, precise, and simpler.

Some instruments measure the wavefront deformations and some others measure the transverse aberrations. Next, we will describe the mathematical representations used for the wavefront deformations and the transverse aberrations and also the relations between them.

1.1 Mathematical Wavefront Representations

The simplest and natural representation of the wavefront deformations in an optical system can be written in Cartesian coordinates, up to the Kth power as:

$$W(x, y) = \sum_{i=0}^{K} \sum_{j=0}^{i} a_{i,j} x^i y^{i-j}$$
$$= a_{00} + a_{10}x + a_{11}y + a_{20}x^2 + a_{21}xy + a_{22}y^2 + \cdots + a_{KK}y^K \qquad (1)$$

We may observe that there are $n+1$ terms for each power n. For example, there are three terms for power two, four terms for the power three, and so on.

However with the exception of the piston term and the tilts, the well-known Seidel aberrations are represented by the combination of two or more monomials, they cannot be represented by just a monomial term. Kingslake (1925–1926) proposed representing the Seidel aberrations by

$$W(x, y) = A + Bx + Cy + D(x^2 + y^2) + E(x^2 - y^2) \\ + F(x^2 + y^2)y + G(x^2 + y^2)^2 \quad (2)$$

where these aberrations are:
A = piston or constant term;
B = tilt about the y-axis (image displacement along the x-axis);
C = tilt about the x-axis (image displacement along the y-axis);
D = defocusing;
E = astigmatism;
F = coma; and
G = spherical aberration.

A limitation of this expression is that this wavefront is for axially symmetric optical systems, whose wavefront has symmetry about the z-axis. A more general expression is convenient for optical testing. Also, since most pupils have a circular shape, it seems more natural to write the expression for the aberrations in polar coordinates. Then, the first characteristic to be defined is if the angle θ is measured from the x- or y-axes. If only axially symmetric systems are considered, the wavefronts exiting from the system would have symmetry about the z-axis. In this case, the common practice in optical design was to measure the angle from the y-axis. However in optical testing, this symmetry is not common and thus it is more natural to measure the angle counter-clockwise from the x-axis. Thus, we have:

$$x = \rho\cos\theta \quad \text{and} \quad y = \rho\sin\theta \quad (3)$$

The terms can be of the form $\rho^n \cos^m\theta$ and $\rho^n \sin^m\theta$ or alternatively, of the form $\rho^n \cos m\theta$ and $\rho^n \sin m\theta$. The two forms are frequent, but here, the second form will be used. On the other hand, Hopkins (1950) has pointed out that when using polar coordinates, in order to have a single valued function the following conditions must be satisfied in polar coordinates: (a) the value of m should be smaller than or equal to the value of n and (b) the sum $n + m$ should be even. In other words, n and m should be both odd or both even. These aberration terms, up to the fourth power are in Table 1.

Table 1 Aberration Terms $V(\rho,\theta)$

n	m	r	l	Polar Coordinates	Cartesian Coordinates	Name
0	0	0	0	1	1	Piston
1	0	1	1	$r\sin\theta$	y	Tilt about x-axis
1	1	2	1	$r\cos\theta$	x	Tilt about y-axis
2	0	3	0	r^2	x^2+y^2	Defocusing
2	1	4	2	$r^2\sin 2\theta$	$2xy$	Astigmatism, axis at ± 45 degree
2	2	5	2	$r^2\cos 2\theta$	x^2-y^2	Astigmatism, axis at 0 degree or 90 degree
3	0	6	1	$r^3\sin\theta$	$(x^2+y^2)y$	Coma, along y-axis
3	1	7	1	$r^3\cos\theta$	$(x^2+y^2)x$	Coma, along x-axis
3	2	8	3	$r^3\sin 3\theta$	$(3x^2-y^2)y$	Triangular astigmatism, semiaxes at 30, 150, and 270 degree
3	3	9	3	$r^3\cos 3\theta$	$(x^2-3y^2)x$	Triangular astigmatism, semiaxes at 0, 120, and 240 degree
4	0	10	0	r^4	$(x^2+y^2)^2$	Spherical aberration
4	1	11	2	$r^4\sin 2\theta$	$2(x^2+y^2)xy$	Fifth-order astigmatism at ± 45 degree
4	2	12	2	$r^4\cos 2\theta$	x^4-y^4	Fifth-order astigmatism at 0 degree or 90 degree
4	3	13	4	$r^4\sin 4\theta$	$4(x^2-y^2)xy$	Ashtray at 22.5 degree or 67.5 degree
4	4	14	4	$r^4\cos 4\theta$	$(x^2-y^2)^2-4x^2y^2$	Ashtray at 0 degree or 45 degree

The angle θ is measured counter-clockwise from the x-axis. The triangular astigmatism is also called trefoil aberration in several books and articles.

The aberration terms are all of the form:

$$V(\rho,\theta) = \rho^n \begin{bmatrix} \cos \\ \sin \end{bmatrix} (l\theta) \qquad (4)$$

where $l=0, \ldots, n$ with the condition that $n+l$ should be even. Thus, the values of m in this table have to be equal to $m=0, 1, 2, \ldots, n$ in order to satisfy the conditions just mentioned. A general consecutive index r for all these aberration terms is given by:

$$r = \frac{n(n+1)}{2} + m + 1 \tag{5}$$

so that the maximum number of r, represented by L, is equal to the total number of aberration terms, for a power equal to K, given by:

$$L = \frac{(K+1)(K+2)}{2} \tag{6}$$

In the expression for the functions $V(\rho,\theta)$ the selection of the trigonometric functions, sine or cosine, and the value of l are determined as in Table 2.

The value of l can also be calculated as:

$$l = m - \frac{1-(-1)^{n+m}}{2} \tag{7}$$

The values of n and m can be calculated from the value of the consecutive index r, as follows:

$$n = \text{next integer greater than } \frac{-3+(1+8r)^{1/2}}{2} \quad \text{and} \quad m = r - \frac{n(n+1)}{2} - 1 \tag{8}$$

Thus, the expression for the wavefront deformation is the following:

$$\begin{aligned} W(\rho,\theta) &= \sum_{n=0}^{K}\sum_{m=0}^{n} A_l \rho^n \binom{\cos}{\sin} l\theta \\ &= A_0 + A_1 \rho \sin\theta + A_2 \rho \cos\theta + A_3 \rho^2 + A_4 \rho^2 \sin 2\theta \\ &\quad + A_5 \rho^2 \cos 2\theta + A_6 \rho^3 \sin\theta + A_7 \rho^3 \cos\theta + A_8 \rho^3 \sin 3\theta \\ &\quad + A_9 \rho^3 \cos 3\theta + \cdots \end{aligned} \tag{9}$$

Table 2 Determination of the Trigonometric Function sin or cos and the Value of l

$n+m$ Parity	Value of l	Trigonometric Function
Even	m	Cos
Odd	$m-1$	Sin

1.2 Transverse Aberrations

The transverse aberrations are the ray deviations, measured in a plane near the focal plane, of the rays coming from the wavefront, with respect to their ideal intersections with that plane, as illustrated in Fig. 1.

The mathematical relations between the wavefront deformations and the transverse aberration can be illustrated with the help of Fig. 2. The exact expressions relating these two quantities have been given by Rayces (1964) as follows:

$$\frac{\partial W(\rho, \theta)}{\partial x} = \frac{TA_x}{r_w - W(x, y)} \quad \text{and} \quad \frac{\partial W(\rho, \theta)}{\partial y} = \frac{TA_y}{r_w - W(x, y)} \qquad (10)$$

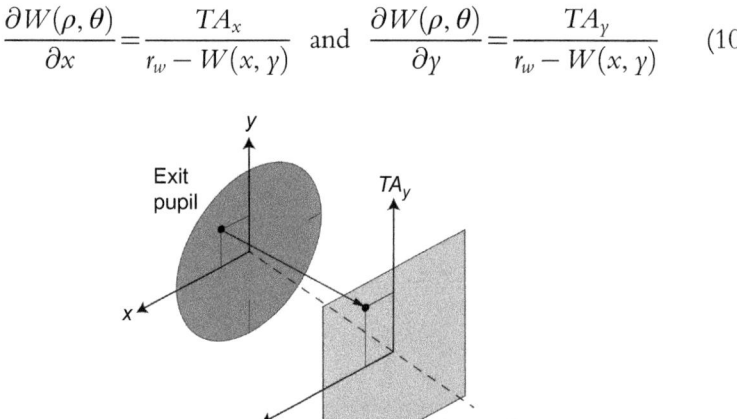

Fig. 1 Transverse aberration measured at an observation plane near the focal plane.

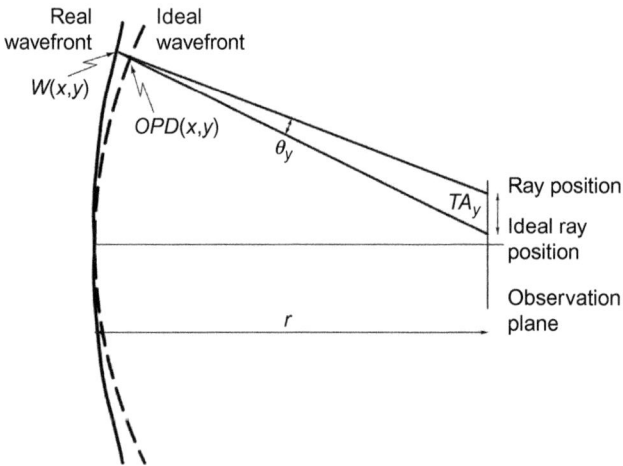

Fig. 2 Relation between wavefront deformations and transverse aberrations.

where r_w is the radius of curvature of the wavefront if the transverse aberrations are measured at the radius of curvature, otherwise, it is the distance from the exit pupil to the plane was the transverse aberrations are measured.

The value of $W(x,y)$ is so small compared with r that an approximate version of these relations taking only the value of r in the denominator of the right-hand side of these equations is normally used, retaining a high accuracy. Then, we can write:

$$TA_\rho = -r_w \frac{\partial W(\rho, \theta)}{\partial \rho}$$
$$= -r_w \begin{pmatrix} A_1 \cos\theta + A_2 \sin\theta + 2A_3\rho + 2A_4\rho\cos 2\theta + 2A_5\rho\sin 2\theta \\ +3A_6\rho^2 \cos\theta + 3A_7\rho^2 \sin\theta + 3A_8\rho^2 \cos 3\theta + 3A_9\rho^2 \sin 3\theta \end{pmatrix} \quad (11)$$

and

$$TA_\theta = -\frac{r_w}{\rho} \frac{\partial W(\rho, \theta)}{\partial \theta}$$
$$= -r_w \begin{pmatrix} -A_1 \sin\theta + A_2 \cos\theta - 2A_4\rho\sin 2\theta + 2A_5\rho\cos 2\theta \\ -A_6\rho^2 \sin\theta + A_7\rho^2 \cos\theta - 3A_8\rho^2 \sin 3\theta + 3A_9\rho^2 \cos 3\theta \end{pmatrix} \quad (12)$$

where r_w is the radius of curvature of the reference sphere, or the distance from the exit pupil to the observation plane, as explained earlier.

1.3 Least Squares Fitting

Let us assume that the wavefront of an optical system can be represented by a linear combination of L linearly independent polynomials $G_r(x,y)$, up to degree K, as follows:

$$W(x, y) = \sum_{r=0}^{L} a_r G_r(x, y) \quad (13)$$

where L is the number of polynomials, given in Eq. (6).

If we sample the wavefront at an array of N sparse or regularly arranged points, as described, and then fit an analytical function represented by this expression to these sampling points, we can write a *variance* or fitting error ε to be minimized as:

$$\varepsilon = \frac{1}{N}\sum_{n=1}^{N}\left[W(x_n, y_n) - W'_n\right]^2$$

$$= \frac{1}{N}\sum_{n=1}^{N}\left[\sum_{r=0}^{L}\alpha_r G_r(x_n, y_n) - W'_n\right]^2 \qquad (14)$$

The minimization is achieved if the variance ε is minimized with respect to the α_r by making the following L derivatives equal to zero:

$$\frac{\partial \varepsilon}{\partial \alpha_k} = \frac{2}{N}\sum_{n=1}^{N}\left[\sum_{r=0}^{L}\alpha_r G_r(x_n, y_n) - W'_n\right]G_m(x_n, y_n) = 0 \qquad (15)$$

Thus obtaining:

$$\sum_{n=1}^{N}\left[\sum_{r=0}^{L}\alpha_r G_r(x_n, y_n) G_m(x_n, y_n)\right] = \sum_{n=1}^{N}W'_n G_m(x_n, y_n) \qquad (16)$$

or

$$\sum_{r=0}^{L}\alpha_r\left[\sum_{n=1}^{N}G_r(x_n, y_n) G_m(x_n, y_n)\right] = \sum_{n=1}^{N}W'_n G_m(x_n, y_n) \qquad (17)$$

This is a system of L linear equations with N unknown values of α_r.

If the functions $G_r(x,y)$ are not polynomials, but L monomials equal to $x^n y^{m-n}$ with $n = 0, \ldots, m$, and $m = 0, \ldots, K$, such that:

$$W(x, y) = \sum_{m=0}^{K}\sum_{n=0}^{m}\alpha_{nm}x^n y^{m-n} = \sum_{r=0}^{L}\alpha_r x^n y^{m-n} \qquad (18)$$

where there is a corresponding value of n and a value m for each value of r, the matrix is of size L and the system cannot be easily solved. However, the matrix becomes diagonal if the functions $G_r(x,y)$ satisfy the orthogonality condition:

$$\sum_{n=1}^{N}G_r(x_n, y_n) G_m(x_n, y_n) = \delta_{r,m} \qquad (19)$$

as shown by Forsythe (1957). The system matrix is diagonal and thus it does not need to be inverted. The coefficients α_r would be given by:

$$\alpha_r = \frac{\sum_{n=1}^{N} W'_n G_m(x_n, y_n)}{\sum_{n=1}^{N} G_m^2(x_n, y_n)} \tag{20}$$

However, this orthogonality condition in general cannot be satisfied unless these functions are constructed by so-called Gram–Schmidt orthogonalization. The functions thus obtained are orthogonal only for that particular sampling point set for which they were constructed and not for any other.

1.4 Gram–Schmidt Orthogonalization

If we want to obtain the polynomials $G_m(x,y)$ that satisfy the orthogonality condition in Eq. (18), we start by taking as the initial seeds a set of linearly independent, but not orthogonal polynomials, for example, the monomial aberrations in Eq. (1) or the aberrations in Table 1 and Eq. (9). Let us represent these seed polynomials in ascending order r by $U_r(x,y)$. Then, we assume that each orthogonal polynomial is equal to the seed polynomial plus a linear combination of all orthogonal preceding polynomials, as follows:

$$\begin{aligned} G_1 &= U_1 \\ G_2 &= U_2 + D_{21} G_1 \\ G_3 &= U_3 + D_{31} G_1 + D_{32} G_2 \\ &\vdots \\ G_r &= U_r + D_{r1} G_1 + D_{r2} G_2 + \cdots + D_{r,r-1} G_{r-1} \end{aligned} \tag{21}$$

In a general manner, this expression can be written as:

$$G_r = U_r + \sum_{s=1}^{r-1} D_{rs} G_s \tag{22}$$

where $r = 0, 1, 2, \ldots, L$. Now, since G_r has to be orthogonal with G_m, if we multiply this expression by G_m, we have:

$$\sum_{i=0}^{N} G_r G_m = \sum_{i=0}^{N} U_r G_m + D_{rm} \sum_{i=0}^{N} G_m^2 = 0 \tag{23}$$

and thus, D_{rm} can be written as:

$$D_{rm} = \frac{\sum_{n=1}^{N} U_r G_m}{\sum_{n=1}^{N} G_m^2} \qquad (24)$$

with $r = 0, 1, 2, \ldots, L$ and $m = 0, 1, 2, 3, \ldots, r-1$.

1.5 Zernike Polynomials

We have seen in the preceding section, the generation of orthogonal polynomials in a discrete base (random or regular) set of N data points. These orthogonal polynomials become the well-known Zernike polynomials when this set of data points satisfies the following two conditions:
(a) the data points are regularly spaced and N approaches infinity and
(b) the data points are inside a circle with a unit radius.
Now, we will describe some properties of this polynomial representation (Kim & Shannon, 1987; Mahajan, 2007). The Zernike polynomials are frequently represented in polar coordinates by the product of two functions, one of them a function of the radial distance ρ in the pupil and the other a function of the angle θ. Here, it is important to point out that in some older publications the angle is measured clockwise from the y-axis, but here and in more recent papers it is measured counter-clockwise from the x-axis.

The angular function is sometimes written as a complex function, real for the term containing $\cos \theta$ and imaginary for the terms containing $\sin \theta$, but an alternative manner is to separate this angular function as the sum of an antisymmetric function and one symmetric function. In this manner the Zernike polynomials will be represented by $Z(\rho,\theta)$. The Zernike polynomials are orthogonal in the unit circle. Frequently the orthogonality condition has a constant in front of the Dirac delta. In this case the polynomials are identical, just multiplied by a constant. Then, these polynomials are said to orthonormals. Instead of the orthogonality condition in Eq. (19), the new one is:

$$\int_0^1 \int_0^{2\pi} Z_n^l Z_{n'}^{l'} \rho \, d\rho \, d\theta = \delta_{nn'} \delta_{ll'} \qquad (25)$$

where in order to normalize the pupil size we define $\rho = S/S_{max}$, n is the degree of the polynomial and l is the harmonic number of the angular function. The orthonormal polynomials can be obtained by just multiplying the orthogonal polynomials by the appropriate constant.

As the monomial aberrations described earlier, the Zernike polynomials are presented with two indices n and l, on which they are dependent. The index n is the degree of the radial function of the polynomial and l is the index of the angular function. The numbers n and l are both even or both odd, making $n-l$ always even. There are $(1/2)(n+1)(n+2)$ linearly independent polynomials of degree $\leq n$, one for each pair of numbers n and l. As pointed out before, these polynomials are the product of two functions, one depending only on the radius ρ and the other being dependent only on the angle θ, as follows:

$$Z_n^l = R_n^l \begin{bmatrix} \sin \\ \cos \end{bmatrix} l\theta \qquad (26)$$

In a centered optical system, all terms with the sine function are zero and only the cosine terms remain. The degree of the radial polynomial $R_n^l(\rho)$ is n and $0 \leq m \leq n$. It may be shown that $|l|$ is the minimum exponent of these polynomials R_n^l. For this ordering of the polynomials (Singer, Totzek, & Gross, 2005), the radial polynomial is given by

$$R_n^l(\rho) = \sum_{s=0}^{n-l} (-1)^s \frac{(n-s)!}{s!\left(\frac{n+l}{2}-s\right)!\left(\frac{n-l}{2}-s\right)!} \rho^{n-2s} \qquad (27)$$

As the previously described wavefront aberration polynomials, Zernike polynomials $Z_n(\rho)$ may also be represented with a single index r, as defined before.

The difference between the monomial polynomials and the Zernike polynomials is that as a consequence of their orthogonality, they are minimized by adding to each of them the proper amount of a low order component, as piston term, tilt, and defocusing. The great advantage of minimizing each aberration polynomial with respect to the reference sphere is that the whole wavefront as well is also minimized. Subtracting or adding any of them preserves the minimization.

Table 3 illustrates the first orthogonal (not orthonormal, since any orthonormalizing constant has been ignored) Zernike polynomials, up to fourth degree. For a list of the orthonormal Zernike polynomials, see Mahajan (2007). Please notice that the ordering of the polynomials for a given power n is not the same in all publications.

The advantage of expressing the wavefront by a linear combination of orthogonal polynomials is that the wavefront deviation represented by each

Table 3 First Zernike Polynomials up to Fourth Degree, Orthogonal (Not Orthonormal) in a Unit Radius Pupil

n	m	r	l	Polar Coordinates	Cartesian Coordinates	Name
0	0	0	0	1	1	Piston
1	0	1	1	$r \sin\theta$	y	Tilt about x-axis
	1	2	1	$r \cos\theta$	x	Tilt about y-axis
2	0	3	0	$2r^2 - 1$	$-1 + 2x^2 + 2y^2$	Defocusing
	1	4	2	$r^2 \sin 2\theta$	$2xy$	Astigmatism, axis at ±45 degree
	2	5	2	$r^2 \cos 2\theta$	$x^2 - y^2$	Astigmatism, axis at 0 degree or 90 degree
3	0	6	1	$(3r^3 - 2\rho) \sin\theta$	$(-2 + 3x^2 + 3y^2)y$	Coma, along y-axis
	1	7	1	$(3r^3 - 2\rho) \cos\theta$	$(-2 + x^2 + 3y^2)x$	Coma, along x-axis
	2	8	3	$r^3 \sin 3\theta$	$(3x^2 - y^2)y$	Triangular astigmatism, semiaxes at 30, 150, and 270 degree
	3	9	3	$r^3 \cos 3\theta$	$(x^2 - 3y^2)x$	Triangular astigmatism, semiaxes at 0, 120, and 240 degree
4	0	10	0	$6r^4 - 6r^2 + 1$	$(x^2 + y^2)^2$	Spherical aberration
	1	11	2	$(4r^4 - 3r^2) \sin 2\theta$	$-4x^4 + 4y^4 + 3x^2 - 3y^2$	Fifth-order astigmatism at ±45 degree
	2	12	2	$(4r^4 - 3r^2) \cos 2\theta$	$(-6 + 8y^2 + 8x^2)xy$	Fifth-order astigmatism at 0 degree or 90 degree
	3	13	4	$r^4 \sin 4\theta$	$4(x^2 - y^2)xy$	Ashtray at 22.5 degree or 67.5 degree
	4	14	4	$r^4 \cos 4\theta$	$(x^2 - y^2)^2 - 4x^2 y^2$	Ashtray at 0 degree or 45 degree

The angle θ is measured from the y-axis.

term is a best fit. Then, any combination of these terms must also be a best fit. In other words, the mean wavefront deformation for all Zernike polynomials is zero, with the exception of the piston term.

The first 15 Zernike polynomials are shown in Table 3. Some Zernike polynomials are illustrated in the isometric plots in Fig. 3.

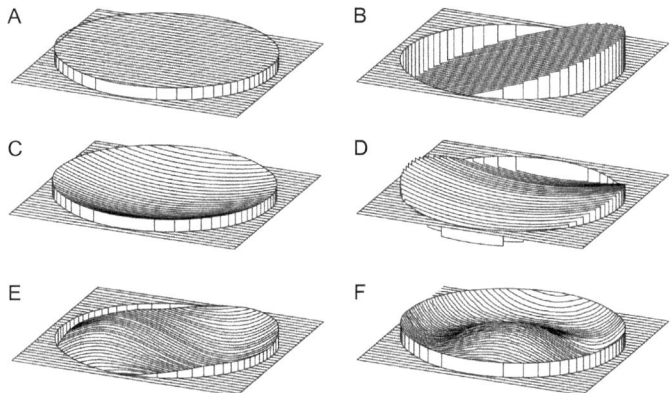

Fig. 3 Isometric plots of some Zernike polynomials: (A) Piston term, (B) tilt, (C) defocusing, (D) astigmatism, (E) coma, and (F) spherical aberration.

1.6 Aspheric Optical Surface Representation

We have described at the beginning of this chapter, the mathematical representation of a wavefront with any kind of aspheric deformations. This representation is for aspheric surfaces with rotational symmetry. However, optical surfaces may be nonrotationally symmetric.

To understand and test aspherical surfaces, the first task is to mathematically specify the surface shape. The testing of aspherical surfaces is extremely important and has been studied and described by many authors, for example, by Schulz (1988). Since, as we pointed out, the most common aspheric surfaces have rotational symmetry, they may be defined by means of the following relation, taking the z-axis as the axis of revolution:

$$z = \frac{cS^2}{1 + [1 - (K+1)c^2 S^2]^{1/2}} + A_1 S^4 + A_s S^6 + A_3 S^8 + A_4 S^{10} \qquad (28)$$

where $S^2 = x^2 + y^2$ and $c = 1/r = 1/\text{radius of curvature}$. Also, $A_1, A_2, A_3,$ and A_4 are the aspheric deformation constants and K is a function of the eccentricity of a conic surface ($K = -e^2$), called the conic constant. We can notice that a simpler expression for the sagittal of a sphere can be found, but this is better for automatic calculations that avoid singularities when the curvature is zero, i.e., for plane surfaces. A sphere touching the optical surface at its vertex and with the same radius of curvature is called an osculating sphere, as illustrated in Fig 4. If the coefficients A_i are all zero, the surface is a conic surface of revolution, according to Table 4.

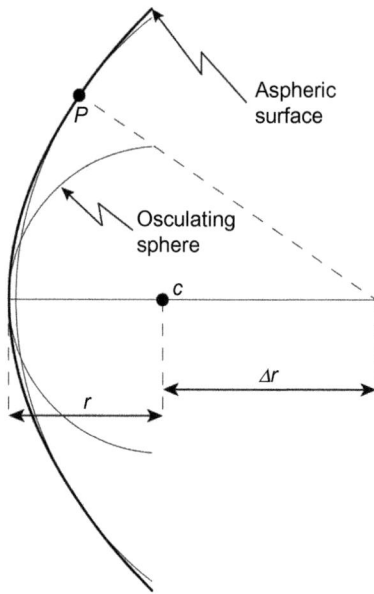

Fig. 4 Aspheric optical surface showing the osculating sphere and a sphere touching the aspheric surface along a ring passing through the point P.

Table 4 Values of Conic Constants for Conic Surfaces

Type of Conic	Conic Constant Value
Hyperboloid	$K < -1$
Paraboloid	$K = -1$
Prolate spheroid or Ellipsoid: (Ellipse rotated about its major axis)	$-1 < K < 0$
Sphere	$K = 0$
Oblate spheroid (Ellipse rotated about its minor axis)	$K > 0$

In an ellipsoid with rotational symmetry about the z-axis the eccentricity e is defined as:

$$e = \frac{\sqrt{a^2 - b^2}}{a} \tag{29}$$

where a is the major semiaxis, along the z-axis and b is the minor semiaxis, along the y-axis. For the case of the oblate spheroid the same definition holds but then a is the minor semiaxis, along the z-axis and b is the major semiaxis, along the y-axis. Thus, the eccentricity is an imaginary number and $K>0$, since $a<b$.

Sometimes it is convenient to consider an aspheric or conic optical surface as the sum of its osculating sphere plus some deformation terms, as follows:

$$z = \frac{cS^2}{1+[1-c^2S^2]^{1/2}} + B_1 S^4 + B_2 S^6 + B_3 S^8 + B_4 S^{10} \tag{30}$$

where

$$\begin{aligned} B_1 &= A_1 + \frac{[(K+1)-1]c^3}{8} \\ B_2 &= A_2 + \frac{[(K+1)^2-1]c^5}{16} \\ B_3 &= A_3 + \frac{5[(K+1)^3-1]c^7}{128} \\ B_4 &= A_4 + \frac{7[(K+1)^4-1]c^9}{256} \end{aligned} \tag{31}$$

The radial slopes of an aspherical wavefront are to be given by:

$$\frac{dz}{dS} = \frac{cS}{1+[1-(k-1)c^2S^2]^{1/2}} + 4B_1 S^3 + 6B_2 S^5 + 8B_3 S^7 + 10B_4 S^9 \tag{32}$$

The second radial derivative is given by:

$$\frac{d^2z}{dS^2} = \frac{c}{1+[1-(k-1)c^2S^2]^{3/2}} + 12B_1 S^2 + 30B_2 S^4 + 56B_3 S^6 + 90B_4 S^8 \tag{33}$$

hence, the principal curvatures, sagittal, and tangential, are given by:

$$c_t = \frac{\dfrac{d^2z}{dS^2}}{\left[1+\left(\dfrac{dz}{dS}\right)^2\right]^{3/2}} \quad \text{and} \quad c_s = \frac{\dfrac{dz}{dS}}{S\left[1+\left(\dfrac{dz}{dS}\right)^2\right]^{1/2}} \tag{34}$$

In an aspheric surface, we can imagine a sphere tangent at the surface along a circle concentric with the optical surface, touching the aspheric surface at a point P as in Fig. 4. We will define the local radius of curvature of the conic as the radius of curvature of this sphere and the local center of curvature is the center of this sphere. However, this is not the same as the tangential curvature previously defined. The local radius of curvature is equal to the radius of curvature r plus the aberration of the normals, represented by Δr. This aberration of the normals can be obtained from Eq. (32), since the slope of the line going from the point P to the local center of curvature is equal to the first derivative or slope of the aspheric surface, obtaining:

$$\Delta r = \frac{S}{\left(\dfrac{dz}{dS}\right)} + z - r \tag{35}$$

which, for conic surfaces becomes:

$$\Delta r = -Kz \tag{36}$$

An spherocylindrical surface, without aspheric deformation, with its cylindrical axis oriented in a vertical or horizontal direction, can be represented by (Menchaca & Malacara, 1986):

$$z = \frac{c_x x^2 + c_y y^2}{1 + \left[1 - \dfrac{c_x^2 x^2 + c_y^2 y^2}{x^2 + y^2}\right]^{1/2}} \tag{37}$$

This is not the only manner to represent an aspheric surface. If a high-order aspheric coefficient is slightly modified during the process of designing the optical system, only the edge of the circular surface is modified, and a very small variation introduces large surface deviations. When evaluating the surface and making numerical computations the same effect appears, introducing large numeric errors. To solve this problem, Forbes (2007) has introduced a new method where the surface deviations are not represented with respect to the tangential plane sphere, but with respect to the closest sphere. The conic constant is removed, since it can be included in the aspheric constants. These deviations with respect to the closest sphere are written as a sum of polynomials specially chosen so that none can be even approximately represented by a linear combination of the others, by making them orthogonal. Surfaces without rotational symmetry are also represented with this method (Forbes, 2011).

2. TESTS THAT MEASURE WAVEFRONT DISTORTIONS

Interferometers measure the wavefront deformations using the interference of light, by producing the interference of two wavefronts (Malacara-Hernández, 2007). One of these wavefronts is usually flat and is used as a reference, and the other is a distorted wavefront whose distortions are to be measured. With this method, small wavefront deformations, smaller than a small fraction of the wavelength of light, can be measured.

When testing an optical system, in most interferometric tests we illuminate it with a point light source, frequently from a gas laser. Then, the entrance pupil has a constant irradiation over its aperture and also has a constant phase, in other words, the wavefront is perfectly flat or spherical. Since the exit pupil is conjugate to the entrance pupil, the irradiance is also constant over its aperture. However, the wavefront may be distorted due to the presence of aberrations in the optical system under test.

In any optical test, the goal is to measure the wavefront deformations at the exit pupil. So the interferogram under analysis and the image detector must be located at the exit pupil or at one of its real images. Frequently extra optical elements must be used to image the exit pupil at the image detector.

A two-wave interferogram with a flat wavefront, and also one whose deformations are given by $W(x,y)$, are schematically represented in Fig. 5.

The amplitude $E_1(x,y)$ in the observing plane is obtained by the sum of the amplitudes of the two waves, with their corresponding relative phase $A_1(x,y)$ and $A_2(x,y)$, given by

$$E_1(x, y) = A_1(x, y) \exp i(kW(x, y)) + A_2(x, y) \exp i(kx\sin\theta) \quad (38)$$

with $k = 2\pi/\lambda$. The irradiance $I(x,y)$ is thus given by

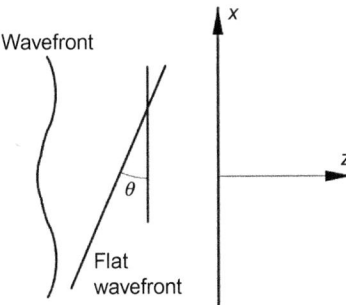

Fig. 5 Two interfering wavefronts.

$$E_1(x,y) \cdot E_1^*(x,y) = A_1^2(x,y) + A_2^2(x,y)$$
$$+ 2A_1(x,y)A_2(x,y)\cos k[x\sin\theta - W(x,y)] \quad (39)$$

where the symbol * denotes the complex conjugate.

2.1 Newton Interferometer

This is the simplest interferometer we can construct to test an optical surface. Two optical surfaces with nearly opposite shapes, one concave and one convex, or most commonly, two flats are placed on top each other as in Fig. 6, and interference fringes are observed. To understand this configuration, let us assume that the reference surface, as well as the surface under test, is flat and a tilt is present, then, the fringes look straight and parallel. If the surface to be measured is irregular the fringes are also irregular. A ray of light coming from an extended light source arrives to the two optical surfaces in close contact. Then, it is reflected at both of the surfaces, to go to the observer's eye.

In this interferometer the paths travelled before interference by the two beams are almost the same, the only difference being the air gap between the two surfaces. If this air gap has a thickness t, the optical path difference is approximately $2t$. Hence, the phase difference $\delta\phi$ is

$$\delta\phi(x,y) = 2t(x,y) \quad (40)$$

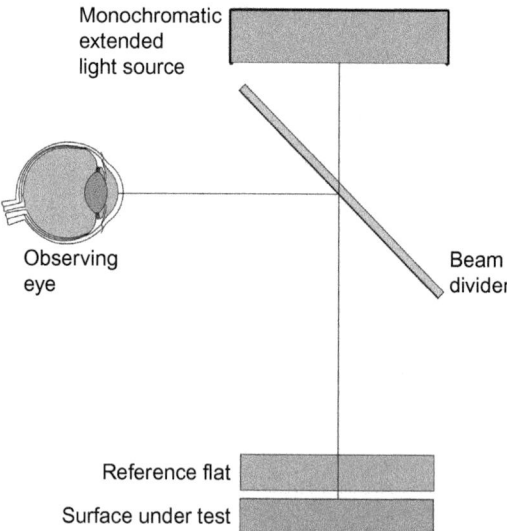

Fig. 6 Newton interferometer.

Optical Testing and Interferometry

The monochromatic extended light source can be any lamp with gas or vapor discharge, like sodium or mercury. A highly monochromatic light source is not required. Even a fluorescent lamp with a green filter can be used. However, a gas or vapor lamp without the fluorescent coating to avoid the continuum spectrum is better in order to obtain more contrast.

The air gap thickness $t(x,y)$ is related to the shape $z(x,y)$ of the surface under test by

$$t(x, y) = z(x, y) + Bx + Cy \tag{41}$$

The last two terms are any possible tilts between the two surfaces. The accuracy of a plane surface tested against a flat reference can be estimated as in Fig. 7.

The well-known Newton rings appear when the two tilts are zero and the surface and the surface under test are spherical with a long radius of curvature as shown in Fig. 8.

When a small tilt is introduced by pushing one on the surfaces on the edge, the rings are displaced in the direction of the finger pushing the edge, and in the opposite direction if the surface is convex, as in Fig. 9.

Spherical convex as well as concave surfaces can also be tested if the reference surface has the opposite curvature, as illustrated in Fig. 10.

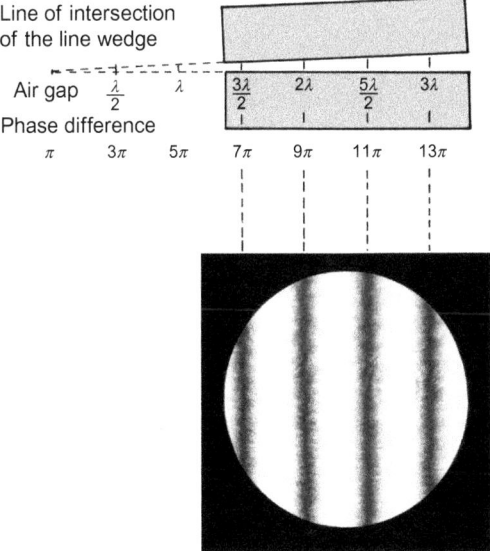

Fig. 7 Straight fringes formed by placing one flat against another in a Newton interferometer.

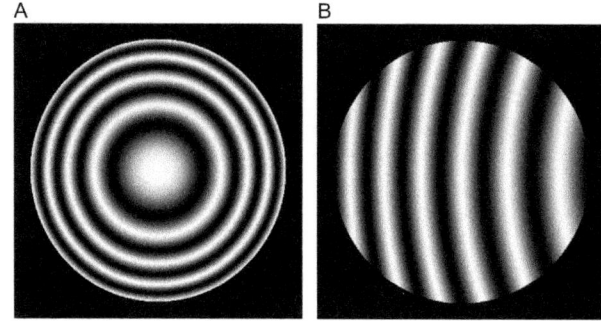

Fig. 8 (A) Concentric circular fringes when a spherical nearly plane surface is on top of a reference flat and (B) arcs formed when there is a small tilt between the two surfaces.

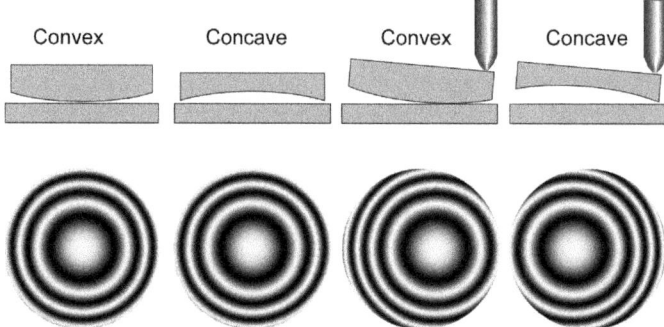

Fig. 9 Determination of the convexity or concavity of a spherical nearly flat surface on top of a reference plane.

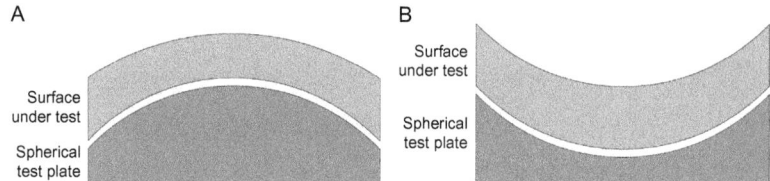

Fig. 10 A pair of spherical surfaces, one convex and one concave, one on top of the other. (A) Testing a convex surface and (B) testing a concave surface.

If the curvature is too strong, the upper surface of the glass with the surface under test has to be curved in order to observe the whole surface without the observer having to move his head.

The air gap is so small that the light source does not need to be small and collimated. A disadvantage of this interferometer is that the surface under test may be scratched when placed over the reference flat surface if the operator is not sufficiently careful.

2.2 Fizeau Interferometer

A Fizeau interferometer closely resembles a Newton interferometer, as shown in Fig. 11. The fundamental difference is that the two surfaces forming the interfering beams, which are the surface under test and the reference surface, are not in close contact. They can have a large separation.

This large separation between the interferometer and the surface under test requires that the illuminating beam is well collimated in order to have a flat wavefront. Thus, since, the light source must have a high spatial coherence, we need to have a point light source and a collimator.

Owing to the large separation between the two surfaces the light source must also be highly monochromatic. For surface separations of a few millimeters a vapor or a gas electrical discharge light source with a colored filter is sufficient but for very large separations a gas laser is necessary.

As in the Newton interferometer, if a light ray is reflected at the two surfaces of the interferometer separated by a distance of d the optical path difference OPD, as illustrated in Fig. 12, is given by

$$OPD = AB + BC - AD = 2d\cos\theta \tag{42}$$

thus, for an on-axis collimated light beam and perpendicular incidence, the optical path difference OPD_0 is

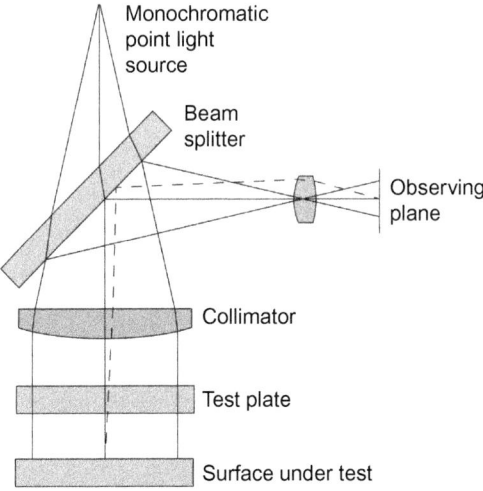

Fig. 11 Fizeau interferometer.

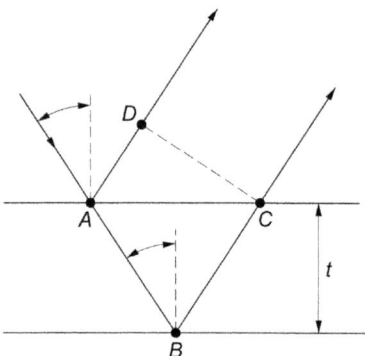

Fig. 12 Optical path difference in a Fizeau interferometer between the two nearly parallel interfering rays.

$$OPD_0 = 2d \qquad (43)$$

If the light beam is not perfectly collimated because the light source is not a point or because the collimator is defocused or because it has spherical aberration, a change in the optical path difference is introduced by an amount given by

$$\Delta OPD = OPD_0 - OPD = 2d(1 - \cos\theta) \approx d\theta^2 \qquad (44)$$

If the precision of the interferometer has to be λ/N, the maximum allowed value of θ is given by:

$$\theta = \sqrt{\frac{\lambda}{Nd}} \qquad (45)$$

Thus, the larger the distance d is, the smaller the tolerance in the angle θ is. This sets a maximum possible value for the size of the point light source or the spherical aberration of the collimator. If we have a small extended light source with semidiameter s the fringes have a reasonably high contrast as long as

$$\Delta OPD = \frac{ds^2}{f} \leq \frac{\lambda}{4} \qquad (46)$$

We can see that the light source can increase its size s if the air gap thickness d is reduced.

When the collimator lens has spherical aberration, the refracted wavefront is not perfectly flat. The maximum transverse aberration TA in this lens can be interpreted as the semidiameter s of the light source. The conclusion is that the quality requirements for the collimator lens increase as the optical path difference is also increased. If the optical path difference is zero, the collimator lens can have any magnitude of spherical aberration. If d is very small the tolerance in the angle θ is quite large, which is the case of the Newton interferometer.

A Fizeau interferometer can have different configurations to test plane glass plates as in Fig. 13. If the reference surface and the surface under test are in the same glass plate we can measure the angle between the two plane surfaces. In this second case, also illustrated in Fig. 13, the optical path difference OPD for nearly normal incidence, would be given by:

$$OPD = 2nd \qquad (47)$$

Thus, we can test the glass homogeneity if we are sure that the surfaces are flat or, the flatness and parallelism of the surfaces if we are sure about the homogeneity of the glass.

Concave and convex surfaces can also be tested, using both, a plane or a concave reference surface. Fig. 14 shows the testing of a concave surface, using a flat or a concave reference surface.

Fig. 15 illustrates the testing arrangement of a convex surface, also using a flat or a concave reference surface.

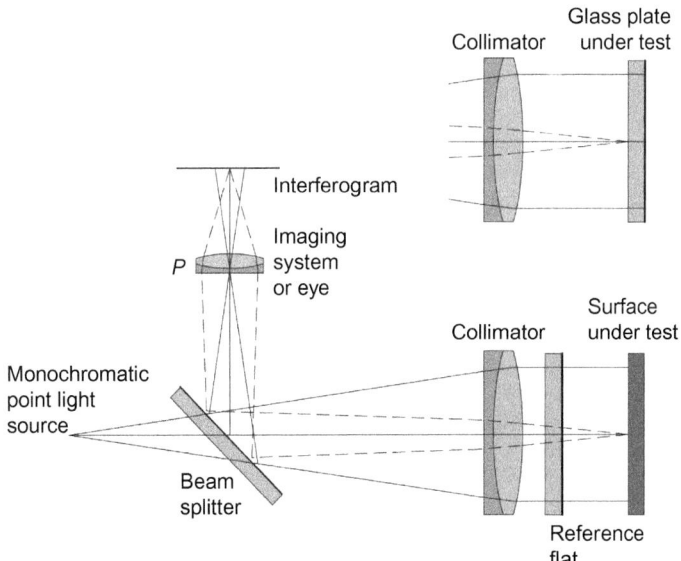

Fig. 13 Fizeau interferometer configuration to test an optical surface or a glass plate.

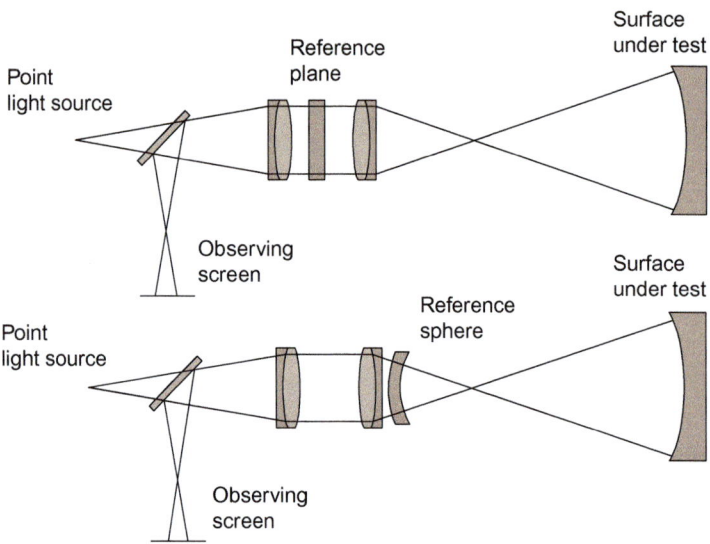

Fig. 14 Fizeau interferometer configurations to test concave surfaces.

2.3 Twyman–Green Interferometer

The Michelson interferometer is the basis for the Twyman–Green (Twyman, 1918) interferometer, as shown in Fig. 16. Instead of an extended light source, it has a point light source with a collimator to produce a single flat wavefront that arrives to the beam splitter as in the Fizeau interferometer.

To observe the interference pattern a convergent lens is used to project the two interfering beams at the exit pupil of the system or optical element under test, to the pupil of the eye of the observer. If the wavefront deformations are small and a convergent lens is not used, an observing screen as in Fig. 16 can be used to observe the interference fringes.

The beam splitter must have extremely flat surfaces and its material must be highly homogeneous. The nonreflecting surface must not reflect any light, to avoid spurious interference fringes. The nonreflecting face must be coated with an antireflection multilayer coating. Otherwise, we may have an incidence angle on the beam splitter with a magnitude equal to the Brewster angle and properly polarizing the incident light beam.

Another configuration possibility is illustrated in Fig. 17, where the two beams are separated with a cube polarizing beam splitter. The incident beam has to be linearly polarized at 45 degree. A half-wave phase plate has its slow or fast axis in the vertical direction, in order to produce circularly polarized light. This phase plate can be rotated a small angle to equate the irradiances of

Optical Testing and Interferometry

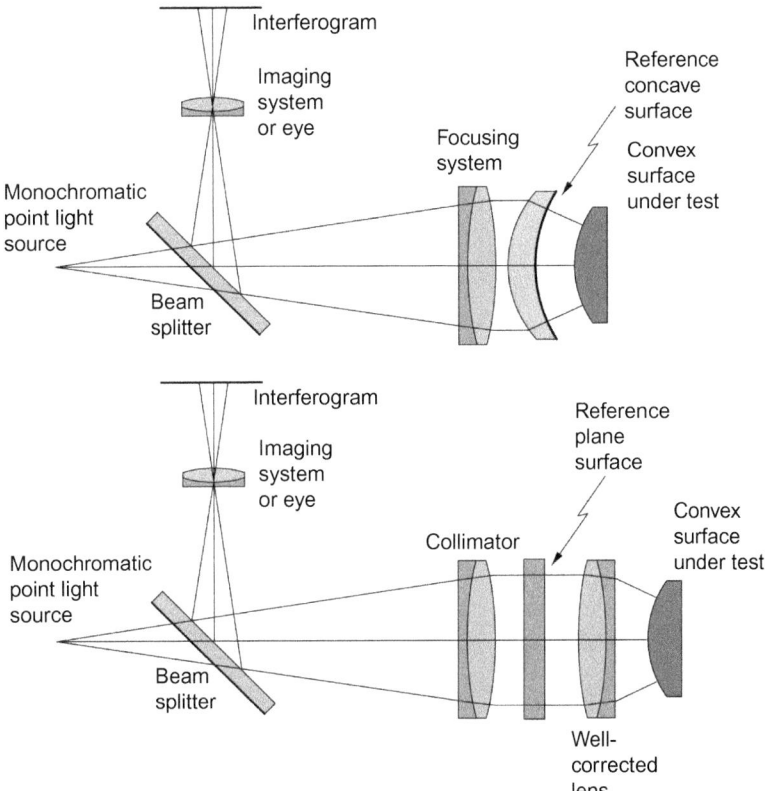

Fig. 15 Fizeau interferometer configurations to test convex surfaces.

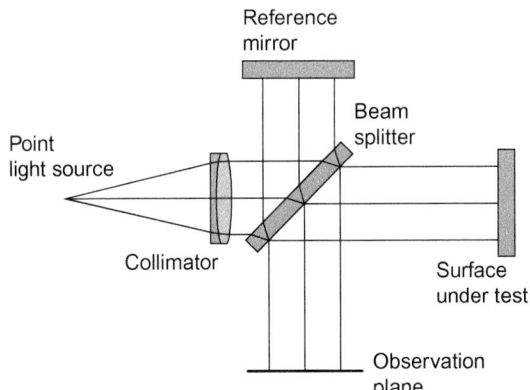

Fig. 16 Twyman–Green interferometer.

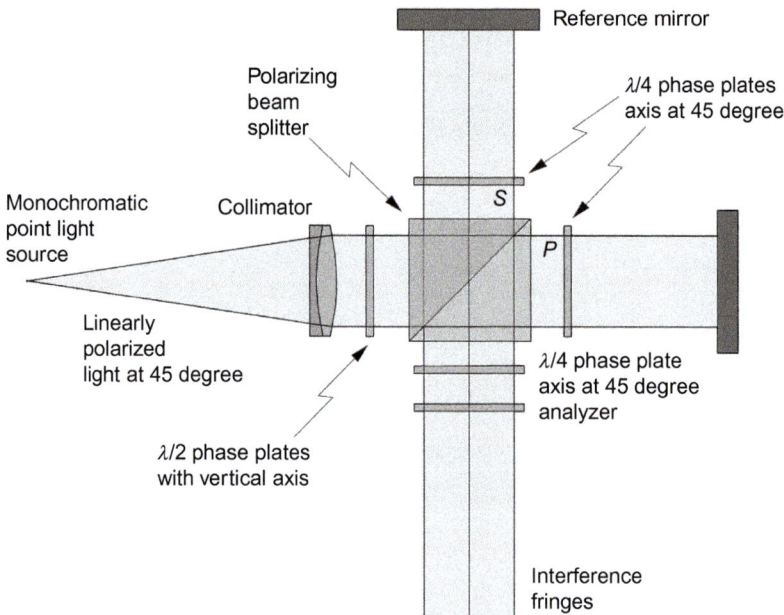

Fig. 17 Twyman–Green interferometer with a cube polarizing beam splitter.

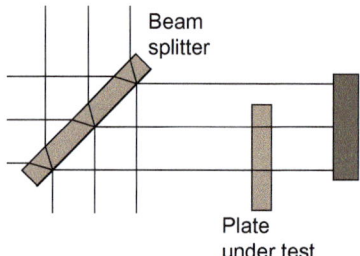

Fig. 18 Testing a glass plate in a Twyman–Green interferometer.

the two interfering beams. The beam splitter produces two beams, one linearly polarized in the horizontal direction and the other in the vertical direction. Two quarter wave phase plates at 45 degree are traversed twice making the two beams linearly polarized in the proper directions so that the two beams are directed to the interfering plane.

When testing a glass plate, inserted on the testing arm of the interferometer, as illustrated in Fig. 18, the optical path difference OPD is given by

$$OPD = (n-1)\,t \tag{48}$$

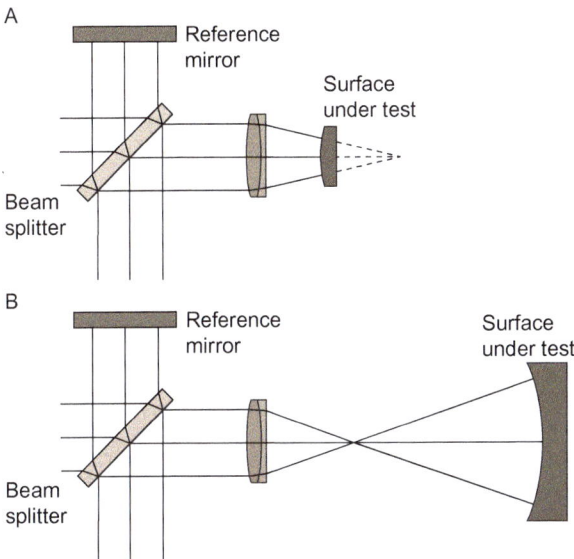

Fig. 19 Testing a lens or convex and concave surfaces in a Twyman–Green interferometer. (A) A convex surface and (B) a concave surface.

where n is the refractive of the glass plate and t is its thickness. When no fringes are present, we can conclude that $(n-1)t$ is a constant, but not n or t independently. If we compare this expression with the equivalent for the Fizeau interferometer, in Eq. (47), we see that n and t can independently be measured if both Fizeau and Twyman–Green interferometers are used.

A convex spherical mirror with its center of curvature at the focus of the lens is used to test lenses with long focal lengths, and a concave spherical mirror to test lenses with short focal lengths as shown in Fig. 19.

The convergent lens can also be tested with these configurations if the concave or convex surfaces are assumed to be perfect. A small, flat mirror at the focus of the lens can also be employed. The small region being used on the flat mirror is so small that its surface does not need to be very accurate. However, the wavefront is rotated 180 degree making the spatial coherence requirements (a gas laser is necessary) higher and canceling odd aberrations like coma.

Large astronomical mirrors can also be tested with a unequal path interferometer as described by Houston, Buccini, and O'Neill (1967). However, it must be remembered that if the collimator is not producing a perfectly flat wavefront, the optical path difference cannot be very large, as in the Fizeau interferometer. Many other optical components like prisms or diffraction gratings can be tested using this interferometer.

When testing a concave surface with a long radius of curvature the optical path difference becomes quite large. Then, the light source has to be highly monochromatic. A gas laser with a stabilized simple longitudinal mode is appropriate in this case. The Twyman–Green and Fizeau interferograms produce the same interferograms for the same aberration. Fig. 20 shows the interferogram produced by the primary aberrations.

It should be remembered that if the wavefront deformation is large or the wavefront is aspheric, the exit pupil of the system under test must be imaged on the detector with some extra optics elements that do not introduce extra wavefront deformations.

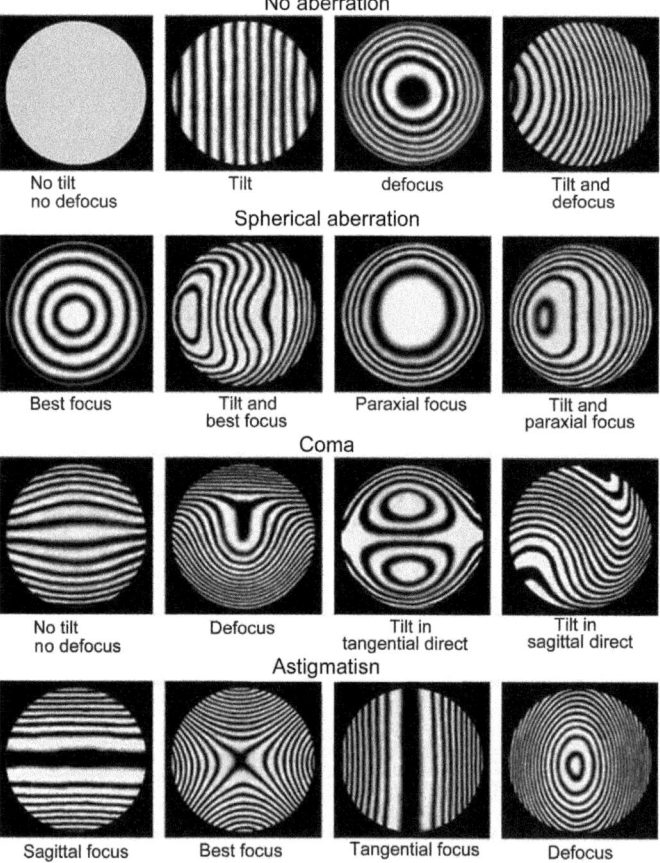

Fig. 20 Interferograms for the primary aberrations.

2.4 Common Path Interferometers

Common path interferometers are named in this manner because the wavefront under test as well as the reference wavefront travel along the same path. There are many different configurations, where the two interfering wavefronts are generated using diffraction, scattering, reflection, double refraction, or polarization phenomena.

Two common and useful examples are based on scattering on a ground and semipolished surface. In Fig. 21, an interferometer designed by Burch (1953, 1962) using a scatter plate is illustrated. A light source formed by a small lamp with a lens forms a small image of the lamp over the surface of a concave surface under test, passing through the scattering plate. Since the scattering plate is on one side of the center of curvature of the concave surface, an image of this scattering plate is formed on a second scattering surface located on the opposite side of the center of curvature. The first scattering surface lets go through it some unscattered light to form the image of the lamp at the center of the mirror, but some light is also scattered and illuminates the whole surface of the concave mirror.

The second scattering plate should be identical to the first one, but rotated 180 degree, so that the image of the first one exactly overlaps the second scattering surface. Then, to the second surface two different beams of light arrive, the light unscattered in the first plate, which is reflected at the center of the mirror, and also the beam scattered in the first plate. In this manner, the following four beams enter the observer's eye:

(a) A beam unscattered in both the first and the second scattering plates. This produces a bright small spot at the center of the observed concave surface.

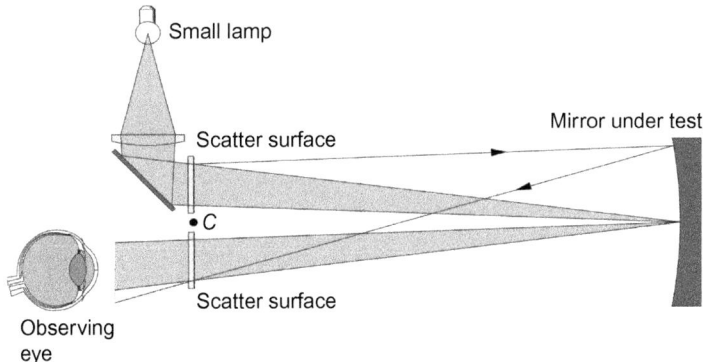

Fig. 21 Scatter plate interferometer with two identical scatter plates.

(b) A beam scattered in both plates. This is very dim, almost unobservable; it only reduces the fringes contrasts a little bit.
(c) The beam that goes through the first plate without being scattered and it is scattered at the second plate, where it is scattered in a wide angle. It carries no information about the mirror.
(d) The fourth beam is scattered in the first plate, so that it covers the whole aperture of the concave mirror. Thus, it carries information about the mirror shape. Then, it goes through the second plate without being scattered.

The last two beams interfere after going out from the second scattering plate, producing a Twyman–Green or Fizeau like fringe pattern.

We have pointed out that the two scattering plates have to be identical, but one of them rotated 180 degree with respect to the other. This is necessary so that the scattering in both plates are identical point to point. There are several procedures reported in the literature to fabricate these scattering plates. One of them is by making a double exposure photographic contact copy of a scattering plate, but rotating the original 180 degree between exposures. Another method is by using a diffracting plate instead of scattering, but with rotational symmetry.

Murty (1963) has reported a system like this with a pair of identical Fresnel zone plates. The great advantage is its full rotational symmetry and that it is simpler to make them. The useful aperture of the zone plate should be smaller than the pupil of the eye. Murty recommends an aperture of about 3.0 mm. If the interferometer is used to test a mirror with an f-number ($f/\#$) the zone plate must have a focal length equal to $1.5/(f/\#)$ and an aperture of 3 mm.

To avoid the need for a scattering or diffractive plate with the required symmetry at 180 degree, an interferometer like in Fig. 22 has been designed. The two scattering steps take place in the same scattering plate. A small flat mirror is used to reflect the light back to the scattering plate, after being reflected twice at the concave mirror. This system has some characteristics that make it different from the first system:

(a) The sensitivity to symmetric aberrations, like defocusing and spherical aberration is doubled due to the double pass through the concave mirror under test.
(b) The sensitivity to antisymmetrical aberrations, like tilt and coma in the x direction, is cancelled out. So tilt in this direction cannot be introduced.
(c) Owing to the insensitivity to antisymmetric aberrations, the fringe patterns are also more insensitive to vibrations.

Optical Testing and Interferometry 103

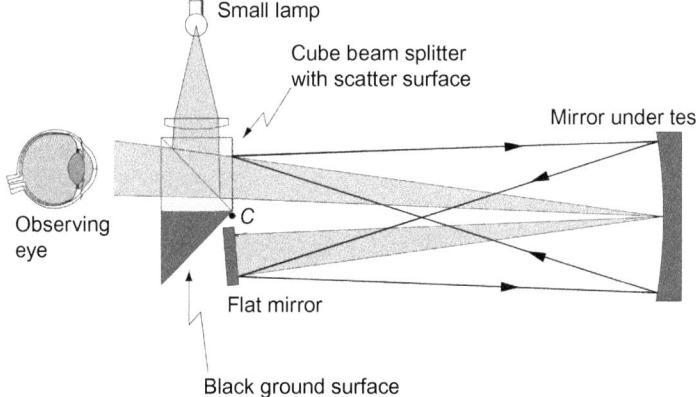

Fig. 22 Scatter plate interferometer with only one scatter plate.

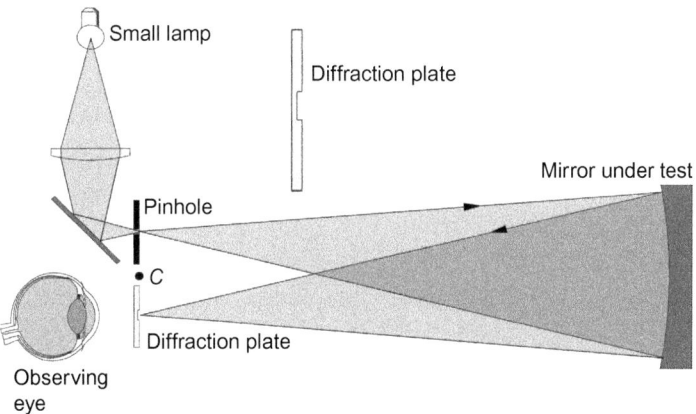

Fig. 23 Point diffraction interferometer.

(d) The mirror under test has to be coated with aluminum or silver because the double reflection at the mirror reduces the light and hence the interference pattern visibility too much.

Another popular common path interferometer first proposed by Linnik in 1933 and later by Smart and Steel (1975) is the point diffraction interferometer in Fig. 23. This interferometer has a specially designed phase plate at the point of convergence of the returning beam. This phase plate has a small circular phase step at the center. This small region can have a higher or a lower optical path than the rest of the plate, by increasing or decreasing its thickness inside this circle. Its diameter has to be around the size of the Airy disc, so that the light diffracted in this optical path discontinuity produces a spherical wavefront that acts as the interferometer reference wavefront. The rest of the plate transmits the aberrated wavefront. The central circle can have a different transmittance than the rest of the plate in order to optimize the contrast of the fringe pattern.

3. TESTS THAT MEASURE TRANSVERSE ABERRATIONS

Some tests measure the transverse aberrations, which represent the slopes of the wavefront (Malacara-Hernández, 2007). If knowledge of the wavefront is desired, an integration of the transverse aberrations has to be performed.

3.1 Foucault or Knife-Edge Test

This is probably the simplest and oldest method to evaluate the quality of a concave spherical surface, invented by Foucault (1858) in France. It is described in detail by Ojeda-Castañeda (2007). This method is so simple and easy to implement that is it has been used by millions of amateur telescope makers for more than a century. In this method, a concave spherical surface is illuminated with a very small light source placed near the center of the spherical surface, called the center of curvature. Then, all the light is reflected back to the center of curvature. To separate the point light source from the reflected image of this light source, this is laterally displaced as shown in Fig. 24.

The observer places an eye with its pupil as close to the reflected image as possible, so that all the reflected light enters to the eye. Under these conditions, the concave surface appears completely and evenly illuminated.

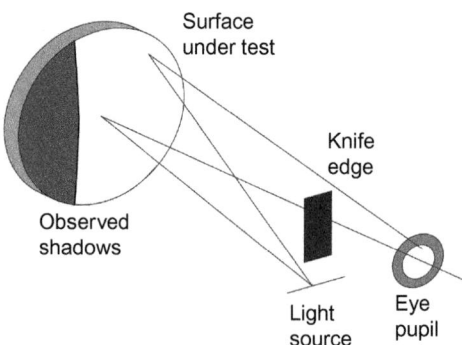

Fig. 24 Schematics of the Foucault knife-edge test.

Fig. 25 Foucault test pattern for a spherical surface and the knife edge outside of focus.

After this, a straight knife edge is placed in front of the eye so that this blocks off some of the returning light beam. Then a shadow on the concave surface is observed as illustrated in Fig. 25. The shape of the shadow depends on the surface deformations as well as on the position of the knife edge with respect to the image.

If the optical surface being examined is perfectly spherical and the knife edge is inside the image of the light source, closer to the optical surface, the shadow appears straight and on the same side as the knife. If the knife edge is now placed outside of the image of the light source, the straight shadow is on the opposite side of the knife. If the knife is exactly on the plane of the image of the light source the optical surface is either completely illuminated or completely dark, depending on the lateral position of the knife.

If the surface is not spherical the rays will not be reflected back to a common point close to the center of curvature. Then, the knife, depending on its position, will block out some rays. Under these conditions the surface being tested appears unevenly illuminated, with shadows in some regions, depending on the shape of the surface. These shadows and illuminated zone for a paraboloidal mirror are shown in Fig. 26.

In Fig. 27, we see three Foucault images of an aspheric mirror with the knife edge at three different local centers of curvature. The brightness along the circles touching the sphere is constant.

Amateur telescope makers frequently use this test to test their paraboloids. They only have to measure the difference between the paraxial radius of curvature and the point where the marginal normals to the surface join with the optical axis. From Eq. (36), we can see that this distance Δr has to be:

Fig. 26 Foucault test pattern for a paraboloidal concave surface.

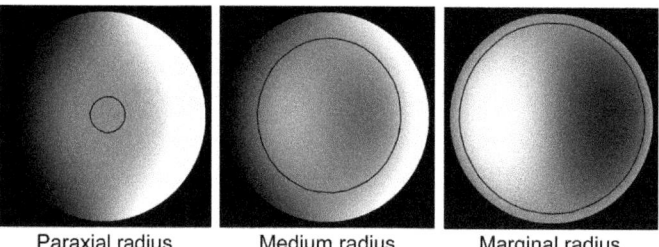

Paraxial radius   Medium radius   Marginal radius

Fig. 27 Three Foucault images with the knife edge at three different local centers of curvature. The knife edge is exactly at the optical axis.

$$\Delta r = \frac{S^2}{2r} \qquad (49)$$

where S is the distance of the point being considered to the optical axis, which is equal to the semidiameter of the mirror for the marginal rays and r is the radius of curvature of the osculating sphere.

The deviation of the reflected ray from the ideal point image as measured in the image plane is defined as the transverse aberration. Then, if the knife edge is parallel to the y-axis, the knife edge intercepts all reflected rays whose transverse aberration component in the x direction is greater than the x coordinate for this edge. This test gives the observer an intuitive and simple idea of the surface deformations. A surface deformation that is as small as a fraction of the wavelength of light can be easily detected.

If the light source is replaced by a long and thin light source (a slit illuminated from behind), parallel to the knife edge, each point of the light source would produce the same Foucault pattern, but it would be much

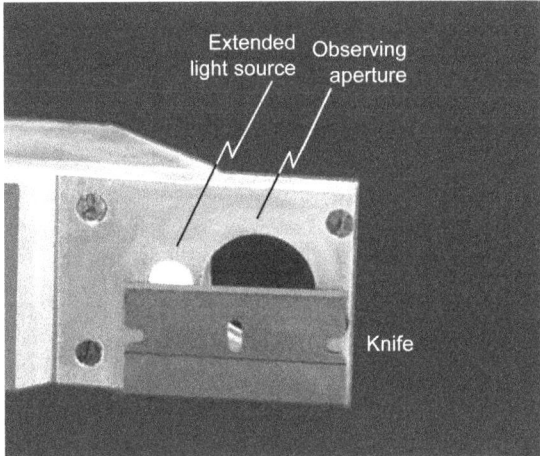

Fig. 28 Foucault or knife-edge test with an extended light source.

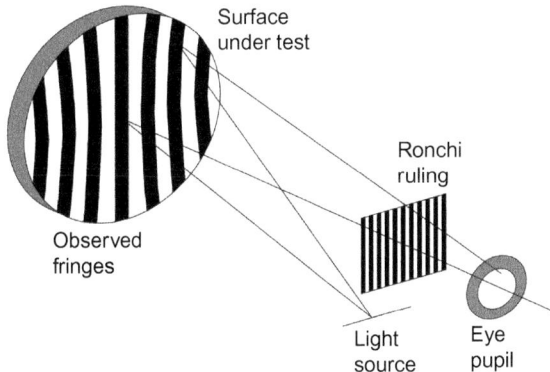

Fig. 29 Ronchi test schematics.

brighter. In practice, the arrangement in Fig. 28 can be used with an extended light source and the knife edge covering the light source as well as the observing eye. This is equivalent to having just a slit, but much simpler to ensure the parallelism of the light source and the knife edge. The reason is that a real image of the extended light source is formed over the knife at the observing aperture and only one slit of light goes through this aperture.

3.2 Ronchi Test

The Ronchi (1923) test, described in detail by Cornejo-Rodriguez (2007), uses a ruling instead of a knife with a set of straight, parallel, and equidistant lines, as shown in Fig. 29, in front of the reflected light beam. Then, the shadows of the lines on the ruling are observed over the optical surface being tested. It is important to point out that a Ronchi ruling is much coarser than a diffraction grating, with spatial frequencies as low as 50–100 lines per inch.

It has been proved that the Ronchi patterns are really lateral shear interferograms and that the ruling can also be considered as a diffraction grating. Both models, geometrical, interpreting the fringes as shadows and the physical, interpreting the fringes as interference, predict the correct shape of the fringes.

The Ronchi patterns produced by a spherical surface with the ruling inside and outside of focus have straight and parallel fringes as illustrated in Fig. 30.

However, if the surface is not spherical the fringes are not straight, as shown in the case of a paraboloidal surface in Fig. 31.

Let us assume that the surface deformations with respect to the ideal sphere are given by $z(x,y)$, thus, the reflected wavefront has distortions $W(x,y)$ with respect to the ideal spherical shape given by

$$W(x, y) = 2z(x, y) \qquad (50)$$

Then, the shape of the fringes is given by the relation

$$\frac{\partial W(x, y)}{\partial x} = -\frac{TA_x(x, y)}{r_w} = -\frac{md}{r_w} \qquad (51)$$

where m is the fringe number and d is the period or separation between two lines on the ruling. In order to find the Ronchi patterns for the primary

Fig. 30 Ronchi patterns for a spherical concave surface, with the ruling outside or inside of focus.

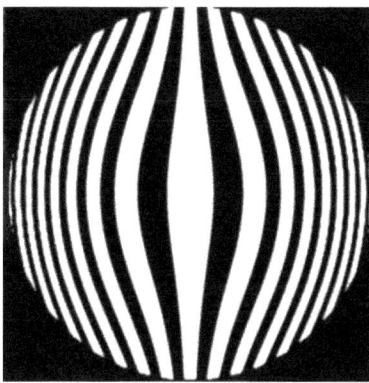

Fig. 31 Ronchi patterns for a paraboloidal concave surface with the ruling inside of focus.

aberrations we can use the wavefront Eq. (2), proposed by Kingslake for the primary aberrations of a centered optical system, obtaining:

$$\frac{\partial W(x, y)}{\partial x} = B + 2Dx + 2Ex + 2Fxy + 4G(x^2 + y^2)x = -\frac{md}{r_w} \quad (52)$$

where the tilt B about the y-axis and the tilt C about the x-axis just shift the fringes along the x- or y-axis, respectively. There is a Ronchi fringe for each value of the integer m.

We can notice that this test is sensitive to defocusing, astigmatism, coma in the direction perpendicular to the fringes, and spherical aberration. There is no sensitivity along the direction of the lines in the Ronchi ruling. To have it, we need two crossed Ronchi rulings or two Ronchi patterns in orthogonal directions.

Fig. 32 shows the Ronchi patterns for defocus, coma along the y-axis, coma along the x-axis, and spherical aberration at the paraxial, medium, and marginal foci. The Ronchi pattern for astigmatism is identical to the pattern for defocus, but it can be detected if the Ronchi ruling is rotated, because the separation between the fringes changes.

A null test for paraboloids or aspheric telescope mirrors can be made by designing a Ronchi ruling with curved lines, instead of straight, with the required shape to obtain straight fringes, as shown by Malacara and Cornejo (1974).

As in the case of the Foucault test, if the light source is replaced by a long and thin light source (a slit illuminated from behind), parallel to the lines in the Ronchi ruling, each point of the light source would produce the same

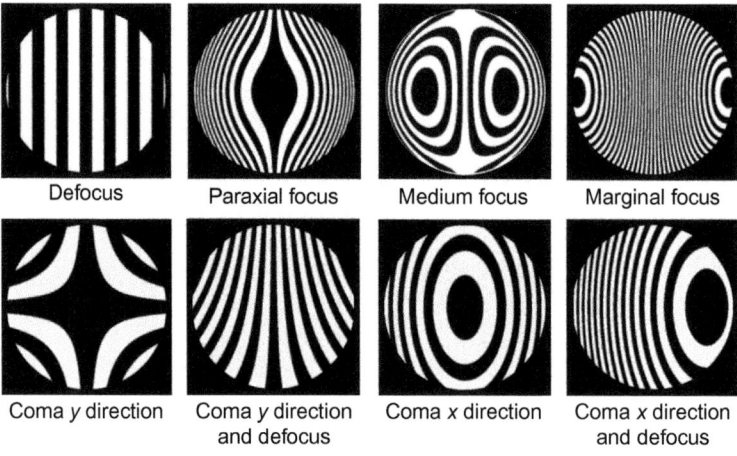

Fig. 32 Ronchi patterns for the primary aberrations.

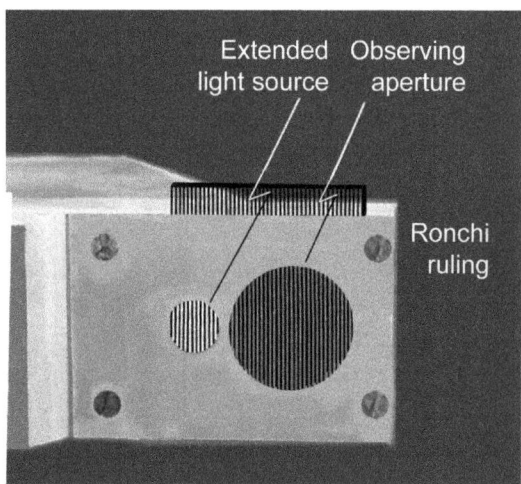

Fig. 33 Ronchi test with an extended light source.

Ronchi pattern, but it would be much brighter. The arrangement in Fig. 33 can be used, with an extended light source and the Ronchi ruling covering the light source as well as the observing eye. This is equivalent to having just a slit, but much simpler to ensure the parallelism of the light source and the knife edge. The reason is that the extended light source with the Ronchi ruling strips is formed over the Ronchi ruling at the observing aperture with unit magnification.

Optical Testing and Interferometry 111

3.3 Hartmann Test

In the Hartmann test (1900), described by Malacara-Doblado and Ghozeil (2007), an opaque screen with a rectangular array of sides is placed close to the surface under test. A photographic plate near the image plane records the light spots corresponding to each of the holes on the screen. To fully separate these spots and to identify them correctly, this plate has to be slightly shifted either inside or outside of focus as in Fig. 34. This test, as well as the Ronchi test can be interpreted in terms of geometrical optics.

When the surface under test is spherical the spots registered on the photographic plate form a rectangular an array of spots resembling the array of holes on the Hartmann screen with holes. In the case of nonspherical surfaces the array of spots is distorted, as in the case of a paraboloid, shown in Fig. 35.

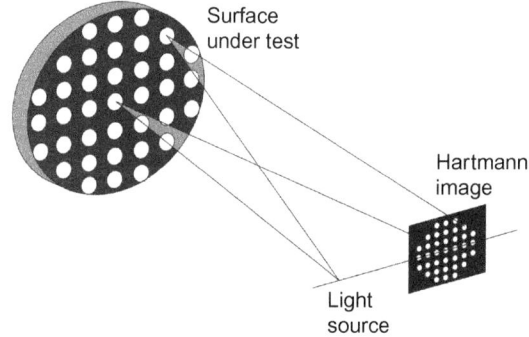

Fig. 34 Hartmann test schematics.

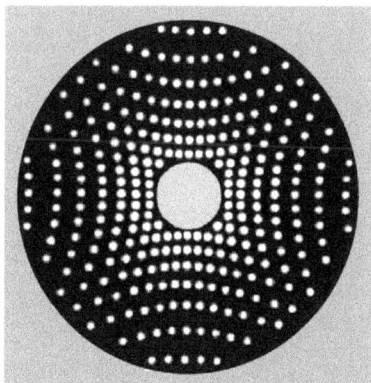

Fig. 35 Hartmann test image for a concave paraboloid with the observation plane inside of focus.

The distances of the actual positions of the spots in the photographic plate from the ideal positions if the surface is spherical are the ray transverse aberrations. If these transverse aberrations are measured, the real shape of the optical surface under test can be calculated as will be described next. Let us represent the transverse aberrations in the x direction by $TA_x(x,y)$ and the transverse aberrations in the y direction by $TA_y(x,y)$. The aberrations are related to the optical surface slopes with respect to the ideal spherical surface in the same direction by

$$TA_x(x, y) = 2r\frac{\partial W(x, y)}{\partial x} \tag{53}$$

and

$$TA_y(x, y) = 2r\frac{\partial W(x, y)}{\partial y} \tag{54}$$

where r is the radius of curvature and $W(x,y)$ is the function representing the surface deformations with respect to the ideal sphere with radius of curvature r. Alternatively, by integration of these expressions we may obtain:

$$W(x, y) = \frac{1}{2r}\int_0^x TA_x(x, y)\mathrm{d}x = \frac{1}{2r}\int_0^y TA_y(x, y)\mathrm{d}y \tag{55}$$

To be able to carry out this integration the one to one correspondence between the sampling points on the exit pupil of the system and the Hartmann spots on the observation plane must be known. When the wavefront is highly aspherical, this is almost impossible, or at least quite difficult if the observation plane is inside the caustic zone. For this reason the observation plane must be either inside or outside of focus, outside of the caustic limits.

Many modern refinements of this test have been made. It is so powerful that it is still being used quite extensively to test astronomical mirrors.

3.4 Shack–Hartmann and Other Modified Hartmann Tests

There are many variations of the classic Hartmann test, as described by Malacara-Hernández and Malacara-Doblado (2015). An extremely interesting and useful modification of the Hartmann test was proposed by Platt and Shack (1971), in order to measure collimated (flat wavefronts), as in Fig. 30. The Shack–Hartmann test illustrated in Fig. 36 was devised to allow the

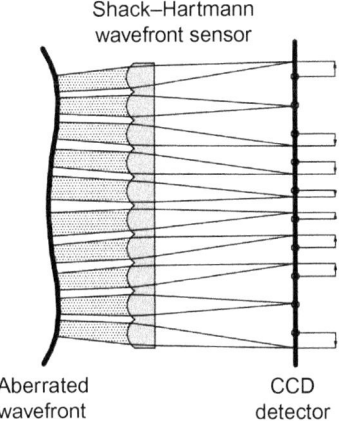

Fig. 36 Shack–Hartmann test.

measuring of small collimated wavefronts with a large precision. The basic sensor consists of a bidimensional array of lenslets molded in plastic or glass.

With this arrangement we can measure collimated wavefronts in lenses or telescope mirrors, or even the wavefront aberrations of the human eye, as described by many researchers, for example, by Hoffer, Artal, Singer, Aragon, and Williams (2001). The main differences with the basic Hartmann test are that a collimated wavefront with a small diameter can be measured and that the spots are focused and their diameters are determined by diffraction. If the Shack–Hartmann lens array is a square array, as in most cases, we have a null test only for flat or slightly spherical wavefronts.

The Shack–Hartmann test has been applied with success to the measurement of the aberrations of the human eye, as reported, for example, by Prieto, Vargas-Martín, Goelz, and Artal (2000) and Canovas and Ribak (2007).

Another possible Hartmann-like configuration is created by placing a small aperture imaging lens at the point conjugate to the virtual object position, as in Fig. 5. In front of the lens, near the front focal plane, an array of light sources is located. This array of light sources can be a LED array or even a nonluminous object with a reticle drawn on it. This arrangement can be imagined as a modification of the previous one, based on the Ronchi test, where the sampling plane (the array of light sources) is at the front of the entrance pupil of the system and not at the back of the system, after the exit pupil. The observing eye (Ronchi ruling) or camera should focus on the lens or on the exit pupil of the optical system under test. Then, as a consequence, the spots are not focused on the measuring plane, as in the original Hartmann test. The small imaging lens and the virtual object are at conjugate positions, where the lens is assumed to be working.

As before, the array of light sources can be replaced by a screen with a rectangular array of lines, and the imaging lens by the naked eye as in the testing of a lens in Fig. 37, as described by López-Ramírez, Doblado, and Malacara-Hernández (2000). This is a useful tool for aspheric lenses, as in Fig. 38. A null test configuration is obtained when the virtual object position and the imaging lens are at proper aberration-free conjugate positions for which the lens was designed.

Other possible modified arrangements for null tests of concave paraboloidal mirrors or for concave spherical mirrors with ruled nonluminous screens are illustrated in Fig. 39 and described by Malacara-Doblado and Ghozeil (2007).

With a similar configuration all the images of the light sources on the ellipsoid are in a plane, very close to the strong convex surface being measured. Since the diameter of the convex surface is small compared with the length of the ovoid, the virtual image positions of the light sources can be

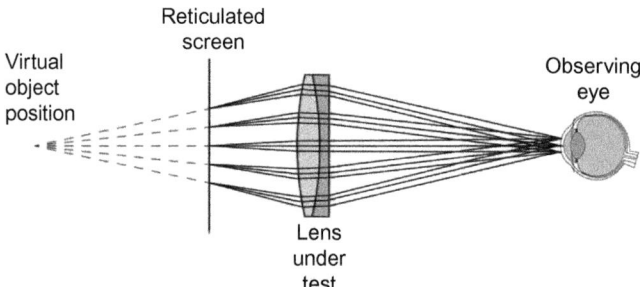

Fig. 37 Testing with an array of small light sources.

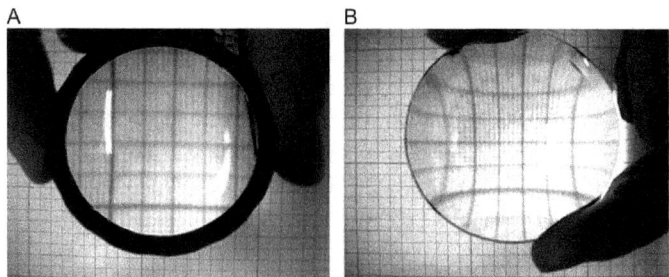

Fig. 38 Observation of a ruled screen through a lens (A) without spherical aberration and (B) with spherical aberration.

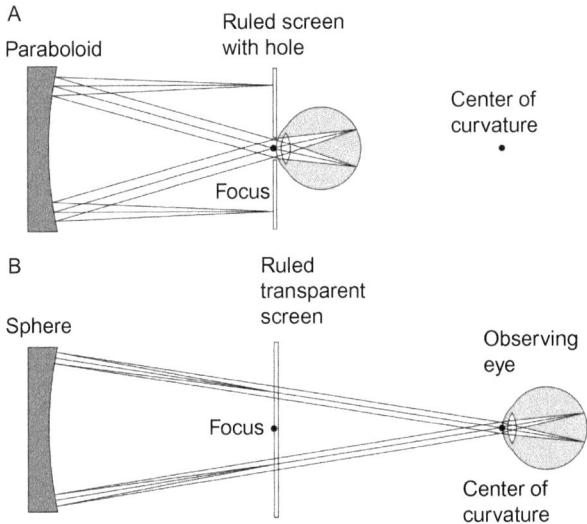

Fig. 39 Modified Hartmann test observing a ruled screen. (A) With a ruled screen with a hole and (B) with a ruled transparent screen.

considered as being at this convex surface. An additional advantage of this configuration is that the spots are focused on the detector as observing plane. A strong convex surface, for example the corneal surface of the eye can be measured as in Fig. 40A. If the virtual observations surface with the images of the light sources is desired to be in a flat surface, the array of light sources has to be over an ovoidal surface, as shown by Mejía-Barbosa and Malacara-Hernández (2001). This configuration has been used to measure the corneal topography of the eye. Instead of the ellipsoidal surface with the light sources, a cylindrical surface with the light sources or a dark screen with an array of lines can be used, as shown in Fig. 40B. The spots in the image are not exactly in a plane and not all the sampling surface with the light sources or lines array is in perfect focus. But since the center of gravity of the spots does not move, the accuracy of the instrument is not seriously affected as shown by Díaz-Uribe and Campos-García (2000).

The essential difference between the Hartmann test and the Ronchi test is that in the Hartmann test the sampling is performed at the exit pupil of the system and the transverse aberration measurements are made at the observation plane near the point of convergence of the rays. On the other hand, in the Ronchi test the sampling is performed at the observation plane near the point where the rays converge and the transverse aberration measurements are made at the exit pupil of the system.

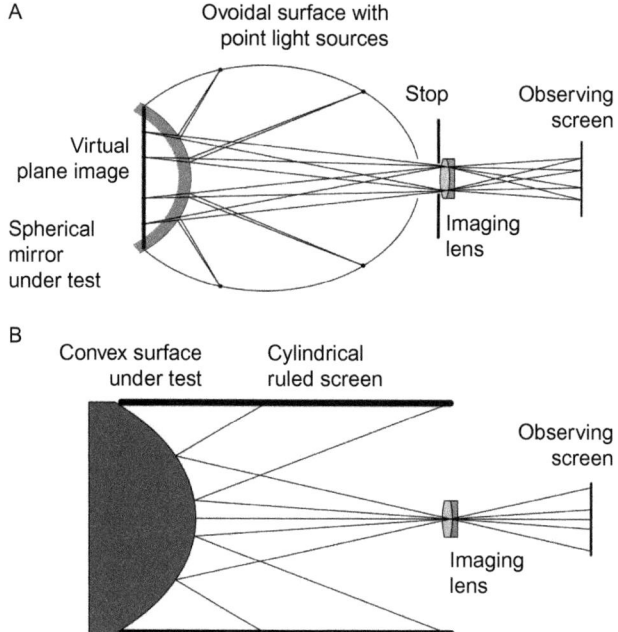

Fig. 40 Observation of a ruled screen plotted in a tube after reflection in a convex surface. (A) An ovoidal surface and (B) a cylindrical surface.

There is another possibility, as was reported by Roddier and Roddier (1992). The sampling screen can be placed near the point light source. The interpretation is simple in terms of physical optics, assuming that the sampling screen produces diffraction. This sampling screen can be one-dimensional, with one Ronchi ruling, or two-dimensional, with two crossed Ronchi rulings, or a small screen with a two-dimensional array of holes. When this light source with the diffracting screen in front of it is observed from the concave mirror under test, multiple light sources will appear, forming a two-dimensional array of virtual light sources. The images of these light sources will appear at the conjugate plane, near the observation plane, close to the center of curvature of the mirror. There is a plane in the vicinity of these images where a Talbot autoimage of the sampling screen appears. However, any aspheric aberration or surface distortions on the mirror under test will distort the rectangular two-dimensional array in this image.

3.5 Lateral Shearing Interferometers

In the Fizeau as well as in the Twyman–Green interferometers the wavefront under test is interfering with a perfectly flat reference wavefront.

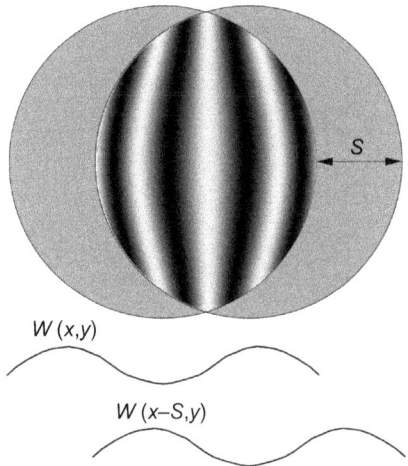

Fig. 41 Wavefronts in a lateral shear interferometer.

In lateral shear interferometers, the wavefront to be measured is interfering with an identical copy, laterally displaced with respect to it, as illustrated in Fig. 41. Several authors reported different types of lateral shearing interferometers before the advent of the laser, for example Saunders (1961). In these interferometers the optical path difference is given by

$$OPD(x, y) = OPD_0 + W(x, y) - W(x - S, y) \qquad (56)$$

where OPD_0 is a constant and S is the relative lateral displacement or shear between the two wavefronts. The interferogram produces interference fringes due to this $OPD(x,y)$. The wavefront deformation cannot be retrieved as easily as in a Twyman–Green interferometer. It has to be calculated by integration.

If the lateral shear S is small compared with the wavefront diameter, the optical path difference can be approximated by

$$OPD(x, y) \approx OPD_0 + S \frac{\partial W(x, y)}{\partial x} \qquad (57)$$

Thus, we can see that the interferogram in this case gives us information about the wavefront slope in the direction of the shear (x-axis). A full wavefront retrieval in the direction of the shear (x-axis) can be obtained only if two interferograms with lateral shears in two orthogonal directions are obtained and measured.

The lateral shear interferograms for the primary aberrations may be obtained with the expression for the primary aberrations, as will now be described. From Eq. (55), the optical path difference, taking OPD_0 as zero, is given by:

$$OPD(x, y) = BS + 2DSx + 2ESx + 2FSxy + 4GS(x^2 + y^2)x = m\lambda \quad (58)$$

The interferogram with a defocused wavefront is given by

$$2DSx = m\lambda \quad (59)$$

This is a system of straight, parallel, and equidistant fringes. These fringes are perpendicular to the lateral shear direction. When the defocusing is large, the spacing between the fringes is small. If there is no defocus, there are no fringes in the field. In the case of spherical aberration the interferogram is given by:

$$4GS(x^2 + y^2)x = m\lambda \quad (60)$$

If this aberration is combined with defocus we have:

$$[4G(x^2 + y^2)x + 2Dx]S = m\lambda. \quad (61)$$

Then, the interference fringes are cubic along the y-axis, but they are symmetric about the x-axis. This pattern looks like coma in the Twyman–Green interferometer.

If coma aberration is present, the interferogram is:

$$2FxyS = m\lambda \quad (62)$$

when the lateral shear is S in the sagittal direction x. It can be shown that if the lateral shear is T in the tangential direction y, the fringes are given by

$$F(x^2 + 3y^2)T = m\lambda \quad (63)$$

The fringe pattern for defocus and astigmatism are identical, making the interferometer insensitive to astigmatism. It can be detected only with two orthogonal fringe patterns. The fringes are straight and parallel as in the case of defocus, but with a different separation for both interferograms.

The experimental optical configuration of a lateral shear interferometer can adopt many geometries but the simplest system, invented by Murty (1964), is illustrated in Fig. 42. Since this interferometer is not compensated, a laser light source is required. The two interfering beams are generated at the two faces of a single glass plate, making the fringe pattern extremely stable, and insensitive to vibrations. With this configuration the collimating lens is also the lens under test.

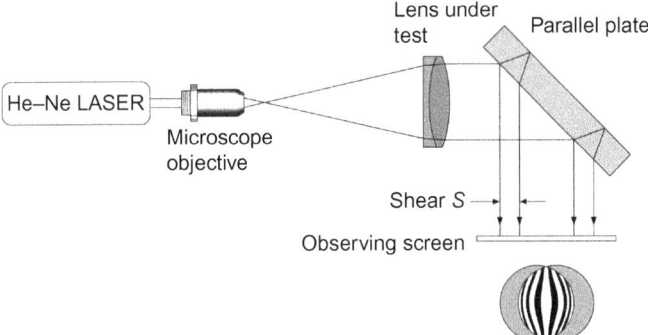

Fig. 42 Murty's lateral shear interferometer.

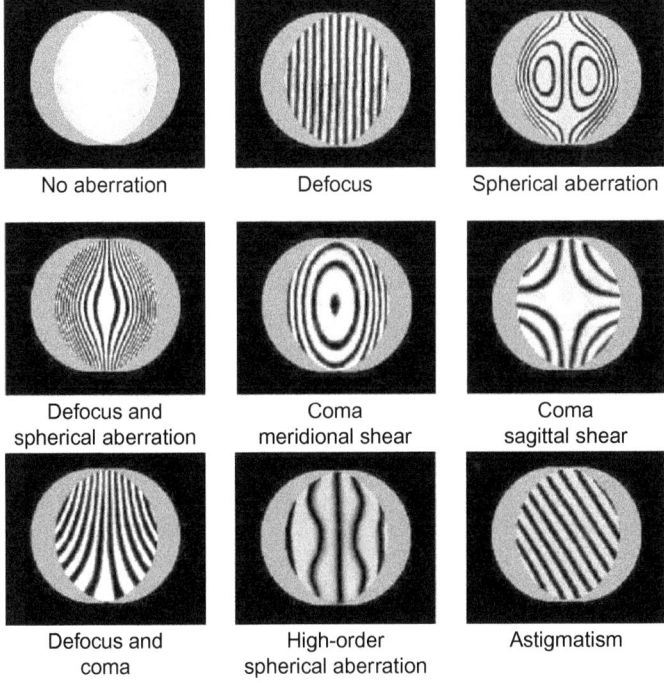

Fig. 43 Lateral shearing interferograms for the primary aberrations.

Fig. 43 shows the lateral shearing interferometer patterns obtained for the primary aberrations.

If the shear is not small enough, the interferogram do not represent the wavefront slope. Then, a procedure to obtain the wavefront deformation has to be employed. One of these possible methods has been proposed by Saunders (1961) and it is described in Fig. 44.

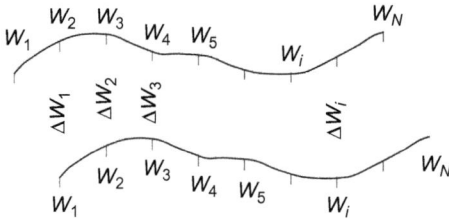

Fig. 44 Saunders method to find the wavefront in a lateral shear interferometer.

To begin, let us assume that $W_1 = 0$. Then, we may write

$$W_1 = 0$$
$$W_2 = \Delta W_1 + W_2$$
$$W_3 = \Delta W_2 + W_2 \qquad (64)$$
$$\vdots$$
$$W_n = \Delta W_{n-1} + W_{n-1}$$

A disadvantage of this method is that the wavefront is evaluated only at points separated by a constant distance S. Intermediate values are not measured and have to be interpolated.

There are many other methods that can be used for retrieving the wavefront. One of them is the Rimmer–Wyant method (Rimmer, 1974; Rimmer & Wyant, 1975), using a polynomial least squares fitting to the measured points.

4. TESTS THAT MEASURE CURVATURE

There are some tests that measure the local curvatures instead of the wavefront deformations or the transverse aberrations. Basically, with different variations, they are two different approaches. They will be briefly described.

4.1 Hartmann Test With Four Sampling Points

Using the Hartmann test, the local slopes, which are the first derivatives of the wavefront deformations, are measured at every sampling point. Then, the second derivative, or curvature, is represented by the separation between the Hartmann spots (Hernández-Gómez, Malacara-Hernández, & Malacara-Hernández, 2014).

With four sampling points in a square we can determine the wavefront tilts coefficients A_1 and A_2, the defocus term A_3, and the astigmatism terms A_4 and A_5, using the expressions:

$$A_1 = -\frac{1}{4r_w}\sum_{n=1}^{4}\left[TA'_\rho \cos\theta - TA'_\theta \sin\theta\right] = -\frac{1}{4r_w}\sum_{n=1}^{4}\left[TA'_x\right]$$

$$A_2 = -\frac{1}{4r_w}\sum_{n=1}^{4}\left[TA'_\rho \sin\theta + TA'_\theta \cos\theta\right] = -\frac{1}{4r_w}\sum_{n=1}^{4}\left[TA'_y\right]$$

$$A_3 = -\frac{1}{8\rho_0 r_w}\sum_{n=1}^{4}\left[TA'_\rho\right] = -\frac{1}{8\rho_0 r_w}\sum_{n=1}^{4}\left[TA'_x \cos\theta + TA'_y \sin\theta\right]$$

$$A_4 = -\frac{1}{8\rho_0 r_w}\sum_{n=1}^{4}\left[TA'_\rho \cos 2\theta - TA'_\theta \sin 2\theta\right] \quad (65)$$

$$= -\frac{1}{8\rho_0 r_w}\sum_{n=1}^{4}\left[TA'_x \cos\theta - TA'_y \sin\theta\right]$$

$$A_5 = -\frac{1}{8\rho_0 r_w}\sum_{n=1}^{4}\left[TA'_\rho \sin 2\theta + TA'_\theta \cos 2\theta\right]$$

$$= -\frac{1}{8\rho_0 r_w}\sum_{n=1}^{4}\left[TA'_x \sin\theta + TA'_y \cos\theta\right]$$

where ρ_0 is the distance from the center to the corners of the square cell. For simplicity the subscript n has been omitted. Thus, in a square cell in the wavefront deformation space, we have a surface representing five aberration terms, i.e., two tilts, defocus and two primary astigmatism terms, with axis at ±45 degree and at 0 degree or 90 degree (see Fig. 45).

Then, with a full rectangular array of sampling spots, the curvatures at each of the squares can be determined.

4.2 Irradiance Transport Equation

The observation of defocused stellar images, known as a star test, has been used for many years as a sensitive method to detect small wavefront deformations. The principle of this method is based on the fact that the illumination in a defocused image is not homogeneous if the wavefront has deformations. These deformations produce also variations in the local slopes and curvatures of the wavefront. If the focus is shortened, the light energy will be concentrated at a shorter focus and vice versa. An obvious consequence is that the illumination at two observing planes, located

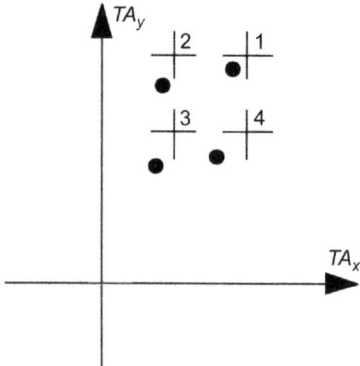

Fig. 45 Curvature measurement with an array of four Hartmann sampling points. The separations from the ideal positions are shown.

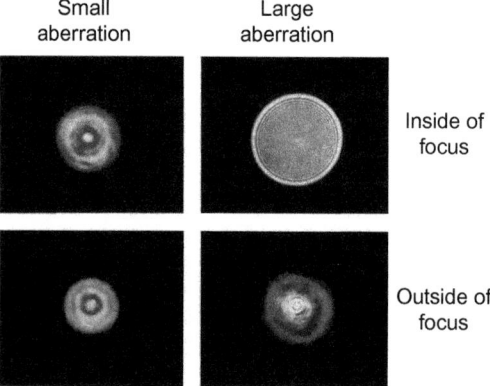

Fig. 46 Images taken at two different defocused planes.

symmetrically with respect to the focus, has different illumination densities (irradiances). However, this test remained for a long time only a qualitative visual test. An example of this effect is illustrated in Fig. 46.

As we mentioned at the beginning of Section 2, when we test an optical system, the entrance pupil is illuminated with a constant irradiation over its whole aperture and has a constant phase. Since the exit pupil is conjugate to the entrance pupil, the irradiance is also constant at this plane, but the wavefront may be distorted due to the aberrations in the optical system.

The principle of the method to be described is more easily understood if we consider a pupil or diffracting screen as in Fig. 47, where the irradiance is a constant, but the wavefront may have deformations. Before arriving to the pupil, the light beam passed through the plane P_1, where neither the

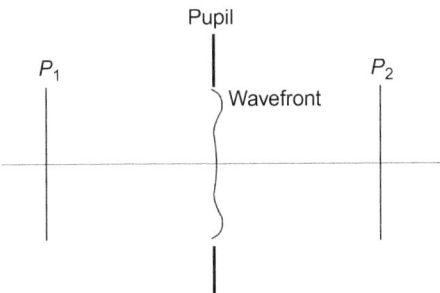

Fig. 47 Exit pupil of an optical system with a wavefront and two symmetrically placed observing planes.

irradiance nor the phase are constants. After passing through the pupil, the light beam finally arrives to plane P_2, where again, neither the phase nor the irradiance are constants.

If we measure the irradiance distributions at the two planes P_1 and P_2, we can prove that the wavefront deformations at the pupil can be calculated. To make the mathematical analysis simpler, the irradiance measurements have to be taken at two planes symmetrically located with respect to the pupil. In Fig. 47, one plane is real because it is located after the pupil, but the other plane is virtual, because it is located before the pupil. In practice this problem has several possible solutions.

One example of a useful arrangement is reported by Shomali, Darudi, and Nasiri (2012) and illustrated in Fig. 48. The irradiances are measured at the planes 1 and 2, symmetrically located with respect to a real image of the entrance pupil.

Fig. 49 illustrates how the measurements can be taken in a telescope with focal length f. Ideally, any two measurement planes P_1 and P_2 for the irradiances should be symmetrically located with the pupil (objective). The plane P'_2 is conjugate to plane P_2. Then, the plane P'_1 is conjugate to P_1. These two planes P'_1 and P'_2 are symmetrically placed with respect to the focus of the objective, at equal distances l from the focus, thanks to the presence of the lens with focal length $f/2$.

The distances Δz and l are related by

$$\Delta z = \frac{f(f-l)}{l} \tag{66}$$

Roddier and Roddier (1991a, 1991b) have pointed out that the small lens with length $f/2$ is not necessary if l is small compared with f. We must

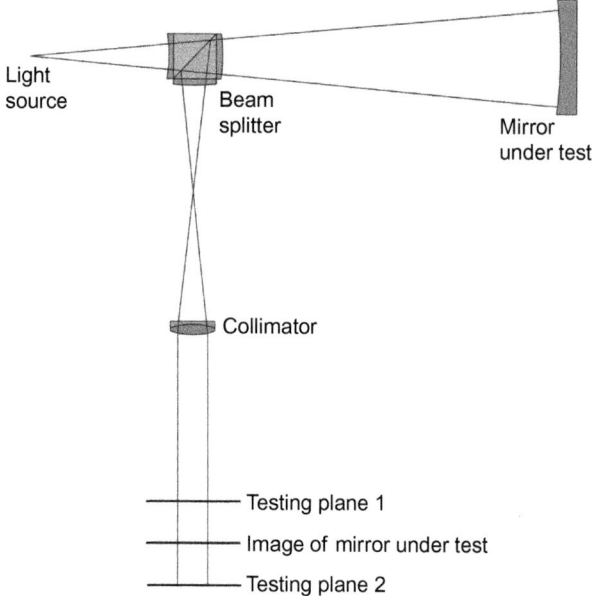

Fig. 48 Testing an aspherical mirror with two irradiance measurements in planes P_1 and P_2.

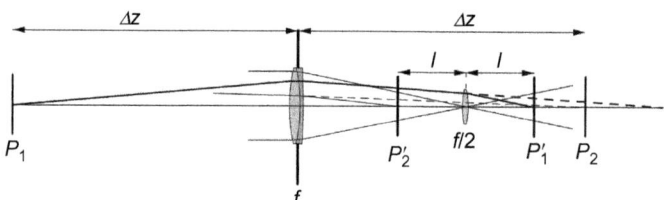

Fig. 49 Location of the planes where the irradiance measurements has to be taken in a telescope.

take into account, that one defocused image is rotated 180 degree respect to the other and also any possible difference in the magnification of the two images. The important consideration is that the subtracted and added irradiances in the two measured images must correspond to the same point (x,y) on the pupil.

Roddier (1988) and Roddier, Roddier, and Roddier (1988) proposed this quantitative wavefront evaluation method to measure the wavefront local curvatures. The local curvatures c_x and c_y of a nearly flat wavefront in the x and y directions are given by the second partial derivatives of this wavefront, as

$$c_x = \frac{\partial^2 W(x,y)}{\partial x^2} \quad \text{and} \quad c_y = \frac{\partial^2 W(x,y)}{\partial y^2} \tag{67}$$

hence, the Laplacian, defined by:

$$\nabla^2 W(x,y) = \frac{\partial^2 W(x,y)}{\partial x^2} + \frac{\partial^2 W(x,y)}{\partial y^2} = 2\rho(x,y) \tag{68}$$

is twice the value of the average local curvature $\rho(x,y)$. We must point out that the curvature at any point is different for different directions. This whole expression is known as the Poisson equation. If we measure by any means the average local curvatures, solving this equation we can find the wavefront deformations. Then to solve the Poisson equation in order to obtain the wavefront deformations $W(x,y)$ we need to have:

(a) the average local curvatures distribution $\rho(x,y)$, which is a scalar field, since no direction is involved, like in the wavefront slopes and
(b) the radial wavefront slopes at the edge of the circular pupil, to be used as Neumann boundary conditions.

As described by Roddier et al. (1988), the simplest method to solve the Poisson equation once the Laplacian has been determined, is the Jacobi iteration algorithm. On the other hand, Noll (1978) has shown that Jacobi's method is essentially the same as that derived by Hudgin (1977) to find the wavefront from slope measurements.

Let us consider a light beam propagating with an average direction along the z-axis, after passing through a diffracting aperture (pupil) on the x–y plane. The irradiance as well as the wavefront shape travels continuously changing its shape along the trajectory. As proved by Teague (1983), the wave disturbance $u(x,y,z)$ at a point (x,y,z) can be found with good accuracy, even with a diffracting aperture with sharp edges, using the Huygens–Fresnel diffraction theory if a paraxial approximation is taken. This approximation considers the Huygens wavelets as emitted in a narrow cone, with a parabolic approximation for the wavefront shape of each wavelet. This is a geometrical optics approximation. Teague (1985) and Steibl (1984) have shown that if a wide diffracting aperture, much larger than the wavelength is assumed, the disturbance at any plane, with any value of z may be found with the differential equation (Papoulis, 1968):

$$\nabla^2 u(x,y,z) + 2k^2 u(x,y,z) + 2ik\frac{\partial u(x,y)}{\partial z} = 0 \tag{69}$$

where $k = 2\pi/\lambda$. We consider a solution to this equation of the form:

$$u(x, y, z) = I^{1/2}(x, y, z) \exp(ikW(x, y, z)) \qquad (70)$$

where $I(x,y,z)$ is the irradiance. If we substitute this disturbance expression into the differential equation, we can obtain after some algebraic steps (Teague, 1983, 1985), a complex function that should be made equal to zero. After that, equating real and imaginary parts to zero, we obtain two equations:

$$\frac{\partial W}{\partial z} = 1 + \frac{1}{4k^2 I}\nabla^2 I - \frac{1}{2}\nabla W \cdot \nabla W - \frac{1}{8k^2 I^2}\nabla I \cdot \nabla I \qquad (71)$$

and

$$\frac{\partial I}{\partial z} = -\nabla I \cdot \nabla W - I\nabla^2 W \qquad (72)$$

where the (x,y,z) dependence has been omitted for notational simplicity and the Laplacian ∇^2 and the gradient ∇ operators work only on the lateral coordinates x and y.

The first expression is the phase transport equation that can be used to find the wavefront shape at any point along the wavefront trajectory. The second expression is the irradiance transport equation, which expresses how the irradiance changes as the light beam propagates along the z-axis.

Ichikawa, Lohmann, and Takeda (1988) reported experimental demonstration of phase retrieval based on this equation. Following him, each term in the terms in these two expressions can be interpreted as follows:

(a) The gradient ∇W is the direction and magnitude of the local slope of the wavefront, which is the light ray direction and ∇I is the direction in which the irradiance changes with maximum speed. Thus, their scalar product $\nabla I \cdot \nabla W$ is the irradiance variation along the optical axis z, due to the local wavefront tilt. Ichikawa et al. (1988) call this a prism term.
(b) The term $I\nabla^2 W$ can be interpreted as the irradiance variation as the beam propagates along the z-axis, caused by the local wavefront average curvature. Ichikawa et al. (1988) called this a lens term.

As explained before, the transport equation is a geometrical optics approximation, valid away from sharp apertures and as long as the aperture is large enough compared with the wavelength.

Roddier, Roddier, Stockton, and Pickles (1990) used the transport equation just described to measure the wavefront deformation in a telescope. Let $P(x,y)$ be the transmittance of the pupil, which is equal to one inside the pupil and zero outside. Furthermore, we assume that the illumination at the pupil's plane ($z=0$) is uniform and equal to a constant I_0 inside the pupil. Hence, the irradiance gradient $nI=0$ everywhere except at the edge of the pupil, where it abruptly goes to zero, thus we can write:

$$\nabla I(x, y, 0) = -I_0 n \delta_c \tag{73}$$

where δ_c is the Dirac delta distribution around the edge of the pupil and n is a unit vector perpendicular to the edge and pointing outward. Substituting this gradient into the irradiance transport equation one obtains:

$$\left(\frac{\partial I}{\partial z}\right)_{z=0} = -I_0 \bullet \left(\frac{\partial W}{\partial n}\right)_{z=0} \delta_c - I_0 P \nabla^2 W \tag{74}$$

where the derivative on the right-hand side of this expression is the wavefront derivative in the outward direction, perpendicular to the pupil's edge. Curvature sensing consists of taking the difference between the irradiance I_1 and I_2 observed in two planes, symmetrically located with respect to the pupil. Thus, the measured irradiances at these two planes are:

$$\begin{aligned} I_1 &= I_0 + \left(\frac{\partial I}{\partial z}\right)_{z=0} \Delta z \\ I_2 &= I_0 - \left(\frac{\partial I}{\partial z}\right)_{z=0} \Delta z \end{aligned} \tag{75}$$

When the wavefront is perfectly flat at the pupil, the Laplacian at all points inside the pupil and the radial slope at the edge of the pupil are both zero. Then, I_2 is equal to I_1. Having obtained these data, one may form the so-called sensor signal, which is a function of x and y as:

$$s = \frac{I_1 - I_2}{I_1 + I_2} = \frac{1}{I_0}\left(\frac{\partial I}{\partial z}\right)_{z=0} \Delta z \tag{76}$$

Substituting Eq. (74) into Eq. (76) gives:

$$\frac{I_1 - I_2}{I_1 + I_2} = \left(\frac{\partial W}{\partial n}\delta_c - P\nabla^2 W\right)\Delta z \tag{77}$$

Thus with the irradiance I_1 and I_2 in two planes symmetrically located with respect to the pupil ($z=0$), we obtain the left term of this expression. This gives us the Laplacian of $W(x,y)$ (average local curvature) for all points inside the aperture and the wavefront slope $\partial W/\partial n$ around the pupil's edge with transmittance $P(x,y)$ as a Neumann boundary condition, to be used when solving Poisson's equation.

5. INTEROGRAM ANALYSIS

Given an interferogram from a Twyman–Green or Fizeau interferometer, the wavefront deformations can be obtained by several possible procedures, as will now be described.

5.1 Sparse Sampling of the Fringes

If the interferogram is not digitized in real time but a picture is taken, the irradiance at each pixel might be wrong due to a nonlinear response of the detector, especially if the image is captured on photographic paper as was done some years ago. In this case only the maxima (or minima) of the fringes can be measured with some acceptable accuracy. Another problem is that the order of interference can be easily assigned if the fringes do not form closed loops, as in Fig. 50A. The orders are in consecutive starting from the number one. The order can also start at a large number and then decrease. The difference is only a piston term and the sign of the retrieved wavefront deviations. If the fringes form closed loops we must take into account that the order of interference is the same for all points along the fringe, as illustrated in Fig. 50B. Then, after the order numbers has been assigned, the wavefront deformation is just the product of the order number multiplied by the wavelength of the light.

Having found the wavefront deformation at many points over the aperture, a least squares fitting of those points to a polynomial is performed, to

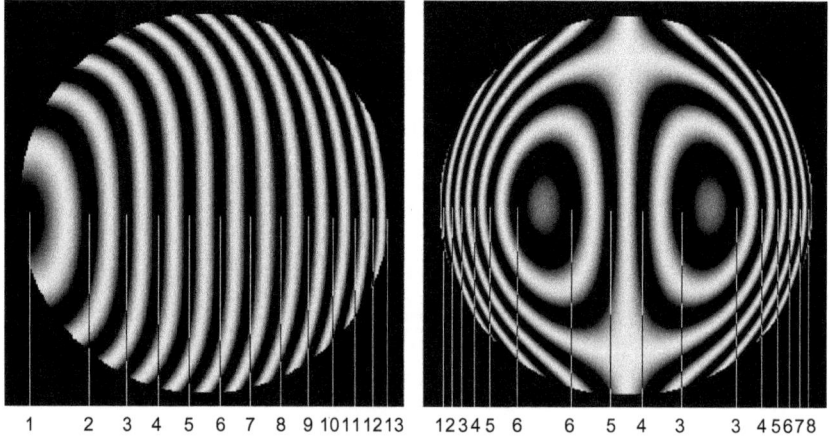

Fig. 50 Assigning the order of interference of the fringes.

obtain the whole wavefront in an analytical manner. The disadvantage is that small deformations between the fringes are not measured.

5.2 Digital Interferometry

If only one interferogram picture is taken and the asphericity is not very strong, several interferometric methods can be used to evaluate the fringe pattern (Malacara-Hernández, Servín, & Malacara-Hernández, 1998). However, if the asphericity is strong and the fringe spacing is not larger than twice the pixels separation at the detector (Nyquist condition) the interferogram may become impossible to analyze. Analysis is possible only with a procedure described by Greivenkamp (1987) and only when the following conditions are satisfied:

(a) The pixel size is smaller than the pixel separation. Then, spurious fringes will appear where the Nyquist condition is not satisfied, as illustrated in Fig. 51.
(b) The wavefront's general shape is known.
(c) The expected wavefront is smooth. The problem is then solved by proper phase unwrapping until the retrieved wavefront and its slopes are continuous.

With phase-shifting techniques, a series of Fizeau or Twyman–Green interferograms can be used without the introduction of tilt, even if closed loop fringes are formed. A series of a minimum of three interferograms has to be taken with different values of the constant term for the phase, also called piston term. Then, the interpretation of the wavefront evaluation becomes relatively simple from a mathematical point of view, but it is much more complicated from the experimental view point. When measuring an

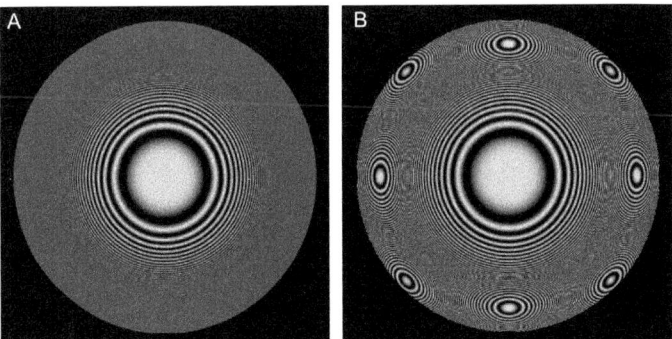

Fig. 51 Interferogram digitalization (A) with enough pixel density and (B) without enough pixel density and thus the Nyquist condition is not satisfied.

aspheric wavefront with phase shifting, the defocusing term has to be properly chosen so that the fringe spacing is a minimum, as described before. As pointed out before the limitation is that by using the optimum focus setting the Nyquist condition is not fulfilled.

If the wavefront has a strong deviation from a sphere, even phase shifting techniques become impossible. Another possibility under these conditions is to test the wavefront by dividing the complete aperture in small regions where the Nyquist condition is not violated. In other words, in all small regions the fringe spacing should be larger than twice the pixels separation. This technique, sometimes referred to as a wavefront stitching technique, will be described in this chapter.

5.3 Single Interferogram Analysis With a Spatial Carrier

A spatial carrier can be introduced in the interferogram by means of a large tilt (linear carrier) or a large defocusing (circular carrier) and the phase demodulation to obtain the wavefront is performed (Malacara-Hernández et al., 1998). Fig. 52 shows three interferograms of the same wavefront, without and with the linear carrier.

The method to obtain the wavefront with a large tilt was described by Takeda, Ina, and Kobayashi (1982) and Kujawinska (1993). It requires the introduction of a large linear carrier in the x direction. The minimum magnitude of this carrier is such that the phase increases (or decreases) in a monotonic manner with x. This condition is necessary to avoid closed loop fringes. This is possible if a tilt is introduced so that W' is always positive. Then, the minimum slope is zero. We know that the tilt is larger than the minimum required when the orders 1 and -1 in the Fourier plane are completely separated from the zero order, as illustrated in Fig. 53.

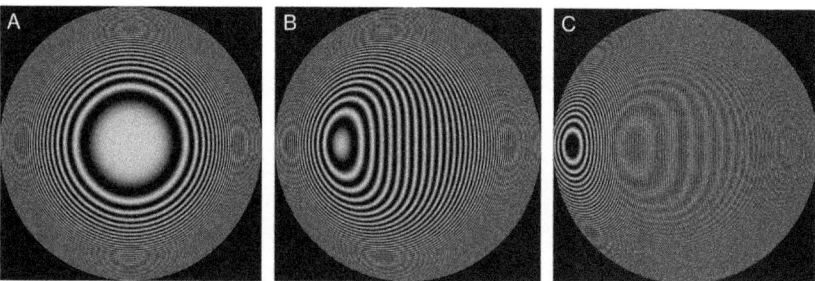

Fig. 52 Interferograms of an aspherical wavefront with three different tilt settings. (A) Without the spatial carrier, (B) with a linear spatial carrier, and (C) with a larger frequency linear carrier.

Fig. 53 Fourier transform in an interferogram with a linear carrier.

The ideal amount of defocusing to have the maximum possible local minimum fringe period, with the maximum possible asphericity is at the paraxial focus, not at the best focus as in phase shifting.

To demodulate the fringe pattern, following the method initially proposed by Ichioka and Inuya (1972), the irradiance values along a line crossing the interferogram perpendicularly to the fringes are measured. If the two interfering beams in the fringe pattern have the same constant amplitude A, the irradiance can be expressed as in Eq. (39) by:

$$I(x,y) = 2A^2[1 + \cos(kx\sin\theta - \phi(x,y))] \quad (78)$$

These irradiance values are multiplied by sinusoidal orthogonal (sine and cosine) reference functions with the same average spatial frequency as the measured function, $\cos(kx\sin\theta)$ and $\cos(kx\sin\theta)$, thus obtaining:

$$\begin{aligned}S(x,y) &= 2A^2[1 + \cos(kx\sin\theta - \phi(x,y))]\sin(kx\sin\theta)\\ C(x,y) &= 2A^2[1 + \cos(kx\sin\theta - \phi(x,y))]\cos(kx\sin\theta)\end{aligned} \quad (79)$$

which can be transformed into:

$$S(x,y) = 2A^2\left[\sin(kx\sin\theta) + \frac{1}{2}\sin\phi(x,y) + \frac{1}{2}\sin(2kx\sin\theta - \phi(x,y))\right]$$

$$C(x,y) = 2A^2\left[\cos(kx\sin\theta) + \frac{1}{2}\cos\phi(x,y) + \frac{1}{2}\cos(2kx\sin\theta - \phi(x,y))\right]$$

$$(80)$$

These multiplications produced two sampled functions with the same fundamental frequency as the two reference functions but with some higher harmonic frequencies plus a low-frequency component. Owing to the signal phase modulation, the low-frequency component contains the desired information about the phase.

To continue the procedure, the two sampled signals are low-passed filtered to remove the spatial frequencies equal and higher than the frequency of the reference functions, obtaining:

$$\bar{S}(x, y) = A^2 \sin\phi(x, y) \\ \bar{C}(x, y) = A^2 \cos\phi(x, y) \qquad (81)$$

Finally, the ratio of the values of these two functions point to point gives the tangent of the desired phase:

$$\tan\phi(x, y) = \frac{\bar{S}(x, y)}{\bar{C}(x, y)} \qquad (82)$$

Many other methods exist for the wavefront retrieval from interferograms of many types, even when the fringes are closed. These can be found in many publications, for example in the books by Malacara-Hernández, Servín, and Malacara-Hernández (1998) and Servín, Quiroga, and Padilla (2014).

6. PHASE SHIFTING INTERFEROMETRY

Phase shifting procedures began with the pioneering work by Bruning et al. (1974) and are well described in the scientific literature, for example by Greivenkamp and Bruning (1992). In phase shifting procedures, the interferogram image is divided in pixels by means of a two-dimensional light detector, for example, by a CCD television camera. Then, the irradiance $I(x,y)$ at each pixel is measured for several values of the phase difference $\varphi(x,y)$ between the two wavefronts.

6.1 Instrumentation

Phase shifting interferometry can be carried out in any two beam interferometer, including lateral shearing interferometers and the Ronchi test. The basic condition is that the phase between the two interfering waves has to be changed in steps with constant phase separation. The irradiance at all pixels in the interferogram pattern is then measured in real time. An example is

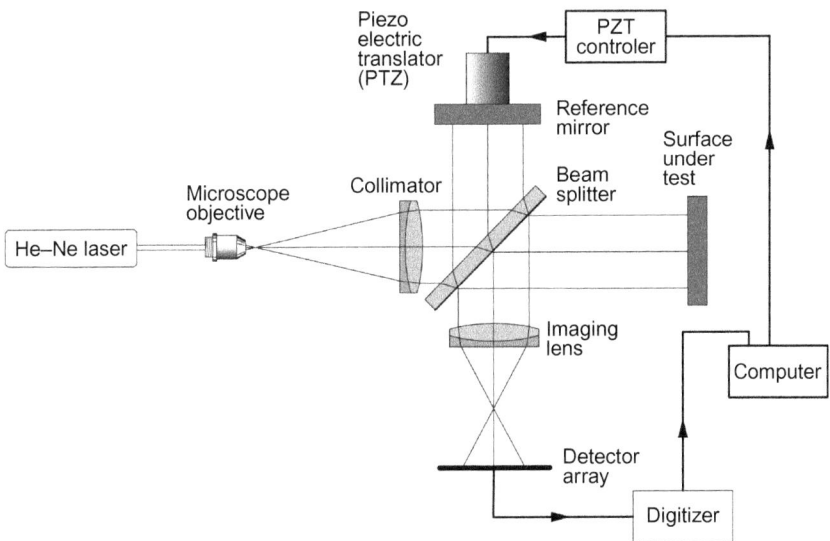

Fig. 54 Phase shifting in a Twyman–Green interferometer with a piezoelectrical crystal.

moving the reference mirror along the light path, with a piezoelectric crystal, as in Fig. 54.

Many different methods to shift the phase of the two interfering beams exist, for example, those illustrated in Fig. 55.

The first method to shift the phase in this figure is by moving one of the mirrors with any method, for example, a magnetic transductor or a piezoelectric crystal. The second method is by rotating a glass plane parallel plate, thus changing the optical path difference in one of the interfering beams. The disadvantage of this method is that the phase shift is not linear with the angle.

The third method is laterally moving a diffraction grating. The zero order of diffraction does not change its frequency, but the first orders of diffraction change their frequency in opposite directions, one increasing it and the other decreasing it. A similar effect can be produced with running fringes in a Bragg cell. The advantage with respect to the moving diffraction grating is that the zero and the other diffracted orders are not present.

Finally, if the light source is linearly polarized in an interferometer, we can insert in the two interfering beams, two $\lambda/4$ phase plates, one of them fixed and the other one rotating, as illustrated in the figure. The first phase plate has its slow and fast axes at 45 degree with respect to the polarization plane of the light beam, producing circularly polarized light. When this

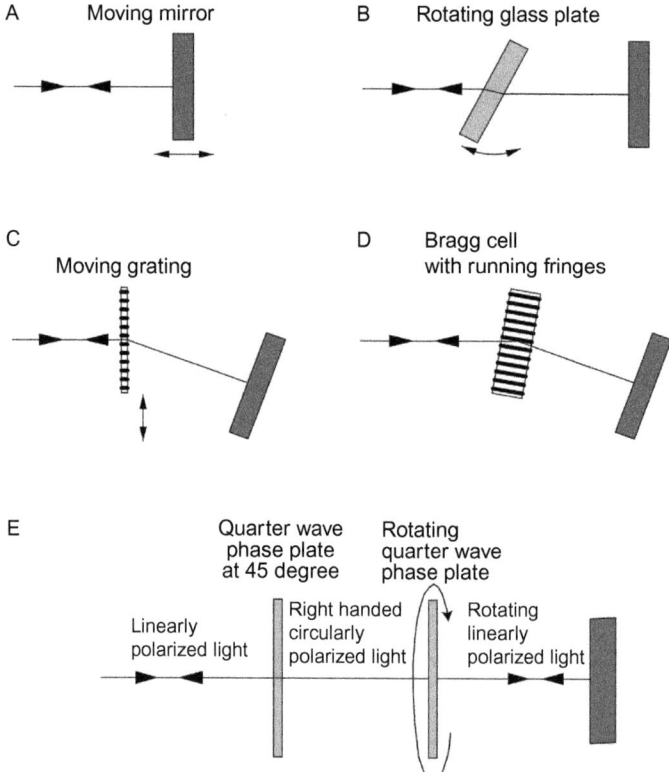

Fig. 55 Some methods to shift the phase of one of the interfering beams.

circularly polarized light enters the second phase plate, the light at the output is plane polarized at 45 degree with respect to the axes of the phase plate. Thus, the plane of polarization will be rotating as the second phase plate rotates. Upon reflection at the mirror, the rotating linearly polarized light goes back to the rotating phase plate, again at 45 degree, since the traveling time was extremely short. Therefore, it becomes circularly polarized, but in opposite sense to the light beam, traveling in the opposite direction. The light is traveling twice, back and forth, through the phase plate and this is equivalent to traveling only once through a half-wave phase plate. Finally after the light passes the first phase plate for a second time, the exiting beam is linearly polarized. However, due to the rotation of the second phase plate, the frequency of the light beam is changed. In other words, the phase of the light beam is shifted by twice the angle rotated by the phase plate.

In phase shifting methods no tilt is necessary but the focus may be adjusted at any desired value. At the best focus we obtain the maximum

Optical Testing and Interferometry 135

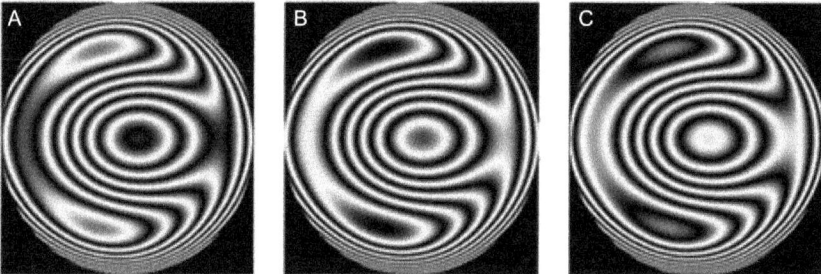

Fig. 56 Three interferograms with different phase to perform phase shifting interferometry. (A) Initial phase difference, (B) second phase difference, and (C) final phase difference.

possible local minimum fringe period. Thus, at this focus the maximum possible degree of asphericity can be tested.

The signal $I(x,y)$ is sampled for at least three different values of the unknown phase $\varphi(x,y)$ at known intervals. Then, three or more interferograms of the same wavefront with different phases of the reference plane are obtained, as illustrated in Fig. 56. From these measurements, the tangent of the phase is calculated using one of many possible well-known (Malacara-Hernández et al., 1998) algorithms.

It is important to realize that the calculated phase is obtained modulus 2π. That is known as a wrapped phase. To unwrap the phase, the proper phase multiples of 2π have to be added to the calculated phase. This is a simple problem if the Nyquist limit has not been exceeded. Otherwise, it becomes almost impossible.

The N different phase steps to obtain the N interferograms can be obtained sequentially with a certain time gap between them. This is the most common method, but sometimes they are measured all simultaneously. The advantage is that all possible vibration effects are eliminated. An example is described by Millerd et al. (2005).

6.2 Algorithms

In order to retrieve the wavefront deformations, the phase has to be sampled at a minimum of three different phase values, not necessarily equidistant. This gives us an infinite number of possibilities, all with different properties, advantages, and disadvantages, depending on the interferometer being used and many other practical environmental conditions.

A theory to find the different characteristics and properties of almost any algorithm has been developed by Freischlad and Koliopoulos (1986). Any algorithm with N equidistant phase steps can be written in a general form as:

$$\tan\phi = \pm\left(\frac{g_1(x)}{g_2(x)}\right) = \pm\left(\frac{\sum_{n=1}^{N} s(x_n)W_{1n}}{\sum_{n=1}^{N} s(x_n)W_{2n}}\right) = \pm\left(\frac{\sum_{n=1}^{N} s(x_n)\sin(2\pi f_r x_n)}{\sum_{n=1}^{N} s(x_n)\cos(2\pi f_r x_n)}\right) \tag{83}$$

The theory developed by Freischlad and Koliopoulos and later described and further developed by several researchers (Malacara, Servín, etc.) is based on the analysis of the Fourier transforms of the functions g_1 and g_2 in Eq. (83). Next, we will describe some of the most popular algorithms.

The wavefront deviation at any pixel in the image is equal to:

$$W = k\phi = \frac{2\pi}{\lambda}\phi \tag{84}$$

The amplitudes of the Fourier transforms of the so-called *sampling functions* $g_1(x)$ and $g_2(x)$ are $G_1(f)$ and $G_2(f)$. The frequency f is the normalized sampling frequency, which is the ratio of the real sampling frequency to the ideal or reference sampling frequency f_r. Thus, a normalized sampling frequency equal to one means that there is no error in the experimental real frequency. The reason is that it is equal to the ideal reference frequency.

An algorithm has no error in its phase calculation when the following four conditions are fulfilled:

(a) The sampling function should have no bias (no additive constant). This happens when at the frequency $f = 0$, the Fourier transforms are equal to zero as follows:

$$G_1(0) = G_2(0) = 0 \tag{85}$$

(b) At the real sampling frequency, we have the same amplitudes of the Fourier transforms as follows:

$$Am[G_1(f)] = Am[G_2(f)] \tag{86}$$

(c) The sampling functions should be orthogonal. This happens when $G_1(f)$ is the complex conjugate of $G_2(f)$. A common particular case is when the sampling functions $g_1(x)$ and $g_2(x)$ are such that one of them is symmetrical and the other antisymmetrical. The Fourier transform of the symmetrical function would be real and the Fourier transform of the antisymmetrical function would be imaginary.

(d) To avoid harmonics interference, the Fourier transforms of the sampling functions must be zero at all harmonics:

$$G_1(n) = G_2(n) = 0 \tag{87}$$

for $n = 2, 3, 4, 5, \ldots$

6.2.1 Three Steps Separated 120 Degree

This algorithm has the minimum number of sampling points, equal to three, with 120 degree separation between them, and the first point at 60 degree, as illustrated in Fig. 57, where G_{ij} are the Fourier transforms of g_{ij}.

We may observe that the sampling function $g_1(x)$ is antisymmetrical and that the sampling function $g_2(x)$ is symmetrical. It can be proved that they satisfy the orthogonality condition at all frequencies. The algorithm to find the phase is then given by:

$$\tan\phi = -\sqrt{3}\frac{s_1 - s_3}{s_1 - 2s_2 + s_3} \tag{88}$$

Examining the amplitudes of the Fourier transforms of the sampling points, in Fig. 58, we observe that at the normalized frequency both amplitudes have the same value, but a small deviation (detuning) introduces an error. On the other hand, if the signal being detected has harmonic frequencies 3, 6, 9, etc., there is no error introduced, but harmonic frequencies 2, 4, 5, 7, may introduce errors.

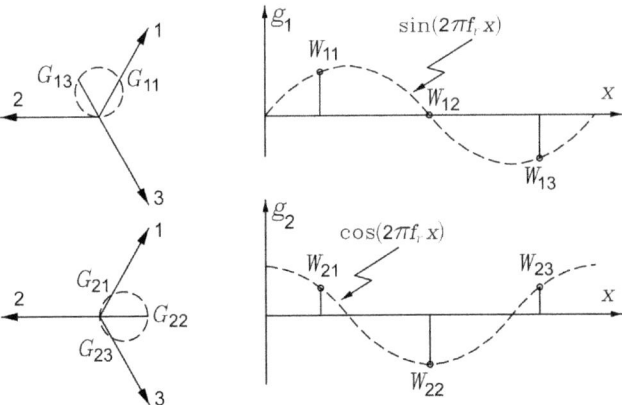

Fig. 57 Three steps algorithm to perform phase shifting interferometry.

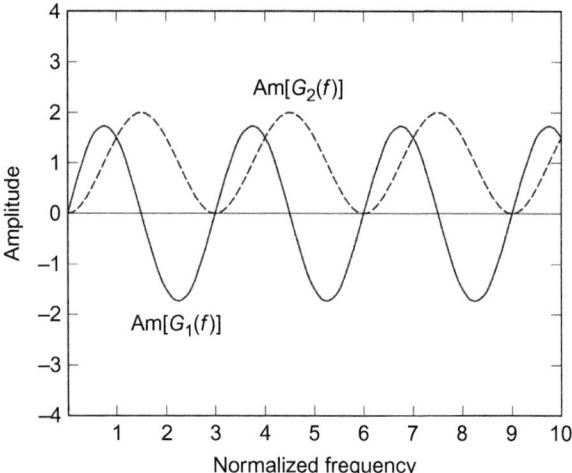

Fig. 58 Fourier analysis of the three steps algorithm.

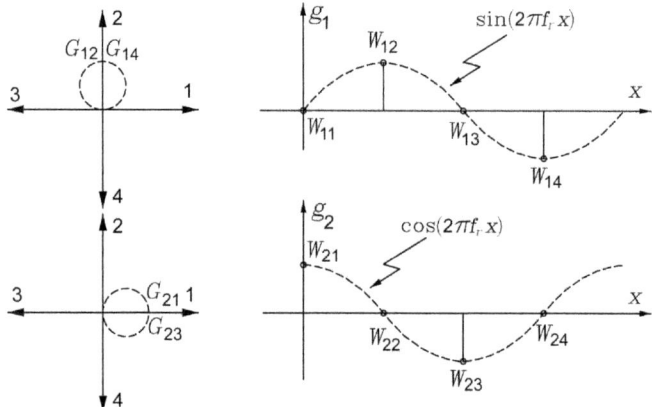

Fig. 59 Four steps algorithm.

6.2.2 Four Steps Separated 90 Degree

This algorithm has four steps separated by 90 degree and the first measurement at 0 degree, as illustrated in Fig. 59, whereas in the preceding algorithm, G_{ij} are the Fourier transforms of the sampling functions g_{ij}.

It can be proved that if there is a deviation of the sampling frequency from the reference frequency (normalized frequency different from one) the orthogonality between the two sampling functions is lost, as illustrated in Fig. 60. The algorithm to find the phase is given by:

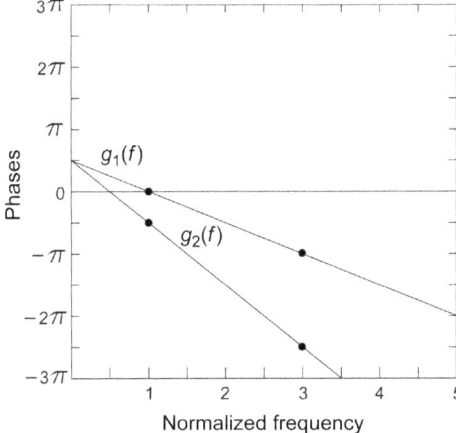

Fig. 60 Phases of sampling functions of the four steps algorithm.

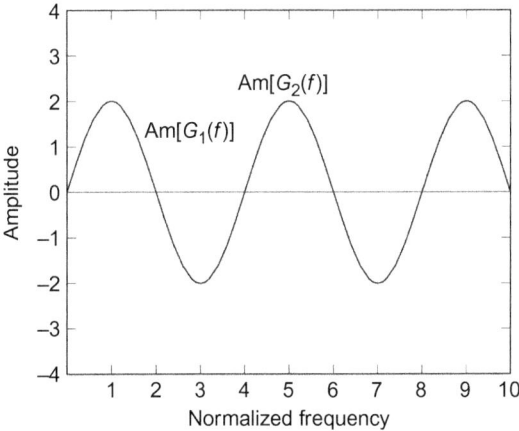

Fig. 61 Fourier analysis of the four steps algorithm.

$$\tan\phi = -\frac{s_2 - s_4}{s_1 - s_3} \qquad (89)$$

In Fig. 60, we have orthogonality when the phase difference between the two sampling functions is equal to $n90$ degree, where n is an odd integer. Thus, we can see that at all even harmonics the orthogonality is lost. Thus, errors can be introduced by the presence of these harmonics in the signal to be detected. At all odd harmonics the orthogonality is preserved. This makes the algorithm insensitive to odd harmonics.

The amplitudes of their Fourier transforms are the same at all frequencies, as shown in Fig. 61.

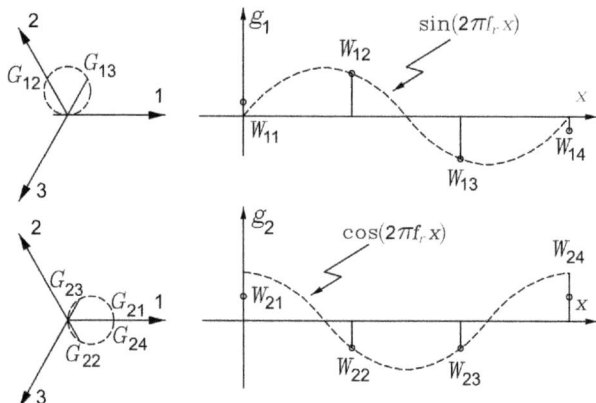

Fig. 62 Four steps (3+1) algorithm.

6.2.3 Four Steps (3+1) Separated 120 Degree

This algorithm has four steps separated by 120 degree and the first measurement at 0 degree, as illustrated in Fig. 62. As in all preceding algorithms, G_{ij} are the Fourier transforms of the sampling functions g_{ij}.

The last sampling point is at 360 degree from the first one. For this reason it is frequently called the (3+1) steps algorithm.

As in the three steps algorithm, the sampling function $g_1(x)$ is antisymmetrical and that the sampling function $g_2(x)$ is symmetrical. They satisfy the orthogonality condition at all frequencies. The algorithm to find the phase is then given by:

$$\tan\phi = -\sqrt{3}\frac{s_2 - s_3}{s_1 - s_2 - s_3 + s_4} \tag{90}$$

Examining the amplitudes of the Fourier transforms of the sampling points, in Fig. 63, we see that at the normalized frequency both amplitudes have the same value, but now a small deviation (detuning) does not introduce any error. This is so because the two curves are tangent to each other at the normalized frequency equal to one. On the other hand, if the signal being detected, has harmonic frequencies 3, 6, 9, etc., there is no error introduced, but harmonic frequencies 2, 4, 5, 7, may introduce errors.

6.2.4 Five Steps (4+1) Separated 90 Degree

This algorithm has five steps separated by 90 degree and the first measurement at 0 degree, as illustrated in Fig. 64. As in all preceding algorithms, G_{ij} are the Fourier transforms of the sampling functions g_{ij}.

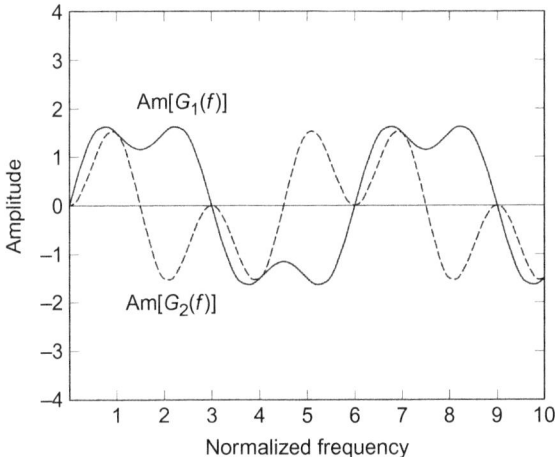

Fig. 63 Fourier analysis of the four steps (3+1) algorithm.

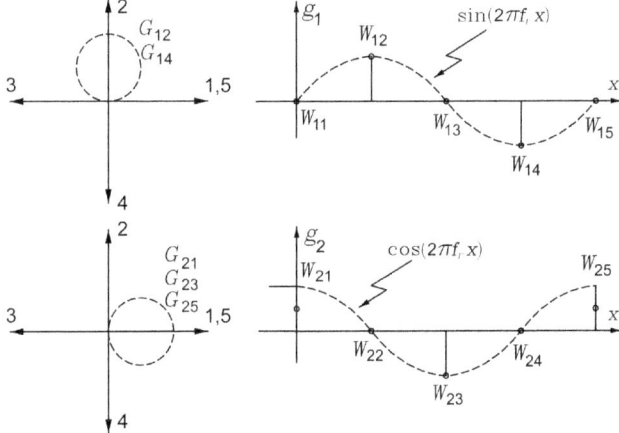

Fig. 64 Five steps (4+1) algorithm.

As in the (3=1) steps algorithm, the last sampling point is at 360 degree from the first one. For this reason it is frequently called the (4+1) steps algorithm.

The sampling function $g_1(x)$ is antisymmetrical and the sampling function $g_2(x)$ is symmetrical. They satisfy the orthogonality condition at all frequencies. The algorithm to find the phase is then given by:

$$\tan\phi = -\left(\frac{s_2 - s_4}{\frac{1}{2}s_1 - s_3 + \frac{1}{2}s_5}\right) \tag{91}$$

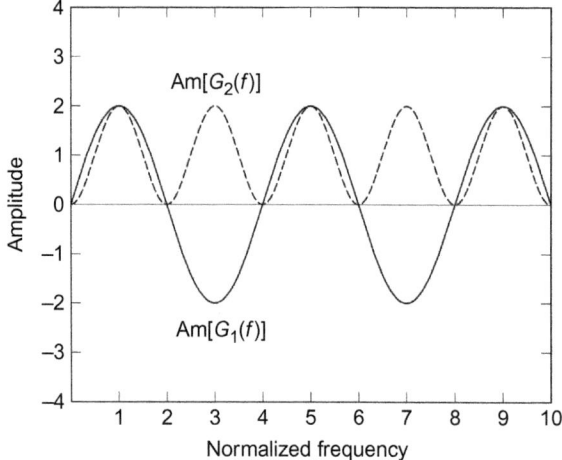

Fig. 65 Fourier analysis of the five steps (4 + 1) algorithm.

We can observe the amplitudes of the Fourier transforms of the sampling points, in Fig. 65, where we see that at the normalized frequency, as in the (3 + 1) steps algorithm, both amplitudes have the same value and that a small deviation (detuning) does not introduce any error. From these plots we can also see that at all harmonic frequencies, the amplitudes are the same and there is also insensitivity to small detuning at these frequencies.

These are not the only algorithms that have been designed. Many others with different properties and characteristics have been devised. A complete review of these algorithms can be found in several publications and books (Malacara et al., 2005; Schreibert & Bruning, 2007).

7. TESTING OF ASPHERICAL SURFACES

To test and measure the shape of an aspheric surface is one of the classical difficult problems in optical testing. There are a large number of different tests, instruments, and configurations, but all of them have different applications, advantages, and disadvantages as described by Schwider (1999) and Malacara, Creath, Schmidt, and Wyant (2007).

Aspheric wavefronts are produced by optical systems, with spherical aberration, using spherical as well as aspherical surfaces. Aspheric surfaces are used in optical systems in order to improve aberration correction and frequently to decrease the number of optical elements needed to make this correction satisfactorily. However, if these surfaces are tested while isolated from the rest of the optical system to which they belong, they frequently produce aspherical wavefronts. The interferometric testing and measurement of aspherical wavefronts are not as simple as in the case of spherical or flat wavefronts.

7.1 Autocompensating Configurations

When testing optical surfaces, the first thing to determine is if the surface has a pair of conjugate points such that if the point light source is located at one of those points, an aberration-free point image is formed at the other point. A typical example is the testing of a refracting objective of doublet. Since normally these optical systems are designed to have its object at an infinite distance, an autocollimating configuration as in Fig. 66 can be used, using a reference flat mirror.

A concave spherical mirror can be tested with the configurations in Fig. 67, where the object light source and the testing points are both at the center of curvatures. If the spherical mirror is independently tested, a flat mirror can be placed at an angle of 45 degree in front of it in order to test it.

When testing a spherical wavefront by interference in a Twyman–Green or Fizeau interferometer the fringes can be made to disappear if the tilts and the defocus are set equal to zero. This is called a *null test*. Or, the fringes can be made straight and parallel, with a separation as wide as desired. On the other hand, if the wavefront is aspheric, when the defocusing is set equal to zero, that is, if it is adjusted at the paraxial focus, by making the axial curvatures of the reference wavefront and the wavefront under test, many fringes will appear at the edge of the pupil, as in Fig. 68A. By introducing some defocus by changing the curvature of any of the two interfering wavefronts, the number of fringes can be reduced, as in Fig. 68B and C.

To test aspherics often a null test is required. As explained before, a null test is that which produces a fringe-free field when the defocus and the tilts

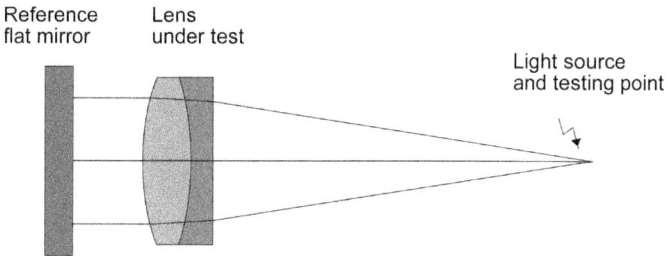

Fig. 66 Testing a doublet without spherical aberration by autocollimation.

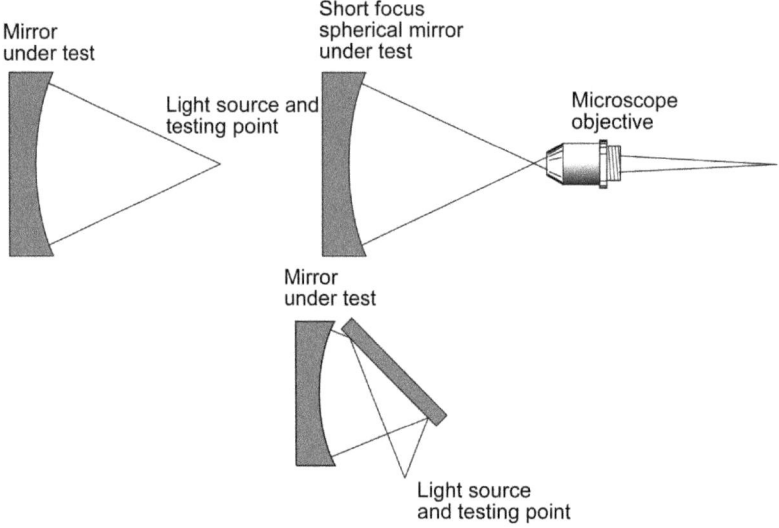

Fig. 67 Testing a concave spherical and flat surface.

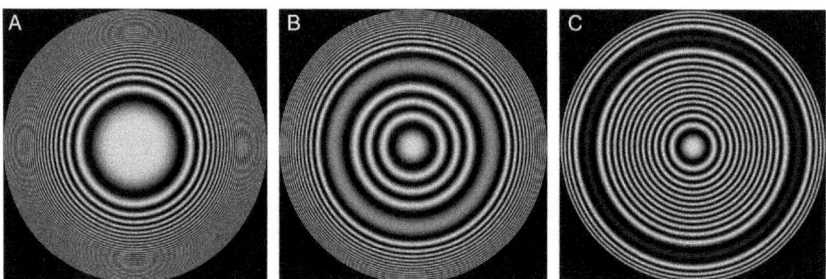

Fig. 68 Interferograms of an aspherical wavefront with three different focus settings. (A) Paraxial focus, (B) an intermediate focus near the paraxial focus, and (C) an intermediate focus near the marginal focus.

are set equal to zero and the desired wavefront is obtained. If a small tilt is introduced, any deviation from straightness of the fringes is a graphical representation of the wavefront deformation. With a null test the wavefront is easily identified and measured with a high accuracy. There are several methods to obtain this null test, but sometimes this is not simple. However, there are some simple cases where special configurations can be used to obtain a null test. These are called autocollimating configurations.

In conic surfaces with symmetry of revolution there is always a pair of conjugate foci that are free of spherical aberration, as illustrated in Fig. 69. The ellipsoid in Fig. 69A is obtained by rotating the ellipse about

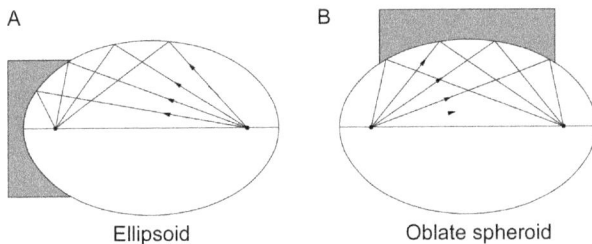

Fig. 69 Testing of (A) ellipsoid and (B) oblate astigmatic spheroids at their conjugate foci.

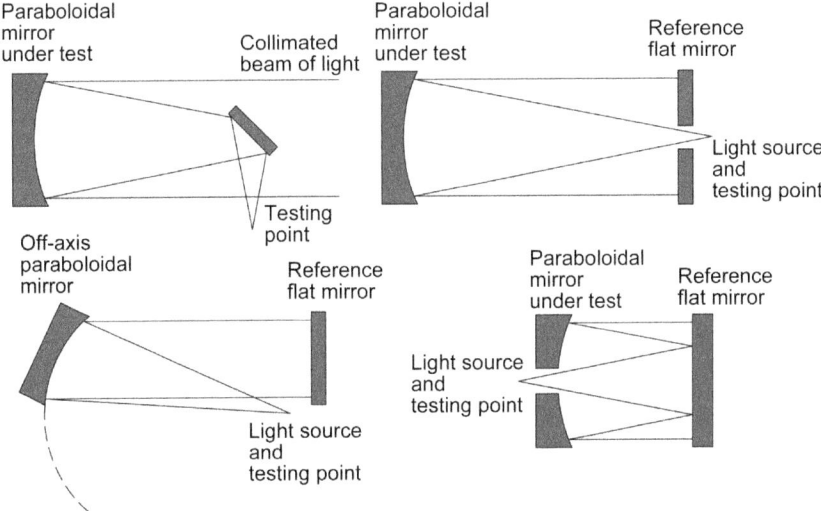

Fig. 70 Testing paraboloidal surfaces.

its major axis. Then all the light emitted by a point light source in one of the foci will be reflected on the internal surface of the ellipsoid, toward the other focus. If the ellipsoid in Fig. 69B is also rotated about its major axis, the generated surface will be like a ring or like the internal surface of a tire. Again, all the light from one focus goes to the other. If the ellipsoid in Fig. 66 is rotated about its minor axis, the generated surface will have rotational symmetry but the foci generate a thin ring and thus a single pair of conjugate foci will not exist. A point light source at any point on the thin ring generated by the foci will produce an image with astigmatism at the other focus.

Several null configurations are possible with these ellipsoids, concave, as well as convex. Some configurations are frequently used to test conical, i.e., paraboloidal, elliptical, or hyperboloidal surfaces, as illustrated in Fig. 70,

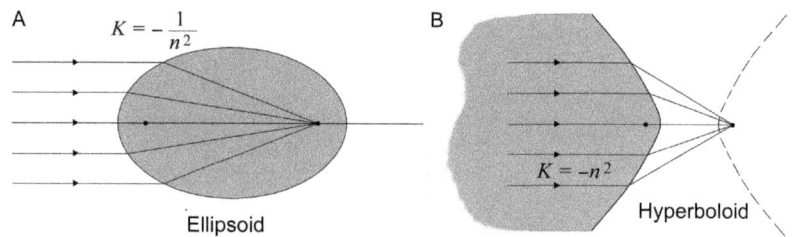

Fig. 71 Testing aplanatic refractive (A) ellipsoid and (B) hyperboloids.

mainly for astronomical telescope mirrors. These are the so-called autocollimating configurations.

Conic refractive conic surfaces can also be free of spherical aberration, but only if the refractive index has a certain value, which is a function of the conic constant of the ellipsoid or hyperboloid, as illustrated in Fig. 71. One of the conjugates is at an infinite distance. In this case the image is also corrected for coma. For this reason these surfaces are said to be aplanatic. Given the conic constant of a surface to be manufactured, the refractive index of the glass has to be selected as close as possible to the required value, but this is not always possible.

7.2 Compensators to Test Aspherical Surfaces

Typically, if a quantitative retrieval of the wavefront is desired, the interferogram is imaged onto a CCD detector. Then, the straightness of the fringes for a perfect wavefront is useful but not absolutely necessary. However, the minimum fringe spacing should be larger than twice the pixel size in the detector. This is the well-known Nyquist condition, which may be impossible to satisfy if the wavefront has a strong asphericity.

Some additional optical elements can be added to the testing system to compensate for the spherical aberration of the wavefront reflected from the aspheric surface. Then, an auxiliary optical system is designed so that, in combination with the aspheric surface, it forms a stigmatic image of a point source. The auxiliary optical system is called a null corrector or null compensator.

Using the basic interferometric configurations described before, many optical elements can be tested, including concave or convex surfaces, prisms, etc. However, the great challenge has always been the testing of aspheric surfaces. The reason is that when an aspheric surface is illuminated with a spherical wavefront nearly always the reflected wavefront is aspheric. Then when the aspheric wavefront interferes with a reference spherical wavefront, the number of nonstraight interference fringes is so large that their interpretation and measurement may become extremely difficult. The number of fringes cannot be reduced below a certain limit by adjusting the radius of curvature of the reference sphere.

To properly test the aspheric surface, the reflected aspheric wavefront must first be transformed into a spherical wavefront. In principle, this can be achieved with a properly designed combination of one or more reflective or refractive optical elements. Ideally, these optical elements must be spherical, so that they can be easily tested.

An autocollimating configuration is not always possible. Then, an optical system called a compensator has to be designed to transform the aspheric wavefront into a spherical or flat wavefront. Many different compensators have been designed, the simplest and one of the first is the Dall compensator in Fig. 72 (Dall, 1947), but the most popular is the Offner (1963) compensator illustrated in Fig. 73.

There are compensators made with mirrors. Their basic principles are the same as in refractive compensators. An example is the reflective Offner compensator. This compensator is widely used to test concave conic (paraboloidal or hyperboloidal) astronomical mirrors. A study of the methods for the design of compensators has been published by Sasian (1988).

7.2.1 Testing Hyperboloids With Autocollimating Configurations

Testing a convex hyperboloid for the secondary mirror of a Cassegrain of Ritchey–Chretien telescope is a particularly interesting problem. A review is presented by Parks and Shao (1988). One interesting method proposed by Meinel and Meinel (1983a, 1983b) tests the hyperboloid from the back of a lens, as if it were a concave mirror, as shown in Fig. 74. The disadvantage is that the glass has to be transparent and highly homogeneous.

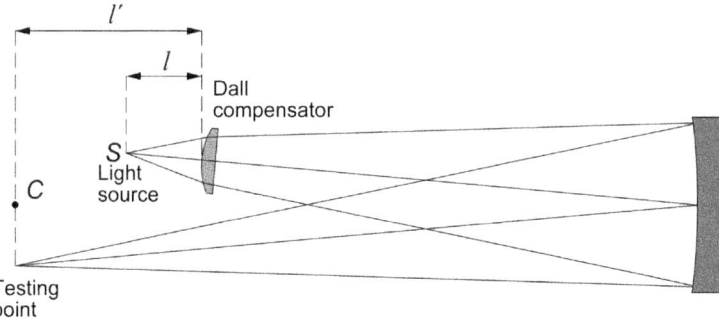

Fig. 72 Dall compensator to test a paraboloid.

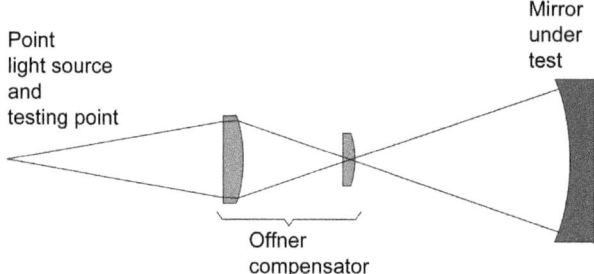

Fig. 73 Offner compensator to test a concave paraboloidal or hyperboloidal mirror.

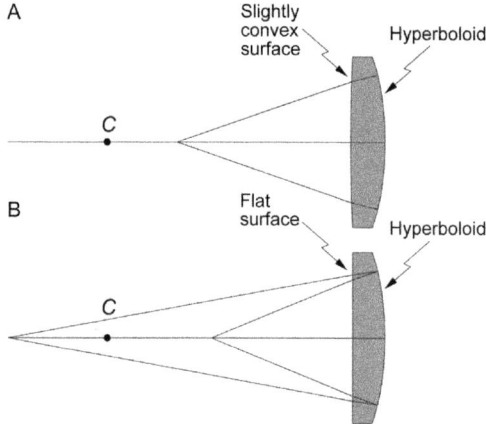

Fig. 74 Testing a hyperboloid on the back surface of a lens. (A) With equal conjugates and a convex spherical surface and (B) with different conjugates and flat surface.

The other surface of the lens has to be polished perfectly spherical, which is not difficult and can be easily tested with traditional methods. The spherical surface introduces in the system the proper amount of spherical aberration to compensate the asphericity of the hyperboloid. There are two possible configurations. One is with equal conjugates, so that the testing point and the light source are at the same place. The spherical surface has to be slightly convex.

If the point light source and the testing point can be at different conjugates, the back surface of the lens can be flat.

Another possibility for testing a hyperboloid is by refraction and a dummy lens on the back, as shown in Fig. 75, designed to compensate for the spherical aberration of the hyperboloidal surface. A positive lens can also be placed in front of the reflecting convex hyperboloid to obtain the null test, as shown by Bruns (1983).

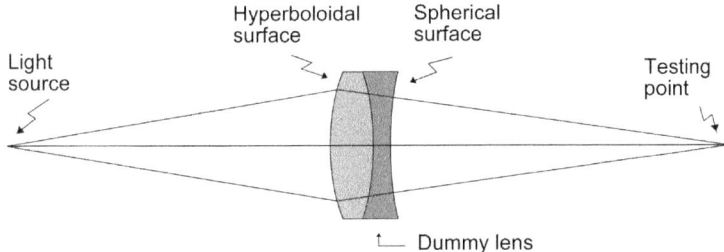

Fig. 75 Testing a hyperboloid by refraction and an auxiliary lens.

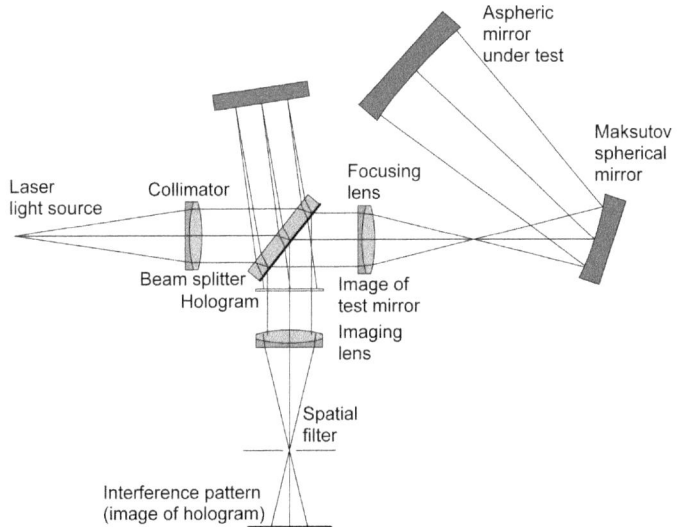

Fig. 76 Arrangement to test a spherical surface in a Twyman–Green interferometer using a computer generated holographic compensator.

7.2.2 Synthetic Hologram Compensators

When master optics are not available to make a real hologram, a computer generated (or synthetic) hologram (CGH) can be made (Beyerlein, Lindlein, & Schwider, 2002; Caulfield, Mueller, Dvore, Epstein, & Loomis, 1981; Dörband & Tiziani, 1985; Lee, 1970, 1974; MacGovern & Wyant, 1971; Ono & Wyant, 1984; Pastor, 1969; Pruss, Reichelt, Tiziani, & Osten, 2004; Reichelt, Pruss, & Tiziani, 2004; Schwider & Burow, 1976; Wyant & Bennett, 1972). A CGH is a binary representation of the actual interferogram (hologram) that would be obtained if the ideal wavefront from the test system is interfered with a tilted plane wavefront. A typical setup is illustrated in Fig. 76. CGHs are an alternative to null optics when testing aspheric optical components.

Fig. 77 Computer generated hologram designed by Wyant and Bennett (1972) and Wyant and O'Neill (1974).

The computer generated hologram, designed by Wyant and Bennett (1972) and Wyant and O'Neill (1974), to compensate the spherical aberration of the surface under test in a Twyman–Green interferometer, as in Fig. 76, is shown in Fig. 77.

7.3 Wavefront Stitching

When the wavefront is strongly aspheric and even with zero tilt in the reference wavefront the minimum fringe spacing is too small, an option is to measure the wavefront by segmenting the complete aperture in small regions where the Nyquist condition is not violated, so that the minimum fringe spacing is larger than twice the pixel spacing. There are many approaches to dividing the aperture, but they can be classified into three broad categories.

One obvious procedure is to use several different defocusing values, as described by Liu, Lawrence, and Koliopoulos (1988) and Melozzi, Pezzati, and Mazzoni (1993). Then, several rings where the fringe spacing never exceeds the Nyquist limit are obtained as illustrated in Fig. 78.

Another method to measure an aspherical surface by stitching is to divide the aperture in many circular zones where the tilt as well as the defocusing is optimized to maximize the minimum fringe spacing. For example, the evaluation of an aspheric surface can be made with many small glass test plates, each one optimized for a different region (Jensen, Chow, & Lawrence, 1984).

This method is also useful when measuring large optical surfaces whose size is much larger than the interferometer aperture (Sjödahl & Oreb, 2002), for example, when testing an extremely large plane (Negro, 1984). Special

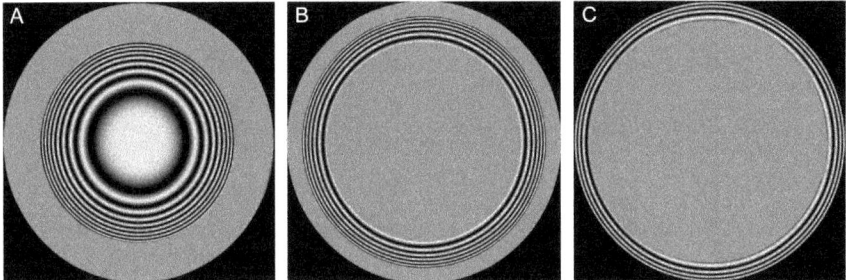

Fig. 78 Ring fringes formed by an aspherical wavefront for three different focus settings. (A) Paraxial focus, (B) an intermediate focus, and (C) marginal focus.

techniques must be used to ensure the continuity of the different apertures, for example with some overlapping and polynomial fitting of the apertures to join them.

A slightly different method using a modified Twyman–Green interferometer has been reported (Garbusi & Osten, 2009; Garbusi, Priss, & Osten, 2008) to test strong aspheric surfaces. A beam stop is used at the Fourier plane of the observation plane in order to limit the spatial frequency of the observed fringe pattern. Also, a two-dimensional array of point light sources is used, instead of a single point light source. This arrangement generates different tilted wavefronts, which in turn produce fringes only when the fringe spacing is smaller than that permitted by the stop at the Fourier plane.

7.4 Two Wavelengths Measurements

The surface under test must be so aspheric that it is not close enough to perform a null test. Even if a null test is attempted, the resulting interferogram contains too many fringes to analyze. If a high accuracy is not needed, a longer wavelength light source could be used in the interferometer to reduce the number of fringes. Another solution is to use two-wavelengths and multiple-wavelengths techniques to synthesize a long effective wavelength using visible light to obtain an interferogram identical to the one that would be obtained if a longer wavelength source were used as shown by several authors, among others by Wyant (1971), Wyant, Oreb, and Hariharan (1984), Cheng and Wyant (1984, 1985), and Creath, Cheng, and Wyant (1985).

Two-wavelengths holography is performed by first registering the image of the fringe pattern obtained by testing an optical element using a wavelength λ_1 in a Twyman–Green or Fizeau interferometer. This recording of the fringe pattern (hologram) is then used to transfer its image to a transparent medium (a slide). This slide is placed in the interferometer at the exact position it occupied during exposure and illuminated with the fringe pattern obtained by testing the optical element using a different wavelength λ_2. The resulting two-wavelengths fringes can either be thought of as the Moiré between the interference fringes stored in the hologram (recorded at λ_1 and replayed at λ_2) and the live interference fringes (at λ_2) or as the secondary interference between the test wavefront stored in the hologram and the live test wavefront. These fringes are identical to those that would be obtained if the optical element was tested using a long effective wavelength given by Wyant (1971):

$$\lambda_e = \frac{\lambda_1 \lambda_2}{|\lambda_1 - \lambda_2|} \tag{92}$$

Using a dye laser, a large range of equivalent wavelengths can be obtained. Tunable helium–neon lasers with four or five distinct wavelengths ranging from green to red are also available.

REFERENCES

Beyerlein, M., Lindlein, N., & Schwider, J. (2002). Dual-wave-front computer-generated holograms for quasi-absolute testing of aspherics. *Applied Optics*, *41*, 2440–3106.

Bruning, J. H., Herriot, D. R., Gallagher, J. E., Rosenfeld, D. P., White, A. D., & Brangaccio, D. J. (1974). Digital wavefront measuring interferometer for testing optical surfaces and lenses. *Applied Optics*, *13*, 2693–2703.

Bruns, D. G. (1983). Null test for hyperbolic convex mirrors. *Applied Optics*, *22*, 12–13.

Burch, J. M. (1953). Scatter fringes of equal thickness. *Nature*, *171*, 889–890.

Burch, J. M. (1962). Scatter-fringe interferometry. *Journal of the Optical Society of America*, *52*, 600.

Canovas, C., & Ribak, E. N. (2007). Comparison of Hartmann analysis methods. *Applied Optics*, *46*, 1–6.

Caulfield, H. J., Mueller, P., Dvore, D., Epstein, A., & Loomis, J. S. (1981). Computer holograms for optical testing. *Proceedings of SPIE*, *306*, 154–157.

Cheng, Y.-Y., & Wyant, J. C. (1984). Two-wavelength phase shifting interferometry. *Applied Optics*, *23*, 4539–4543.

Cheng, Y.-Y., & Wyant, J. C. (1985). Multiple-wavelength phase shifting interferometry. *Applied Optics*, *24*, 804–807.

Cornejo-Rodriguez, A. (2007). Ronchi test. In D. Malacara (Ed.), *Optical shop testing*. (3rd ed.), New York: John Wiley and Sons.

Creath, K., Cheng, Y.-Y., & Wyant, J. C. (1985). Contouring aspheric surfaces using two-wavelength phase-shifting interferometry. *Optica Acta*, *32*, 1455–1464.

Dall, H. E. (1947). A null test for paraboloids. *The Journal of the British Astronomical Association*, *57*, 201–205.

Díaz-Uribe, R., & Campos-García, M. (2000). Null screen testing of fast convex aspheric surfaces. *Applied Optics*, *39*, 2670–2677.

Dörband, B., & Tiziani, H. J. (1985). Testing aspheric surfaces with computer generated-holograms: Analysis of adjustment and shape errors. *Applied Optics*, *24*, 2604–2611.

Forbes, G. W. (2007). Shape specification for axially symmetric optical surfaces. *Optics Express*, *15*, 5218–5226.

Forbes, G. W. (2011). Characterizing the shape of freeform optics. *Optics Express*, *20*, 2483–2499.

Forsythe, G. E. (1957). Generation and use of orthogonal polynomials for data fitting on a digital computer. *Journal of the Society for Industrial and Applied Mathematics*, *5*, 74–88.

Foucault, L. M. (1858). A description des procédés employés pour reconnaitre la configuration des surfaces optiques. *Comptes Rendus de l'Académie des Sciences*, *47*, 958–959.

Freischlad, K. R., & Koliopoulos, C. L. (1986). Modal estimation of a wavefront from difference measurements using the discrete Fourier transform. *Journal of the Optical Society of America A*, *3*, 1852–1861.

Garbusi, E., & Osten, W. (2009). Perturbation methods in optics: Application to the interferometric measurement of surfaces. *Journal of the Optical Society of America A*, *26*, 2538–2549.

Garbusi, E., Priss, C., & Osten, W. (2008). Interferometer for precise and flexible asphere testing. *Optics Letters*, *33*, 2973–2975.

Greivenkamp, J. E. (1987). Sub-nyquist interferometry. *Applied Optics*, *26*, 5245–5258.

Greivenkamp, J. E., & Bruning, J. H. (1992). Phase shifting interferometers. In D. Malacara (Ed.), *Optical shop testing*. New York: John Wiley and Sons.

Hernández-Gómez, G., Malacara-Hernández, Z., & Malacara-Hernández, D. (2014). Hartmann tests to measure the spherical and cylindrical curvatures and the axis orientation of astigmatic lenses o optical surfaces. *Applied Optics*, *53*, 1191–1199.

Hoffer, H., Artal, P., Singer, B., Aragon, J. L., & Williams, D. (2001). Dynamics of the eye's wave aberration. *Journal of the Optical Society of America*, *18*, 497–506.

Hopkins, H. H. (1950). *Wave theory of aberrations* (p. 48). Oxford: Clarendon Press.

Houston, J. B., Jr., Buccini, C. J., & O'Neill, P. K. (1967). A laser unequal path interferometer for the optical shop. *Applied Optics*, *6*, 1237–1242.

Hudgin, R. H. (1977). Wave-front reconstruction for compensated imaging. *Journal of the Optical Society of America*, *67*, 375–378.

Ichikawa, K., Lohmann, A. W., & Takeda, M. (1988). Phase retrieval based on the irradiance transport equation and the Fourier transport method: Experiments. *Applied Optics*, *27*, 3433–3436.

Ichioka, Y., & Inuya, M. (1972). Direct phase detecting system. *Applied Optics*, *11*, 1507–1514.

Jensen, S. C., Chow, W. W., & Lawrence, G. N. (1984). Subaperture testing approaches: A comparison. *Applied Optics*, *23*, 740–744.

Kim, C.-J., & Shannon, R. (1987). Catalog of Zernike polynomials. In R. Shannon & J. C. Wyant (Eds.), *Applied optics and optical engineering:* Vol. 10. New York: Academic Press. Chap. 4.

Kingslake, R. (1925–1926). The inteferometer patterns due to the primary aberrations. *Transactions of the Optical Society*, *27*, 94–104.

Kujawinska, M. (1993). Spatial phase measurement methods. In D. W. Robinson & G. T. Reid (Eds.), *Interferogram analysis* (p. 294). Bristol, Philadelphia: Institute of Physics.

Lee, W. H. (1970). Sampled Fourier transform hologram generated by computer. *Applied Optics*, *9*, 639–643.

Lee, W. H. (1974). Binary synthetic holograms. *Applied Optics*, *13*, 1677–1682.

Liu, Y.-M., Lawrence, G. N., & Koliopoulos, C. L. (1988). Subaperture testing of aspheres with annular zones. *Applied Optics*, *27*, 4504–4513.

López-Ramírez, D., Doblado, M., & Malacara-Hernández, D. (2000). New simple geometrical test for aspheric lenses and mirrors. *Optical Engineering, 39,* 2143–2148.

MacGovern, A. J., & Wyant, J. C. (1971). Computer generated holograms for testing optical elements. *Applied Optics, 10,* 619–624.

Mahajan, V. N. (2007). Zernike polynomial and wavefront fitting. In D. Malacara (Ed.), *Optical shop testing* (3rd ed.). New York: John Wiley and Sons, Inc.

Malacara, D., & Cornejo, A. (1974). Null ronchi test of aspherical surfaces. *Applied Optics, 13,* 1778–1780.

Malacara, D., Creath, K., Schmidt, J., & Wyant, J. C. (2007). Testing of aspheric wavefronts and surfaces. In D. Malacara (Ed.), *Optical shop testing* (3rd ed.). New York: John Wiley and Sons, Inc.

Malacara-Doblado, D., & Ghozeil, I. (2007). Hartmann test. In D. Malacara (Ed.), *Optical shop testing* (3rd ed.). New York: John Wiley and Sons, Inc.

Malacara-Hernández, D. (2007). Optical shop testing. In D. Malacara (Ed.), (3rd ed.). New York: John Wiley and Sons, Inc.

Malacara-Hernández, D., & Malacara-Doblado, D. (2015). What is a Hartmann test. *Applied Optics, 54,* 2296–2300.

Malacara-Hernández, D., Servín, M., & Malacara-Hernández, Z. (1998). *Interferogram analysis for optical testing.* New York: Marcel Dekker.

Meinel, A. B., & Meinel, M. P. (1983a). Self-null corrector test for telescope hyperbolic secondaries. *Applied Optics, 22,* 520–521.

Meinel, A. B., & Meinel, M. P. (1983b). Self-null corrector test for telescope hyperbolic secondaries: Comments. *Applied Optics, 22,* 2405.

Mejia-Barbosa, Y., & Malacara-Hernández, D. (2001). Object surface for applying a modified Hartmann test to measure corneal topography. *Applied Optics, 40,* 5778–5786.

Melozzi, M., Pezzati, L., & Mazzoni, A. (1993). Testing aspheric surfaces using multiple annular interferograms. *Optical Engineering, 32,* 1073–1079.

Menchaca, C., & Malacara, D. (1986). Toroidal and sphero-cylindrical surfaces. *Applied Optics, 25,* 3008–3009.

Millerd, J., Brock, N., Hayes, J., Kimbrough, B., Novak, M., & North-Morris, M. (2005). Modern approaches in phase measuring metrology. *Proceedings of SPIE, 5856,* 14–22.

Murty, M. V. R. K. (1963). Common path interferometer using Fresnel zone plates. *Journal of the Optical Society of America, 53,* 568.

Murty, M. V. R. K. (1964). The use of a single plane parallel plate as a lateral shearing interferometer with a visible Gas laser source. *Applied Optics, 3,* 531.

Negro, J. E. (1984). Subaperture optical system testing. *Applied Optics, 23,* 1921–1930.

Noll, R. J. (1978). Phase estimates from slope-type wave-front sensors. *Journal of the Optical Society of America, 68,* 139–140.

Offner, A. (1963). A null corrector for paraboloidal mirrors. *Applied Optics, 2,* 153–156.

Ojeda-Castañeda, J. (2007). Foucault, wire, and phase modulation tests. In D. Malacara (Ed.), *Optical shop testing.* (3rd ed.), New York: John Wiley and Sons.

Ono, A., & Wyant, J. C. (1984). Plotting errors measurement of CGH using and improved interferometric method. *Applied Optics, 23,* 3905–3910.

Papoulis, A. (1968). *Systems and transforms with applications in optics.* New York: McGraw-Hill.

Pastor, J. (1969). Hologram interferometry and optical technology. *Applied Optics, 8,* 525–531.

Parks, R. E., & Shao, L. Z. (1988). Testing large hyperbolic secondary mirrors. *Optical Engineering, 27,* 1057–1062.

Platt, B. C., & Shack, R. V. (1971). Lenticular Hartmann screen. *Optical Sciences Center Newsletter, 5,* 15–16.

Prieto, P. M., Vargas-Martín, F., Goelz, S., & Artal, P. (2000). Analysis of the performance of the Hartmann–Shack sensor in the human eye. *Journal of the Optical Society of America A*, *17*, 1388–1398.

Pruss, C., Reichelt, S., Tiziani, H. J., & Osten, W. (2004). Computer-generated holograms in interferometric testing. *Optical Engineering*, *43*, 2534–2540.

Rayces, J. L. (1964). Exact relation between wave aberration and ray aberration. *Optica Acta*, *11*, 85–88.

Reichelt, S., Pruss, C., & Tiziani, H. J. (2004). Absolute testing of aspheric surfaces. *Proceedings of SPIE*, *5252*, 252–263.

Rimmer, M. P. (1974). Method for evaluating lateral shearing interferometer. *Applied Optics*, *13*, 623–629.

Rimmer, M. P., & Wyant, J. C. (1975). Evaluation of large aberrations using a lateral shear interferometer having variable shear. *Applied Optics*, *14*, 142–150.

Roddier, F. (1988). Curvature sensing and compensation: A new concept in adaptive optics. *Applied Optics*, *27*, 1223–1225.

Roddier, C., & Roddier, F. (1991a). Reconstruction of the Hubble space telescope mirror figure from out-of-focus stellar images. *Proceedings of SPIE*, *1494*, 11–17.

Roddier, F., & Roddier, C. (1991b). Wavefront reconstruction using iterative Fourier transforms. *Applied Optics*, *30*, 1325–1327.

Roddier, C., & Roddier, F. (1992). New optical testing methods developed at the university of Hawaii: Results on ground-based telescopes and Hubble space telescope. *Proceedings of SPIE*, *1531*, 37–43.

Roddier, F., Roddier, C., & Roddier, N. (1988). curvature sensing: A new wavefront sensing method. *Proceedings of SPIE*, *976*, 203–209.

Roddier, C., Roddier, F., Stockton, A., & Pickles, A. (1990). Testing of telescope optics: A new approach. *Proceedings of SPIE*, *1236*, 756–766.

Ronchi, V. (1923). La frange di combinazioni nello Studio delle superficie e dei sistemi ottici. *Rev. Ottica Mecc. Precis.*, *2*, 9–35.

Sasian, J. M. (1988). Design of null correctors for the testing of astronomical optics. *Optical Engineering*, *27*, 1051–1056.

Saunders, J. B. (1961). Measurement of wavefronts without a reference standard: The wavefront shearing interferometer. *Journal of Research of the National Bureau of Standards*, *65B*, 239–244.

Schreibert, H., & Bruning, J. (2007). Phase shifting interferometry. In D. Malacara (Ed.), *Optical shop testing* (3rd ed.). New York: John Wiley and Sons, Inc.

Schulz, G. (1988). Aspheric surfaces. In E. Wolf (Ed.), *Progress in optics*. Vol. XXV, Chapter IV, North Holland, Amsterdam.

Schwider, J. (1999). Interferometric tests for aspherics. *Optical Society of America Trends in Optics and Photonics*, *24*, 103–114.

Schwider, J., & Burow, R. (1976). The testing of aspherics by means of rotational-symmetric synthetic holograms. *Optica Applicata*, *6*, 83.

Servín, M., Quiroga, J. A., & Padilla, J. M. (2014). *Fringe pattern analysis for optical metrology*. Verlag GmbH: Wiley-VCH.

Shomali, R., Darudi, A., & Nasiri, S. (2012). Application of irradiance transport equation in aspheric surface testing. *Optik*, *123*, 1282–1286.

Singer, W., Totzek, M., & Gross, H. (2005). Handbook of optical systems. In H. Gross (Ed.), Vol. 2 (p. 213): Oberkochen, Germany: Wiley-VCH, Verlag GmbH.

Sjödahl, M., & Oreb, B. F. (2002). Stitching interferometric measurement data for inspection of large optical components. *Optical Engineering*, *41*, 403–408.

Smart, R. N., & Steel, W. H. (1975). Theory and application of point diffraction interferometers. *Proc. ICO conference, Tokyo 1974. The Japanese Journal of Applied Physics*, *14*(Suppl. 1), 351.

Steibl, N. (1984). Phase imaging by the transport equation of intensity. *Optics Communication, 49*, 6–10.

Takeda, M., Ina, H., & Kobayashi, S. (1982). Fourier-transform method of fringe-pattern analysis for computer-based topography and interferometry. *Journal of the Optical Society of America, 72*, 156–160.

Teague, M. R. (1983). Deterministic phase retrieval: A Green's function solution. *Journal of the Optical Society of America, 73*, 1434–1441.

Teague, M. R. (1985). Image formation in terms of the transport equation. *Journal of the Optical Society of America A, 2*, 2019–2026.

Twyman, F. (1918). Correction of optical surfaces. *The Astrophysical Journal, 48*, 256.

Wyant, J. C. (1971). Testing aspherics using two-wavelength holography. *Applied Optics, 10*, 2113–2118.

Wyant, J. C., & Bennett, V. P. (1972). Using computer generated holograms to test aspheric wavefronts. *Applied Optics, 11*, 2833–2839.

Wyant, J. C., & O'Neill, P. K. (1974). Computer generated hologram: Null lens test of aspheric wavefronts. *Applied Optics, 13*, 2762–2765.

Wyant, J. C., Oreb, B. F., & Hariharan, P. (1984). Testing aspherics using two-wavelength holography: Use of digital electronic techniques. *Applied Optics, 23*, 4020–4023.

CHAPTER THREE

Generation of Partially Coherent Beams

Yangjian Cai*,†, Yahong Chen*,†, Jiayi Yu*,†, Xianlong Liu*,†, Lin Liu*,†

*College of Physics, Optoelectronics and Energy & Collaborative Innovation Center of Suzhou Nano Science and Technology, Soochow University, Suzhou, China
†Key Lab of Advanced Optical Manufacturing Technologies of Jiangsu Province & Key Lab of Modern Optical Technologies of Education Ministry of China, Soochow University, Suzhou, China

Contents

1. Introduction	157
1.1 Gaussian Schell-Model Beams	158
1.2 Partially Coherent Beams With Prescribed Phases	160
1.3 Partially Coherent Beams With Prescribed States of Polarization	163
1.4 Partially Coherent Beams With Prescribed Degrees of Coherence	167
2. Characterization and Generation of Various Partially Coherent Beams	170
2.1 GSM Beams	173
2.2 Partially Coherent Beams With Prescribed Phases	177
2.3 Vector Partially Coherent Beams With Prescribed States of Polarization	184
2.4 Partially Coherent Beams With Prescribed Degrees of Coherence	195
3. Summary	212
Acknowledgments	213
References	213

1. INTRODUCTION

Since the invention of the laser, laser beams have been applied in various fields. Laser beams are known for their high spatial coherence and are usually treated as a fully coherent beam. However, light beams with low spatial coherence, so-called partially coherent beams, are preferred in many applications, such as microdensitometry (Kinzly, 1972), linewidth measurement (Nyyssonen, 1977), lithography (Ma & Arce, 2008), holography (Som, Delisle, & Drouin, 1980), optical information processing

(Zhuang & Yu, 1982), inertial confinement fusion (Kato et al., 1984), free-space optical communications (Korotkova, Andrews, & Phillips, 2004a; Ricklin & Davidson, 2002), ghost imaging (Cai & Zhu, 2005), non-interferometric phase imaging (Paganin & Nugent, 1998), quantitative imaging (Gureyev, Paganin, Stevenson, Mayo, & Wilkins, 2004), sub-Rayleigh imaging (Oh, Cho, Scarcelli, & Kim, 2013), optical imaging (Brown & Brown, 2008), solition generation (Akhmediev, Krolikowski, & Snyder, 1998), second-harmonic generation (Ansari & Zubairy, 1986), subsurface diagnostics (Apostol & Dogariu, 2003), collective atomic recoil lasing (Robb & Firth, 2007), lidar (Korotkova, Andrews, & Phillips, 2004b), remote detection (Wu & Cai, 2011), particle trapping (Zhao & Cai, 2011), optical scattering (van Dijk, Fischer, Visser, & Wolf, 2010), and plasmonics (Gan, Gbur, & Visser, 2007). Coherence can be regarded as a consequence of correlations between some components of the fluctuating electric field at two (or more) points. Young's well-known interference experiment (Young, 1802) triggered several studies of the coherence properties of light (Thompson & Wolf, 1957; Wolf, 1954, 1955; Zernike, 1938), and partially coherent beams were rapidly developed following the invention of the laser (Gori, 1998; Mandel & Wolf, 1995; Wolf, 2003). The spatial coherence properties of a scalar beam are generally described by either the mutual intensity in the space–time domain or the cross-spectral density (CSD) in the space–frequency domain (Mandel & Wolf, 1995). The spatial coherence properties of a vector beam are described by either its coherence–polarization matrix (Gori, 1998) in the space–time domain or its CSD matrix (Wolf, 2003, 2007) in the space–frequency domain.

1.1 Gaussian Schell-Model Beams

A Gaussian Schell-model (GSM) beam represents a broad class of partially coherent beams, whose intensity and degree of coherence both satisfy Gaussian distributions. The GSM beam concept was introduced in the 1970s (Collett & Wolf, 1978; Foley & Zubairy, 1976; Wolf & Collett, 1978) and these beams were first generated by De Santis, Gori, Guattari, and Palma (1979) with the help of a rotating ground glass disk (RGGD) and a Gaussian amplitude filter (GAF). Later, Tervonen et al. reported the experimental generation of GSM beams with synthetic acousto-optic holograms (Tervonen, Friberg, & Turunen, 1992). He et al. reported propagation and imaging experiments with GSM beams (He, Turunen, & Friberg, 1988). A GSM beam also can be generated by a suitable superposition of coherent but mutually uncorrelated light beams (Starikov & Wolf, 1982). The quantitative determination of the degree of coherence and the corresponding coherence width of a GSM beam based on measuring of the fourth-order correlation function (FOCF) was reported in Wang, Cai, and He (2006). Experimental observations of the propagation properties of a GSM beam through a fractional Fourier transform optical system, a multimode fiber and an apertured thin lens were reported in Wang and Cai (2007, 2008) and Zhao, Cai, Wang, Lu, and Wang (2008), and it was found that one can shape the beam profile of a GSM beam in those optical systems through varying the initial coherence width. The experimental observation of the focal shift in a focused GSM beam was reported in Wang, Cai, and Korotkovaa (2009a), in which the theoretical prediction that the focal shift increases as the initial coherence width decreases, was verified (Friberg, Visser, Wang, & Wolf, 2001). GSM beams have also been used, instead of entangled photon pairs, to realize ghost imaging (Cai & Zhu, 2005; Valencia, Scarcelli, D'Angelo, & Shih, 2005) and ghost interference (Cai & Zhu, 2004; Vidal, Caetano, Fonseca, & Hickmann, 2009). Experimental observation of nonspecular effects for a GSM beam reflected at an air–glass interface was reported in Merano, Umbriaco, and Mistura (2012). The nonparaxial propagation of a GSM beam in free space and in a uniaxial crystal was investigated in Duan and Lü (2004) and Zhang and Cai (2011), respectively. Propagation properties of a GSM beam in a turbulent atmosphere were explored in Berman, Chumak, and Gorshkov (2007), Gbur and Wolf (2002), Shirai, Dogariu, and Wolf (2003), and Wang and Plonuss (1979), and it was shown that a GSM beam has advantages over a fully coherent Gaussian beam by undergoing reduced turbulence-induced degradation, beam wander, and scintillation. This was verified in several experiments (Dogariu & Amarande, 2003; Liu, Wang, Wei, & Cai, 2014). This shows that GSM beams are promising for application in free-space optical communications (Korotkova, Andrews, et al., 2004a; Ricklin & Davidson, 2002) and lidar system (Korotkova, Andrews, et al., 2004b). In Wu and Cai (2011), it was shown that a GSM beam is useful for remote detection, e.g., one can obtain information about the target by measuring the beam widths and the coherence widths at the source plane and at the receiver plane. The interaction of a GSM beam with a nonlinear crystal was investigated in Ansari and Zubairy (1986), and it was found that the conversion efficiency of the second-harmonic beam can be significantly increased by degrading the coherence of the GSM beam. The theory of a GSM beam scattered by a small homogeneous sphere and the effects of spatial coherence on the angular distribution of the intensity of the scattered fields were explored in van Dijk et al. (2010). Propagation properties of a GSM beam passing through polarization gratings were studied in Piquero, Borghi, and Santarsiero (2001), and it was shown that one can obtain any desirable value of the degree of polarization of the output beam by a suitable choice of the period of the grating. More recently, an experimental determination of the radius of curvature of a GSM beam was reported in Zhu, Chen, and Cai (2013), and the shape–invariant difference between two GSM beams was discussed in Borghi, Gori, Guattari, and Santarsiero (2015a).

Apart from GSM beams, numerous studies have been dedicated to partially coherent beams with prescribed phases (Chen, Liu, & Cai, 2016), states of polarization (Cai et al., 2013; Wolf, 2007), and degrees of coherence (Cai, Chen, & Wang, 2014), especially in the past several years due to the fact that such beams display many extraordinary propagation properties and have advantages over GSM beams in many applications. We will discuss partially coherent beams with prescribed phases, states of polarization, and degrees of coherence in Sections 1.2–1.4.

1.2 Partially Coherent Beams With Prescribed Phases

Partially coherent beams can carry two special kinds of phase (i.e., twist phase and vortex phase) besides the customary quadratic phase. A partially coherent beam with a twist phase is usually called a twisted Gaussian Schell-model (TGSM) beam and was first proposed by Simon and collaborators (Simon & Mukunda, 1993; Simon, Sundar, & Mukunda, 1993; Sundar, Simon, & Mukunda, 1993). The twist phase opens up "a new dimension" in the area of partially coherent fields, and it differs in many aspects from the customary quadratic phase, e.g., unlike the usual phase curvature, the twist phase is bounded in strength due to the fact that the CSD function must be nonnegative definite and it disappears in the limit of full coherence. The twist phase has an intrinsic chiral property and it induces the rotation of the beam spot on propagation (Friberg, Tervonen, & Turunen, 1994a). Friberg et al. reported the first experimental demonstration of a TGSM beam by use of an acousto-optic coherence control technique (Friberg et al., 1994a), and measured the focusing properties of a TGSM beam in Friberg, Tervonen, and Turunen (1994b), where it was found that the presence of the twist phase increases the size of the best focus and moves it toward the image of a completely incoherent source. A TGSM beam can be expressed as an incoherent superposition of ordinary Gaussian beams (Ambrosini, Bagini, Gori, & Santarsiero, 1994) or partially coherent modified Bessel–Gauss beams (Gori & Santarsiero, 2015). The paraxial propagation of a TGSM beam can be studied with the help of the Wigner distribution function (Bastiaans, 2000) or the tensor method (Lin & Cai, 2002a). With the help of the tensor method, propagation properties of a TGSM beam through a paraxial ABCD optical system with and without truncation (Cai & Hu, 2006; Lin & Cai, 2002a), a fractional Fourier transform optical system (Lin & Cai, 2002b), dispersive and absorbing media (Cai, Lin, & Ge, 2002), nonlinear media (Cai & Peschel, 2007), and turbulent atmosphere (Cai & He, 2006; Wang & Cai, 2010; Wang, Cai, Eyyuboğlu, & Baykal, 2012) have been explored in detail. As shown in Cai and He (2006), Wang, Cai, Eyyuboğlu, et al. (2012), the twist phase plays a role in reducing turbulence-induced scintillation besides rotating the beam spot. Ponomarenko introduced TGSM solitons in Ponomarenko (2001a), and he found that the twist phase provides an opportunity for controlling the degree of spatial coherence of solitons without affecting their intensities. Ghost imaging with a TGSM beam was investigated in Cai, Lin, and Korotkova (2009), and it was found that the twist phase enhances the image visibility. The radiation force of a focused TGSM beam on a Rayleigh dielectric sphere was studied in Zhao, Cai, and Korotkovaa (2009), and it was found that it is possible to increase both transverse and longitudinal trapping ranges by raising the absolute value of the twist factor. Tong and Korotkova found that a TGSM beam can serve as an illumination that may produce images with a resolution overcoming the Rayleigh limit (Tong & Korotkova, 2012a). A TGSM beam can carry orbital angular momentum (OAM) (Serna & Movilla, 2001), and one can modulate the OAM of a TGSM beam by a cylindrical thin lens by varying its orientation angle (Cai & Zhu, 2014). The influence of the twist phase on the correlation properties of a partially coherent beam focused by diffractive axicons was explored in Alkelly, Shukri, and Alarify (2012) and Shukri, Alkelly, and Alarify (2012), and a method for synthesizing diffractive axicons based on TGSM beam illumination was proposed in Shukri, Alkelly, and Alarify (2013). More recently, the problem of when a twist phase can be impressed on a partially coherent beam was solved for Schell-model fields endowed with axial symmetry (Borghi, Gori, Guattari, & Santarsiero, 2015b).

A partially coherent beam with a vortex phase is called a partially coherent vortex beam, and such beams display helical wavefronts. It is known that each photon of a vortex beam with a phase term $\exp(il\varphi)$ carries an OAM of $l\hbar$ with l being the topological charge (Allen, Beijersbergen, Spreeuw, & Woerdman, 1992), and such a beam is useful in many applications, such as optical manipulation (Grier, 2003), super-resolution imaging (Tamburini, Anzolin, Umbriaco, Bianchini, & Barbieri, 2006), quantum information transfer (Nagali et al., 2009), free-space data transmission (Wang et al., 2012), and detection of a spinning object (Lavery, Speirits, Barnett, & Padgett, 2013). Partially coherent Bessel–Gauss beams are a typical kind of partially coherent vortex beam and were first introduced in Zahid and Zubairy (1989). In 1998, Gori and collaborators introduced partially coherent sources whose modes are the simplest type of coherent vortex fields belongings to the class of Laguerre–Gaussian (LG) beams (Gori, Santarsiero, & Borghi, 1998). Ponomarenko introduced a class of partially coherent beams carrying optical vortices which can be represented as an incoherent superposition of fully coherent Laguerre–Gauss modes of arbitrary order with the same azimuthal mode index (Ponomarenko, 2001b). Partially coherent vortex beams with a separable phase were introduced theoretically and demonstrated experimentally in Boggatyryova et al. (2003). Later, different types of partially coherent vortex beams, such as partially coherent LG_{0l} beam (Palacios, Maleev, Marathay, & Swartzlander, 2004), partially coherent LG_{pl} beams (Wang, Cai, & Korotkova, 2009b), and GSM vortex beams (Wang, Zhu, & Cai, 2011) were introduced. Partially coherent vortex beams can be generated through imposing a vortex phase on a partially coherent beam with the help of a spiral phase plate (Palacios et al., 2004; Wang, Zhu, & Cai, 2011) or a spatial light modulator (SLM) (Zhao, Dong, Wang, et al., 2012; Zhao, Wang, Dong, et al., 2012). More recently, the digital generation of partially coherent vortex beams with arbitrary azimuthal index using only a SLM was reported (Perez-Garcia, Yepiz, Hernandez-Aranda, Forbes, & Swartzlander, 2016). The optical coherence of a partially coherent vortex beam was analyzed with the help of a Shack–Hartmann wavefront sensor (Stoklasa, Motka, Rehacek, Hradil, & Sánchez Soto, 2014). One can shape the focused intensity distribution of a partially coherent vortex beam by varying its initial coherence width (Wang, Zhu, & Cai, 2011), which is useful for trapping a Rayleigh particle whose refractive index is larger or smaller than that of the embedding medium (Zhao & Cai, 2011). Imposing a vortex phase on a partially coherent beam also is useful for further reducing turbulence-induced scintillation (Liu, Shen, Liu, Wang, & Cai, 2013). The most exciting properties of a partially coherent vortex beam are that the phase singularities disappear on propagation while correlation singularities (Gbur & Swartzlander, 2008; Maleev, Palacios, Marathay, & Swartzlander, 2004; Maleev & Swartzlander, 2008; Palacios et al., 2004) or coherence singularities (van Dijk & Visser, 2009) appear. Correlation singularities (i.e., ring dislocations) in a partially coherent LG_{01} beam were first demonstrated both theoretically and experimentally in Palacios et al. (2004). It was predicted in Yang, Mazilu, and Dholakia (2012) and Zhao, Wang, Dong, Han, and Cai (2012) that the number of ring dislocations in a partially coherent LG_{0l} beam equals $|l|$, which was verified experimentally in Escalante, Perez-Garcia, Hernandez-Aranda, and Swartzlander (2013), thus one can determine the magnitude of the topological charge of a partially coherent LG_{0l} beam through measuring the ring dislocations in the correlation function or the degree of coherence. For a partially coherent LG_{pl} beam, it was demonstrated both theoretically (Yang et al., 2013) and experimentally (Liu et al., 2016) that the number of ring dislocations equals $2p+|l|$, and one can determine the magnitude of the topological charge of a partially coherent LG_{pl} beam through measuring a double-correlation function (Yang & Liu, 2016). More recently, it was demonstrated both theoretically and experimentally that one can determine the sign and the magnitude of the topological charge of a partially coherent LG_{0l} beam simultaneously based on the measuring of the degree of coherence of such a beam after propagating it through a pair of cylindrical lenses (Chen, Liu, Yu, & Cai, 2016).

1.3 Partially Coherent Beams With Prescribed States of Polarization

Polarization is a manifestation of correlations involving components of the fluctuating electric field at a single point and it plays an important role in optical sciences and engineering. Traditionally, coherence and polarization of a light beam were studied separately. In 1994, James found that the degree of polarization of a partially coherent beam may change on propagation in free space (James, 1994). Since then numerous efforts have been devoted to vector partially coherent beams due to their interesting properties and potential applications (Cai et al., 2013; Wolf, 2007). Gori proposed the beam coherence–polarization matrix to characterize the vector beams in the space–time domain (Gori, 1998) and Wolf proposed the CSD matrix to characterize them in the space–frequency domain (Wolf, 2003, 2007). Similar to the classification of vector coherent beams (Zhan, 2009), vector partially coherent beams can be divided into beams with a spatially uniform state of polarization and beams with a spatially nonuniform state of polarization.

A vector partially coherent beam with a uniform state of polarization is usually called partially coherent and partially polarized (Gori, 1998) or a stochastic electromagnetic beam (Wolf, 2003, 2007). The generalized Stokes parameters (Korotkova & Wolf, 2005a) and the polarization ellipse (Korotkova & Wolf, 2005b) were proposed to describe the polarization properties of a stochastic electromagnetic beam. Different definitions of the degree of coherence have been introduced to describe coherence properties of electromagnetic beams (Tervo, Setälä, & Friberg, 2003; Wolf, 2003). The electromagnetic Gaussian Schell-model (EGSM) beam (i.e., a partially coherent and partially polarized GSM beam) is a specific kind of stochastic electromagnetic beam (Gori et al., 2001; Korotkova, Salem, & Wolf, 2004a). The beam conditions and the realizability conditions for an EGSM source were obtained in Gori, Santarsiero, Borghi, and Ramírez-Sánchez (2008), Korotkova, Salem, et al. (2004a), and Roychowdhury and Korotkova (2005). Several methods have been proposed to produce an EGSM source (Basu, Hyde, Xiao, Voelz, & Korotkova, 2014; Ostrovsky, Rodríguez-Zurita, Meneses-Fabián, Olvera-Santamaría, & Rickenstorff-Parrao, 2010) and its properties (Kanseri, Rath, & Kandpal, 2009; Vidal, Fonseca, & Hickmann, 2011; Wang, Wu, Liu, Zhu, & Cai, 2011) were examined both numerically and experimentally. Evolution properties of the degree of polarization and the state of polarization of an EGSM beam in free space (Gori et al., 2001; Korotkova, Visser, & Wolf, 2008; Korotkova & Wolf, 2005b) and in turbulent atmosphere (Korotkova, Salem, Dogariu, & Wolf, 2005; Korotkova, Salem, & Wolf, 2004b; Roychowdhury, Ponomarenko, & Wolf, 2005) have been explored in detail, and it was shown that both the degree of polarization and the polarization ellipse vary on propagation in free space, while in turbulent atmosphere, after propagating over a sufficiently long distance, the degree of coherence regains its initial value, and the polarization ellipse acquires the same shape as in the source plane. Scintillation (i.e., intensity fluctuation) of a stochastic electromagnetic beam propagating in turbulent atmosphere were explored both theoretically (Friberg & Visser, 2015; Korotkova, 2008) and experimentally (Avramov-Zamurovic, Nelson, Malek-Madani, & Korotkova, 2014), and it was found that under suitable conditions the EGSM beam may have reduced scintillation compared to the scalar GSM beam, which makes them attractive for free-space optical communications. Analytic formulas for the CSD matrix of an EGSM beam propagating through a paraxial ABCD optical system in turbulent atmosphere were derived in Cai, Korotkova, Eyyuboğlu, and Baykal (2008), and it can be applied to LIDAR systems (Korotkova, Cai, & Watson, 2009) and for sensing of semirough targets in turbulent atmosphere (Sahin, Tong, & Korotkova, 2010). The resonator theory for stochastic electromagnetic fields was developed in Saastamoinen, Turunen, Tervo, Setala, and Friberg (2005) and Wolf (2006), and the evolution properties of the degree of polarization and propagation factor of an EGSM beam in a resonator were explored with the help of the tensor method (Yao, Cai, Eyyuboğlu, Baykal, & Korotkova, 2008; Zhu & Cai, 2010). Focusing of spatially inhomogeneous partially coherent, partially polarized electromagnetic fields was reported in Foreman and Török (2009). Evolution properties of the degree of coherence, the degree of polarization, and the state of polarization of an EGSM beam propagating through a GRIN fiber were explored in Roychowdhury, Agarwal, and Wolf (2006) and Zhu, Liu, Chen, and Cai (2013). Effects of coherence and polarization on coupling an EGSM beam into a single-mode optical fiber have been investigated both theoretically (Salem & Agrawal, 2009) and experimentally (Zhao, Dong, Wang, et al., 2012; Zhao, Dong, Wu, et al., 2012). Ghost imaging with a stochastic electromagnetic beam was studied both theoretically (Shirai, Kellock, Setala, & Friberg, 2011; Tong, Cai, & Korotkova, 2010) and experimentally (Liu, Wang, Zhang, & Cai, 2015), and it was shown that the source polarization may reduce/enhance the ghost image and its visibility under certain conditions. Correlation singularities in stochastic electromagnetic beams were studied and such singularities have implications for both interference experiments and correlation of intensity fluctuations measurements performed with such beams (Raghunathan, Schouten, & Visser, 2012a, 2012b). The influence of different source parameters on the Hanbury Brown–Twiss effect of a stochastic electromagnetic beam was investigated in Wu and Visser (2014), and it was found that higher-order correlations behave quite differently than the lower-order amplitude-phase correlations. Scattering theories of stochastic electromagnetic fields were developed in Tong and Korotkova (2010) and Wang and Zhao (2010), and the influence of the degree of polarization on the scattering properties was investigated in detail. Stochastic electromagnetic beam carrying a twist phase was introduced in Cai and Korotkova (2009), and its statistical properties and OAM in a uniaxial crystal were explored in Liu, Chen, Guo, and Cai (2015) and Liu, Huang, Chen, Guo, and Cai (2015). More recently, generation and propagation of an EGSM beam with a vortex phase were reported (Liu, Wang, Liu, Zhao, & Cai, 2015).

Vector beams with a nonuniform state of polarization, so-called cylindrical vector beams, such as radially polarized beams and azimuthally polarized beams, exhibit unique tight focusing properties, e.g., a strong longitudinal electric field appears and a much smaller focused beam spot can be formed, which is useful in many applications, such as microscopy, lithography, electron acceleration, proton acceleration, material processing, optical data storage, high-resolution metrology, super-resolution imaging, free-space optical communications, plasmonic focusing, and laser machining (Zhan, 2009). The cylindrical vector partially coherent beam concept was introduced as a natural extension of the coherent cylindrical vector beam, based on the unified theory of coherence and polarization in Dong, Cai, Zhao, and Yao (2011), and nonparaxial cylindrical vector partially coherent beams were introduced in Dong, Feng, Chen, Zhao, and Cai (2012). As typical examples of cylindrical vector partially coherent beams, a radially polarized partially coherent beam (Wang, Cai, Dong, & Korotkova, 2012) and an azimuthally polarized partially coherent beam (Dong, Wang, Zhao, & Cai, 2012) were generated, and their coherence and polarization properties were measured (Wu, Wang, & Cai, 2012). It was shown in Dong, Wang, et al. (2012) and Wang, Cai, Dong, et al. (2012) that one can shape the focused beam profile of a radially or azimuthally polarized partially coherent beam, which is useful for particle trapping and material thermal processing, and it was demonstrated in Wu et al. (2012) that the degree of polarization of a radially polarized partially coherent beam equals one in the source plane and it decreases on propagation in free space, whereas the state of polarization of such a beam remains invariant on propagation. Experimental generation of a polychromatic radially polarized partially coherent beam was reported in Zhu, Zhu, Liu, Wang, and Cai (2013), and it was found that the behavior of the spectral changes of a polychromatic partially coherent RP beam is different from that of a scalar polychromatic GSM beam. Young's interference experiment with a radially polarized partially coherent beam was reported in Zhu, Wang, Chen, Li, and Cai (2014), and it was found that the statistical properties in the interference pattern were affected by not only the coherence width but also by the state of polarization of the incident beam. In Wang, Liu, Liu, Yuan, and Cai (2013), it was demonstrated experimentally that a radially polarized partially coherent beam has advantage over a linearly polarized partially coherent beam for reducing turbulence-induced scintillation. More recently, it was demonstrated both theoretically and experimentally that the vortex phase which is imposed on a radially polarized partially coherent beam will induce changes of both the degree of polarization and the state of polarization of such a beam on propagation besides rotation of the beam spot (Guo, Chen, Liu, Liu, & Cai, 2016), and a radially polarized partially coherent beam with a twist phase was also introduced in Wu (2016).

1.4 Partially Coherent Beams With Prescribed Degrees of Coherence

The degree of coherence of most partially coherent beams in the literature satisfies a Gaussian distribution. Recently, a great deal of attention has been paid to partially coherent beams whose degree of coherence does not satisfy a Gaussian distribution, so-called partially coherent beams with non-conventional correlation functions, because of their extraordinary propagation properties (Cai et al., 2014). In 1987, Gori and Guattari first introduced a J_0-correlated Schell-model source whose degree of coherence is a Bessel function of zero order and pointed out that such a beam can be synthesized with the help of a thin, annular incoherent source, a thin lens, and a spatial filter (Gori & Guattari, 1987). The propagation properties of a J_0-correlated Schell-model beam in free space were studied in Palma, Borghi, and Cincotti (1996), and it was found that the intensity profile of such a beam has properties analogous to those of the Bessel–Gauss beam but the degree of coherence does not keep the J_0-correlation nor the shift invariance. Visser and collaborators predicted that one can shape the focused intensity distribution of J_0-correlated Schell-model beams and produce an intensity minimum at the geometric focus rather than a maximum through varying the spatial coherence width (Gbur & Visser, 2003; van Dijk, Gbur, & Visser, 2008), which can be used in novel optical trapping schemes, to selectively manipulate particles with either a low or high index of refraction, and they reported experimental generation of a J_0-correlated Schell-model beam and verified their predictions in Raghunathan, van Dijk, Peterman, and Visser (2010). Modal expansions for scalar and electromagnetic J_0-correlated

Schell-model sources were investigated in Gori, Santarsiero, and Borghi (2008), and the beam conditions for radiation generated by an electromagnetic J_0-correlated Schell-model source were discussed in Wu et al. (2008). In Gu and Gbur (2010), it was demonstrated that pseudo-Bessel correlated beams have lower scintillation than comparable fully coherent beams in both weak and strong turbulence, which is useful in free-space optical communications.

Since Gori and Santarsiero discussed the sufficient condition for devising a genuine correlation function of a partially coherent beam (Gori & Santarsiero, 2007), numerous efforts have been devoted to partially coherent beams with prescribed degrees of coherence (Cai et al., 2014). Gori et al. explored the sufficiency condition for devising a genuine CSD matrix of an electromagnetic stochastic beam (Gori, Sanchez, Santarsiero, & Shirai, 2009). Martínez-Herrero and collaborators derived the necessary and sufficient nonnegative definiteness conditions for the CSD and the CSD matrix (Martínez-Herrero & Mejías, 2009; Martínez-Herrero, Mejías, & Gori, 2009). A partially correlated azimuthal vortex beam was used as illumination for a simple video microscope and it was found that the image contrast was improved (Brown & Brown, 2008). The generation of structured partially coherent beams of arbitrary spatial coherence by incoherently superposing fully coherent fields was reported in Macías-Romero, Lim, Foreman, and Török (2011), and the creation of a beam with locally varying spatial coherence was reported in Waller, Situ, and Fleischer (2012). In the past several years, a variety of partially coherent beams with prescribed degrees of coherence were introduced (Cai et al., 2014), such as scalar and electromagnetic nonuniformly correlated beams (Lajunen & Saastamoinen, 2011; Tong & Korotkova, 2012b), scalar and electromagnetic nonuniformly correlated pulses (Ding et al., 2013; Lajunen & Saastamoinen, 2013), multi-Gaussian correlated Schell-model (MGCSM) beams with circular or rectangular or elliptical symmetry (Korotkova, 2014; Sahin & Korotkova, 2012; Zhang & Cai, 2014), multi-Gaussian correlated vortex beams (Zhang, Liu, Zhao, & Cai, 2014), generalized MGCSM beams (Wang, Liang, Yuan, & Cai, 2014), LG correlated Schell-model (LGCSM) beams of circular or elliptical or rectangular symmetry (Chen, Liu, Wang, Zhao, & Cai, 2014; Chen, Yu, Yuan, Wang, & Cai, 2016; Mei & Korotkova, 2013a), LGCSM vortex beams (Chen, Wang, Zhao, & Cai, 2014), cosine-GSM beams with circular or rectangular symmetry (Liang, Wang, Liu, Cai, & Korotkova, 2014; Mei & Korotkova, 2013b), vector cosine-Gaussian correlated beams with radial polarization (Zhu et al., 2015), specially correlated partially coherent vector beams (Chen, Liu, et al., 2014), scalar and vector Hermite–Gaussian correlated beams (Chen, Gu, Wang, & Cai, 2015; Chen, Wang, Yu, Liu, & Cai, 2016), optical coherence lattices (Ma & Ponomarenko, 2014, 2015), random sources for optical frames (Korotkova & Shchepakina, 2014), and for beams with azimuthal intensity variation (Wang & Korotkova, 2016). Such beams with prescribed degrees of coherences exhibit many extraordinary properties, such as a self-focusing effect, a lateral shift of the intensity maximum, self-shaping, self-splitting as well as periodicity reciprocity. In 2013, Wang et al. proposed a method for generating partially coherent beams with prescribed degrees of coherence and reported the experimental generation of a circular LGCSM beam with the help of a RGGD and a SLM (Wang, Liu, Yuan, & Cai, 2013). Other beams with prescribed degrees of coherence, such as a multi-Gaussian correlated Schell beam, a cosine-Gaussian correlated beam, a Hermite-Gaussian correlated beam, and a specially correlated partially coherent vector beam were generated subsequently. Modulating the spatial coherence of light in Young's interference experiment by surface plasmons was predicted in Gan et al. (2007). The experimental generation of a spatial coherence comb using a Dammann grating was reported in Vinu, Sharma, Singh, and Senthilkumaran (2014). Numerical modeling of partially coherent sources with arbitrary far-field patterns was described in Voelz, Xiao, and Korotkova (2015), and the experimental generation of such sources for a desired far-field mean irradiance pattern using a single phase-only liquid-crystal SLM was reported in Hyde, Basu, Voelz, and Xiao (2015) and Hyde, Basu, Xiao, and Voelz (2015). In Lehtolahti, Kuittinen, Turunen, and Tervo (2015), deterministic rotating diffusers fabricated by lithographic techniques were used to modulate the coherence properties of the incident beam and the corresponding diffraction pattern. In Nixon, Redding, Friesem, Cao, and Davidson (2013) and Chriki et al. (2015), manipulation of the spatial coherence of a laser source in a modified degenerate cavity laser was demonstrated by changing the size of circular aperture mask or the geometry of the mask placed inside the cavity. Propagation properties of some partially coherent beams with prescribed degrees of coherence in turbulent atmosphere have been studied both theoretically (Gu & Gbur, 2013; Liu, Yu, Cai, & Ponomarenko, 2016; Tong & Korotkova, 2012c; Yu, Chen, Liu, Liu, & Cai, 2015; Yuan et al., 2013) and experimentally (Avramov-Zamurovic, Nelson, Guth, & Korotkova, 2016; Avramov-Zamurovic, Nelson, Guth, Korotkova, & Malek-Madani, 2016). It was found that modulating the structure of the degree of coherence provides a novel way for reducing turbulence-induced deformation and scintillation. Owing to the fact that modulating the structure of degree of coherence provides a novel way for beam shaping, partially coherent beams with prescribed degrees of coherence are useful for particle trapping (Chen & Cai, 2014; Liu & Zhao, 2015). Modulating the structure of the degree of coherence also provides a possible way for controlling the degree of paraxiality of a partially coherent beam (Guo, Chen, Liu, Yao, & Cai, 2016). More recently, generation of optical coherence lattices carrying information was reported in Chen, Ponomarenko, and Cai (2016), and such lattices will be useful for image transmission and optical encryption.

After introducing these developments concerning partially coherent beams, we will introduce the theoretical models for those beams and the methods for generating them in Section 2. In Section 2.1, we introduce the model for a GSM beam and methods for their generation. In Section 2.2, we introduce models for partially coherent beams with prescribed phases (e.g., a TGSM beam and a partially coherent vortex beam) and the methods for generating such beams. In Section 2.3, we introduce models for vector partially coherent beams with prescribed states of polarization (e.g., an EGSM beam and a cylindrical vector partially coherent beam) and the methods for producing them. In Section 2.4, we discuss partially coherent beams with prescribed degrees of coherence and the methods for generating such beams. In Section 3, a brief summary is presented.

 2. CHARACTERIZATION AND GENERATION OF VARIOUS PARTIALLY COHERENT BEAMS

It is known that a scalar partially coherent beam can be characterized by either the mutual intensity in the space–time domain or the CSD in the space–frequency domain (Mandel & Wolf, 1995). The CSD of the field at the source plane is defined as a two-point correlation function

$$W(\mathbf{r}_1, \mathbf{r}_2) = \langle E^*(\mathbf{r}_1) E(\mathbf{r}_2) \rangle, \tag{1}$$

where E denotes the field fluctuating in a direction perpendicular to the z-axis, the asterisk denotes the complex conjugate and the angular brackets denote an ensemble average. The angular frequency dependence of all the quantities of interest is omitted in this review.

To be a mathematically genuine CSD, the CSD must correspond to a nonnegative definite kernel, which is fulfilled if the function can be written in the following form (Gori & Santarsiero, 2007)

$$W(\mathbf{r}_1, \mathbf{r}_2) = \int p(\mathbf{v}) H^*(\mathbf{r}_1, \mathbf{v}) H(\mathbf{r}_2, \mathbf{v}) d^2\mathbf{v}, \tag{2}$$

where H is an arbitrary kernel and p is an arbitrary nonnegative function.

The average intensity and the degree of coherence of a scalar partially coherent beam are defined as (Mandel & Wolf, 1995)

$$I(\mathbf{r}) = W(\mathbf{r}, \mathbf{r}), \quad \mu(\mathbf{r}_1, \mathbf{r}_2) = \frac{W(\mathbf{r}_1, \mathbf{r}_2)}{\sqrt{W(\mathbf{r}_1, \mathbf{r}_1) W(\mathbf{r}_2, \mathbf{r}_2)}}. \tag{3}$$

The CSD of a scalar partially coherent beam generated by a Schell-model source usually can be expanded in the following form

$$W(\mathbf{r}_1, \mathbf{r}_2) = \sqrt{I(\mathbf{r}_1) I(\mathbf{r}_2)} \mu(\mathbf{r}_1, \mathbf{r}_2) f(\varphi_1, \varphi_2), \tag{4}$$

where $f(\varphi_1, \varphi_2)$ represents the phase term. Partially coherent beams with different $I(\mathbf{r})$, $\mu(\mathbf{r}_1, \mathbf{r}_2)$, and $f(\varphi_1, \varphi_2)$ are named differently.

A vector partially coherent beam can be characterized by either the beam coherence–polarization matrix (Gori, 1998) in the space–time domain or the CSD matrix (Wolf, 2003, 2007) in the space–frequency domain. The CSD matrix of the field at two position vectors $\mathbf{r}_1$ and $\mathbf{r}_2$ at the source plane is defined as

$$\overleftrightarrow{W}(\mathbf{r}_1, \mathbf{r}_2) = \begin{bmatrix} W_{xx}(\mathbf{r}_1, \mathbf{r}_2) & W_{xy}(\mathbf{r}_1, \mathbf{r}_2) \\ W_{yx}(\mathbf{r}_1, \mathbf{r}_2) & W_{yy}(\mathbf{r}_1, \mathbf{r}_2) \end{bmatrix}, \tag{5}$$

with elements

$$W_{\alpha\beta}(\mathbf{r}_1, \mathbf{r}_2) = \langle E_\alpha^*(\mathbf{r}_1) E_\beta(\mathbf{r}_2) \rangle, \quad (\alpha = x, y; \beta = x, y). \tag{6}$$

Here E_x and E_y denote the components of the random electric vector, along two mutually orthogonal x and y directions perpendicular to the z-axis.

To be a genuine CSD matrix, the CSD matrix should satisfy the condition of nonnegative definiteness, which is fulfilled if the elements of the CSD matrix have the following integral form (Gori et al., 2009)

$$W_{\alpha\beta}(\mathbf{r}_1, \mathbf{r}_2) = \int p_{\alpha\beta}(\mathbf{v}) H_\alpha^*(\mathbf{r}_1, \mathbf{v}) H_\beta(\mathbf{r}_2, \mathbf{v}) d^2\mathbf{v}, \quad (\alpha = x, y; \beta = x, y), \tag{7}$$

where H_x and H_y are arbitrary kernels and $p_{\alpha\beta}$ are the elements of the following weighting matrix

$$\hat{p}(\mathbf{v}) = \begin{pmatrix} p_{xx}(\mathbf{v}) & p_{xy}(\mathbf{v}) \\ p_{xy}^*(\mathbf{v}) & p_{yy}(\mathbf{v}) \end{pmatrix}. \qquad (8)$$

The elements of the weighting matrix should satisfy the following conditions for any $\mathbf{v}$

$$p_{xx}(\mathbf{v}) \geq 0, \quad p_{yy}(\mathbf{v}) \geq 0, \quad p_{xx}(\mathbf{v})p_{yy}(\mathbf{v}) - |p_{xy}(\mathbf{v})|^2 \geq 0. \qquad (9)$$

The average intensity of a vector partially coherent beam at point $\mathbf{r}$ is given by the formula (Wolf, 2007)

$$\langle I(\mathbf{r}) \rangle = \mathrm{Tr}\overleftrightarrow{W}(\mathbf{r},\mathbf{r}) = W_{xx}(\mathbf{r},\mathbf{r}) + W_{yy}(\mathbf{r},\mathbf{r}), \qquad (10)$$

where Tr denotes the trace of the matrix.

There are two definitions of the degree of coherence for a vector partially coherent beam. According to Wolf (2003), the degree of coherence μ of a vector partially coherent beam at a pair of transverse points with position vectors $\mathbf{r}_1$ and $\mathbf{r}_2$ is defined as follows

$$\mu(\mathbf{r}_1,\mathbf{r}_2) = \frac{\mathrm{Tr}\overleftrightarrow{W}(\mathbf{r}_1,\mathbf{r}_2)}{\sqrt{\mathrm{Tr}\overleftrightarrow{W}(\mathbf{r}_1,\mathbf{r}_1)\mathrm{Tr}\overleftrightarrow{W}(\mathbf{r}_2,\mathbf{r}_2)}}. \qquad (11)$$

According to Tervo et al. (2003), the degree of coherence μ of a vector partially coherent beam is defined by the expression

$$\mu^2(\mathbf{r}_1,\mathbf{r}_2) = \frac{\mathrm{Tr}\left[\overleftrightarrow{W}^{\dagger}(\mathbf{r}_1,\mathbf{r}_2)\overleftrightarrow{W}(\mathbf{r}_1,\mathbf{r}_2)\right]}{\mathrm{Tr}\left[\overleftrightarrow{W}(\mathbf{r}_1,\mathbf{r}_1)\right]\mathrm{Tr}\left[\overleftrightarrow{W}(\mathbf{r}_2,\mathbf{r}_2)\right]}. \qquad (12)$$

Here the symbol "$\dagger$" denotes the Hermitian adjoint.

The degree of polarization of a vector partially coherent beam at point $\mathbf{r}$ is defined by the expression (Wolf, 2003)

$$P(\mathbf{r}) = \sqrt{1 - \frac{4\mathrm{Det}\overleftrightarrow{W}(\mathbf{r},\mathbf{r})}{\left[\mathrm{Tr}\overleftrightarrow{W}(\mathbf{r},\mathbf{r})\right]^2}}, \qquad (13)$$

where Det stands for the determinant of the matrix.

The state of polarization of a vector partially coherent beam can be characterized by the generalized Stokes parameters or the polarization ellipse.

The generalized Stokes parameters which depend on two spatial variables are introduced in Korotkova and Wolf (2005a), and defined as

$$S_0(\mathbf{r}_1, \mathbf{r}_2) = W_{xx}(\mathbf{r}_1, \mathbf{r}_2) + W_{yy}(\mathbf{r}_1, \mathbf{r}_2), \quad (14)$$

$$S_1(\mathbf{r}_1, \mathbf{r}_2) = W_{xx}(\mathbf{r}_1, \mathbf{r}_2) - W_{yy}(\mathbf{r}_1, \mathbf{r}_2), \quad (15)$$

$$S_2(\mathbf{r}_1, \mathbf{r}_2) = W_{xy}(\mathbf{r}_1, \mathbf{r}_2) + W_{yx}(\mathbf{r}_1, \mathbf{r}_2), \quad (16)$$

$$S_3(\mathbf{r}_1, \mathbf{r}_2) = i\left[W_{yx}(\mathbf{r}_1, \mathbf{r}_2) - W_{xy}(\mathbf{r}_1, \mathbf{r}_2)\right]. \quad (17)$$

According to Korotkova and Wolf (2005b), the CSD matrix of a vector partially coherent vector beam at point $\mathbf{r}$ can be locally represented as a sum of completely polarized portion and a completely unpolarized portion, and the state of polarization of the completely polarized portion can be characterized by the polarization ellipse. The major and minor semiaxes of the ellipse, A_1 and A_2, as well as its degree of ellipticity, ε, and its orientation angle, θ, can be related directly to the elements of the CSD matrix with the help of the expressions

$$A_{1,2}(\mathbf{r}, \mathbf{r}) = \frac{1}{\sqrt{2}}\left[\sqrt{(W_{xx}(\mathbf{r}, \mathbf{r}) - W_{yy}(\mathbf{r}, \mathbf{r}))^2 + 4|W_{xy}(\mathbf{r}, \mathbf{r})|^2} \\ \pm \sqrt{(W_{xx}(\mathbf{r}, \mathbf{r}) - W_{yy}(\mathbf{r}, \mathbf{r}))^2 + 4\left[\mathrm{Re}\,W_{xy}(\mathbf{r}, \mathbf{r})\right]^2}\right]^{1/2}, \quad (18)$$

$$\varepsilon(\mathbf{r}, \mathbf{r}) = \frac{A_2(\mathbf{r}, \mathbf{r})}{A_1(\mathbf{r}, \mathbf{r})}, \quad (19)$$

$$\theta(\mathbf{r}, \mathbf{r}) = \frac{1}{2}\arctan\left(\frac{2\mathrm{Re}\,W_{xy}(\mathbf{r}, \mathbf{r})}{W_{xx}(\mathbf{r}, \mathbf{r}) - W_{yy}(\mathbf{r}, \mathbf{r})}\right). \quad (20)$$

In Eq. (18) signs "+" and "−" between the two square roots correspond to A_1 and A_2, respectively.

In Sections 2.1–2.4, we will discuss theoretical models for various scalar and vector partially coherent beams and the methods for their generation.

2.1 GSM Beams

The CSD of a GSM beam in the source plane is defined as (Collett & Wolf, 1978; Foley & Zubairy, 1976; Wolf & Collett, 1978)

$$W(\mathbf{r}_1, \mathbf{r}_2) = C_0 \exp\left[-\frac{\mathbf{r}_1^2 + \mathbf{r}_2^2}{4\sigma_{I0}^2} - \frac{(\mathbf{r}_1 - \mathbf{r}_2)^2}{2\sigma_{g0}^2}\right], \quad (21)$$

where C_0 is a constant, σ_{I0} and σ_{g0} denote the transverse beam width and coherence width, respectively. The degree of coherence of the GSM beam is of the form

$$\mu(\mathbf{r}_1, \mathbf{r}_2) = \exp\left[-\frac{(\mathbf{r}_1 - \mathbf{r}_2)^2}{2\sigma_{g0}^2}\right], \qquad (22)$$

It was shown in Wolf and Collett (1978) that two GSM sources will generate fields with identical far-zone intensity distributions, if the beam width σ_{I0} and the coherence width σ_{g0} are such that each of the quantities

$$\Delta^2 = 1/(k\sigma_{g0})^2 + 1/(2k\sigma_{I0})^2, \qquad (23)$$

$$J(0) = (\sigma_{I0}/\Delta)^2 C_0, \qquad (24)$$

are the same for both sources. Furthermore, it was also shown in Wolf and Collett (1978) that any GSM source whose parameters σ_{I0}, σ_{g0}, and C_0 satisfy the relations

$$1/\sigma_{g0}^2 + 1/(2\sigma_{I0})^2 = 1/(2\delta_L)^2, \qquad (25)$$

$$C_0 = (\delta_L/\sigma_{I0})^2 C_L, \qquad (26)$$

will generate a light beam with the same radiant intensity distribution as a coherent laser source whose intensity distribution at the output mirror is

$$I_l(\mathbf{r}) = C_L \exp\left(-\mathbf{r}^2/2\delta_L^2\right). \qquad (27)$$

It was shown in De Santis et al. (1979) that a GSM beam can be produced starting with a spatially incoherent source and using a collimating lens and an amplitude filter. Fig. 1 shows the experimental setup. A Gaussian laser spot is imaged the lens L_1 on a RGGD. Provided that the spot diameter on the RGGD is large compared with the inhomogeneity scale of the RGGD, the light transmitted by the RGGD can be regarded as spatially incoherent. The RGGD is located in the focal plane of lens L_2 which, in turn, is followed by a GAF F. Using the van Cittert–Zernike theorem and the propagation

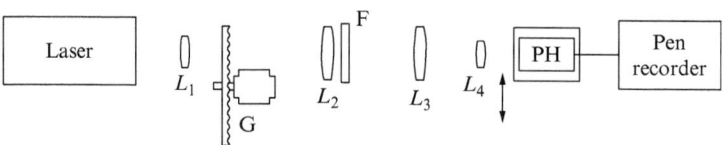

Fig. 1 Experimental setup for generating a GSM beam (De Santis et al., 1979).

law for the CSD, one can show that the light behind the filter is a GSM beam. The far-zone intensity distribution can be examined through the optical system formed by the lenses L_3 and L_4 and the photodetector PH. The coherence width of the generated GSM beam can be controlled through controlling the beam spot size on the RGGD. Tervonen et al. (1992) reported the generation of an anisotropic GSM beam through conversion of a Gaussian laser beam with the help of a synthetic acousto-optic hologram. In principle, a GSM beam also can be generated by appropriately superimposing coherent, but mutually uncorrelated, light beams (Starikov & Wolf, 1982).

The intensity distribution of a GSM beam can be measured by a photodetector or a charge-coupled device (CCD) or a beam profile analyzer. The degree of coherence and the corresponding coherence width of a GSM beam can be quantitatively measured through measuring the FOCF (Wang & Cai, 2007; Wang et al., 2006). As shown in Fig. 2, the generated GSM beam from the GAF first passes through a thin lens with focal length f_1, then is split by a beam splitter (BS) into two distinct imaging optical paths (i.e., $L_1 = 2f_1$). The transmitted beam and reflected beam arrive at single-photon detectors D_1 and D_2, which scan the transverse planes of u_1 and u_2, respectively. The FOCF of the beam at the detector planes is the same as the generated GSM beam just after the GAF. The output signals from D_1 and D_2 are sent to an electronic coincidence circuit to measure the normalized FOCF between two detectors (i.e., the FOCF of the generated GSM beam just after the GAF), which is expressed as

$$g^{(2)}(\mathbf{u}_1, \mathbf{u}_2, \tau) = \frac{\langle I(\mathbf{u}_1, t) I(\mathbf{u}_2, t+\tau) \rangle}{\langle I(\mathbf{u}_1, t) \rangle \langle I(\mathbf{u}_2, t+\tau) \rangle}, \tag{28}$$

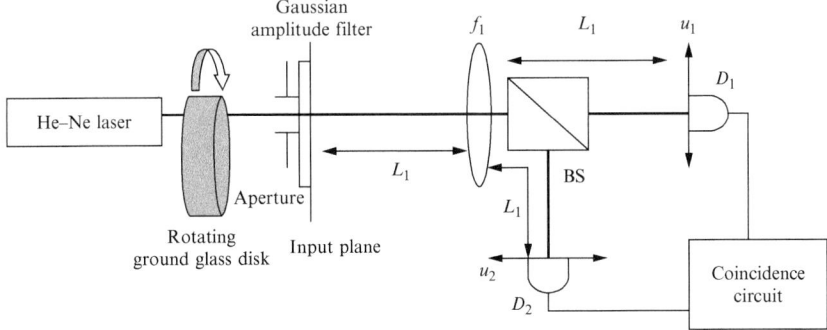

Fig. 2 Experimental setup for measuring the degree of coherence of a GSM beam (Wang & Cai, 2007).

where $I(\mathbf{u}_1, t)$ and $I(\mathbf{u}_2, t+\tau)$ are the instantaneous intensities at D_1 and D_2, τ denotes the delay time of the photon flux of two optical paths. Applying the Gaussian moment theorem (Mandel & Wolf, 1995), the FOCF with $\tau = 0$ can be simplified as

$$g^{(2)}(\mathbf{u}_1, \mathbf{u}_2, 0) = 1 + |\mu(\mathbf{u}_1, \mathbf{u}_2)|^2. \tag{29}$$

Thus, one can measure the square of the modulus of the degree of coherence of the generated GSM beam through measuring its FOCF. Fig. 3 shows the experimental results of the square of the modulus of the degree of coherence of the generated GSM beam, and the corresponding result of the coherence width was obtained by a Gaussian fit through the experimental data.

In principle, the electronic coincidence circuit can be used to measure the two-dimensional distribution of the degree of coherence if D_1 and D_2 scan the entire planes of u_1 and u_2, but this is a slow procedure. In Cai et al. (2014) and Chen, Liu, et al. (2014), a new method is developed to measure the two-dimensional distribution of the degree of coherence. The setup is similar to Fig. 2, while only one imaging optical system is used, i.e., the generated GSM beam passes through a 2f-imaging system and then arrives at a CCD, which is used to measure the instantaneous intensity. The output signal from the CCD is sent to a personal computer to measure

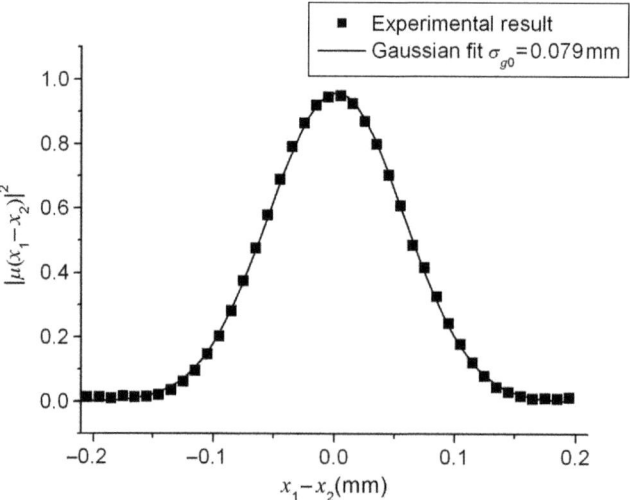

Fig. 3 Experimental result of the square of the modulus of the degree of coherence (along x_1-x_2) for the generated GSM beam just after the GAF (Wang & Cai, 2007).

the normalized FOCF which is given by Eq. (29). In the experiment, the CCD records 2000 or more pictures in total, and each picture denotes one realization of the beam cross section. Each realization can be presented by a matrix $I^{(m)}(u_x, u_y)$ with u_x and u_y being pixel spatial coordinates. Here m denotes each realization and ranges from 1 to 2000 or more. The square of the modulus of the degree of coherence is obtained as

$$|\mu(\mathbf{u}_1, \mathbf{u}_2 = 0)|^2 = \frac{\frac{1}{M}\sum_{m=1}^{M} I^{(m)}(u_{1x}, u_{1y}) I^{(m)}(0, 0)}{\bar{I}(u_{1x}, u_{1y}) \bar{I}(0, 0)} - 1, \quad (30)$$

where

$$\bar{I}(u_{1x}, u_{1y}) = \sum_{m=1}^{M} I^{(m)}(u_{1x}, u_{1y})/M, \quad \bar{I}(0, 0) = \sum_{m=1}^{M} I^{(m)}(0, 0)/M. \quad (31)$$

Here $\bar{I}(u_{1x}, u_{1y})$ and $\bar{I}(0, 0)$ denote the average intensity of all realizations and the average intensity at the central point, respectively.

The above methods can be applied to measure the degree of coherence of various partially coherent beams generated from a Schell-mode source. Once the degree of coherence and the corresponding coherence width are measured, one can simulate the propagation of the generated beam numerically and explore the effect of spatial coherence in various applications quantitatively in experiments.

2.2 Partially Coherent Beams With Prescribed Phases

In this section, we present theoretical models for TGSM beams and partially coherent vortex beams, and the methods for generating such beams.

The CSD of a TGSM beam in the source plane is written as (Simon & Mukunda, 1993)

$$W(\mathbf{r}_1, \mathbf{r}_2) = \exp\left[-\frac{\mathbf{r}_1^2 + \mathbf{r}_2^2}{4\sigma_{I0}^2} - \frac{(\mathbf{r}_1 - \mathbf{r}_2)^2}{2\sigma_{g0}^2} - \frac{ik\mu_0}{2}(\mathbf{r}_1 - \mathbf{r}_2)^T \mathbf{J}(\mathbf{r}_1 + \mathbf{r}_2)\right], \quad (32)$$

where $k = 2\pi/\lambda$ is the wave number with λ being the wavelength, μ_0 is a scalar real-valued twist factor with the dimension of an inverse distance, limited by the double inequality $0 \leq \mu_0^2 \leq \left[k^2 \sigma_{g0}^4\right]^{-1}$ due to the nonnegativity requirement of Eq. (32). In the coherent limit, $\sigma_{g0} \to \infty$, the twist factor μ_0 disappears. $\mathbf{J}$ is an antisymmetric matrix given by

$$\mathbf{J} = \begin{pmatrix} 0 & 1 \\ -1 & 0 \end{pmatrix}. \tag{33}$$

When $\mu_0 = 0$, the TGSM beam reduces to a GSM beam. Owing to the presence of the twist phase, the two-dimensional CSD cannot be split into a product of two one-dimensional CSDs, and one can treat its propagation with the help of the Wigner distribution function (Bastiaans, 2000) or the tensor method (Lin & Cai, 2002a). The twist phase imposes OAM on the beam (Serna & Movilla, 2001) and will induce a rotation of the beam spot on propagation (Friberg et al., 1994b).

In Friberg et al. (1994a), Friberg et al. interpreted the TGSM beams in physical-optics terms by decomposition of such beams into a weighted superposition of overlapping, mutually uncorrelated but spatially coherent component fields, and the authors proposed an astigmatic optical lens system for converting an anisotropic GSM beam into a TGSM beam (see Fig. 4). In Fig. 4, a group of three cylindrical lenses were used, two of which form a standard $4f_y$ imaging system in one transverse coordinate and the third with focal length $f_x = 2f_y$ performs a Fourier transform in the orthogonal direction. Based on this proposed optical system, Friberg et al. reported the generation of a TGSM beam by use of an acousto-optic coherence control technique. The acousto-optic coherence control technique was used to convert a coherent Gaussian beam into an anisotropic GSM beam which is expressed as superposition of coherent but mutually uncorrelated Gaussian beams. Fig. 5 shows theoretical simulation and experimental observation of the rotation of the TGSM beam. In principle, a TGSM beam also can be produced through incoherent superposition of ordinary Gaussian beams (Ambrosini et al., 1994) or partially coherent modified Bessel–Gauss beams (Gori & Santarsiero, 2015).

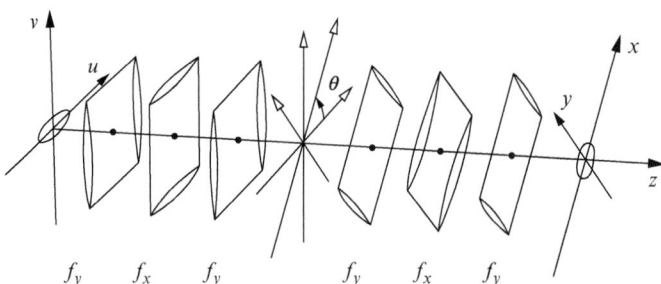

Fig. 4 Astigmatic optical lens system for converting an anisotropic GSM beam into a TGSM beam (Friberg et al., 1994a).

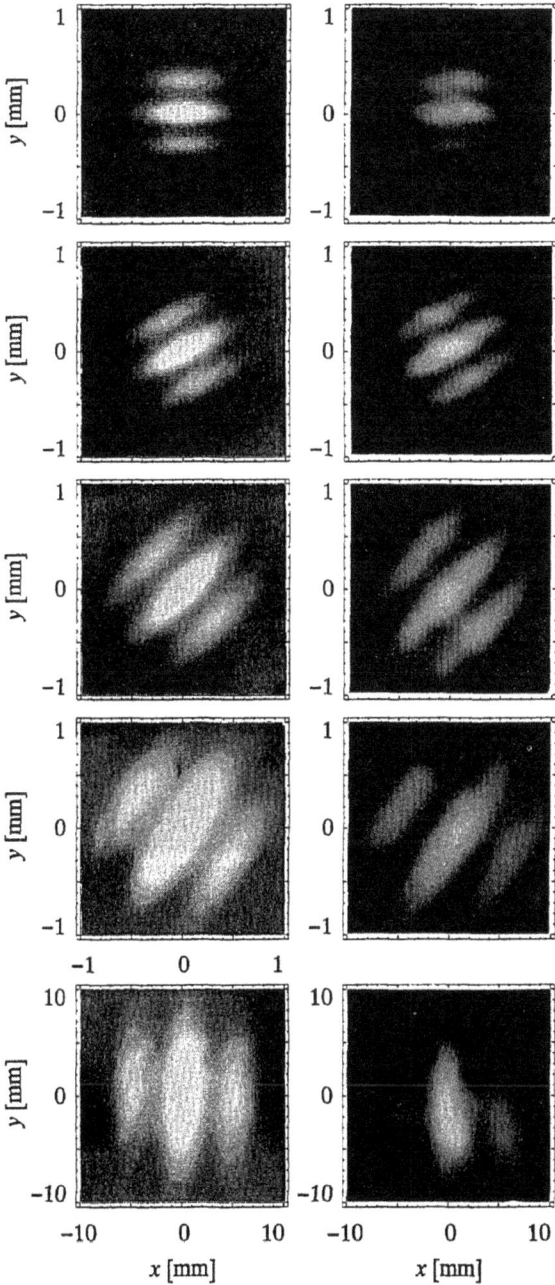

Fig. 5 Theoretical simulation (*left*) and experimental demonstration (*right*) of the rotation of the twisted GSM beam: three individual components in the coherent beam decomposition are shown on propagation in planes (from *top* to *bottom*) $z=0$, 150, 300, 450, and 5000 mm (Friberg et al., 1994a).

Next we discuss partially coherent vortex beams. There are different kinds of partially coherent vortex beams, and here we only treat Gaussian Schell-model vortex (GSMV) beams and partially coherent LG beams.

For a GSMV beam, in cylindrical coordinates, its CSD in the source plane is defined as (Wang, Zhu, & Cai, 2011)

$$W(r_1, r_2, \varphi_1, \varphi_2) = \exp\left[-\frac{r_1^2 + r_2^2}{4\sigma_{I0}^2} - \frac{r_1^2 + r_2^2 - 2r_1 r_2 \cos(\varphi_1 - \varphi_2)}{2\sigma_{g0}^2} + il(\varphi_1 - \varphi_2)\right], \quad (34)$$

where r and φ are the radial and azimuthal (angle) coordinates, respectively, and l denotes the topological charge. When $\sigma_{g0} = \infty$, the GSMV beam reduces to a coherent Gaussian vortex beam. A GSMV beam can be produced through imposing a vortex phase on a GSM beam with the help of a spiral phase plate (see Fig. 6). One can shape the focused intensity distribution of a GSMV beam through varying its initial coherence width (see Fig. 7), which is useful for trapping particles whose refractive index is larger or smaller than that of their embedding (Zhao & Cai, 2011).

In a cylindrical coordinate system, the CSD of a partially coherent LG_{pl} beam in the source plane is expressed as (Wang et al., 2009b)

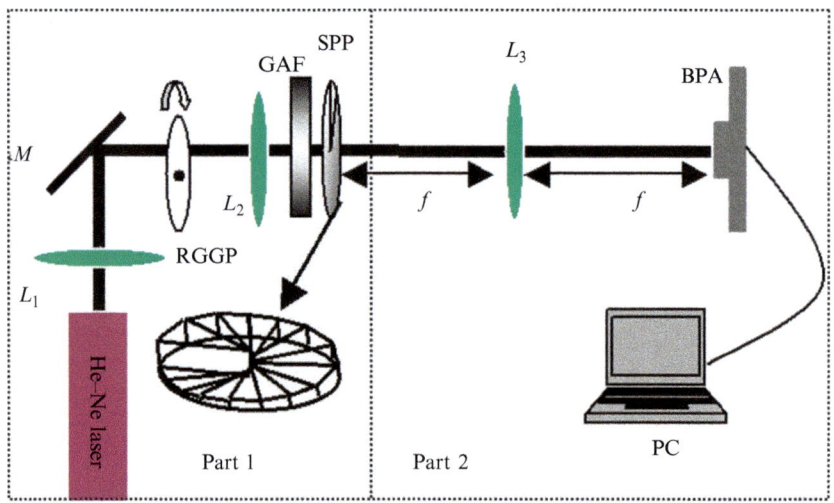

Fig. 6 Experimental setup for generating a GSMV beam and measuring its focused intensity. *BPA*, beam profile analyzer (Wang, Zhu, & Cai, 2011); *GAF*, Gaussian amplitude filter; L_1, L_2, L_3, thin lenses; *M*, mirror; *RGGP*, rotating ground glass plate; *SPP*, spiral phase plate.

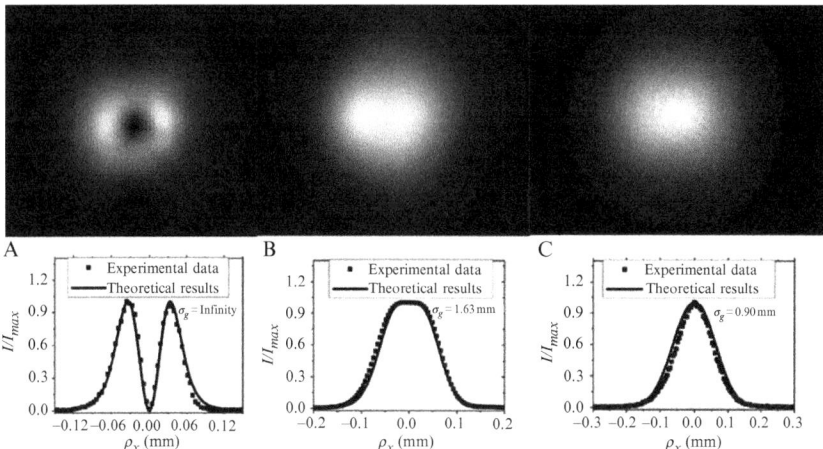

Fig. 7 Experimental results of the focused intensity distribution and the corresponding cross line (*dotted curve*) of the generated GSMV beam for different coherence widths. The *solid curves* are calculated by theoretical formulae (Wang, Zhu, & Cai, 2011). (A) $\sigma_g =$ infinity; (B) $\sigma_g = 1.63$ mm; (C) $\sigma_g = 0.90$ mm.

$$W(r_1, r_2, \varphi_1, \varphi_2) = \left(\frac{\sqrt{2}r_1}{\omega_0}\right)^l \left(\frac{\sqrt{2}r_2}{\omega_0}\right)^l L_p^l\left(\frac{2r_1^2}{\omega_0^2}\right) L_p^l\left(\frac{2r_2^2}{\omega_0^2}\right) \exp\left(-\frac{r_1^2 + r_2^2}{\omega_0^2}\right)$$
$$\times \exp(il\varphi_1 - il\varphi_2) \times \exp\left[-\frac{r_1^2 + r_2^2 - 2r_1 r_2 \cos(\varphi_1 - \varphi_2)}{2\sigma_{g0}^2}\right],$$
(35)

where L_p^l denotes the Laguerre polynomial with mode orders p and l. When $p=0$, Eq. (35) reduces to the CSD of a partially coherent LG_{0l} beam (Palacios et al., 2004). When $p=0$ and $l=0$, Eq. (35) reduces to the CSD of a GSM beam.

The generation of a partially coherent LG_{01} beam was demonstrated in Palacios et al. (2004), where the light from a halogen bulb first passes through two separate apertures and an imaging system, then illuminates a vortex phase mask, producing a partially coherent LG_{01} beam. Here the two apertures were used to control the coherence of the light. In Zhao, Dong, Wang, et al. (2012) and Zhao, Dong, Wu, et al. (2012), a partially coherent LG_{0l} beam was generated through conversion of a GSM beam by a SLM (see Fig. 8), where the generated GSM beam goes toward a SLM, which acts as a grating with fork pattern designed by the method of computer-generated holograms. The first-order diffraction pattern of the beam

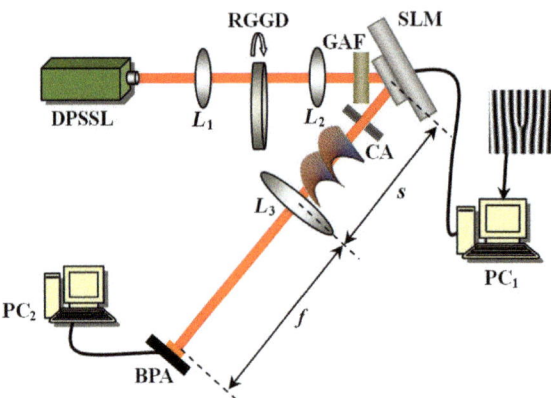

Fig. 8 Experimental setup for generating a partially coherent LG_{0l} beam. *BPA*, beam profile analyzer (Zhao, Dong, Wang, et al., 2012; Zhao, Dong, Wu, et al., 2012); *CA*, circular aperture; *GAF*, Gaussian amplitude filter; L_1, L_2, L_3, thin lenses; *RGGD*, rotating ground glass disk; *SLM*, spatial light modulator.

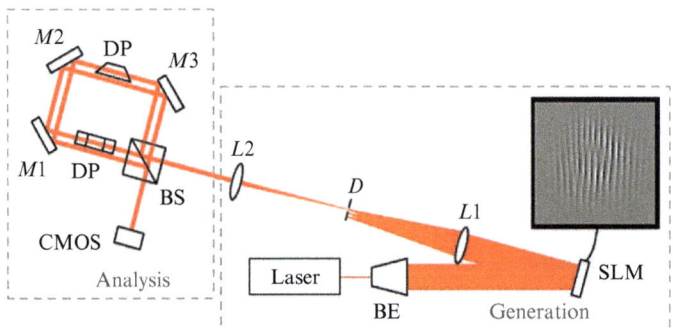

Fig. 9 Experimental setup for digital generation of partially coherent vortex beams. *BE*, beam expander; *BS*, beam splitter; *CMOS*, camera (Perez-Garcia et al., 2016); *D*, iris diaphragm; *DP*, dove prism; *He–Ne*, laser; *L1–L2*, lenses; *M1–M2*, mirrors; *SLM*, spatial light modulator.

reflected from the SLM is regarded as a partially coherent LG_{0l} beam and is selected out by a circular aperture. The experimental setup shown in Fig. 8 also can be applied to generate a partially coherent LG_{pl} beam with a designed grating (Liu, Wang, et al., 2016). More recently, in Perez-Garcia et al. (2016), Perez-Garcia et al. reported the digital generation of partially coherent vortex beams (see Fig. 9), where partially coherent vortex beams with arbitrary azimuthal index were generated using only a SLM. This approach is based on digitally simulating the intrinsic randomness of broadband light passing through a spiral phase plate.

In Palacios et al. (2004), it was demonstrated that a correlation singularity (i.e., a ring dislocation) exists in the cross-correlation function of a partially coherent LG_{01} beam in the far-field plane. Later, it was shown both theoretically (Yang et al., 2012; Zhao, Wang, et al., 2012) and experimentally (Escalante et al., 2013) that the number of ring dislocations of a partially coherent LG_{0l} beam in the cross-correlation function or the degree of coherence in the far field or in the focal plane equals $|l|$ (see Fig. 10), which means that one can determine the magnitude of the topological charge of a partially coherent LG_{0l} beam through measuring its cross-correlation function or degree of coherence. More recently, it was shown in Chen, Liu, et al. (2016) that the sign and the magnitude of the topological charge of a partially coherent LG_{0l} beam can be determined simultaneously by measuring the modulus of the degree of coherence of such a beam propagating through a pair of cylindrical lenses (see Fig. 11). It is found from Fig. 12 that the distribution of the modulus of the degree of coherence becomes anisotropic and it rotates anticlockwise (or clockwise) during propagation when the sign of the topological charge is positive (or negative), furthermore, the modulus

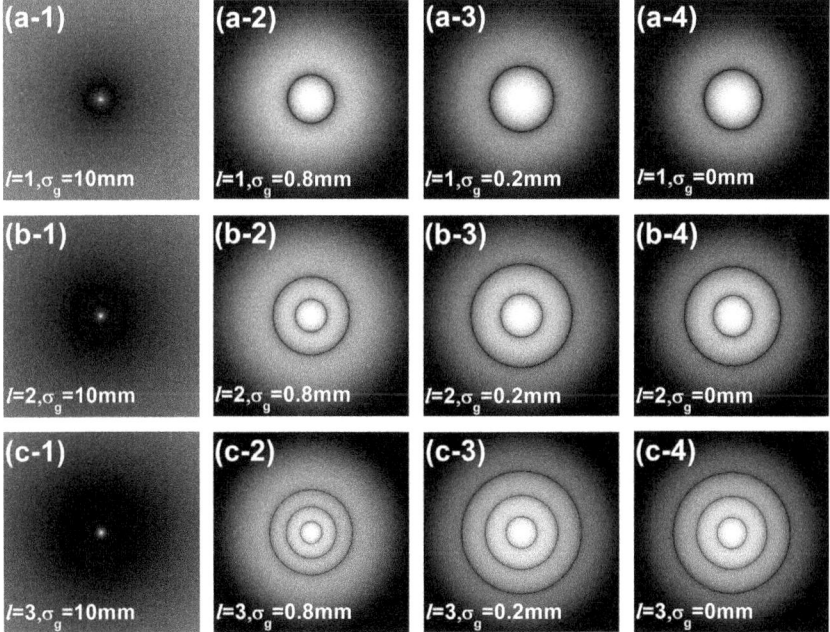

Fig. 10 Distribution of $\log[1 + |\mu(\rho_1, \theta_1, 0, 0)|]$ of a focused LG_{0l} beam in the focal plane for different coherence widths and topological charges (Zhao, Wang, et al., 2012).

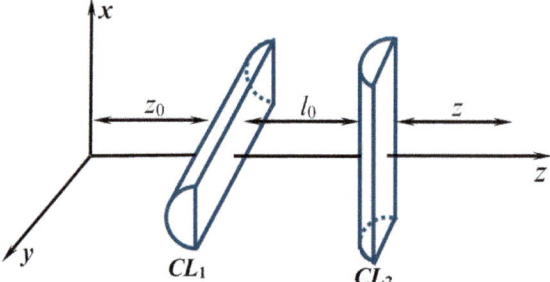

Fig. 11 Scheme for an axially nonsymmetric ABCD optical system consisting of a pair of cylindrical lenses and free space. CL_1 and CL_2 represent two cylindrical lenses. The direction of curvature of CL_1 is along the y-axis and the direction of curvature of CL_2 is along the x-axis (Chen, Liu, et al., 2016).

of the degree of coherence displays a fringe distribution within certain propagation distances and the number of the bright fringes equals $2|l|+1$. For a partially coherent LG_{pl} beam, it was demonstrated both theoretically (Yang et al., 2013) and experimentally (Liu, Wang, et al., 2016) that the number of ring dislocations in the far field or in the focal plane equals $2p+|l|$.

2.3 Vector Partially Coherent Beams With Prescribed States of Polarization

In this section, we present some typical examples of vector partially coherent beams with a spatially uniform state of polarization, or a spatially nonuniform state of polarization, i.e., EGSM beams (partially coherent and partially polarized GSM beams), radially polarized and azimuthally polarized partially coherent beams.

The EGSM beam is a vector partially coherent beams with a spatially uniform state of polarization (i.e., a stochastic electromagnetic beam), whose elements of the CSD matrix in the source plane are expressed as (Gori et al., 2001; Korotkova, Salem, et al., 2004a)

$$W_{\alpha\beta}(\mathbf{r}_1,\mathbf{r}_2) = A_\alpha A_\beta B_{\alpha\beta} \exp\left[-\frac{\mathbf{r}_1^2}{4\sigma_\alpha^2} - \frac{\mathbf{r}_2^2}{4\sigma_\beta^2} - \frac{(\mathbf{r}_1-\mathbf{r}_2)^2}{2\delta_{\alpha\beta}^2}\right], \qquad (36)$$

where A_x and A_y are the amplitudes of x and y components of the electric field, respectively. σ_i is the r.m.s. width of the intensity distribution along i direction, δ_{xx}, δ_{yy}, and δ_{xy} are the r.m.s. widths of the autocorrelation functions of the x component of the field, of the y component of the field, and of

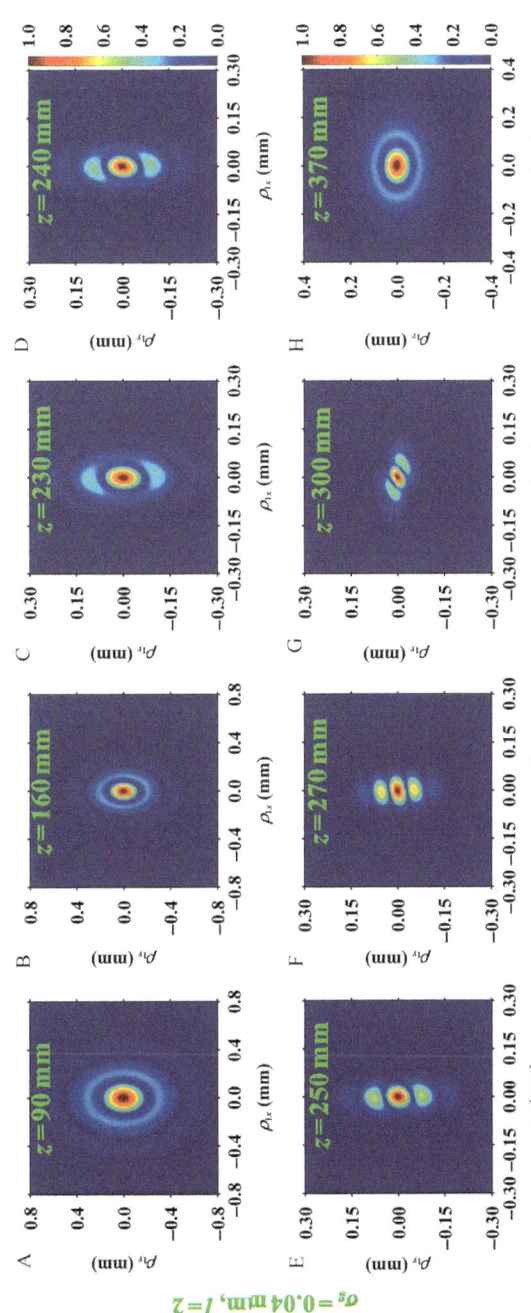

Fig. 12 Normalized distribution of $\log\left[1+|\mu(\rho_{1x},\rho_{1y},0,0)|\right]$ of a partially coherent LG_{0l} beam after passing through a couple of cylindrical lenses at different propagation distances with $l=2$ and $\sigma_g=0.04$ mm (Chen, Liu, et al., 2016). (A) $z=90$ mm; (B) $z=160$ mm; (C) $z=230$ mm; (D) $z=240$ mm; (E) $z=250$ mm; (F) $z=270$ mm; (G) $z=300$ mm; (H) $z=370$ mm.

the mutual correlation function of the x and y field components, respectively. $B_{xx} = B_{yy} = 1$, $B_{xy} = |B_{xy}| \exp\left(i\phi_{xy}\right)$ is the complex correlation coefficient between the x and y components of the electric field with ϕ_{xy} being the phase difference between the x and y components.

The conditions for the elements of the CSD matrix of an EGSM source under which the source generates an EGSM beam can expressed as (Korotkova, Salem, et al., 2004a)

$$\frac{1}{4\sigma_x^2} + \frac{1}{\delta_{xx}^2} \leq \frac{2\pi^2}{\lambda^2}, \quad \frac{1}{4\sigma_y^2} + \frac{1}{\delta_{xx}^2} \leq \frac{2\pi^2}{\lambda^2}. \tag{37}$$

The sufficiency condition on the choice of the parameters needed to describe a physically realizable EGSM source can be expressed as (Gori et al., 2001; Roychowdhury & Korotkova, 2005)

$$\max\{\delta_{xx}, \delta_{yy}\} \leq \delta_{xy} \leq \min\left\{\frac{\delta_{xx}}{\sqrt{|B_{xy}|}}, \frac{\delta_{yy}}{\sqrt{|B_{xy}|}}\right\}. \tag{38}$$

The necessary conditions for the parameters needed to describe a physically realizable EGSM source are (Roychowdhury & Korotkova, 2005)

$$\frac{A_x^2 \sigma_x^4 \delta_{xx}^2}{\delta_{xx}^2 + 4\sigma_x^2} - \frac{2A_x A_y |B_{xy}| \sigma_x^2 \sigma_y^2 \delta_{xy}^2}{\delta_{xy}^2 + 2\sigma_x^2 + 2\sigma_y^2} + \frac{A_y^2 \sigma_y^4 \delta_{yy}^2}{\delta_{yy}^2 + 4\sigma_y^2} \geq 0, \tag{39}$$

$$\frac{2\sigma_x^2 \delta_{xx}^2}{\delta_{xx}^2 + 4\sigma_x^2} - \frac{\left(\sigma_x^2 + \sigma_y^2\right)\delta_{xy}^2}{\delta_{xy}^2 + 2\sigma_x^2 + 2\sigma_y^2} + \frac{2\sigma_y^2 \delta_{yy}^2}{\delta_{yy}^2 + 4\sigma_y^2} \leq 0. \tag{40}$$

The state of polarization of an EGSM beam is uniform in the source plane (see Fig. 13), i.e., the polarization ellipse at every point across the source plane is the same. Both the degree of polarization and the state of polarization of an EGSM beam vary on propagation in free space (Gori et al., 2001; Korotkova & Wolf, 2005b), while in a turbulent atmosphere, the degree of polarization of regains its initial value and the polarization ellipse in the far field acquires the same shape as in the source plane (Korotkova et al., 2005; Roychowdhury et al., 2005).

An EGSM source was generated using a Mach–Zehnder interferometer in Piquero et al. (2002). The experimental setup is shown in Fig. 14. A linearly polarized beam emitted by an Argon laser enters the Mach–Zehnder interferometer. A half-wave plate rotates the polarization axis of

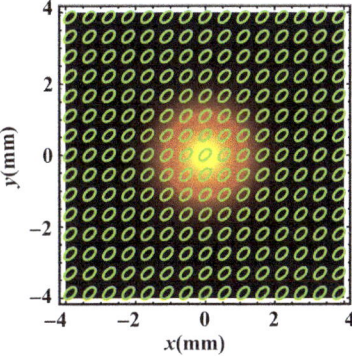

Fig. 13 Intensity distribution and state of polarization of an EGSM beam in the source plane with $A_x = A_y = 1$, $B_{xy} = 0.1 \exp(i\pi/3)$, $\sigma_x = \sigma_y = 1$ mm, $\delta_{xx} = \delta_{yy} = \delta_{xy} = 0.2$ mm.

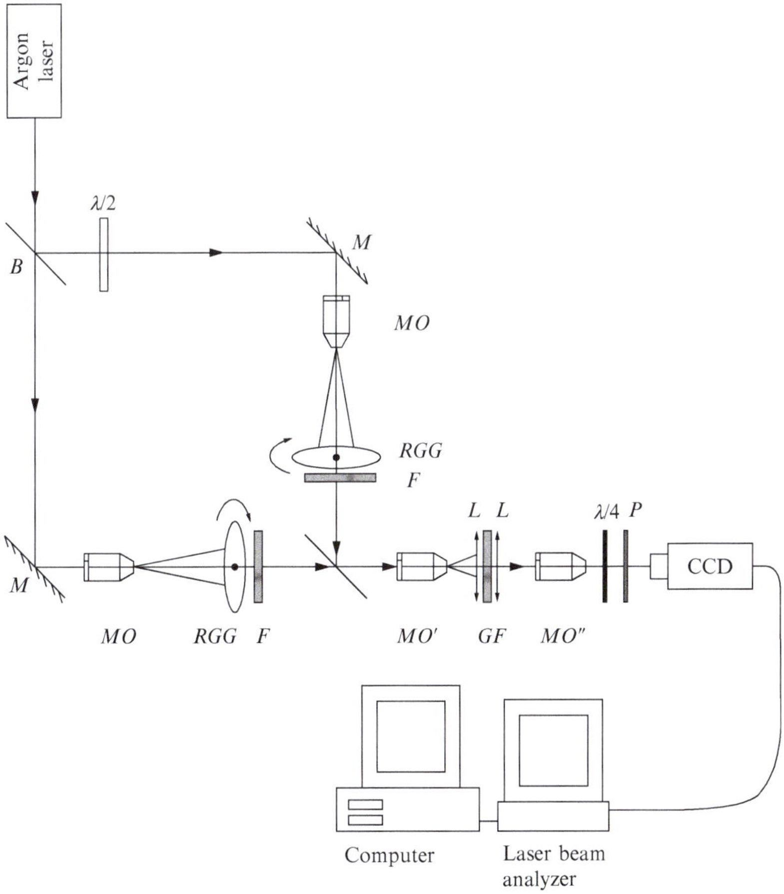

Fig. 14 Experimental setup for generating an EGSM source. B, beam splitter; GF, Gaussian filter; L, lens with f = 10 cm; M, mirror; MO, microscope objective 40×; MO', microscope objective 40×; MO", microscope objective 10×; λ/4 and λ/2, quarter-wave plates and half-wave plates, respectively (Piquero et al., 2002).

the beam in one of the arms by 90 degree. Two microscope objectives (MOs) and two rotating ground glass plates (RGGPs) are used to produce two orthogonally polarized spatially incoherent sources. Two suitably shaped intensity filters cover each of these incoherent sources, which are superimposed at the output of the interferometer and then imaged by another microscope objective (MO′). The image thus obtained represents (approximately) a spatially incoherent source with a Gaussian intensity profile. By means of a lens (L) placed at the distance f (corresponding to its focal length) from the above image, and using a Gaussian intensity filter (GF) just after the lens, an EGSM beam source is obtained whose CSD matrix does not have antidiagonal elements. A rotation by 45 degree of the reference frame in the transverse plane leads to a CSD matrix with both diagonal and antidiagonal elements. The beam generated by such a source is analyzed in the far field by means of a CCD camera with the help of another lens (L) and a microscope objective MO″. The degree of polarization of the generated EGSM beam in the far field displays a Gaussian distribution as was predicted (see Fig. 15).

Later, different methods have been proposed to generate an EGSM source with the help of SLMs (Basu et al., 2014; Ostrovsky et al., 2010; Shirai, Korotkova, & Wolf, 2005). The generalized Stokes parameters of an EGSM beam were determined directly from the usual Stokes parameters in experiments (Basu et al., 2014). In Vidal et al. (2011), the influence of the coherence coefficients of the source on the degree of coherence of an EGSM

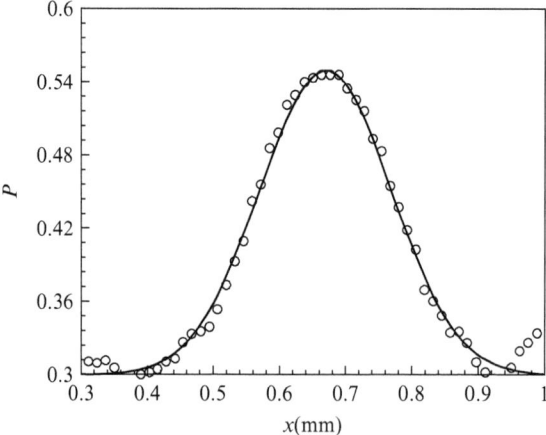

Fig. 15 Local degree of polarization of an EGSM beam vs x for $y=0$ in the far field (Piquero et al., 2002).

beam on propagation in free space was measured, and the results are shown in Fig. 16. One finds from Fig. 16 that the degree of coherence of an EGSM beam during free-space propagation can be controlled by changing the coherence coefficients of the source. In Wang, Wu, et al. (2011), the beam parameters of an EGSM source were measured and the experimental setup for generating an EGSM source and measuring its beam parameters is shown in Fig. 17. The EGSM source is generated with the help of a Mach–Zehnder interferometer, RGGP, and GAF (see Part I of Fig. 17). In the experiment, ϕ_{xy} is assumed to be zero because the eikonals along the two arms of the MZI are equal.

Part 2 of Fig. 17 shows the experimental setup for measuring the parameters of the generated EGSM beam. The generated EGSM beam first passes through a linear polarizer P_2 and a thin lens L_4 with focal length $f=15$ cm, then is split into two beams by a BS. The transmitted and reflected beams going through two separated $2f$-imaging systems will arrive at single photon detectors D_1 and D_2, which scan the transverse planes of **u** and **v**, respectively. Both the distances from the GAF to L_4 and from L_4 to D_1 and D_2 are $2f$. An electronic coincidence circuit is used to measure the FOCF between the two detectors.

First, we adjust linear polarizer P_2 to set its transmission axis along x-axis. In this case, only the element W_{xx} is nonzero behind P_2. By measuring the intensity distribution and its maximum intensity at plane **u** or **v** with the help

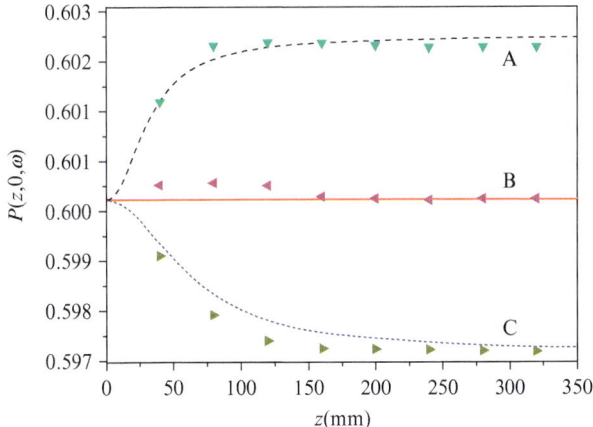

Fig. 16 Degree of polarization as a function of the propagation distance. *Triangles* are experimental points for: (A) $\delta_{xx} > \delta_{yy}$, (B) $\delta_{xx} = \delta_{yy}$, and (C) $\delta_{xx} < \delta_{yy}$. The *continuous curves* denote the theoretical predictions (Vidal et al., 2011).

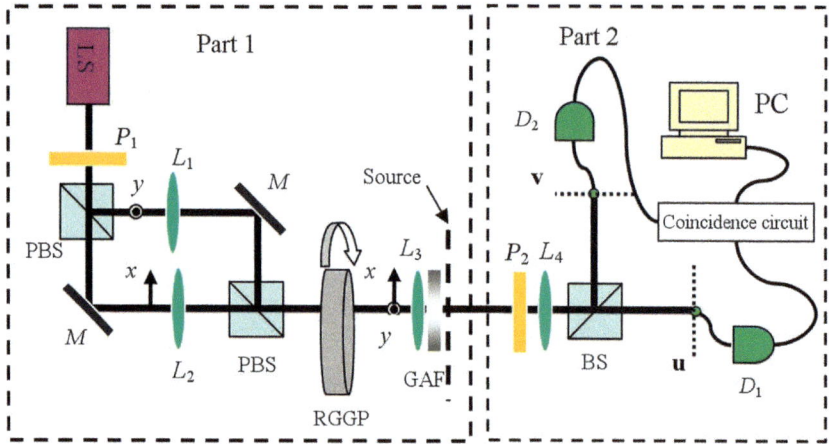

Fig. 17 Experimental setup for generating an EGSM beam and measuring its beam parameters. *BS*, 50:50 beam splitter; D_1, D_2, single photon detector (Wang, Wu, et al., 2011); *GAF*, Gaussian amplitude filter; L_1, L_2, L_3, L_4, thin lens; *M*, reflecting mirror; *LS*, He–Ne laser; P_1, P_2, linear polarizer; *PBS*, polarization beam splitter; *RGGP*, rotating ground glass plate.

of the single photon detector D_1 or D_2, we can obtain the values of the parameters σ_x and A_x. The FOCF between D_1 and D_2 is given by the expression

$$g_{xx}^{(2)}(u_1 - v_1, \tau) = \frac{\langle I_x(u_1, t) I_x(v_1, t+\tau) \rangle}{\langle I_x(u_1, t) \rangle \langle I_x(v_1, t+\tau) \rangle}, \qquad (41)$$

where τ denotes the delay time of the photon flux of two optical paths. Applying the Gaussian moment theorem, $g_{xx}^{(2)}(u_1 - v_1, \tau)$ with $\tau = 0$ can be simplified to

$$g_{xx}^{(2)}(u_1 - v_1, \tau = 0) = 1 + \exp\left[-\frac{(u_1 - v_1)^2}{\delta_{xx}^2}\right]. \qquad (42)$$

We fix D_2 at $\boldsymbol{v} = 0$, and D_1 scans along the plane $\boldsymbol{u}$. The coincidence circuit records the FOCF between D_1 and D_2. Thus we can obtain the distribution of the normalized FOCF $G_{xx}^{(2)}(u_1, \tau = 0)$ with $\tau = 0$. From the curve of the Gaussian fit for the experimental results, we can obtain the value of δ_{xx}. If we adjust P_2 to set its transmission axis along y-axis, only the element W_{yy} is nonzero behind P_2. Then through a similar operation for obtaining σ_x, A_x, and δ_{xx}, we can measure the values of the parameters σ_y, A_y, and δ_{yy}.

Fig. 18 shows the experimental scheme for measuring the parameters $|B_{xy}|$ and δ_{xy}. Different from part 2 of Fig. 17, P_2 is now removed, and

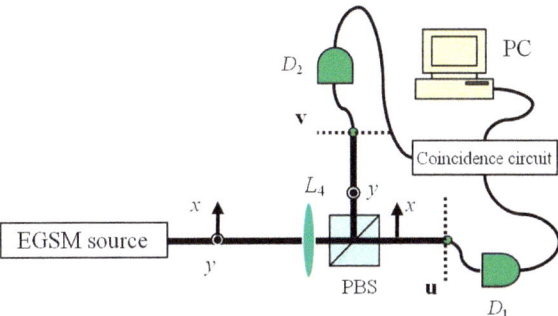

Fig. 18 Experimental scheme for measuring the beam parameters $|B_{xy}|$ and δ_{xy} (Wang, Wu, et al., 2011).

the BS is replaced with a polarization beam splitter (PBS). The x-component and y-component of the field will arrive at D_1 and D_2, respectively. The normalized FOCF between two detectors with $\tau = 0$ can be expressed as

$$g_{xy}^{(2)}(u_1 - v_1, \tau = 0) = \frac{\langle I_x(u_1, t) I_y(v_1, t) \rangle}{\langle I_x(u_1, t) \rangle \langle I_y(v_1, t) \rangle}$$
$$= 1 + |B_{xy}|^2 \exp\left[-\frac{(u_1 - v_1)^2}{\delta_{xy}^2}\right]. \quad (43)$$

Following the same procedure for obtaining δ_{xx}, D_2 is fixed at $v=0$, and D_1 scans along the plane u, and the coincidence circuit records the distribution of the normalized FOCF $G_{xy}^{(2)}(u_1, \tau = 0)$. From the curve of the Gaussian fit for the experimental results, we can obtain the value of δ_{xy}. From Eq. (43), $|B_{xy}|$ can be obtained by the relation $|B_{xy}| = \sqrt{G_{xy}^2(u_1, \tau = 0) - 1}$ under the condition that $u_1 = 0$.

After all beam parameters of an EGSM beam have been measured, we can study the influence of coherence and polarization on the ghost image formed with an EGSM beam (Liu, Wang, et al., 2015), and the effect of coherence and polarization on the efficiency of coupling an EGSM beam into a single-model optical fiber (Zhao, Dong, Wu, et al., 2012).

Next we discuss radially polarized and azimuthally polarized partially coherent beams, which are typical examples of cylindrical vector partially coherent beams with a spatially nonuniform state of polarization. The elements of the CSD matrix of a radially polarized partially coherent beam in the source plane are defined as (Wang, Cai, Dong, et al., 2012; Wu et al., 2012)

$$W_{xx}(x_1, y_1, x_2, y_2)$$
$$= \frac{x_1 x_2}{\omega_0^2} \exp\left[-\frac{x_1^2 + y_1^2 + x_2^2 + y_2^2}{\omega_0^2} - \frac{(x_1 - x_2)^2 + (y_1 - y_2)^2}{2\delta_0^2}\right], \quad (44)$$

$$W_{xy}(x_1, y_1, x_2, y_2)$$
$$= \frac{x_1 y_2}{\omega_0^2} \exp\left[-\frac{x_1^2 + y_1^2 + x_2^2 + y_2^2}{\omega_0^2} - \frac{(x_1 - x_2)^2 + (y_1 - y_2)^2}{2\delta_0^2}\right], \quad (45)$$

$$W_{yx}(x_1, y_1, x_2, y_2) = W_{xy}^*(x_2, y_2, x_1, y_1), \quad (46)$$

$$W_{yy}(x_1, y_1, x_2, y_2)$$
$$= \frac{y_1 y_2}{\omega_0^2} \exp\left[-\frac{x_1^2 + y_1^2 + x_2^2 + y_2^2}{\omega_0^2} - \frac{(x_1 - x_2)^2 + (y_1 - y_2)^2}{2\delta_0^2}\right], \quad (47)$$

where ω_0 and δ_0 denote the transverse beam size and spatial coherence width, respectively. For an azimuthally polarized partially coherent beam, the elements of its CSD matrix are defined as (Dong, Wang, et al., 2012)

$$W_{xx}(x_1, y_1, x_2, y_2)$$
$$= \frac{y_1 y_2}{\omega_0^2} \exp\left[-\frac{x_1^2 + y_1^2 + x_2^2 + y_2^2}{\omega_0^2} - \frac{(x_1 - x_2)^2 + (y_1 - y_2)^2}{2\delta_0^2}\right], \quad (48)$$

$$W_{xy}(x_1, y_1, x_2, y_2) =$$
$$-\frac{x_2 y_1}{\omega_0^2} \exp\left[-\frac{x_1^2 + y_1^2 + x_2^2 + y_2^2}{\omega_0^2} - \frac{(x_1 - x_2)^2 + (y_1 - y_2)^2}{2\delta_0^2}\right], \quad (49)$$

$$W_{yx}(x_1, y_1, x_2, y_2) = W_{xy}^*(x_2, y_2, x_1, y_1), \quad (50)$$

$$W_{yy}(x_1, y_1, x_2, y_2)$$
$$= \frac{x_1 x_2}{\omega_0^2} \exp\left[-\frac{x_1^2 + y_1^2 + x_2^2 + y_2^2}{\omega_0^2} - \frac{(x_1 - x_2)^2 + (y_1 - y_2)^2}{2\delta_0^2}\right]. \quad (51)$$

In the source plane, the radially polarized or azimuthally polarized partially coherent beams display a dark hollow profile and radial or azimuthal polarization (see Fig. 19). The dark hollow beam profile of a radially or azimuthally polarized partially coherent beam disappears on propagation and finally becomes a flat topped or Gaussian beam profile in the far field or in the focal plane depending on its initial coherence width (Dong, Wang, et al., 2012; Wang, Cai, Dong, et al., 2012). Moreover, the degree of

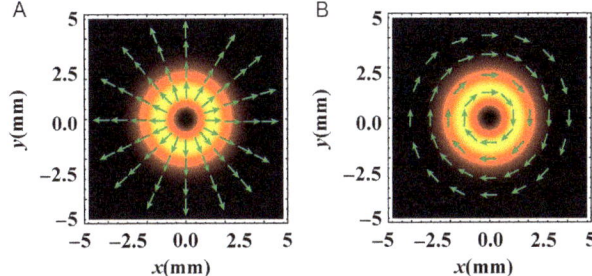

Fig. 19 Intensity distribution and state of polarization of (A) a radially polarized partially coherent beam and (B) an azimuthally polarized partially coherent beam with $\omega_0 = 1$ mm and $\delta_0 = 0.2$ mm.

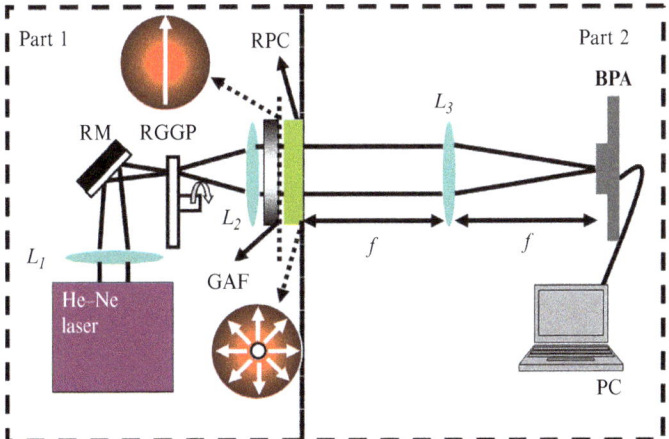

Fig. 20 Experimental setup for generating a partially coherent RP beam and measuring its focused intensity. *BPA*, beam profile analyzer; *GAF*, Gaussian amplitude filter; L_1, L_2, L_3, thin lenses; *PC*, personal computer (Wang, Cai, Dong, et al., 2012); *RGGP*, rotating ground glass plate; *RM*, reflecting mirror; *RPC*, radial polarization converter.

polarization of such a beam is equal to 1 in the source plane and decreases on propagation in free space, which means that the beam depolarizes on propagation, while its state of polarization remains invariant (Wu et al., 2012).

Experimental generation of a radially polarized partially coherent beam was first reported in Wang, Cai, Dong, et al. (2012). The experimental setup is shown in Fig. 20. In the experiment, a radially polarized partially coherent beam is generated through conversion of a generated linearly polarized GSM beam by a radial polarization converter. The focused intensity is measured by the BPA, and the results are shown in Fig. 21. An azimuthally polarized

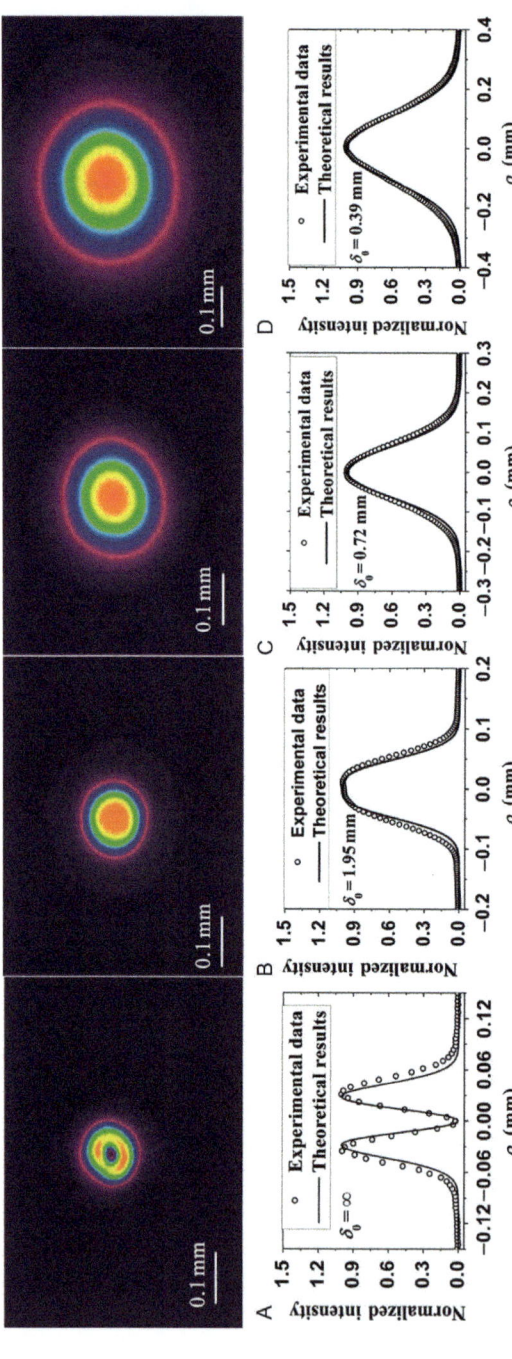

Fig. 21 Experimental results of the focused intensity and the corresponding cross line (*black solid curve*) of the generated radially polarized partially coherent beam for four different values of δ_0. The *black solid curves* are theoretical results (Wang, Cai, Dong, et al., 2012). (A) $\delta_0 = \infty$; (B) $\delta_0 = 1.95$ mm; (C) $\delta_0 = 0.72$ mm; (D) $\delta_0 = 0.39$ mm.

partially coherent beam can be generated in a similar way (Dong, Wang, et al., 2012). It was found that one can shape the beam profile of an azimuthally polarized partially coherent beam focused by a high numerical aperture objective lens by varying its initial spatial coherence width, which is useful for particle trapping.

The generation of a polychromatic radially polarized partially coherent beam was reported in Zhu, Zhu, et al. (2013). Experimental studies of a radially polarized partially coherent beam propagating through Young's two-slit setup, or thermal turbulence were reported in Zhu et al. (2014) and Wang, Liu, et al. (2013), respectively. More recently, vortex phase-induced changes of intensity distribution and state of polarization of a radially polarized partially coherent beam were demonstrated both theoretically and experimentally in Guo, Chen, Liu, et al. (2016).

2.4 Partially Coherent Beams With Prescribed Degrees of Coherence

In this section, we discuss theoretical models for some partially coherent beams with prescribed degrees of coherence, and introduce the methods for manipulating the degree of coherence.

The CSD of the well-known J_0-correlated Schell-model beam in the source plane is defined as (Gori & Guattari, 1987)

$$W(\mathbf{r}_1, \mathbf{r}_2) = T(\mathbf{r}_1) T(\mathbf{r}_2) J_0(\beta |\mathbf{r}_1 - \mathbf{r}_2|), \qquad (52)$$

where J_0 is the Bessel function of the first kind and zero order, β is a real constant and $T(\mathbf{r}_j)$ $(j=1,2)$ is a real valued function whose square gives the optical intensity at the point $\mathbf{r}_j$ $(j=1,2)$. It was demonstrated numerically in van Dijk et al. (2008) and Gbur and Visser (2003) that for high-Fresnel-number focusing systems illuminated by a J_0-correlated Schell-model beam, it is possible to change the intensity distribution and even to produce a local minimum of intensity at the geometrical focus by altering the coherence length. Generation of a J_0-correlated Schell-model beam was reported in Raghunathan et al. (2010). The experimental setup is shown in Fig. 22. The output of a 15 mW He–Ne laser, operating at 632.8 nm, is focused by Lens 1 onto a rotating optical diffuser, producing an incoherent beam. The incoherent beam illuminates a thin annulus of inner radius 1.2 mm and outer radius 1.5 mm. The annulus is positioned in the back focal plane of a 3.7 m lens (Lens 2), which produces a J_0-correlated field in its focal plane. This field is incident on an iris of radius 2.5 mm and focused by a lens

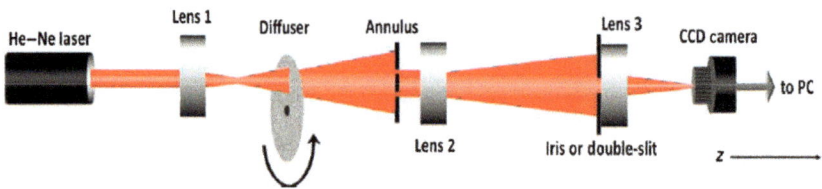

Fig. 22 Experimental setup for generating a J_0-correlated Schell-model beam and measuring its focused intensity (Raghunathan et al., 2010).

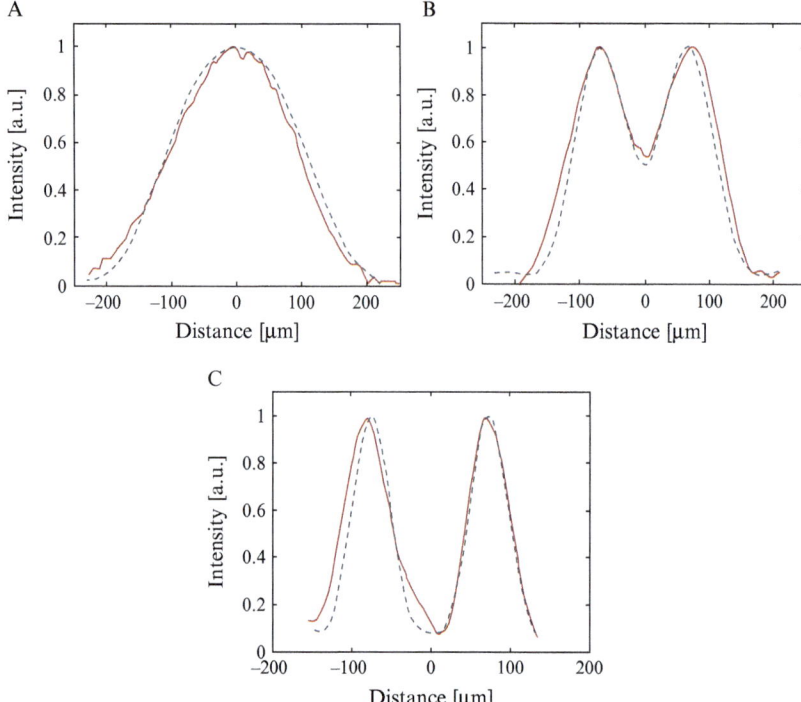

Fig. 23 Transforming the intensity maximum in the focal plane into a minimum by varying the iris radius, (A) 0.25, (B) 0.75, and (C) 1:2 mm (Raghunathan et al., 2010).

of focal length 10.6 cm (Lens 3). The focused image is captured using a CCD camera connected to a PC via a frame grabber. Fig. 23 shows the experimental results of the intensity in the focal plane for different values of the iris radius, which confirms that one can obtain a local minimum of intensity through varying the iris radius.

The CSD of the most recently introduced scalar partially coherent beams with prescribed degree of coherence can be expressed in the following form

$$W(\mathbf{r}_1, \mathbf{r}_2) = \exp\left[-\frac{\mathbf{r}_1^2 + \mathbf{r}_2^2}{4\sigma_0^2}\right]\mu(\mathbf{r}_1, \mathbf{r}_2). \tag{53}$$

Such beams have Gaussian beam profiles with width σ_0 in the source plane.

For a nonuniformly correlated beam, in the one-dimensional case, the degree of coherence in the source plane is defined as follows (Lajunen & Saastamoinen, 2011)

$$\mu(x_1, x_2) = \exp\left\{-\frac{\left[(x_2 - x_0)^2 - (x_1 - x_0)^2\right]^2}{\delta_0^4}\right\}, \tag{54}$$

where δ_0 is a measure of the correlation length and the state of coherence depends on the lateral coordinate, its maximum being located in the region centered at x_0. As shown in Lajunen and Saastamoinen (2011), due to its special distribution of the degree of coherence, the nonuniformly correlated beam displays self-focusing and a lateral shift of the intensity maximum (when $x_0 \neq 0$) on propagation in free space.

For a MGCSM beam, the degree of coherence in the source plane is defined as (Sahin & Korotkova, 2012)

$$\mu(\mathbf{r}_1, \mathbf{r}_2) = \frac{1}{C_0}\sum_{m=1}^{M}\binom{M}{m}\frac{(-1)^{m-1}}{m}\exp\left[-\frac{(\mathbf{r}_1 - \mathbf{r}_2)^2}{2m\delta_0^2}\right], \tag{55}$$

with $C_0 = \sum_{m=1}^{M}\frac{(-1)^{m-1}}{m}\binom{M}{m}$ and $\binom{M}{m}$ stands for binomial coefficients.

The MGCSM beam displays self-shaping on propagation, e.g., the initial Gaussian beam profile evolves into a flat-topped beam profile in the far field or in the focal plane. A generalized MGCSM beam was proposed in Wang et al. (2014), the initial Gaussian beam profile of the generalized MGCSM beam of the first or second kind evolves into a dark hollow or flat-topped beam in the far field.

For an elliptical LGCSM beam, the degree of coherence in the source plane is defined as (Chen, Liu, et al., 2014)

$$\mu(\mathbf{r}_1, \mathbf{r}_2) = L_n^0\left[\frac{(x_1 - x_2)^2}{2\delta_{0x}^2} + \frac{(y_1 - y_2)^2}{2\delta_{0y}^2}\right]\exp\left[-\frac{(x_1 - x_2)^2}{2\delta_{0x}^2} - \frac{(y_1 - y_2)^2}{2\delta_{0y}^2}\right], \tag{56}$$

where L_n^0 denotes the Laguerre polynomial with mode orders n and 0, δ_{0x} and δ_{0y} denote the coherence widths in x- and y-directions, respectively. When $\delta_{0x} = \delta_{0y}$, Eq. (56) reduces to the CSD of a circular LGCSM beam (Mei & Korotkova, 2013a). The circular or elliptical LGCSM beam also displays self-shaping properties, e.g., the initial Gaussian beam profile evolves into a circular or elliptical dark hollow beam profile in the far field or in the focal plane.

For a Hermite-Gaussian correlated Schell-model (HGCSM) beam, the degree of coherence in the source plane is defined as (Chen et al., 2015)

$$\mu(\mathbf{r}_1, \mathbf{r}_2) = \frac{H_{2m}\left[(x_2 - x_1)/\sqrt{2}\delta_{0x}\right] H_{2n}\left[(y_2 - y_1)/\sqrt{2}\delta_{0y}\right]}{H_{2m}(0) \, H_{2n}(0)} \exp\left[-\frac{(x_2 - x_1)^2}{2\delta_{0x}^2} - \frac{(y_2 - y_1)^2}{2\delta_{0y}^2}\right], \quad (57)$$

where δ_{0x} and δ_{0y} are the coherence widths along x- and y-directions, respectively. Here H_m denotes the Hermite polynomial of order. The HGCSM beam displays self-splitting, i.e., the initial Gaussian beam profile is split into two or four beam spots in the far field depending on the initial beam orders m and n.

For a nonuniform multi-Gaussian correlated beam, the degree of coherence in the source plane is defined as (Chen & Cai, 2016)

$$\mu(\mathbf{r}_1, \mathbf{r}_2) = \mu_x(x_1, x_2)\mu_y(y_1, y_2), \quad (58)$$

with

$$\mu_x(x_1, x_2) = \exp\left\{-\frac{\left[(x_2 - x_0)^2 - (x_1 - x_0)^2\right]^2}{\delta_x^4}\right\}, \quad (59)$$

$$\mu_y(y_1, y_2) = \frac{1}{C_0}\sum_{m=1}^{M}\frac{(-1)^{m-1}}{\sqrt{m}}\binom{M}{m}\exp\left[-\frac{(y_1 - y_2)^2}{m\delta_y^2}\right]. \quad (60)$$

The intensity distribution of a nonuniform multi-Gaussian correlated beam exhibits self-focusing and self-shifting in one direction and self-shaping in the other direction on propagation in free space.

In Wang, Liu, et al. (2013), it was shown that a scalar partially coherent beam with a prescribed degree of coherence can be produced from an incoherent source with a prescribed intensity distribution through propagation, and the LGCSM beam was generated as an application example. Later, different kinds of partially coherent beams with prescribed degrees of coherence, such as a cosine Gaussian-correlated Schell-model beam (Liang et al., 2014), an elliptical LGCSM beam (Chen, Liu, et al. (2014), an LGCSM vortex beam (Chen, Wang, Zhao, et al., 2014), an MGCSM beam (Avramov-Zamurovic, Nelson, Guth, & Korotkova, 2016), a generalized MGCSM beam (Wang et al., 2014), and an HGCSM beam (Chen et al., 2015). Optical coherence lattices (Chen, Ponomarenko, et al., 2016) were demonstrated experimentally.

Fig. 24 shows the experimental setup for generating a partially coherent beam with a prescribed degree of coherence and measuring its beam properties. A beam emitted from a He–Ne laser is reflected by a mirror and passes through a beam expander, then it goes toward a SLM, which acts as a phase grating designed by the method of computer-generated holograms. The first-order diffraction pattern generated by an SLM is selected with the help of a circular aperture, and may be regarded as a coherent beam with a prescribed intensity distribution. After passing through a thin lens L, the prescribed coherent beam illuminates the RGGD, producing an incoherent beam with prescribed intensity distribution. After passing through the thin lens L_1 and the GAF, the incoherent beam with prescribed intensity distribution turns into a partially coherent bream with prescribed degree of coherence. The $2f$-imaging system and the CCD are used to measure the degree of coherence and the BPA is used to measure the focused intensity distribution.

Figs. 25 and 26 show the experimental results of the square of the modulus of the degree of coherence of the generated HGCSM beam and the focused intensity distribution of such beams on propagation for different

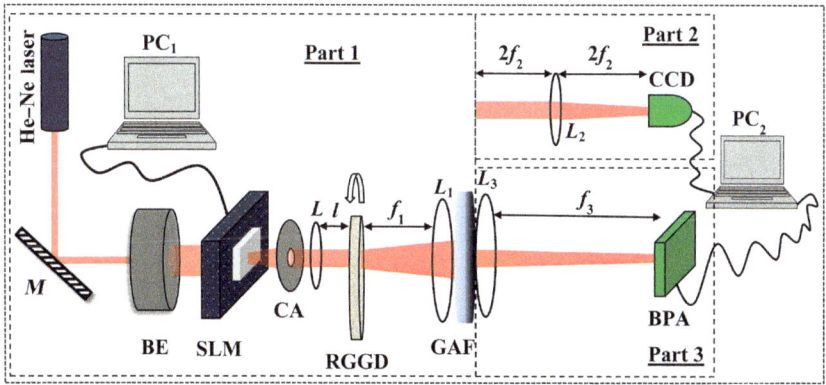

Fig. 24 Experimental setup for generating a partially coherent beam with prescribed degree of coherence, measuring the square of the modulus of its degree of coherence, and its focused intensity. *BE*, beam expander; *BPA*, beam profile analyzer; *CA*, circular aperture; *CCD*, charge-coupled device; *GAF*, Gaussian amplitude filter; L, L_1, L_2, L_3, thin lenses; PC_1, PC_2, personal computers (Cai et al., 2014); *RGGD*, rotating ground glass disk; *RM*, reflecting mirror; *SLM*, spatial light modulator.

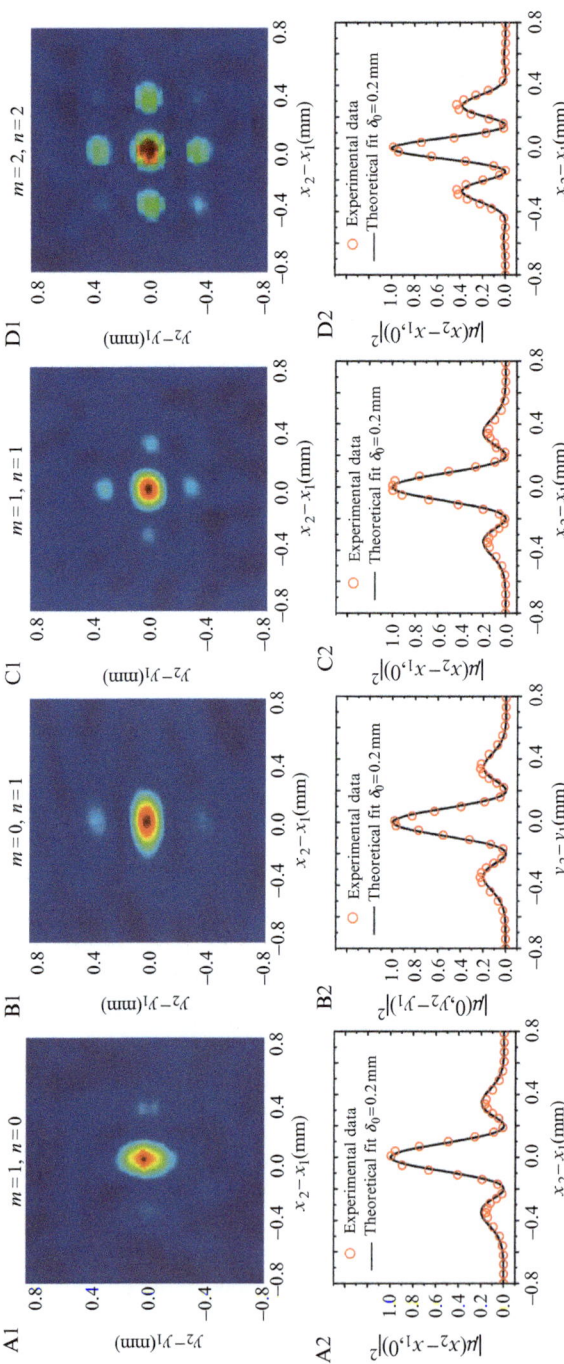

Fig. 25 Experimental results of the square of the modulus of the degree of coherence of the generated HGCSM beam for different values of m and n and the corresponding cross line (*dotted curve*). The *solid line* denotes the theoretical fit of the experimental results with $\delta_0 = 0.2$ mm (Chen et al., 2015). (A1) $m = 1, n = 0$; (B1) $m = 0, n = 1$; (C1) $m = 1, n = 1$; (D1) $m = 2, n = 2$; (A2) $m = 1, n = 0$; (B2) $m = 0, n = 1$; (C2) $m = 1, n = 1$; (D2) $m = 2, n = 2$.

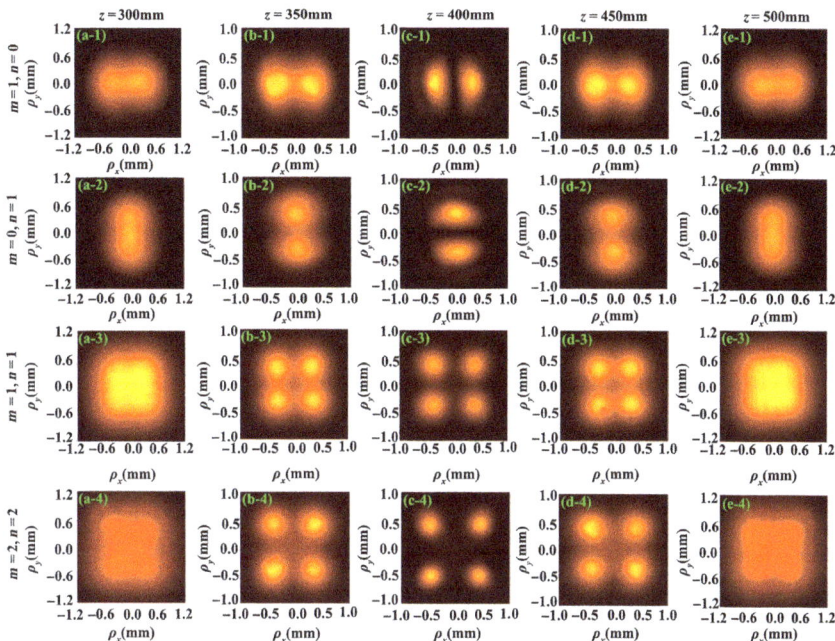

Fig. 26 Experimental results for the intensity distribution of an HGCSM beam focused by a thin lens with $f=400$ mm for different values of m and n with $\sigma_0 = 1$ mm and $\delta_0 = 0.2$ mm at selected propagation distances (Chen et al., 2015).

values of m and n, and the splitting properties of such beams were verified experimentally. Figs. 27 and 28 show the experimental results of the square of the modulus of the degree of coherence of the generated elliptical LGCSM beam ($n=5$) and its intensity in the geometrical focal plane for different values of the coherence widths δ_{0x} and δ_{0y}, and the self-shaping properties were verified experimentally.

In Chen et al. (2014), it was shown that a vector partially coherent beam with prescribed degree of coherence can be produced from a vector incoherent source with prescribed intensity distribution through propagation, and a specially correlated radially polarized (SCRP) beam was generated in this manner. Later, vector cosine-Gaussian correlated beams (Zhu et al., 2015) and vector HGCSM beams (Chen, Wang, et al., 2016) were generated experimentally. Here we only discuss SCRP beams. The elements of the CSD matrix of a SCRP beam are defined as follows (Chen, Wang, Liu, et al., 2014)

$$W_{xx}(\mathbf{r}_1, \mathbf{r}_2) = \exp\left[-\frac{\mathbf{r}_1^2 + \mathbf{r}_2^2}{4\sigma_0^2} - \frac{(\mathbf{r}_1 - \mathbf{r}_2)^2}{2\delta_0^2}\right]\left(1 - \frac{(x_1 - x_2)^2}{\delta_0^2}\right), \quad (61)$$

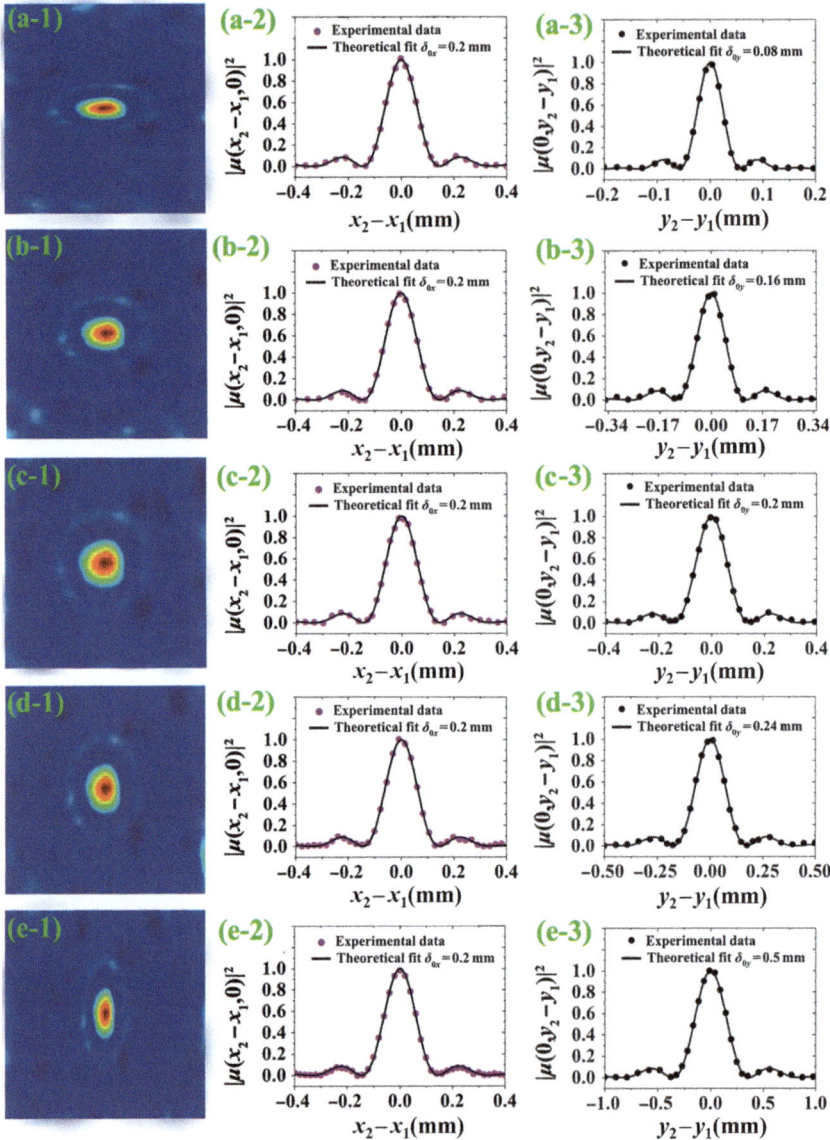

Fig. 27 Experimental results for the squared modulus of the degree of coherence and the corresponding cross lines (*dotted curves*) of the generated elliptical LGCSM beam ($n=5$) for different values of the coherence widths δ_{0x} and δ_{0y}. The *solid curve* is a theoretical fit (Chen, Liu, et al., 2014). (a-1) $\delta_{0x} = 0.2$ mm, $\delta_{0y} = 0.08$ mm; (a-2) $\delta_{0x} = 0.2$ mm; (a-3) $\delta_{0y} = 0.08$ mm; (b-1) $\delta_{0x} = 0.2$ mm, $\delta_{0y} = 0.16$ mm; (b-2) $\delta_{0x} = 0.2$ mm; (b-3) $\delta_{0y} = 0.16$ mm; (c-1) $\delta_{0x} = 0.2$ mm, $\delta_{0y} = 0.2$ mm; (c-2) $\delta_{0x} = 0.2$ mm; (c-3) $\delta_{0y} = 0.2$ mm; (d-1) $\delta_{0x} = 0.2$ mm, $\delta_{0y} = 0.24$ mm; (d-2) $\delta_{0x} = 0.2$ mm; (d-3) $\delta_{0y} = 0.24$ mm; (e-1) $\delta_{0x} = 0.2$ mm, $\delta_{0y} = 0.5$ mm; (e-2) $\delta_{0x} = 0.2$ mm; (e-3) $\delta_{0y} = 0.5$ mm.

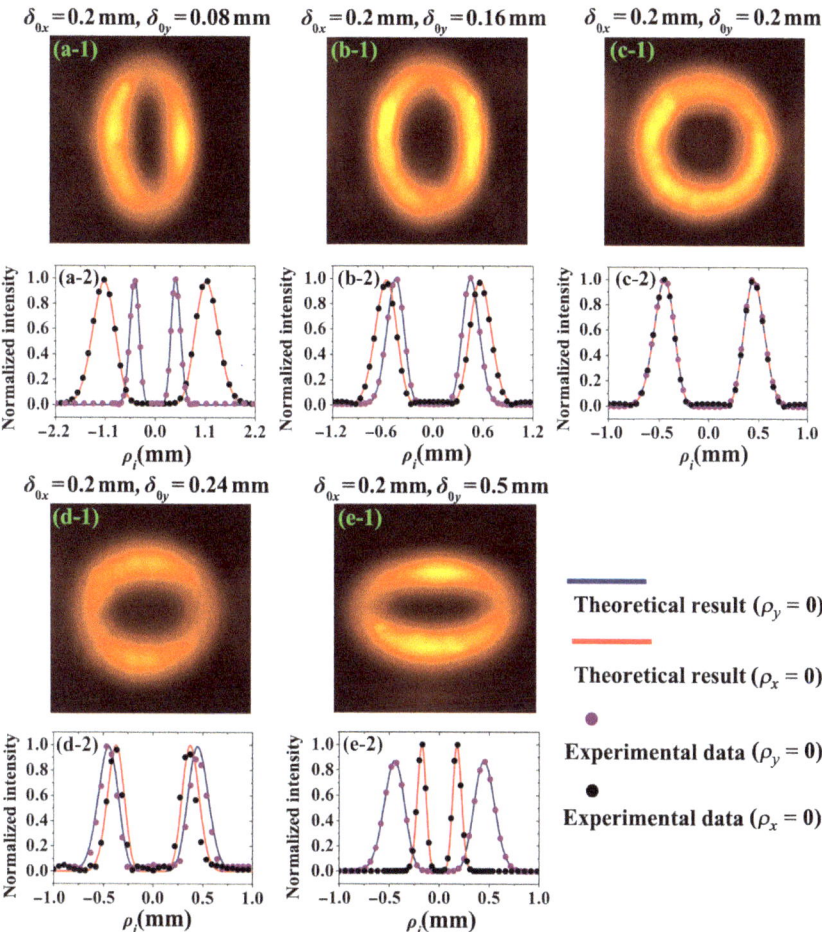

Fig. 28 Experimental results of the intensity distribution of the generated elliptical LGCSM beam ($n=5$) and the corresponding cross lines in the geometrical focal plane for different values of the coherence widths δ_{0x} and δ_{0y}. The *solid curves* denote the theoretical results (Chen, Liu, et al., 2014).

$$W_{yy}(\mathbf{r}_1, \mathbf{r}_2) = \exp\left[-\frac{\mathbf{r}_1^2 + \mathbf{r}_2^2}{4\sigma_0^2} - \frac{(\mathbf{r}_1 - \mathbf{r}_2)^2}{2\delta_0^2}\right]\left(1 - \frac{(y_1 - y_2)^2}{\delta_0^2}\right), \qquad (62)$$

$$W_{xy}(\mathbf{r}_1, \mathbf{r}_2) = -\exp\left[-\frac{\mathbf{r}_1^2 + \mathbf{r}_2^2}{4\sigma_0^2} - \frac{(\mathbf{r}_1 - \mathbf{r}_2)^2}{2\delta_0^2}\right]\frac{(x_2 - x_1)(y_2 - y_1)}{\delta_0^2}, \qquad (63)$$

$$W_{yx}(\mathbf{r}_1, \mathbf{r}_2) = W_{xy}^*(\mathbf{r}_2, \mathbf{r}_1). \qquad (64)$$

Here σ_0 and δ_0 denote the beam width and coherence width, respectively.

For a SCRP beam (Chen, Wang, Liu, et al., 2014), the distribution of the square of its degree of coherence $\mu^2(\mathbf{r}_1, \mathbf{r}_2 = 0)$ and its correlation functions $\mu_{xx}^2(\mathbf{r}_1, \mathbf{r}_2 = 0)$, $\mu_{yy}^2(\mathbf{r}_1, \mathbf{r}_2 = 0)$, and $\mu_{xy}^2(\mathbf{r}_1, \mathbf{r}_2 = 0)$ all have non-Gaussian profiles. Hence such a beam displays interesting propagation properties, e.g., it has a Gaussian beam profile in the source plane and it evolves into a radially polarized beam with a dark hollow beam profile in the far field. Moreover, its degree of polarization equals zero in the source plane while its degree of polarization increases on propagation, which is quite different from the conventional radially polarized partially coherent beam (Wu et al., 2012). Fig. 29 shows the experimental setup for generating an SCRP beam and measuring its beam properties. A linearly polarized He–Ne laser beam with $\lambda = 632.8$ nm passes through a RPC and becomes a radially polarized beam. The generating radially polarized beam illuminates a RGGD producing an incoherent radially polarized beam. After passing through the thin lens L_1 with focal length f_1 and the GAF, the generated incoherent RP beam becomes an SCRP beam, whose CSD matrix elements are given by Eqs. (61)–(64). The BPA is used to measure its focused intensity, and the PBS and the CCD are used to measure their degree of coherence and degree of polarization. Fig. 30 shows the experimental results of the square of the degree of coherence $\mu^2(\mathbf{r}_1, \mathbf{r}_2 = 0)$, the square of the correlation functions

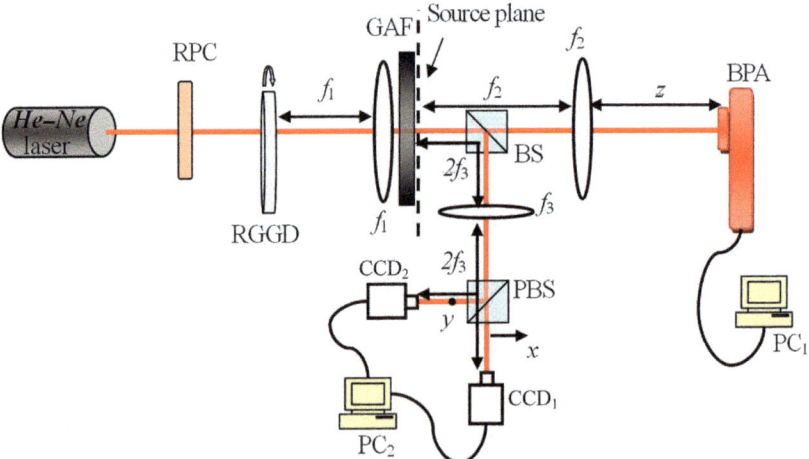

Fig. 29 Experimental setup for generating a SCRP beam and measurement of the degree of coherence, correlation functions, and the focused intensity. *BS*, beam splitter; *BPA*, beam profile analyzer; *CCD$_1$*, *CCD$_2$*, charge-coupled devices (Chen, Wang, Liu, et al., 2014); *GAF*, Gaussian amplitude filter; *L$_1$*, *L$_2$*, *L$_3$*, thin lenses; *PBS*, polarization beam splitter; *RPC*, radial polarization converter; *RGGD*, rotating ground glass disk.

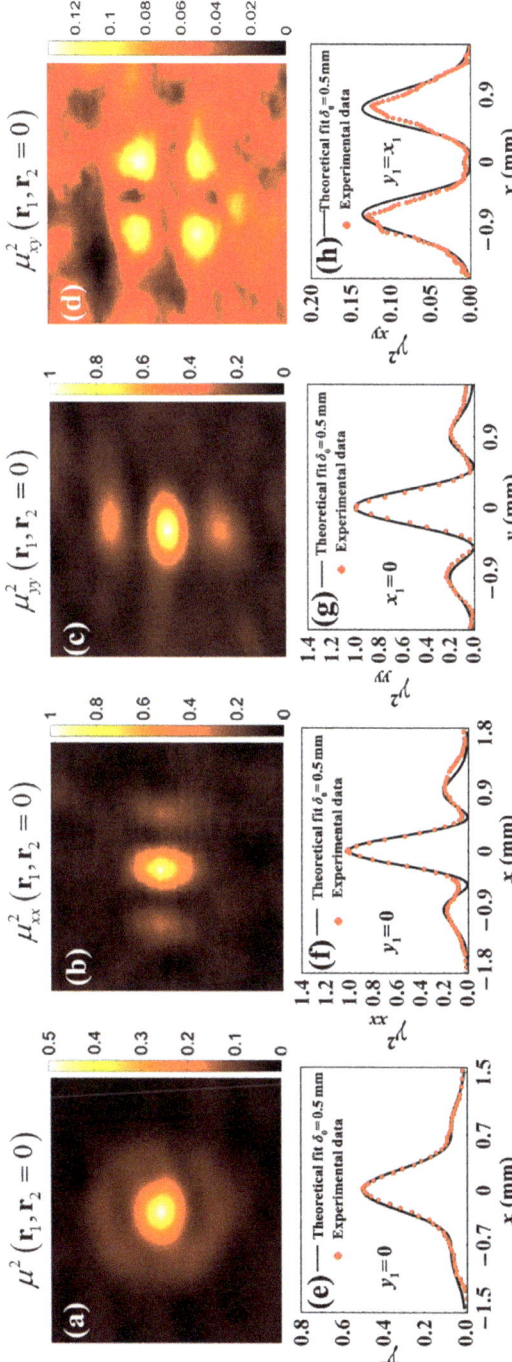

Fig. 30 Experimental results for the square of the degree of coherence $\mu^2(\mathbf{r}_1, \mathbf{r}_2 = 0)$, the square of the correlation functions $\mu_{xx}^2(\mathbf{r}_1, \mathbf{r}_2 = 0)$, $\mu_{yy}^2(\mathbf{r}_1, \mathbf{r}_2 = 0)$, $\mu_{xy}^2(\mathbf{r}_1, \mathbf{r}_2 = 0)$ and the corresponding cross lines (dotted curves) of the generated SCRP beam. The solid curve is a result of the theoretical fit (Chen, Wang, Liu, et al., 2014).

$\mu_{xx}^2(\mathbf{r}_1, \mathbf{r}_2 = 0)$, $\mu_{yy}^2(\mathbf{r}_1, \mathbf{r}_2 = 0)$, $\mu_{xy}^2(\mathbf{r}_1, \mathbf{r}_2 = 0)$ of the generated SCRP beam. Fig. 31 shows the experimental results for the intensity distribution of a focused SCRP beam on propagation. Fig. 32 shows the experimental results of the degree of polarization of the generated SCRP beam after passing a thin lens. The experimental results are consistent with the theoretical predictions.

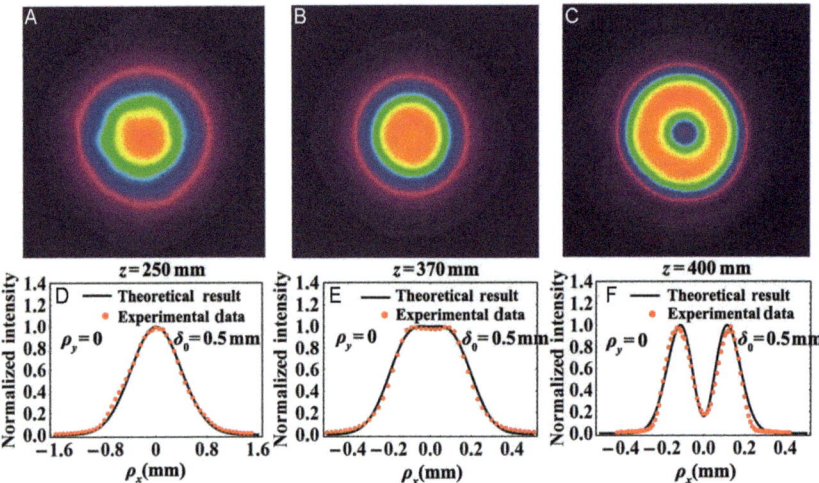

Fig. 31 Experimental results for the intensity distribution and corresponding cross line (*dotted curve*) of the generated SCRP beam focused by a thin lens with focal length $f = 400$ mm at several propagation distances for the case $\delta_0 = 0.5$ mm. The *solid curve* is the theoretical result (Chen, Wang, Liu, et al., 2014). Intensity distribution: (A) $z = 250$ mm; (B) $z = 370$ mm; (C) $z = 400$ mm. Cross line: (D) $A = 250$ mm; (E) $z = 370$ mm; (F) $z = 400$ mm.

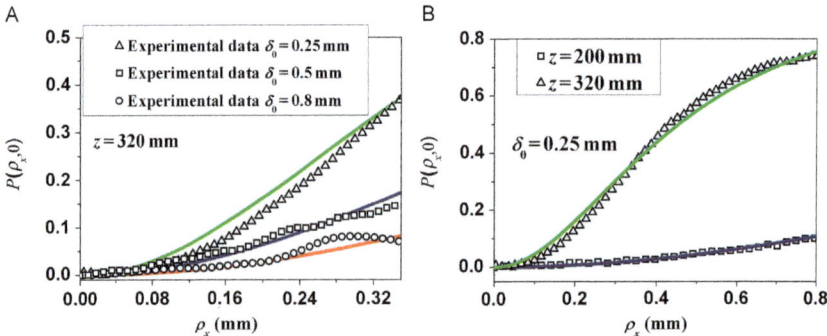

Fig. 32 Experimental results of the degree of polarization of the generated SCRP beam after passing a thin lens vs the transverse coordinate ρ_x ($\rho_y = 0$) for different values of the initial coherence width δ_0 and the propagation distance. The *solid curve* is the theoretical result (Chen, Wang, Liu, et al., 2014). (A) Different coherence widths. (B) Different propagation distances.

In Hyde, Basu, Xiao, et al. (2015) and Hyde, Basu, Voelz, et al. (2015), using a single 512×512 Boulder nonlinear systems model P512-0635 SLM, a new technique is developed to tailor the degree of coherence of a partially coherent Schell-model source, which can produce any desired mean far-field irradiance pattern. Fig. 33 shows the

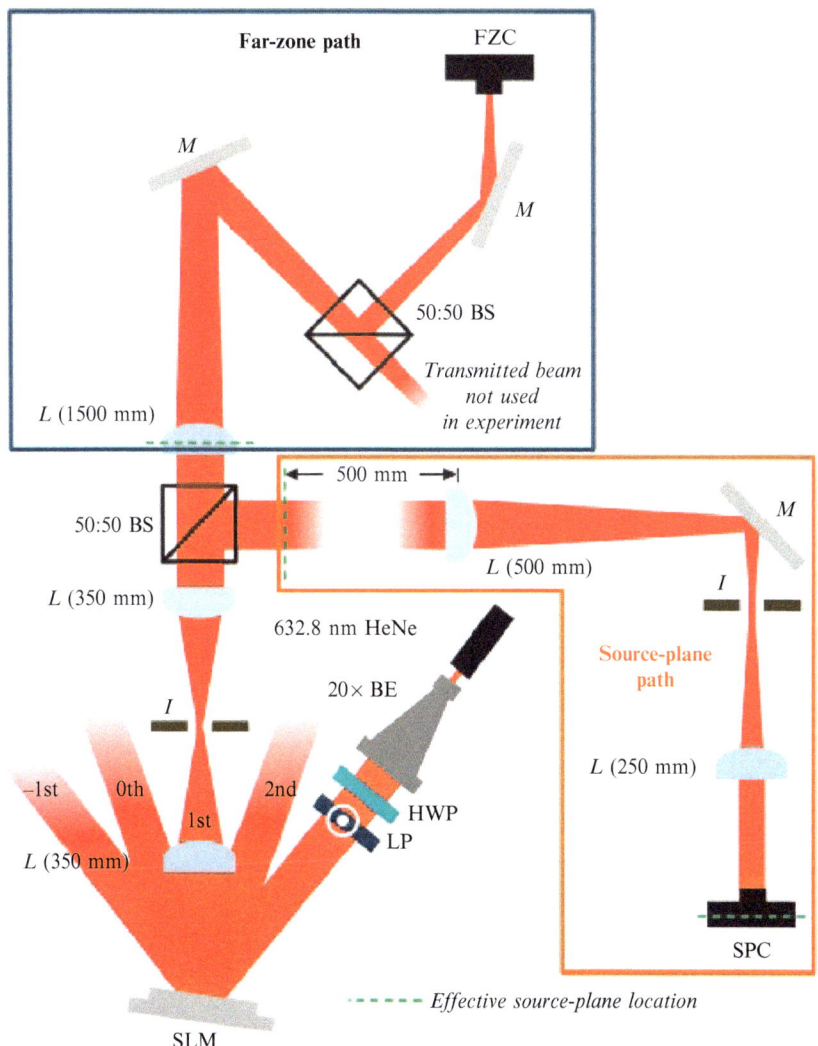

Fig. 33 Schematic of the experimental setup for producing any desired far-field mean irradiance pattern. *BE*, beam expander; *BS*, beam splitter; *FZC*, far-zone camera; *HWP*, half-wave plate; *I*, iris; *L*, lens; *LP*, linear polarizer; *M*, mirror; *SLM*, spatial light modulator; *SPC*, source-plane camera (Hyde, Basu, Voelz, et al., 2015).

schematic of the experimental setup. In Avramov-Zamurovic, Nelson, Guth, and Korotkova (2016), an MGCSM beam was generated with the help of a nematic phase only, reflective SLM, and the scintillation properties of such a beam propagating through a weakly turbulent air channel over a distance of 70 m were explored experimentally, and the reduction in the scintillation index through controlling the coherence width and the summation index were demonstrated. In Lehtolahti et al. (2015), it was shown that the degree of coherence can be tailored by deterministic rotating diffusers (see Fig. 34), when a coherent Gaussian beam illuminates a designed diffuser, the diffraction pattern generated by the fabricated diffuser will display a complicated shape (see Fig. 35). In Chen, Ponomarenko, et al. (2016), optical coherence lattices were generated through the synthesis of multiple partially coherent Schell-model beams. The experimental setup and experimental results are shown in Figs. 36 and 37, and the most interesting finding is that information can be encoded into and, in principle, recovered from the lattice degree of coherence (see Fig. 38).

The works mentioned earlier were limited to manipulation of the degree of coherence of the light beam outside of the laser cavity. In Nixon et al. (2013), it was shown that one can tune the spatial coherence of a modified degenerate laser over a broad range with minimum variation in the total

Fig. 34 The schematic structure of the diffuser. The *dark* and *light gray areas* mean phase delays of zero and π radians (Lehtolahti et al., 2015).

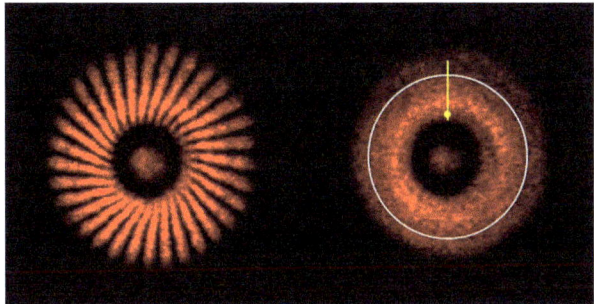

Fig. 35 Diffraction pattern generated by the fabricated diffuser while stationary (*left*) and rotating (*right*). The *vertical line* shows the measurement range for radial modulation with the reference point marked with a *dot*; the *circle* shows the measurement range for azimuthal modulation (Lehtolahti et al., 2015).

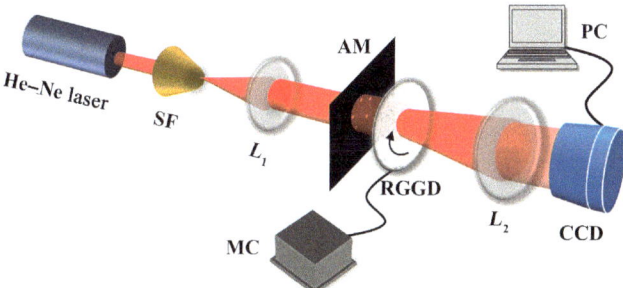

Fig. 36 Experimental setup for generation and measurement of coherence lattices in a partially coherent beam. *AM*, amplitude mask; *CCD*, charge-coupled device (Chen, Ponomarenko, et al., 2016); L_1 and L_2, thin lenses; *MC*, motion controller; *RGGD*, rotating ground glass disk; *SF*, spatial filter assembly.

output power, which is based on varying the diameter of a spatial filter (i.e., a pinhole aperture) inside the cavity. Later, in Chriki et al. (2015), it was demonstrated that global spatial coherence can be manipulated in a modified degenerate cavity laser, which was accomplished with intracavity binary amplitude masks which determine the functional form of the degree of coherence, in accordance to the Van Cittert–Zernike theorem. Fig. 39 shows the experimental arrangements for the modified degenerate cavity laser and for characterizing the spatial coherence properties. Fig. 40 shows the experimental images of the degree of coherence $|\mu|$. As shown in Fig. 40,

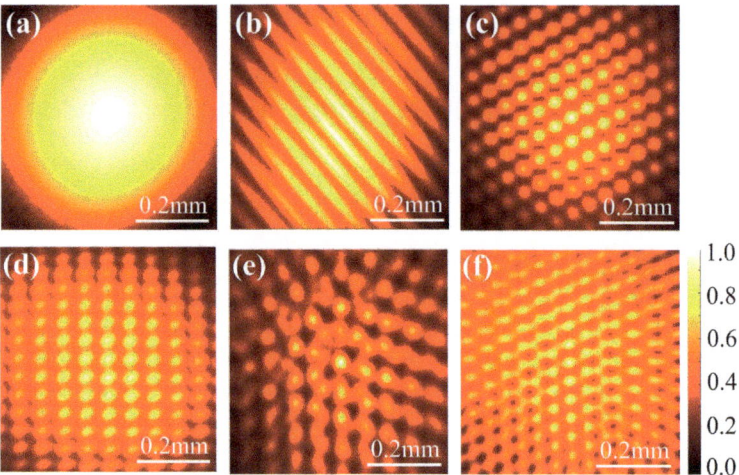

Fig. 37 Experimental results of the lattice degree of coherence magnitude distribution for different numbers of Schell-model beams. (A) $M=1$, (B) $M=2$, (C) $M=3$, (D) $M=4$, (E) $M=5$, and (F) $M=6$ (Chen, Ponomarenko, et al., 2016).

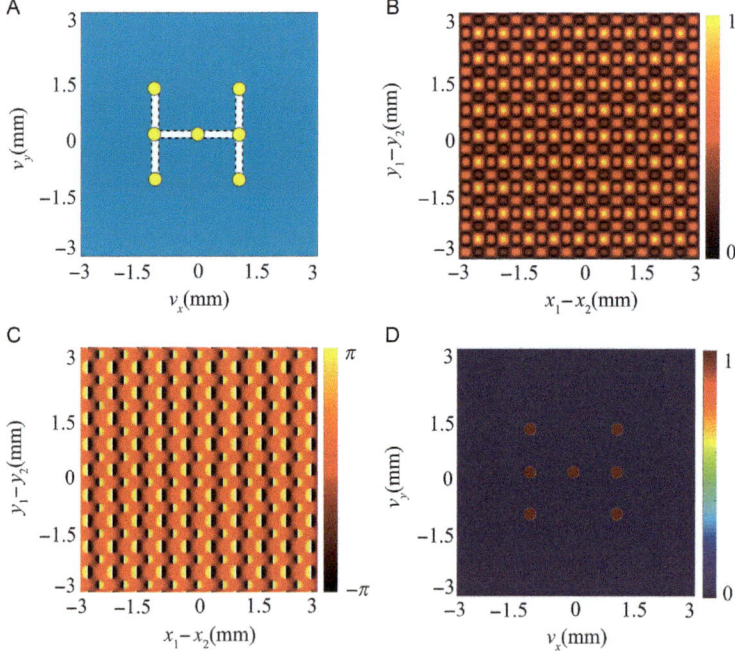

Fig. 38 Recovering information concealed in optical coherence lattices. (A) The image "H" is to be hidden; (B) the lattice degree of coherence magnitude distribution of the image; (C) the lattice degree of coherence phase distribution of the image; and (D) the recovered image (Chen, Ponomarenko, et al., 2016).

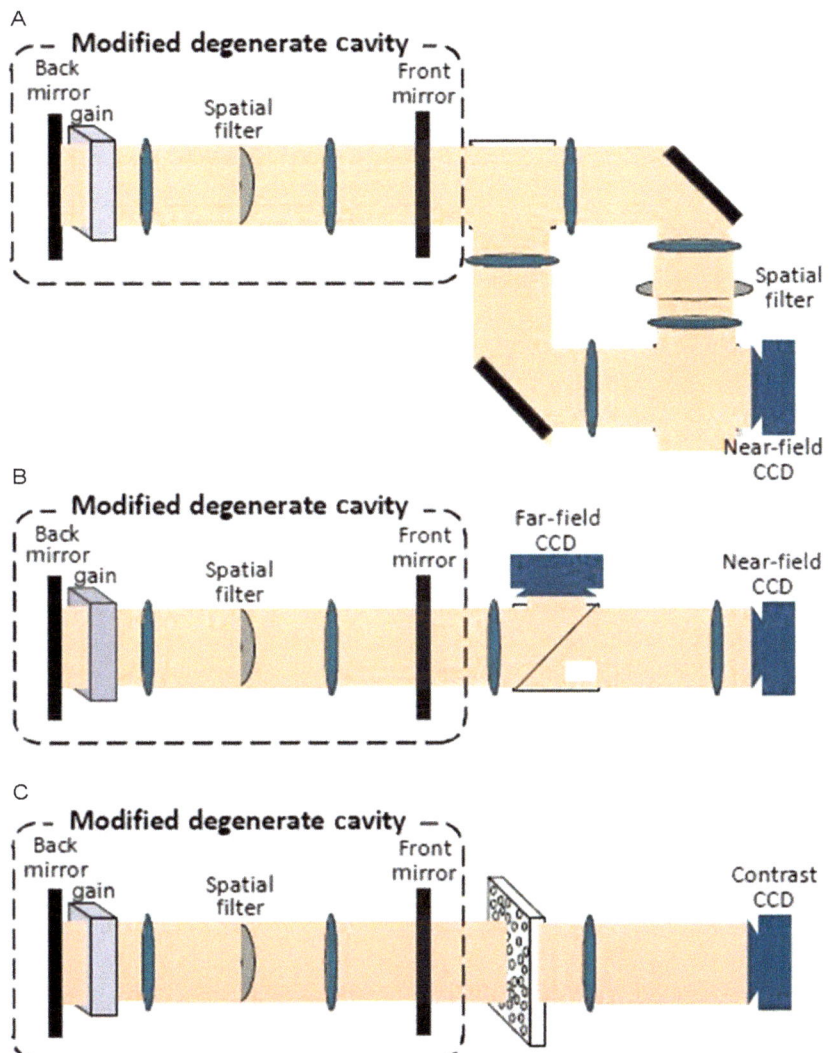

Fig. 39 Experimental arrangements for the modified degenerate cavity laser and for characterizing the spatial coherence properties. (A) The arrangement for measuring the absolute value of the complex coherence factor $|\mu|$. (B) The arrangement for measuring M^2. (C) The arrangement for measuring speckle contrast (Chriki et al., 2015).

a circular aperture, an annular aperture, a double aperture, and an array and circular apertures lead to a jinc coherence function, a zero-order Bessel coherence function, a cosine coherence function, and a comb coherence function, respectively.

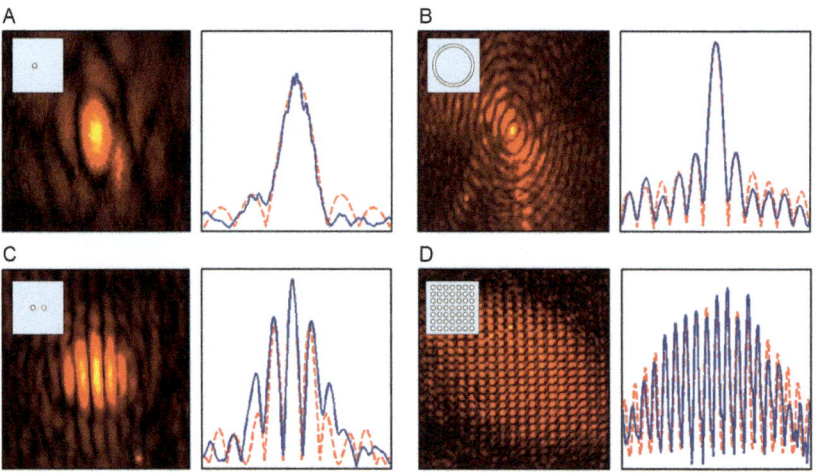

Fig. 40 Experimental images of the absolute value of the degree of coherence $|\mu|$ and their measured (continuous *blue line*) and calculated (*dashed red line*) cross sections for different geometries of the spatial filter. (A) A circular aperture resulting in a jinc coherence function; (B) an annular aperture resulting in a zero order Bessel coherence function; (C) a double aperture resulting in a cosine coherence function; and (D) an array of circular apertures resulting in a comb coherence function (Chriki et al., 2015).

3. SUMMARY

We have presented an overview of the developments of conventional GSM beams and partially coherent beams with prescribed phases, states of polarization, and degrees of coherence in Section 1. In Section 2, we began by introducing the fundamental theories for characterizing scalar and vector partially coherent beams. We then discussed the theoretical models for GSM beams and various partially coherent beams with prescribed phases (e.g., TGSM beams and partially coherent vortex beams), states of polarization (e.g., EGSM beams, radially polarized or azimuthally polarized partially coherent beams), and degrees of coherence (e.g., MGCSM beams, LGCSM beams and HGCSM beams), and the methods for their generation. Conventional partially coherent beams, such as GSM beams, have important applications in various fields, and we have shown that partially coherent beams with prescribed beam properties display many extraordinary propagation properties and have advantages over GSM beams in many applications.

The generation of partially coherent beams with prescribed phases, states of polarization, and degrees of coherence is rapidly developing. Only a few papers have been devoted to manipulation of the coherence of laser beams inside the cavity, and there are still many interesting problems for further research. For example, the design of cavities which generate partially coherent beams with prescribed properties, linear and nonlinear interactions of partially coherent beams with prescribed properties with matter, and novel applications of partially coherent beams with prescribed properties. We hope our review will stimulate further efforts in this area of research.

ACKNOWLEDGMENTS
This work was supported by the National Natural Science Fund for Distinguished Young Scholars (11525418), the National Science Foundation of China (11274005 and 11404234), the Project of the Priority Academic Program Development (PAPD) of Jiangsu Higher Education Institutions, and the Innovation Plan for Graduate Students in the Universities of Jiangsu Province (KYZZ16_0079 and KYLX15_1254).

REFERENCES
Akhmediev, N., Krolikowski, W., & Snyder, A. W. (1998). Partially coherent solitons of variable shape. *Physical Review Letters*, *81*, 4632.
Alkelly, A. A., Shukri, M., & Alarify, Y. S. (2012). Influence of twist phenomenon of partially coherent field with uniform-intensity diffractive axicons. *Journal of the Optical Society of America. A*, *29*, 417.
Allen, L., Beijersbergen, M. W., Spreeuw, R., & Woerdman, J. P. (1992). Orbital angular momentum of light and the transformation of Laguerre–Gaussian laser modes. *Physical Review. A*, *4*, 8185.
Ambrosini, D., Bagini, V., Gori, F., & Santarsiero, M. (1994). Twisted Gaussian Schell-model beams: A superposition model. *Journal of Modern Optics*, *41*, 1391.
Ansari, N. A., & Zubairy, M. S. (1986). Second-harmonic generation by a Gaussian Schell-model source. *Optics Communications*, *59*, 385.
Apostol, A., & Dogariu, A. (2003). Spatial correlations in the near field of random media. *Physical Review Letters*, *91*, 093901.
Avramov-Zamurovic, S., Nelson, C., Guth, S., & Korotkova, O. (2016). Flatness parameter influence on scintillation reduction for multi-Gaussian Schell-model beam propagating in turbulent air. *Applied Optics*, *55*, 3442.
Avramov-Zamurovic, S., Nelson, C., Guth, S., Korotkova, O., & Malek-Madani, R. (2016). Experimental study of electromagnetic Bessel–Gaussian Schell model beams propagating in a turbulent channel. *Optics Communications*, *359*, 207.
Avramov-Zamurovic, S., Nelson, C., Malek-Madani, R., & Korotkova, O. (2014). Polarization-induced reduction in scintillation of optical beams propagating in simulated turbulent atmospheric channels. *Waves in Random and Complex Media*, *24*, 452.
Bastiaans, M. J. (2000). Wigner distribution function applied to twisted Gaussian light propagating in first-order optical systems. *Journal of the Optical Society of America. A*, *17*, 2475.
Basu, S., Hyde, M. W., IV, Xiao, X., Voelz, D. G., & Korotkova, O. (2014). Computational approaches for generating electromagnetic Gaussian Schell-model sources. *Optics Express*, *22*, 31691.
Berman, G. P., Chumak, A. A., & Gorshkov, V. N. (2007). Beam wandering in the atmosphere: The effect of partially coherence. *Physical Review. E*, *76*, 056606.

Boggatyryova, V. G., Felde, V. C., Polyanskii, P. V., Ponomarenko, S. A., Soskin, M. S., & Wolf, E. (2003). Partially coherent vortex beams with a separable phase. *Optics Letters*, *28*, 878.

Borghi, R., Gori, F., Guattari, G., & Santarsiero, M. (2015a). Shape-invariant difference between two Gaussian Schell-model beams. *Journal of the Optical Society of America. A*, *32*, 790.

Borghi, R., Gori, F., Guattari, G., & Santarsiero, M. (2015b). Twisted Schell-model beams with axial symmetry. *Optics Letters*, *40*, 4504.

Brown, D. P., & Brown, T. G. (2008). Partially correlated azimuthal vortex illumination: Coherence and correlation measurements and effects in imaging. *Optics Express*, *16*, 20418.

Cai, Y., Chen, Y., & Wang, F. (2014). Generation and propagation of partially coherent beams with nonconventional correlation functions: A review [Invited]. *Journal of the Optical Society of America. A*, *31*, 2083.

Cai, Y., & He, S. (2006). Propagation of a partially coherent twisted anisotropic Gaussian Schell-model beam in a turbulent atmosphere. *Applied Physics Letters*, *89*, 041117.

Cai, Y., & Hu, L. (2006). Propagation of partially coherent twisted anisotropic Gaussian Schell-model beams through an apertured astigmatic optical system. *Optics Letters*, *31*, 685.

Cai, Y., & Korotkova, O. (2009). Twist phase-induced polarization changes in electromagnetic Gaussian Schell-model beams. *Applied Physics B*, *96*, 499.

Cai, Y., Korotkova, O., Eyyuboğlu, H. T., & Baykal, Y. (2008). Active laser radar systems with stochastic electromagnetic beams in turbulent atmosphere. *Optics Express*, *16*, 15834.

Cai, Y., Lin, Q., & Ge, D. (2002). Propagation of partially coherent twisted anisotropic Gaussian Schell-model beams in dispersive and absorbing media. *Journal of the Optical Society of America. A*, *19*, 2036.

Cai, Y., Lin, Q., & Korotkova, O. (2009). Ghost imaging with twisted Gaussian Schell-model beam. *Optics Express*, *17*, 2450.

Cai, Y., & Peschel, U. (2007). Second-harmonic generation by an astigmatic partially coherent beam. *Optics Express*, *15*, 15480.

Cai, Y., Wang, F., Zhao, C., Zhu, S., Wu, G., & Dong, Y. (2013). Partially coherent vector beams: From theory to experiment. In Q. Zhan (Ed.), *Vectorial optical fields: Fundamentals and applications* (pp. 221–273). Singapore: World Scientific. Chapter 7.

Cai, Y., & Zhu, S. (2004). Ghost interference with partially coherent radiation. *Optics Letters*, *29*, 2716.

Cai, Y., & Zhu, S. (2005). Ghost imaging with incoherent and partially coherent light radiation. *Physical Review. E*, *71*, 056607.

Cai, Y., & Zhu, S. (2014). Orbital angular moment of a partially coherent beam propagating through an astigmatic ABCD optical system with loss or gain. *Optics Letters*, *39*, 1968.

Chen, Y., & Cai, Y. (2014). Generation of a controllable optical cage by focusing a Laguerre–Gaussian correlated Schell-model beam. *Optics Letters*, *39*, 2549.

Chen, Y., & Cai, Y. (2016). Correlation-induced self-focusing and self-shaping effect of a partially coherent beam. *High Power Laser Science and Engineering*, *4*, e20.

Chen, Y., Gu, J., Wang, F., & Cai, Y. (2015). Self-splitting properties of a Hermite-Gaussian correlated Schell-model beam. *Physical Review. A*, *91*, 013823.

Chen, Y., Liu, L., & Cai, Y. (2016). Research progress in phase modulation of partially coherent beam and its fundamental applications. *Chinese Science Bulletin*, *61*, 1952.

Chen, Y., Liu, L., Wang, F., Zhao, C., & Cai, Y. (2014). Elliptical Laguerre–Gaussian correlated Schell-model beam. *Optics Express*, *22*, 13975.

Chen, J., Liu, X., Yu, J., & Cai, Y. (2016). Simultaneous determination of the sign and the magnitude of the topological charge of a partially coherent vortex beam. *Applied Physics B*, *122*, 201.

Chen, Y., Ponomarenko, S. A., & Cai, Y. (2016). Experimental generation of optical coherence lattices. *Applied Physics Letters*, *109*, 061107.

Chen, Y., Wang, F., Liu, L., Zhao, C., Cai, Y., & Korotkova, O. (2014a). Generation and propagation of a partially coherent vector beam with special correlation functions. *Physical Review. A*, *89*, 013801.

Chen, Y., Wang, F., Yu, J., Liu, L., & Cai, Y. (2016). Vector Hermite-Gaussian correlated Schell-model beam. *Optics Express*, *24*, 15232.

Chen, Y., Wang, F., Zhao, C., & Cai, Y. (2014b). Experimental demonstration of a Laguerre–Gaussian correlated Schell-model vortex beam. *Optics Express*, *22*, 5826.

Chen, Y., Yu, J., Yuan, Y., Wang, F., & Cai, Y. (2016). Theoretical and experimental studies of a rectangular Laguerre–Gaussian correlated Schell-model beam. *Applied Physics B*, *122*, 31.

Chriki, R., Nixon, M., Pal, V., Tradonsky, C., Barach, G., Friesem, A. A., et al. (2015). Manipulating the spatial coherence of a laser source. *Optics Express*, *23*, 12989.

Collett, E., & Wolf, E. (1978). Is complete spatial coherence necessary for the generation of highly directional light beams? *Optics Letters*, *2*, 27.

De Santis, P., Gori, F., Guattari, G., & Palma, C. (1979). An example of a Collett–Wolf source. *Optics Communications*, *29*, 256.

Ding, C., Cai, Y., Zhang, Y., Wang, H., Zhao, Z., & Pan, L. (2013). Stochastic electromagnetic plane-wave pulse with non-uniform correlation distribution. *Physics Letters A*, *377*, 1563.

Dogariu, A., & Amarande, S. (2003). Propagation of partially coherent beams: Turbulence-induced degradation. *Optics Letters*, *28*, 10.

Dong, Y., Cai, Y., Zhao, C., & Yao, M. (2011). Statistics properties of a cylindrical vector partially coherent beam. *Optics Express*, *19*, 5979.

Dong, Y., Feng, F., Chen, Y., Zhao, C., & Cai, Y. (2012). Statistical properties of a nonparaxial cylindrical vector partially coherent field in free space. *Optics Express*, *20*, 15908.

Dong, Y., Wang, F., Zhao, C., & Cai, Y. (2012). Effect of spatial coherence on propagation, tight focusing and radiation forces of an azimuthally polarized beam. *Physical Review A*, *86*, 013840.

Duan, K., & Lü, B. (2004). Partially coherent nonparaxial beams. *Optics Letters*, *29*, 800.

Escalante, A. Y., Perez-Garcia, B., Hernandez-Aranda, R. I., & Swartzlander, G. A. (2013). Determination of angular momentum content in partially coherent beams through cross correlation measurements. *Proceedings of SPIE*, *8843*, 884302.

Foley, J. T., & Zubairy, M. S. (1976). The directionality of Gaussian Schell-model beams. *Optics Communications*, *26*, 297.

Foreman, M. R., & Török, P. (2009). Focusing of spatially inhomogeneous partially coherent, partially polarized electromagnetic fields. *Journal of the Optical Society of America. A*, *26*, 2470.

Friberg, A. T., Tervonen, E., & Turunen, J. (1994a). Interpretation and experimental demonstration of twisted Gaussian Schell-model beams. *Journal of the Optical Society of America. A*, *11*, 1818.

Friberg, A. T., Tervonen, E., & Turunen, J. (1994b). Focusing of twisted Gaussian Schell-model beams. *Optics Communications*, *106*, 127.

Friberg, A. T., & Visser, T. D. (2015). Scintillation of electromagnetic beams generated by quasi-homogeneous sources. *Optics Communications*, *335*, 82.

Friberg, A. T., Visser, T. D., Wang, W., & Wolf, E. (2001). Focal shifts of converging diffracted waves of any state of spatial coherence. *Optics Communications*, *196*, 1.

Gan, C. H., Gbur, G., & Visser, T. D. (2007). Surface plasmons modulate the spatial coherence of light in Young's interference experiment. *Physical Review Letters*, *98*, 043908.

Gbur, G., & Swartzlander, G. A. (2008). Complete transverse representation of a correlation singularity of a partially coherent field. *Journal of the Optical Society of America. B*, *25*, 1422.

Gbur, G., & Visser, T. D. (2003). Can spatial coherence effects produce a local minimum of intensity at focus? *Optics Letters, 28*, 1627.

Gbur, G., & Wolf, E. (2002). Spreading of partially coherent beams in random media. *Journal of the Optical Society of America. A, 19*, 1592.

Gori, F. (1998). Matrix treatment for partially polarized, partially coherent beams. *Optics Letters, 23*, 241.

Gori, F., & Guattari, G. (1987). Modal expansion for J_0-correlated Schell-model sources. *Optics Communications, 64*, 311.

Gori, F., Sanchez, V. R., Santarsiero, M., & Shirai, T. (2009). On genuine cross-spectral density matrices. *Journal of Optics A: Pure and Applied Optics, 11*, 085706.

Gori, F., & Santarsiero, M. (2007). Devising genuine spatial correlation functions. *Optics Letters, 32*, 3531.

Gori, F., & Santarsiero, M. (2015). Twisted Gaussian Schell-model beams as series of partially coherent modified Bessel–Gauss beams. *Optics Letters, 40*, 1587.

Gori, F., Santarsiero, M., & Borghi, R. (1998). Partially coherent sources with helicoidal modes. *Journal of Modern Optics, 45*, 539.

Gori, F., Santarsiero, M., & Borghi, R. (2008). Modal expansion for J_0-correlated electromagnetic sources. *Optics Letters, 33*, 1857.

Gori, F., Santarsiero, M., Borghi, R., & Ramírez-Sánchez, V. (2008). Realizability condition for electromagnetic Schell-model sources. *Journal of the Optical Society of America. A, 25*, 1016.

Gori, F., Santarsiero, M., Piquero, G., Borghi, R., Mondello, A., & Simon, R. (2001). Partially polarized Gaussian Schell-model beams. *Journal of Optics A: Pure and Applied Optics, 3*, 1.

Grier, D. G. (2003). A revolution in optical manipulation. *Nature, 424*, 810.

Gu, Y., & Gbur, G. (2010). Scintillation of pseudo-Bessel correlated beams in atmospheric turbulence. *Journal of the Optical Society of America. A, 27*, 2621.

Gu, Y., & Gbur, G. (2013). Scintillation of nonuniformly correlated beams in atmospheric turbulence. *Optics Letters, 38*, 1395.

Guo, L., Chen, Y., Liu, X., Liu, L., & Cai, Y. (2016a). Vortex phase-induced changes of the statistical properties of a partially coherent radially polarized beam. *Optics Express, 24*, 13714.

Guo, L., Chen, Y., Liu, L., Yao, M., & Cai, Y. (2016b). Correlation-induced changes of the degree of paraxiality of a partially coherent beam. *Journal of the Optical Society of America. A, 33*, 251.

Gureyev, T. E., Paganin, D. M., Stevenson, A. W., Mayo, S. C., & Wilkins, S. W. (2004). Generalized eikonal of partially coherent beams and its use in quantitative imaging. *Physical Review Letters, 93*, 068103.

He, Q. S., Turunen, J., & Friberg, A. T. (1988). Propagation and imaging experiments with Gaussian Schell-model beams. *Optics Communications, 67*, 245.

Hyde, M. W., IV, Basu, S., Voelz, D. G., & Xiao, X. (2015a). Experimentally generating any desired partially coherent Schell-model source using phase-only control. *Journal of Applied Physics, 118*, 093102.

Hyde, M. W., IV, Basu, S., Xiao, X., & Voelz, D. G. (2015b). Producing any desired far-field mean irradiance pattern using a partially coherent Schell-model source. *Journal of Optics, 17*, 055607.

James, D. F. V. (1994). Changes of polarization of light beams on propagation in free space. *Journal of the Optical Society of America. A, 11*, 1641.

Kanseri, B., Rath, S., & Kandpal, H. C. (2009). Direct determination of the generalized Stokes parameters from the usual Stokes parameters. *Optics Letters, 34*, 719.

Kato, Y., Mima, K., Miyanaga, N., Arinaga, S., Kitagawa, Y., Nakatsuka, M., et al. (1984). Random phasing of high-power lasers for uniform target acceleration and plasma-instability suppression. *Physical Review Letters, 53*, 1057.

Kinzly, R. E. (1972). Partially coherent imaging in a microdensitometer. *Journal of the Optical Society of America*, *62*, 386.

Korotkova, O. (2008). Scintillation index of a stochastic electromagnetic beam propagating in random media. *Optics Communications*, *281*, 2342.

Korotkova, O. (2014). Random sources for rectangularly-shaped far fields. *Optics Letters*, *39*, 64.

Korotkova, O., Andrews, L. C., & Phillips, R. L. (2004a). Model for a partially coherent Gaussian beam in atmospheric turbulence with application in Lasercom. *Optical Engineering*, *43*, 330.

Korotkova, O., Andrews, L. C., & Phillips, R. L. (2004b). A lidar model for a rough-surface target: Method of partial coherence. *Proceedings of SPIE*, *5237*, 49.

Korotkova, O., Cai, Y., & Watson, E. (2009). Stochastic electromagnetic beams for LIDAR systems operating through turbulent atmosphere. *Applied Physics B*, *94*, 681.

Korotkova, O., Salem, M., Dogariu, A., & Wolf, E. (2005). Changes in the polarization ellipse of random electromagnetic beams propagating through the turbulent atmosphere. *Waves in Random and Complex Media*, *15*, 353.

Korotkova, O., Salem, M., & Wolf, E. (2004a). Beam conditions for radiation generated by an electromagnetic Gaussian Schell-model source. *Optics Letters*, *29*, 1173.

Korotkova, O., Salem, M., & Wolf, E. (2004b). The far-zone behavior of the degree of polarization of electromagnetic beams propagating through atmospheric turbulence. *Optics Communications*, *233*, 225.

Korotkova, O., & Shchepakina, E. (2014). Random sources for optical frames. *Optics Express*, *22*, 10622.

Korotkova, O., Visser, T. D., & Wolf, E. (2008). Polarization properties of stochastic electromagnetic beams. *Optics Communications*, *281*, 515.

Korotkova, O., & Wolf, E. (2005a). Generalized Stokes parameters of random electromagnetic beams. *Optics Letters*, *30*, 198.

Korotkova, O., & Wolf, E. (2005b). Changes in the state of polarization of a random electromagnetic beam on propagation. *Optics Communications*, *246*, 35.

Lajunen, H., & Saastamoinen, T. (2011). Propagation characteristics of partially coherent beams with spatially varying correlations. *Optics Letters*, *36*, 4104.

Lajunen, H., & Saastamoinen, T. (2013). Non-uniformly correlated partially coherent pulses. *Optics Letters*, *21*, 190.

Lavery, M. P. J., Speirits, F. C., Barnett, S. M., & Padgett, M. J. (2013). Detection of a spinning object using light's orbital angular momentum. *Science*, *341*, 537.

Lehtolahti, J., Kuittinen, M., Turunen, J., & Tervo, J. (2015). Coherence modulation by deterministic rotating diffusers. *Optics Express*, *23*, 10453.

Liang, C., Wang, F., Liu, X., Cai, Y., & Korotkova, O. (2014). Experimental generation of cosine Gaussian-correlated Schell-model beams with rectangular symmetry. *Optics Letters*, *39*, 769.

Lin, Q., & Cai, Y. (2002a). Tensor ABCD law for partially coherent twisted anisotropic Gaussian-Schell model beams. *Optics Letters*, *27*, 216.

Lin, Q., & Cai, Y. (2002b). Fractional Fourier transform for partially coherent Gaussian Schell-model beams. *Optics Letters*, *27*, 1672.

Liu, L., Chen, Y., Guo, L., & Cai, Y. (2015). Twist phase-induced changes of the statistical properties of a stochastic electromagnetic beam propagating in a uniaxial crystal. *Optics Express*, *23*, 12454.

Liu, L., Huang, Y., Chen, Y., Guo, L., & Cai, Y. (2015). Orbital angular moment of an electromagnetic Gaussian Schell-model beam with a twist phase. *Optics Express*, *23*, 30283.

Liu, X., Shen, Y., Liu, L., Wang, F., & Cai, Y. (2013). Experimental demonstration of vortex phase-induced reduction in scintillation of a partially coherent beam. *Optics Letters*, *38*, 5323.

Liu, R., Wang, F., Chen, D., Wang, Y., Zhou, Y., Gao, H., et al. (2016). Measuring mode indices of a partially coherent vortex beam with Hanbury Brown and Twiss type experiment. *Applied Physics Letters, 108*, 051107.

Liu, X., Wang, F., Liu, L., Zhao, C., & Cai, Y. (2015a). Generation and propagation of an electromagnetic Gaussian Schell-model vortex beam. *Journal of the Optical Society of America. A, 32*, 2058.

Liu, X., Wang, F., Wei, C., & Cai, Y. (2014). Experimental study of turbulence-induced beam wander and deformation of a partially coherent beam. *Optics Letters, 39*, 3336.

Liu, X., Wang, F., Zhang, M., & Cai, Y. (2015b). Experimental demonstration of ghost imaging with an electromagnetic Gaussian Schell-model beam. *Journal of the Optical Society of America. A, 32*, 910.

Liu, X., Yu, J., Cai, Y., & Ponomarenko, S. A. (2016). Propagation of optical coherence lattices in the turbulent atmosphere. *Optics Letters, 41*, 4182.

Liu, X., & Zhao, D. (2015). Trapping two types of particles with a focused generalized multi-Gaussian Schell model beam. *Optics Communications, 354*, 250.

Ma, X., & Arce, G. R. (2008). PSM design for inverse lithography with partially coherent illumination. *Optics Express, 16*, 20126.

Ma, L., & Ponomarenko, S. A. (2014). Optical coherence gratings and lattices. *Optics Letters, 39*, 6656.

Ma, L., & Ponomarenko, S. A. (2015). Free-space propagation of optical coherence lattices and periodicity reciprocity. *Optics Express, 23*, 1848.

Macías-Romero, C., Lim, R., Foreman, M. R., & Török, P. (2011). Synthesis of structured partially spatially coherent beams. *Optics Letters, 36*, 1638.

Maleev, I. D., Palacios, D. M., Marathay, A. S., & Swartzlander, G. A. (2004). Spatial correlation vortices in partially coherent light: Theory. *Journal of the Optical Society of America. B, 21*, 1895.

Maleev, I. D., & Swartzlander, G. A. (2008). Propagation of spatial correlation vortices. *Journal of the Optical Society of America. B, 25*, 915.

Mandel, L., & Wolf, E. (1995). *Optical coherence and quantum optics.* Cambridge: Cambridge University Press.

Martínez-Herrero, R., & Mejías, P. M. (2009). Elementary-field expansions of genuine cross-spectral density matrices. *Optics Letters, 34*, 2303.

Martínez-Herrero, R., Mejías, P. M., & Gori, F. (2009). Genuine cross-spectral densities and pseudo-modal expansions. *Optics Letters, 34*, 1399.

Mei, Z., & Korotkova, O. (2013a). Random sources generating ring-shaped beams. *Optics Letters, 38*, 91.

Mei, Z., & Korotkova, O. (2013b). Cosine-Gaussian Schell-model sources. *Optics Letters, 38*, 2578.

Merano, M., Umbriaco, G., & Mistura, G. (2012). Observation of nonspecular effects for Gaussian Schell-model light beams. *Physical Review. A, 86*, 033842.

Nagali, E., Sciarrino, F., De Martini, F., Marrucci, L., Piccirillo, B., Karimi, E., et al. (2009). Quantum information transfer from spin to orbital angular momentum of photons. *Physical Review Letters, 103*, 013601.

Nixon, M., Redding, B., Friesem, A. A., Cao, H., & Davidson, N. (2013). Efficient method for controlling the spatial coherence of a laser. *Optics Letters, 38*, 3858.

Nyyssonen, D. (1977). Linewidth measurement with an optical microscope: The effect of operating conditions on the image profile. *Applied Optics, 16*, 2223.

Oh, J. E., Cho, Y. W., Scarcelli, G., & Kim, Y. H. (2013). Sub-Rayleigh imaging via speckle illumination. *Optics Letters, 38*, 682.

Ostrovsky, A. S., Rodríguez-Zurita, G., Meneses-Fabián, C., Olvera-Santamaría, M. A., & Rickenstorff-Parrao, C. (2010). Experimental generating the partially coherent and partially polarized electromagnetic source. *Optics Express, 18*, 12864.

Paganin, D., & Nugent, K. A. (1998). Noninterferometric phase imaging with partially coherent light. *Physical Review Letters, 80*, 2586.

Palacios, D., Maleev, I., Marathay, A., & Swartzlander, G. A. (2004). Spatial correlation singularity of a vortex field. *Physical Review Letters, 92*, 143905.

Palma, C., Borghi, R., & Cincotti, G. (1996). Beams originated by J_0-correlated Schell-model planar sources. *Optics Communications, 125*, 113.

Perez-Garcia, B., Yepiz, A., Hernandez-Aranda, R. I., Forbes, A., & Swartzlander, G. A. (2016). Digital generation of partially coherent vortex beams. *Optics Letters, 41*, 3471.

Piquero, G., Borghi, R., & Santarsiero, M. (2001). Gaussian Schell-model beams propagating through polarizating gratings. *Journal of the Optical Society of America. A, 18*, 1399.

Piquero, G., Gori, F., Romanini, P., Santarsiero, M., Borghi, R., & Mondello, A. (2002). Synthesis of partially polarized Gaussian Schell-model sources. *Optics Communications, 208*, 9.

Ponomarenko, S. A. (2001a). Twisted Gaussian Schell-model solitons. *Physical Review E, 64*, 036618.

Ponomarenko, S. A. (2001b). A class of partially coherent beams carrying optical vortices. *Journal of the Optical Society of America. A, 18*, 150.

Raghunathan, S. B., Schouten, H. F., & Visser, T. D. (2012a). Correlation singularities in partially coherent electromagnetic beams. *Optics Letters, 37*, 4179.

Raghunathan, S. B., Schouten, H. F., & Visser, T. D. (2012b). Topological reactions of correlation functions in partially coherent electromagnetic beams. *Journal of the Optical Society of America. A, 30*, 582.

Raghunathan, S. B., van Dijk, T., Peterman, E. J. G., & Visser, T. D. (2010). Experimental demonstration of an intensity minimum at the focus of a laser beam created by spatial coherence: Application to the optical trapping of dielectric particles. *Optics Letters, 35*, 4166.

Ricklin, J. C., & Davidson, F. M. (2002). Atmospheric turbulence effects on a partially coherent Gaussian beam: Implications for free-space laser communication. *Journal of the Optical Society of America. A, 19*, 1794.

Robb, G. R. M., & Firth, W. J. (2007). Collective atomic recoil lasing with a partially coherent pump. *Physical Review Letters, 99*, 253601.

Roychowdhury, H., Agarwal, G. P., & Wolf, E. (2006). Changes in the spectrum, in the spectral degree of polarization, and in the spectral degree of coherence of a partially coherent beam propagating through a gradient-index fiber. *Journal of the Optical Society of America. A, 23*, 940.

Roychowdhury, H., & Korotkova, O. (2005). Realizability conditions for electromagnetic Gaussian Schell-model sources. *Optics Communications, 249*, 379.

Roychowdhury, H., Ponomarenko, S. A., & Wolf, E. (2005). Change in the polarization of partially coherent electromagnetic beams propagating through the turbulent atmosphere. *Journal of Modern Optics, 52*, 1611.

Saastamoinen, T., Turunen, J., Tervo, J., Setala, T., & Friberg, A. T. (2005). Electromagnetic coherence theory of laser resonator modes. *Journal of the Optical Society of America. A, 22*, 103.

Sahin, S., & Korotkova, O. (2012). Light sources generating far fields with tunable flat profiles. *Optics Letters, 37*, 2970.

Sahin, S., Tong, Z., & Korotkova, O. (2010). Sensing of semi-rough targets embedded in atmospheric turbulence by means of stochastic electromagnetic beams. *Optics Communications, 283*, 4512.

Salem, M., & Agrawal, G. P. (2009). Effects of coherence and polarization on the coupling of stochastic electromagnetic beams into optical fibers. *Journal of the Optical Society of America. A, 26*, 2452.

Serna, J., & Movilla, J. M. (2001). Orbital angular momentum of partially coherent beams. *Optics Letters, 26*, 405.

Shirai, T., Dogariu, A., & Wolf, E. (2003). Mode analysis of spreading of partially coherent beams propagating through atmospheric turbulence. *Journal of the Optical Society of America. A, 20*, 1094.

Shirai, T., Kellock, H., Setala, T., & Friberg, A. T. (2011). Visibility in ghost imaging with classical partially polarized electromagnetic beams. *Optics Letters, 36*, 2880.

Shirai, T., Korotkova, O., & Wolf, E. (2005). A method of generating electromagnetic Gaussian Schell-model beams. *Journal of Optics A: Pure and Applied Optics, 7*, 232.

Shukri, M. A., Alkelly, A. A., & Alarify, Y. S. (2012). Spatial correlation properties of twisted partially coherent light focused by diffractive axicons. *Journal of the Optical Society of America. A, 29*, 2019.

Shukri, M., Alkelly, A. A., & Alarify, Y. S. (2013). Apodized design of diffractive axicons for twisted partially coherent light. *Applied Optics, 52*, 1881.

Simon, R., & Mukunda, N. (1993). Twisted Gaussian Schell-model beams. *Journal of the Optical Society of America. A, 10*, 95.

Simon, R., Sundar, K., & Mukunda, N. (1993). Twisted Gaussian Schell-model beams. I. Symmetry structure and normal-mode spectrum. *Journal of the Optical Society of America. A, 10*, 2008–2016.

Som, S. C., Delisle, C., & Drouin, M. (1980). Holography in partially coherent-light. *Optics Communications, 32*, 370.

Starikov, S., & Wolf, E. (1982). Coherent-mode representation of Gaussian Schell-model sources and their radiation fields. *Journal of the Optical Society of America. A, 72*, 923.

Stoklasa, B., Motka, L., Rehacek, J., Hradil, Z., & Sánchez-Soto, L. L. (2014). Wavefront sensing reveals optical coherence. *Nature Communications, 5*, 3275.

Sundar, K., Simon, R., & Mukunda, N. (1993). Twisted Gaussian Schell-model beams. II. Spectrum analysis and propagation characteristics. *Journal of the Optical Society of America. A, 10*, 2017.

Tamburini, F., Anzolin, G., Umbriaco, G., Bianchini, A., & Barbieri, C. (2006). Overcoming the Rayleigh criterion limit with optical vortices. *Physical Review Letters, 97*, 163903.

Tervo, J., Setälä, T., & Friberg, A. T. (2003). Degree of coherence for electromagnetic fields. *Optics Express, 11*, 1137.

Tervonen, E., Friberg, A. T., & Turunen, J. (1992). Gaussian Schell-model beams generated with synthetic acousto-optic holograms. *Journal of the Optical Society of America. A, 9*, 796.

Thompson, B. J., & Wolf, E. (1957). Two-beam interference with partially coherent light. *Journal of the Optical Society of America, 47*, 895.

Tong, Z., Cai, Y., & Korotkova, O. (2010). Ghost imaging with electromagnetic stochastic beams. *Optics Communications, 283*, 3838.

Tong, Z., & Korotkova, O. (2010). Theory of weak scattering of stochastic electromagnetic fields from. *Physical Review. A, 82*, 033836.

Tong, Z., & Korotkova, O. (2012a). Beyond the classical Rayleigh limit with twisted light. *Optics Letters, 37*, 2595.

Tong, Z., & Korotkova, O. (2012b). Electromagnetic nonuniformly correlated beam. *Journal of the Optical Society of America. A, 29*, 2154.

Tong, Z., & Korotkova, O. (2012c). Non-uniformly correlated light beams in uniformly correlated media. *Optics Letters, 37*, 3240.

Valencia, A., Scarcelli, G., D'Angelo, M., & Shih, Y. (2005). Two-photon imaging with thermal light. *Physical Review Letters, 94*, 063601.

van Dijk, T., Fischer, D. G., Visser, T. D., & Wolf, E. (2010). Effects of spatial coherence on the angular distribution of radiant intensity generated by scattering on a sphere. *Physical Review Letters, 104*, 173902.

van Dijk, T., Gbur, G., & Visser, T. D. (2008). Shaping the focal intensity distribution using spatial coherence. *Journal of the Optical Society of America. A, 25*, 575.

van Dijk, T., & Visser, T. D. (2009). Evolution of singularities in a partially coherent vortex beam. *Journal of the Optical Society of America. A*, *26*, 741.

Vidal, I., Caetano, D. P., Fonseca, E. J. S., & Hickmann, J. M. (2009). Effects of pseudothermal light source's transverse size and coherence width in ghost-interference experiment. *Optics Letters*, *34*, 1450.

Vidal, I., Fonseca, E. J. S., & Hickmann, J. M. (2011). Light polarization control during free-space propagation using coherence. *Physical Review. A*, *84*, 033836.

Vinu, R. V., Sharma, M. K., Singh, R. K., & Senthilkumaran, P. (2014). Generation of spatial coherence comb using Dammann grating. *Optics Letters*, *39*, 2407.

Voelz, D., Xiao, X., & Korotkova, O. (2015). Numerical modeling of Schell-model beams with arbitrary far-field patterns. *Optics Letters*, *40*, 352.

Waller, L., Situ, G., & Fleischer, J. W. (2012). Phase-space measurement and coherence synthesis of optical beams. *Nature Photonics*, *6*, 474.

Wang, F., & Cai, Y. (2007). Experimental observation of fractional Fourier transform for a partially coherent optical beam with Gaussian statistics. *Journal of the Optical Society of America. A*, *24*, 1937.

Wang, F., & Cai, Y. (2008). Experimental generation of a partially coherent flat-topped beam. *Optics Letters*, *33*, 1795.

Wang, F., & Cai, Y. (2010). Second-order statistics of a twisted Gaussian Schell-model beam in turbulent atmosphere. *Optics Express*, *18*, 24661.

Wang, F., Cai, Y., Dong, Y., & Korotkova, O. (2012a). Experimental generation of a radially polarized beam with controllable spatial coherence. *Applied Physics Letters*, *100*, 051108.

Wang, F., Cai, Y., Eyyuboğlu, H. T., & Baykal, Y. (2012b). Twist phase-induced reduction in scintillation of a partially coherent beam in turbulent atmosphere. *Optics Letters*, *37*, 184.

Wang, F., Cai, Y., & He, S. (2006). Experimental observation of the coincidence fractional Fourier transform with a partially coherent beam. *Optics Express*, *14*, 6999.

Wang, F., Cai, Y., & Korotkova, O. (2009a). Experimental observation of focal shifts in focused partially coherent beams. *Optics Communications*, *282*, 3408.

Wang, F., Cai, Y., & Korotkova, O. (2009b). Partially coherent standard and elegant Laguerre–Gaussian beams of all orders. *Optics Express*, *17*, 22366.

Wang, F., & Korotkova, O. (2016). Random sources for beams with azimuthal intensity variation. *Optics Letters*, *41*, 516.

Wang, F., Liang, C., Yuan, Y., & Cai, Y. (2014). Generalized multi-Gaussian correlated Schell-model beam: From theory to experiment. *Optics Express*, *22*, 23456.

Wang, F., Liu, X., Liu, L., Yuan, Y., & Cai, Y. (2013a). Experimental study of the scintillation index of a radially polarized beam with controllable spatial coherence. *Applied Physics Letters*, *103*, 091102.

Wang, F., Liu, X., Yuan, Y., & Cai, Y. (2013b). Experimental generation of partially coherent beams with different complex degrees of coherence. *Optics Letters*, *38*, 1814.

Wang, S. C. H., & Plonus, M. A. (1979). Optical beam propagation for a partially coherent source in the turbulent atmosphere. *Journal of the Optical Society of America*, *69*, 1297.

Wang, F., Wu, G., Liu, X., Zhu, S., & Cai, Y. (2011). Experimental measurement of the beam parameters of an electromagnetic Gaussian Schell-model source. *Optics Letters*, *36*, 2722.

Wang, J., Yang, J., Fazal, I. M., Ahmed, N., Yan, Y., Huang, H., et al. (2012). Terabit free-space data transmission employing orbital angular momentum multiplexing. *Nature Photonics*, *6*, 488.

Wang, T., & Zhao, D. (2010). Scattering theory of stochastic electromagnetic light waves. *Optics Letters*, *35*, 2412.

Wang, F., Zhu, S., & Cai, Y. (2011). Experimental study of the focusing properties of a Gaussian Schell-model vortex beam. *Optics Letters*, *36*, 3281.

Wolf, E. (1954). Optics in terms of observable quantities. *Nuovo Cimento, 12*, 884.
Wolf, E. (1955). A macroscopic theory of interference and diffraction of light from finite sources. II. Fields with a spectral range of arbitrary width. *Proceedings of the Royal Society of London A, 230*, 246.
Wolf, E. (2003). Unified theory of coherence and polarization of random electromagnetic beams. *Physics Letters A, 312*, 263.
Wolf, E. (2006). Coherence and polarization properties of electromagnetic laser modes. *Optics Communications, 265*, 60.
Wolf, E. (2007). *Introduction to the theory of coherence and polarization of light*. Cambridge University Press.
Wolf, E., & Collett, E. (1978). Partially coherent sources which produce the same far-field intensity distribution as a laser. *Optics Communications, 25*, 293.
Wu, G. (2016). Propagation properties of a radially polarized partially coherent twisted beam in free space. *Journal of the Optical Society of America. A, 33*, 345.
Wu, G., & Cai, Y. (2011). Detection of a semi-rough target in turbulent atmosphere by a partially coherent beam. *Optics Letters, 36*, 1939.
Wu, G., Lou, Q., Zhou, J., Guo, H., Zhao, H., & Yuan, Z. (2008). Beam conditions for radiation generated by an electromagnetic J_0-correlated Schell-model source. *Optics Letters, 33*, 2677.
Wu, G., & Visser, T. D. (2014). Hanbury Brown–Twiss effect with partially coherent electromagnetic beams. *Optics Letters, 39*, 2561.
Wu, G., Wang, F., & Cai, Y. (2012). Coherence and polarization properties of a radially polarized beam with variable spatial coherence. *Optics Express, 20*, 28301.
Yang, Y., Chen, M., Mazilu, M., Mourka, A., Liu, Y., & Dholakia, K. (2013). Effect of the radial and azimuthal mode indices of a partially coherent vortex field upon a spatial correlation singularity. *New Journal of Physics, 15*, 113053.
Yang, Y., & Liu, Y. (2016). Measuring azimuthal and radial mode indices of a partially coherent vortex field. *Journal of Optics, 18*, 015604.
Yang, Y., Mazilu, M., & Dholakia, K. (2012). Measuring the orbital angular momentum of partially coherent optical vortices through singularities in their cross-spectral density functions. *Optics Letters, 37*, 4949.
Yao, M., Cai, Y., Eyyuboğlu, H. T., Baykal, Y., & Korotkova, O. (2008). Evolution of the degree of polarization of an electromagnetic Gaussian Schell-model beam in a Gaussian cavity. *Optics Letters, 33*, 2266.
Young, T. (1802). An account of some cases of the production of colours, not hitherto described. *Philosophical Transactions of the Royal Society of London, 92*, 387.
Yu, J., Chen, Y., Liu, L., Liu, X., & Cai, Y. (2015). Splitting and combining properties of an elegant Hermite-Gaussian correlated Schell-model beam in Kolmogorov and non-Kolmogorov turbulence. *Optics Express, 23*, 13467.
Yuan, Y., Liu, X., Wang, F., Chen, Y., Cai, Y., Qu, J., et al. (2013). Scintillation index of a multi-Gaussian Schell-model beam in turbulent atmosphere. *Optics Communications, 305*, 57.
Zahid, M., & Zubairy, M. S. (1989). Directionality of partially coherent Bessel–Gauss beams. *Optics Communications, 70*, 361.
Zernike, F. (1938). The concept of degree of coherence and its application to optical problems. *Physica, 5*, 785.
Zhan, Q. (2009). Cylindrical vector beams: From mathematical concepts to applications. *Advances in Optics and Photonics, 1*, 1.
Zhang, L., & Cai, Y. (2011). Statistical properties of a nonparaxial Gaussian Schell-model beam in a uniaxial crystal. *Optics Express, 19*, 13312.
Zhang, Y., & Cai, Y. (2014). Random source generating far field with elliptical flat-topped beam profile. *Journal of Optics, 16*, 075704.

Zhang, Y., Liu, L., Zhao, C., & Cai, Y. (2014). Multi-Gaussian Schell-model vortex beam. *Physics Letters A, 378*, 750.

Zhao, C., & Cai, Y. (2011). Trapping two types of particles using a focused partially coherent elegant Laguerre–Gaussian beam. *Optics Letters, 36*, 2251.

Zhao, C., Cai, Y., & Korotkova, O. (2009). Radiation force of scalar and electromagnetic twisted Gaussian Schell-model beams. *Optics Express, 17*, 21472.

Zhao, C., Cai, Y., Wang, F., Lu, X., & Wang, Y. (2008). Generation of a high-quality partially coherent dark hollow beam with a multimode fiber. *Optics Letters, 33*, 1389.

Zhao, C., Dong, Y., Wang, Y., Wang, F., Zhang, Y., & Cai, Y. (2012). Experimental generation of a partially coherent Laguerre–Gaussian beam. *Applied Physics B, 109*, 345.

Zhao, C., Dong, Y., Wu, G., Wang, F., Cai, Y., & Korotkova, O. (2012). Experimental demonstration of coupling of an electromagnetic Gaussian Schell-model beam into a single-mode optical fiber. *Applied Physics B, 108*, 891.

Zhao, C., Wang, F., Dong, Y., Han, Y., & Cai, Y. (2012). Effect of spatial coherence on determining the topological charge of a vortex beam. *Applied Physics Letters, 101*, 261104.

Zhu, S., & Cai, Y. (2010). M^2-factor of a stochastic electromagnetic beam in a Gaussian cavity. *Optics Express, 18*, 27567.

Zhu, S., Chen, Y., & Cai, Y. (2013). Experimental determination of the radius of curvature of an isotropic Gaussian Schell-model beam. *Journal of the Optical Society of America. A, 30*, 171.

Zhu, S., Chen, Y., Wang, J., Wang, H., Li, Z., & Cai, Y. (2015). Generation and propagation of a vector cosine-Gaussian correlated beam with radial polarization. *Optics Express, 23*, 33099.

Zhu, S., Liu, L., Chen, Y., & Cai, Y. (2013). State of polarization and propagation factor of a stochastic electromagnetic beam in a gradient-index fiber. *Journal of the Optical Society of America. A, 30*, 2306.

Zhu, S., Wang, F., Chen, Y., Li, Z., & Cai, Y. (2014). Statistical properties in Young's interference pattern formed with a radially polarized beam with controllable spatial coherence. *Optics Express, 22*, 28697.

Zhu, S., Zhu, X., Liu, L., Wang, F., & Cai, Y. (2013). Theoretical and experimental studies of the spectral changes of a polychromatic partially coherent radially polarized beam. *Optics Express, 21*, 27682.

Zhuang, S., & Yu, F. (1982). Apparent transfer function for partially coherent optical information processing. *Applied Physics B, 28*, 359.

CHAPTER FOUR

Optical Models and Symmetries

Kurt Bernardo Wolf
Instituto de Ciencias Físicas, Universidad Nacional Autónoma de México, Cuernavaca, Morelos, Mexico

Contents

1. Introduction — 226
2. The Euclidean Group — 228
3. The Fundamental Object of a Model — 230
4. The Geometric Model — 232
   4.1 Euclidean Group and Coset Parameters — 232
   4.2 Geometric and Dynamic Postulates — 234
   4.3 Hamiltonian Structure and Phase Space — 236
   4.4 Canonical and Optical Transformations — 237
   4.5 Refracting Surfaces and the Root Transformation — 239
5. The Wave and Helmholtz Models — 241
   5.1 Coset Parameters for the Wave Model — 242
   5.2 Euclidean Generators and Casimir Invariants — 243
   5.3 Hilbert Space for Helmholtz Wavefields — 244
6. Paraxial Models — 245
   6.1 Contraction of the Euclidean to the Heisenberg–Weyl Algebra and Group — 246
   6.2 The Heisenberg–Weyl Algebra and Group — 247
7. Linear Transformations of Phase Space — 249
   7.1 Geometric Model — 249
   7.2 Wave Model: Canonical Integral Transforms — 253
8. The Metaxial Regime — 258
   8.1 Aberrations in 2D Systems — 258
   8.2 Axis-Symmetric Aberrations in 3D Systems — 262
9. Discrete Optical Models — 266
   9.1 The Contraction of so(4) to iso(3) — 266
   9.2 The Plane Pixelated Screen — 267
   9.3 The Kravchuk Oscillator States — 268
   9.4 2D Screens and $U(2)_F$ Transformations — 271
   9.5 Square and Circular Pixelated Screens — 275
   9.6 Aberrations of 1D Finite Discrete Signals — 279

Acknowledgment 282
Appendix A. The SU(2) Wigner Function 282
   A.1 The Wigner Operator 283
   A.2 The SU(2) Wigner Matrix 285
   A.3 Closing Remarks 287
References 288

1. INTRODUCTION

It is not difficult to argue that optics offers a richer field for the recognition and use of symmetries than mechanics—classical or quantum. The harvest includes the geometric, wave, and finite (pixelated) models of optics; in turn, the first two encompass the global (4π), paraxial and metaxial (aberration) regimes, while the finite model, when the number and density of pixels increases without bound, limits to the continuous cases.

The *mother symmetry* of all these models is the Euclidean group and algebra of translations and rotations. This statement may appear disappointing at first sight because that is the symmetry of a homogeneous and isotropic vacuum. But quite on the contrary, as we shall show, this symmetry serves as a basis for the construction of phase space in geometric optics, as well as the Hilbert space formulation of wave optics. A *deformation* of this group to that of rotations produces the finite model of pixelated optical systems. Using various techniques of deformation and contraction we succinctly shift between them and their distinct regimes.

The theory of Lie algebras and groups was developed by mathematicians during the second half of the 19th century and applied to find and classify all possible crystallographic lattices in three dimensions—which are but discrete subgroups of the Euclidean group. Early researchers in quantum mechanics found that rotational bands in the spectra of atomic systems were naturally characterized by the underlying geometric symmetry, while systematic level degeneracies were due to hidden, higher symmetries. The second half of the 20th century became the heyday of Lie group theory as nuclear and elementary particle physics presented quantum-number patterns and conservation laws whose origin was understood to be due to symmetries of Hamiltonians that were themselves unknown; yet, they provided conservation laws and sum rules for the observed reaction rates.

The main reason for the statement in our first sentence is that mechanical Hamiltonians basically read $H = p^2/2m + V(q)$, separated into a fixed kinetic term of squared momentum p, and an in-principle arbitrary potential term, normally of position q alone. In optics on the other hand, we come to write and use evolution-generating Hamiltonians of various other forms, according to whether the regime is global, paraxial or metaxial, geometric, wave or finite; and a phase space with symplectic metric follows. Most generators are of *co*-variance, rather than *in*-variance transformations, and belong to finite-dimensional Lie algebras whose representation theory is well established. We shall basically work in $D = 3$ space dimensions, occasionally with the $D = 2$ case for clarity in some figures, but most developments are in principle valid for generic D dimensions.

Since the main preoccupation of early optical research was the faithful formation of images on a plane screen, the angles of incoming rays—their *momenta*—did not really matter, so they were not placed on the same level of interest as ray *positions*. The recent surge of literature on linear canonical transforms for the paraxial regime of geometric and wave optics has highlighted the necessity of treating both position and momentum as coordinates of a phase space where one-dimensional wavefields could be visually displayed on a plane through their Wigner function as a music sheet with time and frequency axes. As we shall show, the phase space and Wigner function constructs fit also some of the other optical models and regimes, albeit with different topologies, but based on purely group-theoretical premises and applicable to other Lie groups beside the Euclidean.

First of all, in Section 2 we introduce the Euclidean group and its structure. The symmetries of the basic objects of the geometric and wave models of light—a line and a plane in space—are presented in Section 3. The reader will appreciate that these are two among a number of other possibilities based on the symmetry of chosen fundamental "objects" within the same Euclidean mother group. The geometric model of light with a dynamical postulate builds a Hamiltonian system whose canonical and optical transformations are the subject of Section 4, while Section 5 builds a Hilbert space analogue for Helmholtz monochromatic wavefields.

The contraction of the Euclidean group along the evolution axis yields in Section 6 the paraxial models of geometric and wave optical models under the Heisenberg–Weyl group of translations in position and paraxial momentum. Linear transformations, obtained through a quadratic extension of this algebra in Section 7, lead to the symplectic group of canonical geometric and wave transformations, the former serving to introduce the unitary Fourier subgroup, and the latter realized by integral transforms.

Higher-order extensions, addressed in Section 8, take us to the metaxial regime where the classification and action of aberrations are set forth.

The Euclidean group of continuous optical models is actually the result of a contraction of a higher compact group: rotations in 4-space. There, the operators of position and momentum have finite spectra of equally spaced eigenvalues; in Section 9 we thus present discrete optical models on linear or rectangular pixelated screens and the unitary transformations under which no information is lost. Finally, when the number and density of pixels grows without bound, one recovers the continuous models based on the Euclidean group and canonical transformations. In Appendix A we review a Wigner function defined on the rotation group that allows us to plot discrete and finite signals on phase space, in particular of aberrated signals; this too contracts to the generally known Wigner function on paraxial phase space.

2. THE EUCLIDEAN GROUP

Consider a 3D (three-dimensional) space whose points are labeled by column vectors[1] $\vec{r} = (x, y, z)^\top \in \mathsf{R}^3$. The rigid transformations of this space are translations and rotations that we indicate, respectively, by 3-vectors $\vec{\tau} = (\tau_x, \tau_y, \tau_z)^\top \in \mathsf{T}_3 = \mathsf{R}^3$, and 3×3 real, special[2] orthogonal matrices $\mathbf{R}(\phi, \theta, \psi) \in \mathsf{SO}(3)$, $\mathbf{R}\,\mathbf{R}^\top = \mathbf{R}^\top \mathbf{R} = \mathbf{1}$, where ϕ, θ, and ψ are the Euler angles of rotation around the z-, x-, and z-axes.[3] The action of these transformations can be written as a 4×4 matrix in $3 + 1$ block-diagonal form,[4]

$$\mathbf{E}(\vec{\tau}, \mathbf{R}) \begin{pmatrix} \vec{r} \\ 1 \end{pmatrix} := \begin{pmatrix} \mathbf{R} & \vec{\tau} \\ 0 & 1 \end{pmatrix} \begin{pmatrix} \vec{r} \\ 1 \end{pmatrix} = \begin{pmatrix} \mathbf{R}\vec{r} + \vec{\tau} \\ 1 \end{pmatrix}. \qquad (1)$$

It is immediate to verify through this 4×4 matrix realization that the set of transformations (1) form a *group*, that is, they satisfy the four axioms:

composition: $\quad \mathbf{E}(\vec{\tau}_1, \mathbf{R}_1)\, \mathbf{E}(\vec{\tau}_2, \mathbf{R}_2) = \mathbf{E}(\vec{\tau}_1 + \mathbf{R}_1 \vec{\tau}_2, \mathbf{R}_1 \mathbf{R}_2), \qquad (2)$

identity: $\quad \mathbf{E}(\vec{0}, \mathbf{1}) = \mathbf{1} \quad (4 \times 4 \text{ unit}), \qquad (3)$

---

[1] We indicate by $^\top$ the transpose of an array.
[2] That is, of unit determinant; thus, reflections across one coordinate, or 3D inversions, are not included in this group.
[3] A more common (if older) parametrization rotates around the z-, y-, and z-axes; ours has the advantage of generalizing easily to D dimensions by rotating successively in the 1-2, 2-3, 3-4, ..., $(D-1)$-D planes.
[4] We use the notation $A := B$ when the symbol A is defined by the expression B.

$$\text{inverse}: \quad \mathbf{E}(\vec{\tau},\mathbf{R})^{-1} = \mathbf{E}(-\mathbf{R}^{-1}\vec{\tau},\mathbf{R}^{-1}), \tag{4}$$

$$\begin{aligned}\text{associativity}: \quad & \mathbf{E}(\vec{\tau}_1,\mathbf{R}_1)(\mathbf{E}(\vec{\tau}_2,\mathbf{R}_2)\mathbf{E}(\vec{\tau}_3,\mathbf{R}_3)) \\ & = (\mathbf{E}(\vec{\tau}_1,\mathbf{R}_1)\mathbf{E}(\vec{\tau}_2,\mathbf{R}_2))\mathbf{E}(\vec{\tau}_3,\mathbf{R}_3). \end{aligned} \tag{5}$$

This is the group of *inhomogeneous special orthogonal* transformations, denoted ISO(3) or, in common parlance, the *Euclidean* group E_3 in three dimensions. Its elements can be factored into translations and rotations, as

$$\mathbf{E}(\vec{\tau},\mathbf{R}) = \mathbf{E}(\vec{\tau},\mathbf{1})\mathbf{E}(\vec{0},\mathbf{R}). \tag{6}$$

The *manifold* of the Euclidean group has six coordinates: three for translations $\vec{\tau} \in \mathsf{R}^3$ and three for rotations through Euler angles $(\psi, \theta, \phi) \in \mathsf{S}^3$, where S^3 is the 3D sphere (in a 4D ambient space). Eqs. (2) and (6) also indicate that, while the group contains the two *subgroups* of translations and of rotations, they are *not* on the same footing since rotations act on translations but not vice versa. The structure of the Euclidean group is that of a *semidirect* product (Gilmore, 1978; Sudarshan & Mukunda, 1974; Wybourne, 1974) indicated

$$\mathsf{E}_3 := \mathsf{ISO}(3) = \mathsf{T}_3 \circledS \mathsf{SO}(3). \tag{7}$$

In such a composition the left factor (translations) is called the *invariant* subgroup, while the right factor (rotations) is the *factor* subgroup.

In (2) we see that the composition functions for the group parameters of a product of elements are *analytic* functions of the parameters of the factors. Thus E_3 is a *Lie* group, whose (local) structure is determined by the infinitesimal neighborhood of the identity element. When we abbreviate all coordinates by $\xi := (\vec{\tau},\mathbf{R})$, the action of a Euclidean group element $E(\xi')$ on functions $f(\xi)$ of the group manifold is $E(\xi') : f(\xi) = f(E(\xi')^{-1}\xi)$.[5] In the Taylor series around the identity element (3), the first derivatives provide the *generators* of the one-parameter subgroup lines and form the Lie *algebra*[6] e_3 of the Euclidean group. They yield the familiar operators of translation and rotation in skew-Hermitian form and on the six coordinates of the E_3 manifold,

---

[5] The *inverse* of $E(\xi')$ in the argument of $f(\xi)$ ensures that the action of two or more group transformations of the manifold $\{\xi\}$ maintain their order of application.

[6] A Lie algebra is the real vector space spanned by the generators of the group, with one extra operation: the *Lie bracket* $\{\hat{A}, \hat{B}\}$. This operation is bilinear $\{\hat{A}, b\hat{B}+c\hat{C}\} = b\{\hat{A}, \hat{B}\} + c\{\hat{A}, \hat{C}\}$, skew-symmetric $\{\hat{A}, \hat{B}\} = -\{\hat{B}, \hat{A}\}$, and satisfies the Jacobi's identity $\{\hat{A}, \{\hat{B}, \hat{C}\}\} + \{\hat{B}, \{\hat{C}, \hat{A}\}\} + \{\hat{C}, \{\hat{A}, \hat{B}\}\} = 0$.

$$\hat{T}_x^E = -\partial_{\tau_x}, \quad \hat{T}_y^E = -\partial_{\tau_y}, \quad \hat{T}_z^E = -\partial_{\tau_z}, \tag{8}$$

$$\hat{J}_x^E = -\tau_y \partial_{\tau_z} + \tau_z \partial_{\tau_y} + \cot\theta \cos\phi \, \partial_\phi + \sin\phi \, \partial_\theta - \frac{\cos\phi}{\sin\theta} \partial_\psi, \tag{9}$$

$$\hat{J}_y^E = -\tau_z \partial_{\tau_x} + \tau_x \partial_{\tau_z} + \cot\theta \sin\phi \, \partial_\phi - \cos\phi \, \partial_\theta - \frac{\sin\phi}{\sin\theta} \partial_\psi, \tag{10}$$

$$\hat{J}_z^E = -\tau_x \partial_{\tau_y} + \tau_y \partial_{\tau_x} - \partial_\phi. \tag{11}$$

The translation generators (8) perform as $\exp(\alpha_i \hat{T}_i^E) f(\tau_i, R) = f(\tau_i - \alpha_i, R)$ and do not affect the rotation parameters R, while the rotation generators (9)–(11) consist of two summands, which act on the translation and on the rotation parameters. Their *commutators*[7] reflect this:

$$[\hat{T}_i, \hat{T}_j] = 0, \quad [\hat{J}_i, \hat{T}_j] = \hat{T}_k, \quad [\hat{J}_i, \hat{J}_j] = \hat{J}_k, \tag{12}$$

where i, j, k are cyclic permutations of 1, 2, 3, respectively. This structure is common to Euclidean generators in all realizations below and characterizes the Euclidean algebra e_3 as a *semidirect sum* of the translation and rotation Lie algebras, $e_3 := iso(3) = t_3 \looparrowleft\oplus so(3)$, following from (7) and written with lowercase letters.

3. THE FUNDAMENTAL OBJECT OF A MODEL

In the geometric model of optics in a 3D vacuum, light rays are idealized as straight directed lines in space, while in the wave model, wavefields are integrated out of directed plane waves, which in turn can be built out of Dirac-δ 2D planes in space (Luneburg, 1964; Wolf, 1989). Through translations T_3 and rotations $SO(3)$, both can be brought to the following *fundamental objects*:

$$\mathcal{O}_G := \text{the } z\text{-axis line}, \tag{13}$$
$$\mathcal{O}_W := \text{the } x\text{–}y \text{ plane}. \tag{14}$$

Fundamental objects are determined by their *symmetry* groups: $\mathcal{O}_G$ is invariant under translations along and rotations around the z-axis (but not inversions $z \leftrightarrow -z$; the line is *directed*), while $\mathcal{O}_W$ is invariant under translations in the x–y plane and rotations around the z-axis (but not inversions).

---

[7] The commutators $[A, B] := AB - BA$ are a realization of Lie brackets.

Indicating by calligraphic font the abstract group elements realized by the 4×4 (boldface) matrices in (1), their respective invariance subgroups are

$$\mathcal{H}_G(s;\psi) : \mathcal{O}_G = \mathcal{O}_G,$$
$$\mathcal{H}_G(s,\psi) := \mathcal{E}\Big((0,0,s)^\top, \mathbf{R}_z(\psi)\Big) \in \mathsf{T}_z \otimes \mathsf{SO}(2)_z \subset \mathsf{E}_3, \qquad (15)$$
$$\mathcal{H}_W(t_x, t_y; \psi) : \mathcal{O}_W = \mathcal{O}_W,$$
$$\mathcal{H}_W(t_x, t_y; \psi) := \mathcal{E}\Big((t_x, t_y, 0)^\top, \mathbf{R}_z(\psi)\Big) \in \mathsf{ISO}(2)_{x,y} \subset \mathsf{E}_3, \qquad (16)$$

where $\otimes$ indicates the direct product of groups. Indeed, *any* subgroup of E_3 can be used as a symmetry group to define a "fundamental object" for some model (useful or not) in crystallography, quantum mechanics, or optics.

Now consider factoring the generic E_3 group element in the following two forms, corresponding with the geometric (13)–(15) and wave (14)–(16) models,

$$\mathcal{E}(\vec{\tau}, \mathbf{R}(\phi,\theta,\psi)) = \mathcal{E}\Big((q_x, q_y, 0)^\top, \mathbf{R}_z(\phi)\mathbf{R}_x(\theta)\Big)\mathcal{H}_G(s;\psi), \qquad (17)$$

$$\mathcal{E}(\vec{\tau}, \mathbf{R}(\phi,\theta,\psi)) = \mathcal{E}\Big((0,0,u)^\top, \mathbf{R}_z(\phi)\mathbf{R}_x(\theta)\Big)\mathcal{H}_W(t_x, t_y; \psi). \qquad (18)$$

While the right factors are elements of the symmetry groups of the geometric and wave fundamental objects $\mathcal{O}_G$ and $\mathcal{O}_W$, the left factors are not. The structure of (17) and (18) is that of a decomposition of E_3 into *cosets* by the *set* of elements in the subgroups $\mathcal{H}_G$ and $\mathcal{H}_W$, respectively.[8] Cosets are subsets of the group; their main properties are that they are *disjoint* (no overlap between any two), and that their union *covers* the group (it provides all elements of the group). The parameters of the *left* factors of (17) and (18) are the coordinates of the *manifold* of cosets in each model, i.e., of all straight lines or all planes in 3-space. When we multiply (17) or (18) on the left by a generic Euclidean transformation $\mathcal{E}(\vec{\tau}', \mathbf{R}') \in \mathsf{E}_3$ and again factor out the symmetry subgroup to the right, we have maps of the space of cosets, i.e., the Euclidean transformations of the manifold of all lines, or of all planes, among each other.

The set of all straight lines in the geometric model of optics is thus a 4D manifold parametrized by $\{q_x, q_y; \theta, \phi\} \in \mathsf{R}^2 \otimes \mathsf{S}^2$, while that of all planes in the wave model is a 3D manifold with coordinates by $\{u; \theta, \phi\} \in \mathsf{R} \otimes \mathsf{S}^2$. For both cases it will be useful to define the unit 3-vector on the sphere

---

[8] These are *right* cosets; if the symmetry group elements were on the left, they would be *left* cosets.

$$\vec{p}(\theta,\phi) = \begin{pmatrix} p_x(\theta,\phi) \\ p_y(\theta,\phi) \\ p_z(\theta) \end{pmatrix} := \mathbf{R}_z(\phi)\,\mathbf{R}_x(\theta) \begin{pmatrix} 0 \\ 0 \\ 1 \end{pmatrix}$$

$$= \begin{pmatrix} \cos\phi & \sin\phi & 0 \\ -\sin\phi & \cos\phi & 0 \\ 0 & 0 & 1 \end{pmatrix} \begin{pmatrix} 1 & 0 & 0 \\ 0 & \cos\theta & \sin\theta \\ 0 & -\sin\theta & \cos\theta \end{pmatrix} \begin{pmatrix} 0 \\ 0 \\ 1 \end{pmatrix} = \begin{pmatrix} \sin\theta\sin\phi \\ \sin\theta\cos\phi \\ \cos\theta \end{pmatrix}, \tag{19}$$

that will be shown below to take the role of a *momentum* vector.

4. THE GEOMETRIC MODEL

In this section we shall introduce the Hamiltonian structure of the geometric model of light, based on the fundamental object $\mathcal{O}_G$ in (13) and the Euclidean transformations that give rise to the manifold of straight directed lines in space. On these we shall then impress a dynamic postulate to describe their change of direction (momentum) due to the inhomogeneity of the medium determined by the gradient of a scalar *refractive index* function over the space of positions (Goodman, 1968).

4.1 Euclidean Group and Coset Parameters

From the composition of the Euclidean 4×4 matrix realization involving (1)–(2)–(6)–(15) we can relate the Euclidean group and coset parameters in (17) through

$$\begin{aligned} \tau_x &= q_x + s\sin\theta\sin\phi, & q_x &= \tau_x - sp_x, \\ \tau_y &= q_y + s\sin\theta\cos\phi, & q_y &= \tau_y - sp_y, \\ \tau_z &= s\cos\theta, & s &= \tau_z/p_z = \tau_z\sec\theta, \end{aligned} \tag{20}$$

where the components of the unit 3-vector $\vec{p}(\theta,\phi)$ are given in (19). The geometric meaning of these coordinates is shown (projected on 2D) in Fig. 1. Each geometric light ray is a coset by $\mathcal{H}_G(s;\psi)$, where $s \in \mathsf{R}$ draws out the line and $\psi \in \mathsf{S}^1$ rotates around it preventing the attachment of any "flag" or "polarization" to this line. The manifold of straight lines is the 4D manifold of cosets

$$\wp := \{q_x, q_y; p_x, p_y; \sigma\} \in \mathsf{R}^2 \otimes \mathsf{D}^2 \otimes \mathsf{Z}_2, \tag{21}$$

where the 2-vector $\mathbf{q} := (q_x, q_y)^\top \in \mathsf{R}^2$ is the intersection of the line with the $z = 0$ plane. The first two components of the ray direction unit

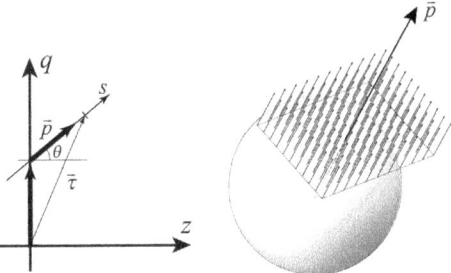

Fig. 1 *Left*: Relation between the 2D Euclidean group parameters $\{\tau_x, \tau_z; \theta\}$, $\vec{p} = (p, p_z)$, and the coset-separated parameters $\{q, p\}$ and $\{s\}$, with respect to the $\{q, z\}$ plane. *Right*: Rendering of the set of 3D geometric rays in the direction of $\vec{p}(\theta, \phi)$ parametrized by translations $\vec{\tau} \perp \vec{p}$.

3-vector $\vec{p}(\theta, \phi)$ form the 2-vector $\mathbf{p} := (p_x, p_y)^\top$; since $|\mathbf{p}| \le 1$, we must use $\sigma := \text{sign}\, p_z \in \{+, -\} =: \mathbb{Z}_2$ to distinguish between the two 2D unit disks, D_+^2 for "forward" rays where s grows in the z direction, and D_-^2 for "backward" rays, where s grows along $-z$. There is a $p_z = 0$ ($\sigma = 0$) circle along the common boundary that stitches together the two disks, but being a 1D submanifold it can and will be ignored.

Beams of geometric light can be described by functions of the $\wp$ manifold of directed lines, $\rho(\mathbf{q}, \mathbf{p}, \sigma)$. The total amount of "geometric light" is then given by $\sum_\sigma \int_\wp d\mu(w)\, \rho(w, \sigma)$, where $w := (\mathbf{q}, \mathbf{p})$; to determine the measure $d\mu(w)$, we start with the invariant measure over the full E_3 group in terms of its six parameters, $\{\vec{\tau}, \mathbf{R}\}$ in (1), putting them in terms of the coordinates used in the coset decomposition $\{\mathbf{q}, \theta, \phi; s, \psi\}$ in (17). The measure can be found then through (20) because it *separates* into two differential forms[9]

$$d^6 E(\vec{\tau}, \mathbf{R}) = d^3\vec{\tau}\, d^3\mathbf{R}(\phi, \theta, \psi) = d^4 w_G(\mathbf{q}, \mathbf{p})\, d^2 h_G(s, \psi), \tag{22}$$

$$d^3 \vec{\tau} = d\tau_x\, d\tau_y\, d\tau_z, \quad d^4 w_G(\mathbf{q}, \mathbf{p}) = dq_x\, dq_y\, dp_x\, dp_y, \tag{23}$$

$$d^3 \mathbf{R}(\phi, \theta, \psi) = d\phi\, d\cos\theta\, d\psi, \quad d^2 h_G(s, \psi) = ds\, d\psi. \tag{24}$$

This measure $d^6 E(\vec{\tau}, \mathbf{R})$ is the unique (up to a numerical factor) *Haar* measure for all Euclidean transformations; it is *invariant* because, for all (fixed) $E' \in \mathsf{E}_3$, $d^6(E'E) = d^6 E = d^6(E E')$. The measure on the space of rays on the z-screen (cosets) is also invariant: $d^4(E' w_G(\mathbf{q}, \mathbf{p})) = d^4 w_G(\mathbf{q}, \mathbf{p})$.

---

[9] Differentials in coordinates that follow from coset decompositions always separate.

This states a Liouville-type conservation law: no light is created nor destroyed under rotations or translations of space.

Written in terms of the parameters of the manifold $\wp$ of directed lines (21) that constitute the geometric model, the generators of the Euclidean algebra of translations (8) and rotations (9)–(11) assume the following form

$$\hat{T}_x^G = -\frac{\partial}{\partial q_x}, \quad \hat{T}_y^G = -\frac{\partial}{\partial q_y}, \quad \hat{T}_z^G = \frac{\sigma}{\sqrt{1-|\mathbf{p}|^2}} \mathbf{p} \cdot \frac{\partial}{\partial \mathbf{q}}, \quad (25)$$

$$\hat{J}_x^G = \sigma\sqrt{1-|\mathbf{p}|^2}\frac{\partial}{\partial p_y} + \frac{\sigma q_y}{\sqrt{1-|\mathbf{p}|^2}} \mathbf{p} \cdot \frac{\partial}{\partial \mathbf{q}}, \quad (26)$$

$$\hat{J}_y^G = -\sigma\sqrt{1-|\mathbf{p}|^2}\frac{\partial}{\partial p_x} - \frac{\sigma q_x}{\sqrt{1-|\mathbf{p}|^2}} \mathbf{p} \cdot \frac{\partial}{\partial \mathbf{q}}, \quad (27)$$

$$\hat{J}_z^G = -\mathbf{q} \times \frac{\partial}{\partial \mathbf{q}} - \mathbf{p} \times \frac{\partial}{\partial \mathbf{p}}. \quad (28)$$

These operators satisfy the same commutation relations (12) that characterize any realization of the Euclidean algebra $\mathbf{e}_3$.

4.2 Geometric and Dynamic Postulates

We have not yet said that $\wp$ is a phase space because from geometry we only showed that $(\mathbf{q}, \mathbf{p})$ is a 4D manifold with a Euclidean-invariant measure. To formulate useful geometric optics we need an extra postulate on its *dynamics*, namely the behavior of the light rays or beams, idealized as cosets $w_G(\mathbf{q}, \mathbf{p}, \sigma)$ or distributions $\rho(\mathbf{q}, \mathbf{p}, \sigma)$, $\sigma = \operatorname{sign} p_z$, as they *evolve* when the reference $z = 0$ screen is translated along $z \in \mathbf{R}$, in a medium that is no longer the homogeneous vacuum assumed above. The rays will then generally not be straight, so we will have a *deformation* of the space of cosets $w_G(\mathbf{q}, \mathbf{p}, \sigma) \in \wp$ that must nevertheless respect the invariance of its measure, with the conservation of its points and integrals (lest light be created or destroyed!).

To describe lines in 3-space that are generally not straight, we use the parametric *ray* 3-vector $\vec{q}(s) = (q_x(s), q_y(s), q_z(s))^\top$ for $s \in \mathbf{R}$, that will be subsequently projected as $\mathbf{q}(s)$ on the standard $z = 0$ screen. The rays $\vec{q}(s)$ are subject to the following two postulates:

- **Geometric postulate.** *Rays are continuous and piecewise differentiable.* This means that, except at points where they *break*, they have a tangent vector (indicated $\vec{p}(s)$), and they never disconnect. Since $|\mathrm{d}\vec{q}(s)| = \mathrm{d}s$, we can write

$$\frac{\mathrm{d}\vec{q}(s)}{\mathrm{d}s} = \frac{\vec{p}(s)}{|\vec{p}(s)|} =: \nabla_{\vec{p}} H(\vec{q}, \vec{p}), \tag{29}$$

where we introduce $H(\vec{q}, \vec{p}) = |\vec{p}| +$arbitrary function of $\vec{q}$.

- **Dynamic postulate.** *The ray direction vector $\vec{p}(s)$ responds linearly to the local 3-space gradient of a real, region-wise differentiable scalar function $n(\vec{q})$ (the refractive index).* This is written as

$$\frac{\mathrm{d}\vec{p}(s)}{\mathrm{d}s} = \nabla_{\vec{q}}\, n(\vec{q}) =: -\nabla_{\vec{q}} H(\vec{q},\, \vec{p}). \tag{30}$$

From the two postulate equations we obtain $H(\vec{q},\, \vec{p}) = |\vec{p}| - n(\vec{q}) +$constant and, using the chain rule, $\mathrm{d}H(\vec{q}(s),\, \vec{p}(s))/\mathrm{d}s = 0$. Incorporating the constant into $n(\vec{q})$ so that $H = 0$, we find the tangent vector $\vec{p}$ to be of length

$$|\vec{p}| = n(\vec{q}) \quad \text{for all } \vec{q} \in \mathsf{R}^3. \tag{31}$$

Thus, to every point of the medium corresponds a sphere of radius $n(\vec{q}) > 0$, —the *Descartes* sphere—that "guides" the ray trajectory, and which proceeds obeying the two above postulates. The geometric and dynamic postulates imply two *conservation* laws: at a point $\bar{s}$ of the trajectory $\vec{q}(s)$, between neighboring points $s_\pm = \bar{s} \pm \varepsilon$ as $\varepsilon \to 0$, and with $\Delta\vec{p}(\bar{s}) = \vec{p}(s_+) - \vec{p}(s_-)$ being parallel to $\nabla_{\vec{q}}\, n(\vec{q}(\bar{s}))$, the two conservation laws are stated as

$$\text{ray continuity}: \quad \vec{q}(s_-) = \vec{q}(s_+), \tag{32}$$
$$\text{refraction law}: \quad \nabla_{\vec{q}}\, n(\vec{q}(\bar{s})) \times \vec{p}(s_+) = \nabla_{\vec{q}}\, n(\vec{q}(\bar{s})) \times \vec{p}(s_-). \tag{33}$$

These, plus piecewise differentiability of $\vec{q}(s)$, imply the two original postulates. In particular (33) yields the well-known equation

$$n_+ \sin \alpha_+ = n_- \sin \alpha_-, \quad \begin{array}{l} n_\pm = |\vec{p}(s_\pm)| \\ \alpha_\pm := \angle\{\vec{p}(s_\pm), \nabla_{\vec{q}}\, n(\vec{q}(s_\pm))\}, \end{array} \tag{34}$$

which holds when the refractive index has a *finite* discontinuity at $\bar{s}$ and $\varepsilon \to 0$. This is of course known as the Ibn Sahl (Rashed, 1990, 1993), Snell, Descartes, and/or sine law of refraction.

4.3 Hamiltonian Structure and Phase Space

Two of the six coordinates $\{\vec{q}(s), \vec{p}(s)\}$ are redundant: the origin $s = 0$, and one of the three components of $\vec{p}(s)$ that lies on a Descartes sphere where we choose to discount the z-component. Noting that the triangle $\triangle(ds, dz, d\mathbf{q})$ is similar and equally oriented with $\triangle(n, p_z, \mathbf{p})$, we can divide the two postulated equations, (29) and (30), by $dz/ds = p_z/n$, to obtain a new pair of Hamilton equations in the essential x, y components of $\vec{q}$ and $\vec{p}$,

$$\frac{d\mathbf{q}}{dz} = \frac{\mathbf{p}}{p_z} =: \frac{\partial h(\mathbf{q}, z; \mathbf{p}, \sigma)}{\partial \mathbf{p}}, \tag{35}$$

$$\frac{d\mathbf{p}}{dz} = \frac{n(\mathbf{q}, z)}{p_z} \frac{\partial n(\mathbf{q}, z)}{\partial \mathbf{q}} =: -\frac{\partial h(\mathbf{q}, z; \mathbf{p}, \sigma)}{\partial \mathbf{p}}, \tag{36}$$

where the *Hamiltonian function* is here

$$h(\mathbf{q}, z; \mathbf{p}, \sigma) := -p_z = -\sigma\sqrt{n(\mathbf{q}, z)^2 - |\mathbf{p}|^2} = -n(\mathbf{q}, z)\cos\theta, \tag{37}$$

and where θ is the angle between the ray direction 3-vector $\vec{p}$ and the z-axis. The 3-vector $\vec{q} = (\mathbf{q}, z)^\top$ thus includes now the evolution parameter z, while $\vec{p} = (\mathbf{p}, -h)^\top$ includes the (minus) Hamiltonian that guides its evolution. At the $z = 0$ screen and on the Descartes sphere of $|\vec{p}|$, the range of coordinates $(\mathbf{q}, \mathbf{p}, \sigma) \in \wp$ form the *phase space* manifold of the geometric model. This is a restricted definition of symplectic phase spaces, but is sufficient for our purposes. In particular, in a homogeneous medium $n(\mathbf{q}, z) = n$, $\partial n/\partial \vec{q} = \vec{0}$, free flight preserves the ray direction and its chart index σ, but shears the position coordinate of $\wp$

$$\begin{aligned} \mathbf{q}(z) &= \mathbf{q}(0) + z\,\mathbf{p}(0)/p_z(0) & \mathbf{p}(z) &= \mathbf{p}(0), \\ &= \mathbf{q}(0) + z\tan\theta, & h(z) &= h(0). \end{aligned} \tag{38}$$

It is time to introduce, for conceptual and computational ease, the *Poisson operator* of a scalar function $f(\mathbf{q}, \mathbf{p})$,

$$\{f, \circ\}_{(\mathbf{q},\mathbf{p})} := \frac{\partial f(\mathbf{q},\mathbf{p})}{\partial \mathbf{q}} \cdot \frac{\partial}{\partial \mathbf{p}} - \frac{\partial f(\mathbf{q},\mathbf{p})}{\partial \mathbf{p}} \cdot \frac{\partial}{\partial \mathbf{q}}. \tag{39}$$

This allows us to write the Hamilton equations (35) and (36) as a $2 \times 2 = 4$-vector equation

$$\frac{d}{dz}\begin{pmatrix} \mathbf{q} \\ \mathbf{p} \end{pmatrix} = \begin{pmatrix} 0 & 1 \\ -1 & 0 \end{pmatrix} \begin{pmatrix} \partial/\partial \mathbf{q} \\ \partial/\partial \mathbf{p} \end{pmatrix} h(\mathbf{q},\mathbf{p},z) = -\{h,\circ\}\begin{pmatrix} \mathbf{q} \\ \mathbf{p} \end{pmatrix}, \qquad (40)$$

where we omit the chart index σ assuming the rays not to "bend over" in the interval of interest. The form (40) is attractive because it shows this system in *evolution* form by identifying $d/dz \leftrightarrow \begin{pmatrix} 0 & 1 \\ -1 & 0 \end{pmatrix} \leftrightarrow -\{h,\circ\}$. Systems governed by a Hamiltonian h in this form are *Hamiltonian systems*, with coordinates $\mathbf{p}$ of ray *momentum* that are *canonically conjugate* to coordinates of ray position $\mathbf{q}$.

4.4 Canonical and Optical Transformations

The invariance we demand of a Hamiltonian system is that if Eq. (40) is valid for the ray coordinates $(\mathbf{q}, \mathbf{p})$ with $h(\mathbf{q},\mathbf{p},z)$ on a plane screen at z, they should continue to be valid on any other screen at z', where they are registered as $(\mathbf{Q}(\mathbf{q},\mathbf{p}), \mathbf{P}(\mathbf{q},\mathbf{p}))$ with $h(\mathbf{Q},\mathbf{P},z')$. From (39) we introduce the *Poisson bracket* of two differentiable functions f, g of $(\mathbf{q}, \mathbf{p})$ defined by $\{f,g\} := \{f,\circ\}g = -\{g,\circ\}f$, and if necessary indicate by a subindex $(\mathbf{q},\mathbf{p})$ the variables with respect to which the derivatives are taken.[10] Then, replacing the differentials and partial derivatives of the new coordinates into (40), we find the conditions of invariance given succinctly by

$$\{Q_i, Q_j\}_{(\mathbf{q},\mathbf{p})} = 0, \quad \{Q_i, P_j\}_{(\mathbf{q},\mathbf{p})} = \delta_{i,j}, \quad \{P_i, P_j\}_{(\mathbf{q},\mathbf{p})} = 0, \qquad (41)$$

for $i,j \in \{x,y\}$. Poisson brackets are a skew-symmetric bilinear form satisfying the conditions of a Lie bracket plus the Leibniz identity,[11] and to whose useful properties we shall return shortly.

The set of all transformations $(\mathbf{q},\mathbf{p}) \leftrightarrow (\mathbf{Q},\mathbf{P})$ that leave (41) invariant form a group because the definition is transitive. Thus we have the functional group of all (linear and nonlinear) *canonical* transformations.[12] Moreover, when it maps the geometric-optical phase space $\wp$ bijectively onto itself (respecting the projections of the two Descartes spheres) we can call it an *optical* transformation. Under optical transformations all admissible rays are conserved. In the group of all canonical transformations some

---
[10] Poisson brackets are also a realization of Lie brackets, where the two partners commute under ordinary multiplication.
[11] The Leibniz identity is $\{fg, h\} = f\{g,h\} + \{f,h\}g$ for functions f, g, h.
[12] It is called *functional*, because its elements are defined by functions $\mathbf{Q}(\mathbf{q},\mathbf{p})$, $\mathbf{P}(\mathbf{q},\mathbf{p})$ that may carry an infinite number of parameters.

order can be established noting that, for any differentiable function $f(\mathbf{q}, \mathbf{p})$, the transformation

$$\exp(\tau\{f, \circ\})\begin{pmatrix}\mathbf{q}\\ \mathbf{p}\end{pmatrix} \mapsto \begin{pmatrix}\mathbf{q}'(\mathbf{q},\mathbf{p};\tau)\\ \mathbf{p}'(\mathbf{q},\mathbf{p};\tau)\end{pmatrix} \quad \text{is canonical,} \quad (42)$$

where the Taylor series expansion of the exponential operator is

$$\exp(\tau\{f, \circ\}) = \sum_{n=0}^{\infty} \frac{(\tau\{f, \circ\})^n}{n!} = 1 + \sum_{n=1}^{\infty} \frac{\tau^n}{n!} \underbrace{\{\ldots\{\{f, \circ\}, \circ\}\ldots, \circ\}, \circ\}}_{n \text{ brackets}}. \quad (43)$$

These exponential operators can "jump into" the arguments of any infinitely differentiable function $F(\mathbf{q}, \mathbf{p})$, as (Steinberg, 1986)

$$e^{\tau\{f,\circ\}} F(\mathbf{q},\mathbf{p}) = F(e^{\tau\{f,\circ\}}\mathbf{q}, e^{\tau\{f,\circ\}}\mathbf{p}). \quad (44)$$

Canonical transformations also preserve the volume element $d^2\mathbf{q} \wedge d^2\mathbf{p}$ of the space of rays/cosets (23), and form one-parameter groups $e^{\tau_1\{f,\circ\}} e^{\tau_2\{f,\circ\}} = e^{(\tau_1+\tau_2)\{f,\circ\}}$.

In homogeneous regions, where the refractive index n is constant, the generator functions of the Euclidean algebra in Poisson bracket form, obtained from (25)–(28), adopt the readily recognized form

$$\hat{T}_x^G = \{p_x, \circ\}, \qquad \hat{T}_y^G = \{p_y, \circ\}, \qquad \hat{T}_z^G = \{\sigma\sqrt{n^2-|\mathbf{p}|^2}, \circ\},$$
$$\hat{J}_x^G = \{q_y \sigma\sqrt{n^2-|\mathbf{p}|^2}, \circ\}, \quad \hat{J}_y^G = -\{q_x \sigma\sqrt{n^2-|\mathbf{p}|^2}, \circ\}, \quad \hat{J}_z^G = \{\mathbf{q}\times\mathbf{p}, \circ\}, \quad (45)$$

where the Hamiltonian h of (37) appears repeatedly. These operators also satisfy the Euclidean algebra $\mathbf{e}_3$ commutation relations (12). The generator functions inside the Poisson bracket yield $\vec{T}^{G\,2} = n^2$, $\vec{T}^G \cdot \vec{J}^G = 0$, and $(\vec{J}^G)^2 = n^2|\mathbf{q}|^2 - (\mathbf{p}\cdot\mathbf{q})^2 = |\vec{q}\times\vec{p}|^2|_\wp$; the last is the square of the Petzval projected on the $z=0$ screen with $\vec{p}$ on the Descartes sphere.

It is not guaranteed that series (42) and (43) will be easy to calculate or compute, nor that it will preserve $\wp$ globally. In general, subgroups of canonical transformations with a *finite* number of parameters are more amenable to ordered discussion.

4.5 Refracting Surfaces and the Root Transformation

The transformation of the manifold of rays due to refraction by a smooth but arbitrary surface $S(x, y, z) = 0$ between a medium with refracting index n and another n' cannot be put cogently in the evolution form (42) since the transformation is "sudden" and discontinuous. Yet it is clearly a most relevant transformation in optics. Note that this is distinct from the "thin lens" approximation in the paraxial regime (to be seen in Section 6) or the "potential jolts" used in quantum mechanics, because the surface S is generally not flat, so it does not act at one given z or time.

We shall assume that the surface S can be described, at least region-wise, by $z = \zeta(\mathbf{q})$, with a well-defined 3D normal vector

$$\vec{\Sigma}(\mathbf{q}) = \begin{pmatrix} \Sigma(\mathbf{q}) \\ -1 \end{pmatrix}, \quad \Sigma(\mathbf{q}) := \begin{pmatrix} \partial \zeta(\mathbf{q})/\partial q_x \\ \partial \zeta(\mathbf{q})/\partial q_y \end{pmatrix}, \tag{46}$$

that takes the place of the gradient of the refractive index, $\nabla_{\vec{q}}\, n(\vec{q})$ in (30) whose magnitude is now infinite, but whose direction is parallel to $\vec{\Sigma}(\mathbf{q})$. We may then resort to the two conservation laws (32) and (33) for position and momentum. As shown in Fig. 2, the coordinates of a ray before and after refraction, $(\mathbf{q}, \mathbf{p})$ and $(\mathbf{q}', \mathbf{p}')$ referred to the *same* screen $z = 0$, meet at point of impact $\bar{\mathbf{q}}$ after free flight (38) by $z = \zeta(\bar{\mathbf{q}})$. There, the component of momentum tangential to the surface, $\vec{\Sigma}(\mathbf{q}) \times \vec{p}$, is conserved. We thus have two 2-vector equations whose members we separate as

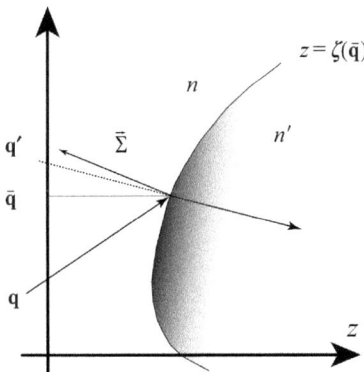

Fig. 2 An interface $z = \zeta(\mathbf{q})$ between homogeneous media n and n'. Referred to the same $z = 0$ screen, the incoming ray crosses it at $\mathbf{q}$, impacts the surface at $\bar{\mathbf{q}}$, refracts, and exits as having crossed the screen at $\mathbf{q}'$.

$$\mathbf{q} + \zeta(\bar{\mathbf{q}})\, \mathbf{p}/p_z = \bar{\mathbf{q}} = \mathbf{q}' + \zeta(\bar{\mathbf{q}})\, \mathbf{p}'/p'_z, \qquad (47)$$

$$\mathbf{p} + \Sigma(\bar{\mathbf{q}})\, p_z =: \bar{\mathbf{p}} = \mathbf{p}' + \Sigma(\bar{\mathbf{q}})\, p'_z, \qquad (48)$$

where we have defined $\bar{\mathbf{p}}$. Also, we note the resulting conservation of $\Sigma(\bar{\mathbf{q}}) \times (\mathbf{p} - \mathbf{p}') = 0$, and hence the coplanarity of $\vec{p}$, $\vec{p}'$ and $\vec{\Sigma}(\mathbf{q})$.

Solving the pair of simultaneous implicit equations (47) and (48) to find explicitly the transformation $\mathcal{S}_{n,n';\zeta}$ produced by the refracting surface $z = \zeta(\mathbf{q})$ between the media n and n',

$$\mathcal{S}_{n,n';\zeta} : \begin{pmatrix} \mathbf{q} \\ \mathbf{p} \end{pmatrix} = \begin{pmatrix} \mathbf{q}'(\mathbf{q},\mathbf{p};\zeta) \\ \mathbf{p}'(\mathbf{q},\mathbf{p};\zeta) \end{pmatrix}, \qquad (49)$$

may seem (and is) daunting. But if we perform it via the intermediate step of using the barred coordinates $(\bar{\mathbf{q}}, \bar{\mathbf{p}})$ in (47) and (48) and define the *root* transformation $\mathcal{R}_{n;\zeta}$ through (Navarro Saad & Wolf, 1986)

$$\bar{\mathbf{q}}(\mathbf{q},\mathbf{p}) = \mathcal{R}_{n;\zeta} : \mathbf{q} = \mathbf{q} + \zeta(\bar{\mathbf{q}})\, \mathbf{p} \Big/ \sqrt{n^2 - |\mathbf{p}|^2}, \qquad (50)$$

$$\bar{\mathbf{p}}(\mathbf{q},\mathbf{p}) = \mathcal{R}_{n;\zeta} : \mathbf{p} = \mathbf{p} + \zeta(\bar{\mathbf{q}}) \sqrt{n^2 - |\mathbf{p}|^2}. \qquad (51)$$

and its inverse,

$$\mathbf{q}'(\bar{\mathbf{q}},\bar{\mathbf{p}}) = \mathcal{R}^{-1}_{n';\zeta} : \bar{\mathbf{q}} = \bar{\mathbf{q}} - \zeta(\bar{\mathbf{q}})\, \mathbf{p}' \Big/ \sqrt{n'^2 - |\mathbf{p}'|^2}, \qquad (52)$$

$$\mathbf{p}'(\bar{\mathbf{q}},\bar{\mathbf{p}}) = \mathcal{R}^{-1}_{n';\zeta} : \bar{\mathbf{p}} = \bar{\mathbf{p}} - \zeta(\bar{\mathbf{q}}) \sqrt{n'^2 - |\mathbf{p}'|^2}, \qquad (53)$$

we will have factorized the refracting surface transformation (49) as

$$\mathcal{S}_{n,n';\zeta} = \mathcal{R}_{n;\zeta}\, \mathcal{R}^{-1}_{n';\zeta} \qquad (54)$$

where each factor depends on the surface ζ and on one medium *only*.[13]

Instead of simultaneous implicit equations in two variables, Eq. (50) is implicit in $\bar{\mathbf{q}}$ only, and of the simpler form $\bar{\mathbf{q}} = \mathbf{q} + f(\bar{\mathbf{q}}; \mathbf{p}; n)$. When solved by repeated replacement (if at all possible), or Taylor expansions of $\zeta(\bar{\mathbf{q}})$ and $\bar{\mathbf{q}}(\mathbf{q},\mathbf{p})$ by powers in an aberration series, this can be now replaced into (51) to find explicitly $\bar{\mathbf{p}}(\mathbf{q},\mathbf{p})$. The inverse transformation (53) is now implicit in $\mathbf{p}'(\bar{\mathbf{q}},\bar{\mathbf{p}})$ and can be similarly solved, and explicitly replaced into (52).

---

[13] It may appear as if the two factors in (54) should be in the opposite order. As can be verified, the property of these operators to "jump into" the argument of functions as in (44), leads to the correct composition $\mathcal{S}_{n,n';\zeta} : \rho(\mathbf{q},\mathbf{p}) = \rho(\mathcal{S}_{n,n';\zeta} : \mathbf{q},\, \mathcal{S}_{n,n';\zeta} : \mathbf{p})$.

Composition of both results then provides the refracting surface transformation (49). With the aid of symbolic computation (Wolf & Krötzsch, 1995) we have worked through aberration expansions up to total order seven in the four components of $\mathbf{q}$, $\mathbf{p}$, for axially symmetric refracting surfaces $\zeta(|\mathbf{q}|)$.

It may seem surprising that the root transformation (50) and (51) is *canonical* (although generally not *optical*); in fact it is only *locally* canonical, because its regions of validity in 4D phase space $\wp$ are bounded by submanifolds of rays that are *tangent* to the surface $z = \zeta(\mathbf{q})$. Beyond, they may be in another region or simply miss the surface. The proof of canonicity is quite simple (Wolf, 2004, chap. 4) when we use one of the four Hamilton *characteristic* functions, $F(\vec{q}', \vec{p})$, of a final position $\vec{q}'$ and initial momentum $\vec{p}$ to determine the remaining coordinates,

$$q_k = \frac{\partial F(\vec{q}', \vec{p})}{\partial p_k}, \quad p'_k = \frac{\partial F(\vec{q}', \vec{p})}{\partial q'_k}, \quad k \in \{x, y, z\}. \tag{55}$$

From here it is easy to see that the basic Poisson brackets (41) are preserved. Now consider the *unit* transformation, whose characteristic function in 6D is $F_{\mathrm{id}}(\vec{q}', \vec{p}) = \vec{q}' \cdot \vec{p}$, and then restrict position $\vec{q}'$ to the surface $q_z = \zeta(\mathbf{q})$ and momentum $\vec{p}$ to the Descartes sphere $|\vec{p}| = n$ in (37). We are left with a 4D transformation whose Hamilton characteristic function is

$$F_{\mathrm{root}}(\bar{\mathbf{q}}, \mathbf{p}) := F_{\mathrm{id}}(\vec{q}', \vec{p})|_{\zeta, n} = \bar{\mathbf{q}} \cdot \mathbf{p} + \sqrt{n^2 - |\mathbf{p}|^2}. \tag{56}$$

When introduced in (55), with $\bar{\mathbf{q}}$ in place of $\vec{q}'$, and $\mathbf{p}$ in place of $\vec{p}$ this becomes the root transformation (50) and (51), proving its canonicity. The root operator $\mathcal{R}_{n;\zeta}$ canonically transforms (regions of) the phase space of rays at the standard screen $z = 0$, to regions of another phase space referred to the warped surface ζ in the medium n (Atzema, Krötzsch, & Wolf, 1997).

5. THE WAVE AND HELMHOLTZ MODELS

We return to the decomposition of the Euclidean group manifold into cosets, but now of the symmetry group $\mathcal{H}_{\mathrm{W}}(t_x, t_y; \psi) = \mathsf{E}_2$ of $\mathcal{O}_{\mathrm{W}}$ in (14) and (16), which determines the "wave" model of 2-planes in 3-space. The coordinates of the *manifold* of cosets are $\{u; \theta, \phi\}$, obtained from (18). The angular parameters $\{\theta, \phi\} \in \mathsf{S}^2$ mark the direction normal to the planes, $\vec{p}(\theta, \phi)$

in (19), which is the same as the Descartes sphere of the geometric model. Thereafter, we shall reduce the wave model to its monochromatic components that are solutions to the Helmholtz equation. There we shall have a map between functions on a sphere and fields (value *and* normal derivative) on the $z = 0$ screen. This map is unitary between the Hilbert space $\mathcal{L}^2(\mathsf{S}^2)$ of square-integrable functions on the sphere, and an interesting Hilbert space $\mathcal{H}_k$ on the screen that is characterized by a *nonlocal* inner product.

5.1 Coset Parameters for the Wave Model

For each direction $\vec{p}(\theta, \phi)$, the planes form a stack characterized by the coset parameter $u \in \mathsf{R}$ as shown in Fig. 3; it will be more convenient to instead use the normal distance from the origin, $s \in \mathsf{R}$. In place of (20) for the geometric model the change of variables is now

$$\begin{aligned}
\tau_x &= t_x \cos\phi + t_y \cos\theta \sin\phi, \\
\tau_y &= -t_x \sin\phi + t_y \cos\theta \cos\phi, \quad s := u\cos\theta, \\
\tau_z &= -t_x \sin\theta + u,
\end{aligned} \qquad (57)$$

As was the case for the geometric model (24), the invariant Haar measure of the mother group E_3 separates into the E_2-invariant measure on each plane (coset), and an invariant measure on the manifold of cosets (planes),

$$d^6 g(\vec{\tau}, \mathbf{R}(\phi, \theta, \psi)) = d^3 w_{\mathsf{W}}(s; \theta, \phi)\, d^3 h_{\mathsf{W}}(t_x, t_y; \psi), \qquad (58)$$
$$d^3 w_{\mathsf{W}}(s; \theta, \phi) = ds\, \sin\theta\, d\theta\, d\phi, \quad d^3 h_{\mathsf{W}}(t_x, t_y; \psi) = dt_x\, dt_y\, d\psi. \qquad (59)$$

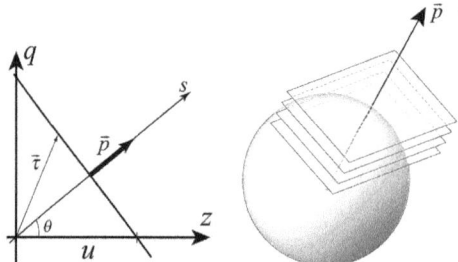

Fig. 3 *Left*: Relation between the 2D Euclidean group parameters $\{\tau_x, \tau_y; \theta\}$, $\vec{p} = (p, p_z)$, and the coset-separated parameters of the wave model, $\{u, \theta\}$ and $\{s\}$, with respect to the ambient $\{q, z\}$ space; cf. Fig. 1. *Right*: Rendering of the set of planes in a 3D space, parametrized by their distance $\{s\}$ to the origin and normal to the direction of $\vec{p}(\theta, \phi)$.

In this model, we have "beams" $\rho(s; \vec{p}(\theta, \phi))$ that stand for plane wavetrains $\rho_{(\theta,\phi)}(s)$ in every direction $\vec{p}(\theta, \phi)$ of the sphere, which integrate into wavefields in 3-space.

5.2 Euclidean Generators and Casimir Invariants

The Euclidean group E_3 presented in (2), is generated by a Lie *algebra* e_3, with three generators of translation $\hat{T}_i = \partial_{\tau_i}$, and three generators of rotations $\hat{J}_i$ as in quantum angular momentum theory, involving derivatives ∂_θ, ∂_ϕ, and ∂_ψ.

As before, when we change variables to coset parameters (57) and eliminate those of the symmetry subgroup, we are left with the generators of transformations in the manifold of this *wavetrain* model,

$$\begin{aligned}
\hat{T}_x^{\mathrm{W}} &= -\sin\theta\sin\phi\,\partial_s, & \hat{J}_x^{\mathrm{W}} &= \cot\theta\sin\phi\,\partial_\phi - \cos\phi\,\partial_\theta, \\
\hat{T}_y^{\mathrm{W}} &= -\sin\theta\cos\phi\,\partial_s, & \hat{J}_y^{\mathrm{W}} &= \cot\theta\cos\phi\,\partial_\phi + \sin\phi\,\partial_\theta, \\
\hat{T}_z^{\mathrm{W}} &= -\cos\theta\,\partial_s, & \hat{J}_z^{\mathrm{W}} &= -\partial_\phi.
\end{aligned} \quad (60)$$

These operators close under commutation into the Lie algebra e_3 with the relations (12), as all its other realizations do. The generator of translations along the s line of the wavetrain along $\vec{p}(\theta,\phi)$ is thus $\vec{p}(\theta,\phi)\,\partial_s$ and, since in vacuum $|\vec{p}| = 1$, the invariant quadratic Casimir operators are

$$\hat{T}^{\mathrm{W}\,2} := \sum_{i=x,y,z} \hat{T}_i^{\mathrm{W}\,2} = \frac{\partial^2}{\partial s^2}, \qquad \sum_{i=x,y,z} \hat{T}_i^{\mathrm{W}} \hat{J}_i^{\mathrm{W}} = 0. \quad (61)$$

Functions in the eigenspace of $\hat{T}^{\mathrm{W}\,2}$ have the form

$$\rho_{(\theta,\phi)}(s) = f_k(\theta,\phi)\,\exp(\mathrm{i}ks), \quad k \in \mathsf{R} - \{0\}, \quad (62)$$

and span subspaces of eigenvalue $-k^2$ that will not mix under Euclidean transformations. These are *monochromatic* beam functions of *wavenumber k*. A monochromatic *wavefield* on $\vec{q} = (q_x, q_y, q_z)^\top \in \mathsf{R}^3$ is the integral of $f_k(\vec{p})$ over $\vec{p}$ in the Descartes sphere,

$$F_k(\vec{q}) = \frac{k}{2\pi} \int_{\mathsf{S}^2} \mathrm{d}^2 S(\vec{p})\, f_k(\vec{p})\,\exp(\mathrm{i}k\vec{p}\cdot\vec{q}), \quad (63)$$

without the now-redundant parameter s, which can be set to any constant (or integrated to a Dirac δ subsequently factored out), its phase attached to $f_k(\vec{p})$, and where we will henceforth omit the subindex with the constant k. The monochromatic wavefields (63) are solutions of the *Helmholtz* equation,

$$\left(\frac{\partial^2}{\partial q_x^2} + \frac{\partial^2}{\partial q_y^2} + \frac{\partial^2}{\partial q_z^2}\right) F(\vec{q}) = -k^2 F(\vec{q}). \tag{64}$$

5.3 Hilbert Space for Helmholtz Wavefields

As in the geometric model, we privilege the description of the objects on the $z = 0$ plane screen, particularly because the 3D Helmholtz wavefields (63) actually depend on beam functions on the 2D surface of the sphere. The integration over the Descartes sphere using the two coordinates $\mathbf{p} = (p_x, p_y)^\top$ of $\vec{p}(\theta, \phi)$ requires $\sigma := \operatorname{sign} p_z$ to distinguish between *two* unit 2D disks D^2 where $|\mathbf{p}| \leq 1$, both having the measure $\sin\theta\, d\theta\, d\phi = dp_x\, dp_y/p_z$, and $p_z = \sigma\sqrt{1 - |\mathbf{p}|^2}$. Denoting by $f_\sigma(\mathbf{p})$ the beam functions in the forward or backward hemispheres, the Helmholtz wavefield on the plane of the screen and its normal derivative on the screen are given by the *wave* transform:

$$F(\mathbf{q}) := F(\vec{q})|_{q_z=0} = \frac{k}{2\pi} \int_{D^2} \frac{d^2\mathbf{p}}{\sqrt{1 - |\mathbf{p}|^2}} (f_+(\mathbf{p}) + f_-(\mathbf{p}))\, e^{ik\mathbf{p}\cdot\mathbf{q}}, \tag{65}$$

$$F_z(\mathbf{q}) := \left.\frac{\partial F(\vec{q})}{\partial q_z}\right|_{q_z=0} = \frac{ik^2}{2\pi} \int_{D^2} d^2\mathbf{p}\, (f_+(\mathbf{p}) - f_-(\mathbf{p}))\, e^{ik\mathbf{p}\cdot\mathbf{q}}. \tag{66}$$

With 2D Fourier analysis, the inverse wave transform between $f_\pm(\mathbf{p})$ on the two disks and the pair $\{F(\mathbf{q}), F_z(\mathbf{q})\}$ on the $z = 0$ screen, is found to be

$$f_\pm(\mathbf{p}) = \frac{k}{4\pi} \int_{R^2} d^2\mathbf{q} \left(\sqrt{1 - |\mathbf{p}|^2}\, F(\mathbf{q}) \pm \frac{1}{ik} F_z(\mathbf{q})\right) e^{-ik\mathbf{p}\cdot\mathbf{q}}. \tag{67}$$

Among wave phenomena, the energy in a wavefield is (proportional) to the integral of the absolute square of the wave function. For the beam functions on the sphere this is given naturally by the usual $\mathcal{L}^2(S^2)$ inner product, $(f, g)_{S^2} := \int_{S^2} d^2\omega\, f(\omega)^* g(\omega)$, $\omega := \{\theta, \phi\}$. Since the wave transform (65)–(67) is closely related with the Fourier transform, which is *unitary*, we can find the inner product for Helmholtz fields $\mathbf{F}(\mathbf{q}) := \{F(\mathbf{q}), F_z(\mathbf{q})\}$ over $\mathbf{q} \in R^2$ on the screen through replacing (67) in $(f, g)_{S^2}$ and performing one of the three resulting integrals. Thus we obtain the *nonlocal* inner product (Steinberg & Wolf, 1981; Wolf, 1989),

$$(\mathbf{F},\mathbf{G})_{\mathcal{H}_k} := \frac{k^2}{4\pi^2} \int_{\mathsf{R}^2} d\mathbf{q} \int_{\mathsf{R}^2} d\mathbf{q}' \qquad (68)$$
$$\times \left(F(\mathbf{q})^* \mu'(|\mathbf{q}-\mathbf{q}'|) \, G(\mathbf{q}') + F_z(\mathbf{q})^* \mu(|\mathbf{q}-\mathbf{q}'|) \, G_z(\mathbf{q}') \right),$$

which characterizes the *Helmholtz* Hilbert space $\mathcal{H}_k$, with the nonlocality measures obtained from the integration,

$$\mu'(v) = \pi \cdot \frac{j_1(v)}{v} = \pi \frac{\sin v - v \cos v}{v^3}, \quad \mu(v) = \frac{\pi}{k^2} j_0(v) = \frac{\pi}{k^2} \frac{\sin v}{v}, \qquad (69)$$

with $v := k|\mathbf{q}-\mathbf{q}'|$, where $j_1(v)$ and $j_0(v)$ are the spherical Bessel functions, and where we note that $\mu'(v) = (k^2/v) \, \partial_v \mu(v)$. For models on N-dimensional screens one has the nonlocality given by Bessel functions of integer or half-integer index for N odd or even. The wave transform (65)–(67) is unitary and the Hilbert spaces $\mathcal{H}_k$ are *unique* for Euclidean-invariant systems, as shown in Steinberg and Wolf (1981). The Helmholtz wavefield energy is $\sim (\mathbf{F},\mathbf{F})_{\mathcal{H}_k}$; in this context, in González-Casanova and Wolf (1995) we have an algorithm to fit the minimum-energy Helmholtz wavefield or normal derivative to the values on a discrete and finite number of sensor points.

6. PARAXIAL MODELS

In the geometric and wave models mothered by the Euclidean group (2)–(5), we have the explicit realizations of the generating Lie algebra $\mathbf{e}_3$ in Eqs. (25)–(28) and (60), respectively. The *paraxial limit* of these models is the regime where their ray directions or plane normals are infinitesimally close to the z-axis. Seen in the group E_3, we concentrate on vanishingly small rotations around the x- and y-axes, while rotations around the z-axis remain as such. We shall follow the structure of the mother Euclidean algebra as it contracts to the Heisenberg–Weyl algebra, whose group will provide the standard objects for classical and wave paraxial phase spaces, and then study their canonical transformations. Essentially, geometric paraxial phase space will be a plane $(\mathbf{q},\mathbf{p}) \in \mathsf{R}^{2D}$ for D-dimensional screens ($D=2$ as we have considered before), while for the wave model $\mathbf{q} \in \mathsf{R}^D$ and a phase will provide the argument for square-integrable wavefields $f(\mathbf{q}) \in \mathcal{L}^2(\mathsf{R}^D)$, compatible with the usual formalism of quantum mechanics. The transformations will correspond, in optics, to free flights, "thin" lenses, as well as harmonic waveguides of quadratic refractive index profile.

6.1 Contraction of the Euclidean to the Heisenberg–Weyl Algebra and Group

For clarity (and without much claim to rigor) we perform the *contraction* of the Euclidean algebra $\mathbf{e}_3$ to the Heisenberg–Weyl algebra $\mathbf{w}_2$ of 2D quantum mechanics, by rescaling the generators T_i, J_j, with the invariant $\sum_i T_i^2 = 1$, and defining

$$T_x^{(\varepsilon)} := \tfrac{1}{\varepsilon} T_x, \quad T_y^{(\varepsilon)} := \tfrac{1}{\varepsilon} T_y, \quad T_z^{(\varepsilon)} := T_z = \sqrt{1 - \varepsilon^2 (T_x^{(\varepsilon)2} + T_y^{(\varepsilon)2})}, \quad (70)$$
$$J_x^{(\varepsilon)} := \varepsilon J_x, \quad J_y^{(\varepsilon)} := \varepsilon J_y, \quad J_z^{(\varepsilon)} := J_z.$$

Now let $\{\circ, \circ\}$ again stand for *Lie brackets*, which are Poisson brackets $\{\circ, \circ\}$ of functions of phase space in the geometric models, or commutators $[\circ, \circ]$ of operators acting on wavefunctions. The Lie bracket relations (12) for the rescaled generators (70) become

$$\{T_i^{(\varepsilon)}, T_j^{(\varepsilon)}\} = 0, \qquad \{J_z^{(\varepsilon)}, J_x^{(\varepsilon)}\} = J_y^{(\varepsilon)}$$
$$\{J_i^{(\varepsilon)}, T_j^{(\varepsilon)}\} = T_k^{(\varepsilon)}, \qquad \{J_z^{(\varepsilon)}, J_y^{(\varepsilon)}\} = -J_x^{(\varepsilon)} \qquad (71)$$
$$\{J_x^{(\varepsilon)}, J_y^{(\varepsilon)}\} = \varepsilon^2 J_z^{(\varepsilon)},$$

with i, j, k cyclic. When $\varepsilon \to 0$ reaches the limit, the structure of the Lie algebra *changes*: the z-translation generator in (70) becomes the unit $T_z^{(0)} = \hat{1}$ that has null Lie brackets with all others. The Lie bracket relations (71), regrouped as convenient, in the limit become

$$\{T_i^{(0)}, T_j^{(0)}\} = 0,$$
$$\{J_i^{(0)}, T_i^{(0)}\} = 0, \qquad \{J_x^{(0)}, T_y^{(0)}\} = \hat{1}, \qquad \{J_z^{(0)}, J_x^{(0)}\} = J_y^{(0)},$$
$$\{J_z^{(0)}, T_x^{(0)}\} = T_y^{(0)}, \qquad \{J_y^{(0)}, T_x^{(0)}\} = -\hat{1}, \qquad \{J_z^{(0)}, J_y^{(0)}\} = -J_x^{(0)}, \qquad (72)$$
$$\{J_z^{(0)}, T_y^{(0)}\} = -T_x^{(0)}, \qquad \qquad \{J_x^{(0)}, J_y^{(0)}\} = 0.$$

Let us now finally change notation to

$$Q_x := J_x^{(0)}, \quad P_x := -\bar{\imath} T_y^{(0)}, \quad R := \bar{\imath} J_z^{(0)},$$
$$Q_y := J_y^{(0)}, \quad P_y := \bar{\imath} T_x^{(0)}, \quad \hat{1} = \bar{\imath} T_z^{(0)}, \qquad (73)$$

where the factor $\bar{\imath}$ is the unit 1 in the *geometric* model of phase space coordinates and Poisson brackets; in the *wave* (or quantum mechanical) realization of the algebra by self-adjoint operators, $\bar{\imath}$ is the imaginary unit i. In these terms, their common Lie brackets are

$$\begin{aligned}&\{Q_i,Q_j\}=0,\\&\{P_i,P_j\}=0, &&\{R,Q_x\}=\bar{\imath}Q_y, &&\{R,P_x\}=\bar{\imath}P_y,\\&\{Q_i,P_j\}=\bar{\imath}\delta_{i,j}\hat{1}, &&\{R,Q_y\}=-\bar{\imath}Q_x, &&\{R,P_y\}=-\bar{\imath}P_x.\end{aligned} \quad (74)$$

This contraction leaves $\{Q_i, P_j, \hat{1}\} \in \mathsf{w}_2$, $i, j, \in \{x, y\}$, in semidirect sum with R, the generator of $\mathsf{so}(2)$ rotations in the x–y plane, i.e., $\mathsf{e}_3 \xrightarrow{\varepsilon \to 0} \mathsf{w}_2 \rtimes \mathsf{so}(2)$, which continues to be a Lie algebra with six generators.

The Heisenberg–Weyl algebra w_2 has five generators, and its exponential is the 5-dimensional group W_2, whose elements can be parametrized as

$$\begin{aligned}w(\boldsymbol{\tau},\boldsymbol{\rho},v) &:= \exp\left(-\mathrm{i}(\boldsymbol{\tau}\cdot\mathbf{Q}+\boldsymbol{\rho}\cdot\mathbf{P}+v\hat{1})\right)\\&= \exp(-\mathrm{i}\boldsymbol{\tau}\cdot\mathbf{Q})\exp(-\mathrm{i}\boldsymbol{\rho}\cdot\mathbf{P})\exp\left(-\mathrm{i}\left(v-\tfrac{1}{2}\boldsymbol{\tau}\cdot\boldsymbol{\rho}\right)\hat{1}\right)\\&= \exp(-\mathrm{i}\boldsymbol{\rho}\cdot\mathbf{P})\exp(-\mathrm{i}\boldsymbol{\tau}\cdot\mathbf{Q})\exp\left(-\mathrm{i}\left(v+\tfrac{1}{2}\boldsymbol{\tau}\cdot\boldsymbol{\rho}\right)\hat{1}\right),\end{aligned} \quad (75)$$

with $\{\boldsymbol{\tau}, \boldsymbol{\rho}, v\} \in \mathsf{R}^4 \otimes \mathsf{S}^1$.[14] The product of two group elements is then

$$w(\boldsymbol{\tau}_1,\boldsymbol{\rho}_1,v_1)\,w(\boldsymbol{\tau}_2,\boldsymbol{\rho}_2,v_2) = w(\boldsymbol{\tau}_1+\boldsymbol{\tau}_2,\,\boldsymbol{\rho}_1+\boldsymbol{\rho}_2,\,v_1+v_2+\tfrac{1}{2}\boldsymbol{\tau}_1\cdot\boldsymbol{\rho}_2), \quad (76)$$

the unit element is $w\,(\mathbf{0},\,\mathbf{0},\,0)$, the inverse of (75) is $w\,(-\boldsymbol{\tau},\,-\boldsymbol{\rho},\,-v)$, and associativity holds.

6.2 The Heisenberg–Weyl Algebra and Group

In the phase space coordinates familiar from mechanics, the paraxial models have momentum generators P_i that stem from the e_3 space translation generators T_i, position generators Q_i that stem from the J_i rotation generators that now generate translations of momentum, and T_z that has become $\bar{\imath}\hat{1}\mapsto 1$, commuting with all.

6.2.1 Geometric Model

In the same way that we selected the symmetry group of a fundamental object in E_3 to parametrize its manifold of cosets, in the W_2 Heisenberg–Weyl group for geometric-optical (and classic mechanical) models we select the fundamental object given by the 1-parameter subgroup $\{e^{v\hat{1}}\}$, so that the space of its cosets is the manifold of phase space points $\{\mathbf{q},\mathbf{p}\}\in\mathsf{R}^4$—without

---

[14] We may instead decree that $v \in \mathsf{R}$ to have the *covering* group of the usually understood Heisenberg–Weyl group.

phases. The action of the group on this manifold is then through the exponentiated Poisson operators,

$$e^{\boldsymbol{\tau}\cdot\{\mathbf{p},\circ\}} f(\mathbf{q},\mathbf{p}) = f(\mathbf{q}-\boldsymbol{\tau},\mathbf{p}), \quad e^{\boldsymbol{\rho}\cdot\{\mathbf{q},\circ\}} f(\mathbf{q},\mathbf{p}) = f(\mathbf{q},\mathbf{p}+\boldsymbol{\rho}), \quad (77)$$

and $e^{v\{1,\circ\}} = 1$. Hence from a fundamental $(\mathbf{0},\mathbf{0})$ point we reach all other points in this phase space by translations. The generator R in (73) rotates jointly $\mathbf{q}$ and $\mathbf{p}$ in their x–y planes.

6.2.2 Wave Model

For the wave/quantum model, the fundamental object has the symmetry subgroup $\{e^{i\boldsymbol{\tau}\cdot\mathbf{Q}}\}$, whose manifold is that of positions $\{\mathbf{q}\}\in\mathbf{R}^2$ and a phase. From here we obtain the realization of the group $\mathbf{W}_2$ on functions $\psi(\mathbf{q})$ of the position manifold,[15]

$$e^{-i\boldsymbol{\tau}\cdot\mathbf{P}}\psi(\mathbf{q}) = \psi(\mathbf{q}-\boldsymbol{\tau}), \ e^{-i\boldsymbol{\rho}\cdot\mathbf{Q}}\psi(\mathbf{q}) = e^{-i\boldsymbol{\rho}\cdot\mathbf{q}}\psi(\mathbf{q}), \ e^{-iv\hat{\mathbf{I}}}\psi(\mathbf{q}) = e^{-iv}\psi(\mathbf{q}), \quad (78)$$

where the generators have the well-known form

$$P_i\psi(\mathbf{q}) = -i\frac{\partial}{\partial q_i}\psi(\mathbf{q}), \quad Q_i\psi(\mathbf{q}) = q_i\psi(\mathbf{q}), \quad i\in\{x,y\}, \quad (79)$$

while $\bar{i}\hat{\mathbf{1}} = i\hat{\mathbf{1}}$.

This model thus realizes $\mathbf{W}_2$ by unitary transformations (78) on $\mathcal{L}^2(\mathbf{R}^2)$, and of $\mathbf{w}_2$ by operators (79) that are essentially self-adjoint. Different choices in the coset decomposition of $\mathbf{W}_2$ yield other realizations of the Heisenberg–Weyl algebra and group (Wolf, 1975). Admittedly, the standard realizations (77) and (78) do not need the "fundamental object" approach for their construction, which we added only for completeness to show that they stem from contraction of the Euclidean models.

---

[15] If we require the physical *units* of the generators: in optics positions q_i have units of distance while momenta p_j have no units; hence $\bar{i}$ also has units of distance and one should introduce the reduced wavelength $\lambda := \lambda/2\pi = 1/k$, to have $\lambda\mathbf{1}$ or $\lambda i\hat{\mathbf{1}}$ in the geometric or wave optical models. In quantum mechanics on the other hand, momentum has units of mass × distance/time, and $\bar{i}$ has units of action, so one has $\hbar i\hat{\mathbf{1}}$ with a *fixed* $\hbar := h/2\pi$.

7. LINEAR TRANSFORMATIONS OF PHASE SPACE

In both the classical geometric and wave/quantum models, the Heisenberg–Weyl algebra is special in having a *center*, i.e., the generator $\hat{1}$ that has null Lie brackets with all others, $\{\hat{1}, \hat{A}\} = 0$, $\hat{A} \in \mathsf{w}_2$. When we introduce the additional operation of *multiplication* between elements of an algebra, we generate its *covering* algebra, whose elements are $\hat{A}\hat{B}$, $\hat{A}\hat{B}\hat{C}$, etc.; this operation is commutative in classical models, and non-commutative, $\hat{A}\hat{B} = \hat{B}\hat{A} + \{\hat{A}, \hat{B}\}$ in wave/quantum models. The Lie bracket of the algebra can be extended to its covering through the Leibniz identity: $\{\hat{A}\hat{B}, \hat{C}\} = \hat{A}\{\hat{B}, \hat{C}\} + \{\hat{A}, \hat{C}\}\hat{B}$. Unit central elements with the property $\hat{1}\hat{A} = \hat{A}\hat{1} = \hat{A}$ allow the *quadratic* extension in the covering algebra to be a Lie algebra in its own right. In this way, out of w_2 we produce the 4D *real symplectic* algebra $\mathsf{sp}(4,\mathsf{R})$ that will generate the group of linear canonical transformations of interest in optics (Goodman, 1968; Kauderer, 1994).

7.1 Geometric Model

In the classical model, whose four w_2 generator functions commute, its quadratic extension contains the following 10 quadratic generator functions

$$\begin{array}{c} q_x^2, \quad q_x q_y, \quad q_y^2, \quad q_x p_x, \quad q_x p_y, \\ p_x^2, \quad p_x p_y, \quad p_y^2, \quad q_y p_x, \quad q_y p_y. \end{array} \quad (80)$$

Linear combinations of these functions belong to a 10D linear vector space where Poisson brackets between any pair give back functions within that set (for example, $\{q_x q_y, q_x p_y\} = q_x^2$); hence this vector space is a Lie *algebra* $\mathsf{sp}(4,\mathsf{R})$, whose name will be justified below. Moreover, Poisson brackets of $\mathsf{sp}(4,\mathsf{R})$ elements with the original w_2 elements return *linear* combinations of elements of the latter (for example, $\{q_x q_y, p_x\} = q_y$). The exponential Taylor series (43) of $\exp(\tau\{a, \circ\})$, $\{a, \circ\} \in \mathsf{sp}(4,\mathsf{R})$ will thus produce finite *linear* transformations of the 4D manifold $(\mathbf{q}, \mathbf{p})$ that will form the Lie *group* $\mathsf{Sp}(4,\mathsf{R})$. These transformations will be *canonical* since they are generated through Poisson bracket operators.

The 4×4 $\mathsf{Sp}(4,\mathsf{R})$ matrices do not exhaust all the 16 independent linear transformations of the 4D manifold $\mathbf{w} := (q_x, q_y, p_x, p_y)^\top \in \mathsf{R}^4$. Canonicity demands that the fundamental Poisson brackets of the w_2 generators written in (41) be respected. These we can write in block matrix form as

$$\{\mathbf{w}^\top, \mathbf{w}\} := \left\{(\mathbf{q}, \mathbf{p}), \begin{pmatrix} \mathbf{q} \\ \mathbf{p} \end{pmatrix}\right\} := \begin{pmatrix} \{q_i, q_j\} & \{p_i, q_j\} \\ \{q_i, p_j\} & \{p_i, p_j\} \end{pmatrix} = \begin{pmatrix} 0 & -\delta_{i,j} \\ \delta_{i,j} & 0 \end{pmatrix}, \quad (81)$$

and demand that when $\mathbf{w} \mapsto \mathbf{M}\mathbf{w}$, with a matrix $\mathbf{M}$, we satisfy

$$\Omega := \{\mathbf{w}^\top, \mathbf{w}\} = \{(\mathbf{M}\mathbf{w})^\top, \mathbf{M}\mathbf{w}\} = \mathbf{M}\Omega\mathbf{M}^\top. \quad (82)$$

In 2×2 block form, with $\mathbf{M} = \begin{pmatrix} \mathbf{a} & \mathbf{b} \\ \mathbf{c} & \mathbf{d} \end{pmatrix}$ and $\Omega = \begin{pmatrix} 0 & -\mathbf{1} \\ \mathbf{1} & 0 \end{pmatrix} = -\Omega^{-1}$, this reads

$$\begin{pmatrix} 0 & -\mathbf{1} \\ \mathbf{1} & 0 \end{pmatrix} = \begin{pmatrix} -\mathbf{a}\mathbf{b}^\top + \mathbf{b}\mathbf{a}^\top & -\mathbf{a}\mathbf{d}^\top + \mathbf{b}\mathbf{c}^\top \\ -\mathbf{c}\mathbf{b}^\top + \mathbf{d}\mathbf{a}^\top & -\mathbf{c}\mathbf{d}^\top + \mathbf{d}\mathbf{c}^\top \end{pmatrix}. \quad (83)$$

From here we conclude six independent conditions:

$$\mathbf{a}\mathbf{b}^\top, \quad \mathbf{c}\mathbf{d}^\top \text{ are symmetric (and also } \mathbf{a}^\top\mathbf{c}, \quad \mathbf{b}^\top\mathbf{d}\text{)}, \quad \mathbf{a}\mathbf{d}^\top - \mathbf{b}\mathbf{c}^\top = \mathbf{1}, \quad (84)$$

$$\Rightarrow \quad \mathbf{M}^{-1} = \Omega \mathbf{M}^\top \Omega^{-1} = \begin{pmatrix} \mathbf{d}^\top & -\mathbf{b}^\top \\ -\mathbf{c}^\top & \mathbf{a}^\top \end{pmatrix} \in \mathsf{Sp}(4, \mathsf{R}). \quad (85)$$

Matrices $\mathbf{M}$ that satisfy (82) with the nondiagonal *metric* matrix Ω are called *symplectic* (Guillemin & Sternberg, 1984; Kauderer, 1994). The product of two symplectic matrices is symplectic, the unit and inverses are symplectic. They form the Lie group $\mathsf{Sp}(4,\mathsf{R})$ of linear canonical transformations of phase space.

There are several schemes to sensibly organize the 10 generators of $\mathsf{sp}(4,\mathsf{R})$ in (80) and their corresponding $\mathsf{Sp}(4,\mathsf{R})$ group parameters. One scheme, which may be called "optical" favors separating the phase space transformations due to three thin anamorphic lens parameters, three free anisotropic flight parameters, and four ideal magnifiers,

$$\exp\left\{\sum_{i \leq j} c_{i,j} q_i q_j, \circ\right\} \begin{pmatrix} \mathbf{q} \\ \mathbf{p} \end{pmatrix} = \begin{pmatrix} \mathbf{1} & 0 \\ \mathbf{c} & \mathbf{1} \end{pmatrix} \begin{pmatrix} \mathbf{q} \\ \mathbf{p} \end{pmatrix}, \quad \mathbf{c} = \begin{pmatrix} c_{x,x} & c_{x,y} \\ c_{x,y} & c_{y,y} \end{pmatrix}, \quad (86)$$

$$\exp\left\{\sum_{i \leq j} b_{i,j} p_i p_j, \circ\right\} \begin{pmatrix} \mathbf{q} \\ \mathbf{p} \end{pmatrix} = \begin{pmatrix} \mathbf{1} & -\mathbf{b} \\ 0 & \mathbf{1} \end{pmatrix} \begin{pmatrix} \mathbf{q} \\ \mathbf{p} \end{pmatrix}, \quad \mathbf{b} = \begin{pmatrix} b_{x,x} & b_{x,y} \\ b_{x,y} & b_{y,y} \end{pmatrix}, \quad (87)$$

$$\exp\left\{\sum_{i,j} a_{i,j} q_i p_j, \circ\right\} \begin{pmatrix} \mathbf{q} \\ \mathbf{p} \end{pmatrix} = \begin{pmatrix} e^{-\mathbf{a}} & 0 \\ 0 & e^{\mathbf{a}^\top} \end{pmatrix} \begin{pmatrix} \mathbf{q} \\ \mathbf{p} \end{pmatrix}, \quad \mathbf{a} = \begin{pmatrix} a_{x,x} & a_{x,y} \\ a_{y,x} & a_{y,y} \end{pmatrix}. \quad (88)$$

Products between elements of the first two subgroups represent all paraxial optical setups of lenses and empty spaces.[16]

Another scheme to organize the generators (80) highlights the structure of the real symplectic algebras and groups by identifying those quadratic functions, linear combinations of (80), which generate *rotations* of phase space. To this end we profit from the 4×4 representation of $\mathbf{M} \in \mathsf{Sp}(4, \mathsf{R})$, to find the representation of the generating *algebra*, $\mathbf{m} \in \mathsf{sp}(4, \mathsf{R})$ for $\mathbf{M} = \exp(\varepsilon \mathbf{m})$ as $\varepsilon \to 0$, so that $\mathbf{M} \approx \mathbf{1} + \varepsilon \mathbf{m}$. Then, from (82),

$$\Omega = \mathbf{M} \Omega \mathbf{M}^\top \quad \Rightarrow \quad \mathbf{m} \Omega = -\Omega \mathbf{m}^\top. \tag{89}$$

Matrices satisfying the last equality represent $\mathsf{sp}(4,\mathsf{R})$ and are called infinitesimal symplectic, or better *Hamiltonian* matrices, since some end up in their own right generating the dynamics of mechanical systems with quadratic Hamiltonians.

Acting on the 4D vector space $\mathbf{w} = (q_x, q_y, p_x, p_y)^\top$, we see that *skew-symmetric* matrices $\mathbf{m} = -\mathbf{m}^\top$ which satisfy (89) generate rotations because $\exp(\theta \mathbf{m}) = \exp(-\theta \mathbf{m}^\top)$ is orthogonal. There are then four linearly independent such matrices, which include two *fractional Fourier transforms* (FrFT) (Mendlovic & Ozaktas, 1993; Ozaktas & Mendlovic, 1993a, 1993b; Ozaktas, Zalevsky, & Kutay, 2001; Sudarshan, Mukunda, & Simon, 1985) that rotate in q_i–p_i planes, cross-rotation (gyrations) in the q_x–p_y and q_y–p_x planes, and rotation in the x–y planes. Their generator functions and representing skew-symmetric matrices are

$$\text{isotropic FrFT:} \quad \ell_0 := \tfrac{1}{4}\left(p_x^2 + p_y^2 + q_x^2 + q_y^2\right) \;\leftrightarrow\; \tfrac{1}{2}\begin{pmatrix} 0 & -1 & 0 & \\ & 0 & & -1 \\ 1 & 0 & & 0 \\ 0 & 1 & & \end{pmatrix}, \tag{90}$$

$$\text{anisotropic FrFT:} \quad \ell_1 := \tfrac{1}{4}\left(p_x^2 - p_y^2 + q_x^2 - q_y^2\right) \;\leftrightarrow\; \tfrac{1}{2}\begin{pmatrix} 0 & -1 & 0 & \\ & 0 & & 1 \\ 1 & 0 & & 0 \\ 0 & -1 & & \end{pmatrix}, \tag{91}$$

$$\text{gyrations:} \quad \ell_2 := \tfrac{1}{2}\left(p_x p_y + q_x q_y\right) \;\leftrightarrow\; \tfrac{1}{2}\begin{pmatrix} 0 & 0 & -1 \\ 0 & -1 & 0 \\ 0 & 1 & 0 \\ 1 & 0 & \end{pmatrix}, \tag{92}$$

---

[16] One would still have to show that *any* $\mathsf{Sp}(4,\mathsf{R})$ can be reached through products of elements in those two subgroups—and if so, with how many elements? In the 1D case of $\mathsf{Sp}(2,\mathsf{R})$ the answer is with up to three lenses and three empty spaces (Wolf, 2004, sect. 10.5), but in D dimensions I believe the question is still open.

$$\text{rotations:} \quad \ell_3 := \tfrac{1}{2}(q_x p_y - q_y p_x) \quad \leftrightarrow \quad \frac{1}{2}\begin{pmatrix} 0 & -1 & & \mathbf{0} \\ 1 & 0 & & \\ & & 0 & -1 \\ \mathbf{0} & & 1 & 0 \end{pmatrix}. \tag{93}$$

Under Poisson brackets, these functions close into a subalgebra of $\mathsf{sp}(4,\mathsf{R})$:

$$\{\ell_i, \ell_j\} = \ell_k, \quad i,j,k \text{ cyclic, and } \{\ell_0, \ell_i\} = 0, \tag{94}$$

that we identify as $\mathsf{u}(2) = \mathsf{u}(1) \oplus \mathsf{su}(2)$, the algebra of 2×2 skew-adjoint matrices, whose center $\mathsf{u}(1)$ is ℓ_0 in (90), the isotropic harmonic oscillator Hamiltonian that generates the fractional Fourier transform, while $\mathsf{su}(2)$ is the well-known angular momentum algebra that generates 3D rotations. The representing matrices satisfy the same algebra (94) under commutation. In Fig. 4 we show the generated $\mathsf{su}(2)$ rotations that include anisotropic Fourier transforms, gyrations, and rotations. This $\mathsf{u}(2)$ is called the *Fourier algebra* $\mathsf{u}_F(2) \subset \mathsf{sp}(4,\mathsf{R})$ of *ortho-symplectic* matrices (Simon & Wolf, 2000).

The finite $\mathsf{Sp}(4,\mathsf{R})$ matrices generated by (90)–(93) are obtained by exponentiation. Since $\exp\tfrac{1}{2}\theta\begin{pmatrix} 0 & -1 \\ 1 & 0 \end{pmatrix} = \begin{pmatrix} \cos\tfrac{1}{2}\theta & -\sin\tfrac{1}{2}\theta \\ \sin\tfrac{1}{2}\theta & \cos\tfrac{1}{2}\theta \end{pmatrix}$, writing $c := \cos\tfrac{1}{2}\theta$ and $s := \sin\tfrac{1}{2}\theta$ for brevity, these are respectively

$$\begin{pmatrix} c\mathbf{1} & -s\mathbf{1} \\ s\mathbf{1} & c\mathbf{1} \end{pmatrix}; \quad \begin{pmatrix} c & 0 & -s & 0 \\ 0 & c & 0 & s \\ s & 0 & c & 0 \\ 0 & -s & 0 & c \end{pmatrix}, \quad \begin{pmatrix} c & 0 & 0 & -s \\ 0 & c & -s & 0 \\ 0 & s & c & 0 \\ s & 0 & 0 & c \end{pmatrix}, \quad \begin{pmatrix} c & -s & & \mathbf{0} \\ s & c & & \\ & & c & -s \\ \mathbf{0} & & s & c \end{pmatrix}. \tag{95}$$

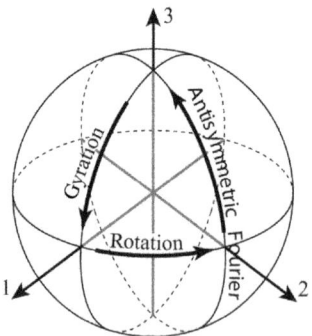

Fig. 4 The SU(2)$_F$ Fourier (sub)-group of anisotropic fractional Fourier transforms generated by (91), gyrations (92), and rotations (93). The isotropic U(1)$_F$ Fourier transforms generated by (90) commute with these and can be visualized as the product circle that closes the S^2 sphere of the figure into a S^3 sphere in a 4D ambient space.

Since the parameter ranges are all finite, $\theta \equiv \theta + 4\pi$, the total *volume* of the group is finite, i.e., it is *compact*; in fact, this $U_F(2)$ is the *maximal* compact subgroup of $Sp(4,R)$. Note that the $SU_F(2)$ factor *covers twice* the normal rotation group $SO(3)$ because of the argument $\tfrac{1}{2}\theta$ in the trigonometric functions, while the central factor $U_F(1)$ can be infinitely covered by R.

The remaining six independent generators of $sp(4,R)$ are represented by symmetric matrices **m** in (89), whose exponentials follow $\exp \tfrac{1}{2}\zeta \begin{pmatrix} 0 & 1 \\ 1 & 0 \end{pmatrix} = \begin{pmatrix} \cosh\tfrac{1}{2}\zeta & \sinh\tfrac{1}{2}\zeta \\ \sinh\tfrac{1}{2}\zeta & \cosh\tfrac{1}{2}\zeta \end{pmatrix}$, and whose elements are unbounded. These and the previous considerations lead us to understand the $Sp(4,R)$ 10D manifold of parameters as a higher-dimensional one-sheeted hyperboloid whose *waist* $U_F(1)$ allows for multiple covers. Its *double* cover is $Mp(4,R)$, the *metaplectic* group.

To study the structure of $Sp(4,R)$ it is helpful to use the accidental homomorphism of the Lie algebra $sp(4,R)$ and the Lie algebra $so(3,2)$ of infinitesimal 5×5 pseudo-orthogonal matrices under the metric $(+,+,+,-,-)$.[17] Using this feature, one can simplify the problem of finding all *inequivalent* optical systems between 2D screens, i.e., the independent matrix conjugation classes in $sp(4,R)$, called *orbits*, obtained from $\alpha\mathbf{MmM}^{-1}$ by letting $\alpha \in R$ and **M** roam over $Sp(4,R)$. For $Sp(2,R)$ this is easy: there are three orbit representatives corresponding to the 1D harmonic oscillator, the repulsive oscillator, and free flight. In $sp(4,R)$ we have 4 continua of orbits, plus 12 points of isolated systems that are inequivalent to each other (Wolf, 2004, chap. 12; Khan & Wolf, 2002).

7.2 Wave Model: Canonical Integral Transforms

One of the most significant extensions in the theory of Fourier analysis during the last few decades is that of linear canonical transforms (Healy, Alper Kutay, Ozaktas, & Sheridan, 2015). This was born as the solution to a rather (now) evident problem in paraxial optics: the transfer function between input and output scalar wavefields, as formulated by Collins in 1970 and, almost

---

[17] This homomorphism is similar to the well-known infinitesimal unitary spin and rotation matrices, $su(2)$ and $so(3)$, as well as between infinitesimal pseudo-unitary, symplectic, and 3D Lorentz Lie algebras: $su(1,1) = sp(2,R) = so(2,1)$. Note carefully that while the Lie algebras are the *same*, their exponentiation to the corresponding Lie *groups* can lead to different *coverings*. Thus $SU(2)$ covers $SO(3)$ *twice*; similarly $SO(2,1)$ is covered twice by $SU(1,1) = Sp(2,R)$, while $Sp(2,R)$ is doubly covered by the group $Mp(2,R)$ of integral transforms (to be seen in the next section), and also has an infinite cover $\overline{Sp}(2,R)$.

simultaneously, as the solution to a question in mathematical physics, investigated by Marcos Moshinsky and Christiane Quesne, on the representation of the group of linear canonical transformations in quantum mechanics (Moshinsky & Quesne, 1971; Quesne & Moshinsky, 1971). The initially separate development of both lines of research and their intertwining provides in itself an interesting foray into the ways scientific knowledge propagates in pure and applied fields that was analyzed in Liberman and Wolf (2015).

7.2.1 Linear Maps of the Heisenberg–Weyl Algebra

Within the context of the optical models based on the quadratic extension of the Heisenberg–Weyl algebra and group, most of the exploratory work on the model of $\mathsf{Sp}(4,\mathsf{R})$ paraxial optical systems has been done above in the geometric realization. In the wave (or quantum) model we have the operators Q_i, P_j in (79) and $i\hat{1}$, which also provide a quadratic extension of their wave W_2 realization.

Following Moshinsky and Quesne, for general dimension D we search for operators $\mathcal{C}_\mathbf{M}$, $\mathbf{M} \in \mathsf{Sp}(2D, \mathsf{R})$ such that, with the matrix inverse to $\mathbf{M} = \begin{pmatrix} \mathbf{a} & \mathbf{b} \\ \mathbf{c} & \mathbf{d} \end{pmatrix}$, namely $\mathbf{M}^{-1}$ given in (85),

$$\mathcal{C}_\mathbf{M} \begin{pmatrix} \mathbf{Q} \\ \mathbf{P} \end{pmatrix} \mathcal{C}_\mathbf{M}^{-1} = \mathbf{M}^{-1} \begin{pmatrix} \mathbf{Q} \\ \mathbf{P} \end{pmatrix} = \begin{pmatrix} \mathbf{d}^\top \mathbf{Q} - \mathbf{b}^\top \mathbf{P} \\ -\mathbf{c}^\top \mathbf{Q} + \mathbf{a}^\top \mathbf{P} \end{pmatrix}. \tag{96}$$

Now the action of this $\mathcal{C}_\mathbf{M}$ on functions $f(\mathbf{q}) \in \mathcal{L}^2(\mathsf{R}^D)$ can be found letting (96) act on some $f(\mathbf{q})$ to obtain 2D simultaneous equations of the form

$$\mathcal{C}_\mathbf{M}(\mathbf{Q}f(\mathbf{q})) = \mathbf{d}^\top \mathbf{Q}\,\mathcal{C}_\mathbf{M}f(\mathbf{q}) - \mathbf{b}^\top \mathbf{P}\,\mathcal{C}_\mathbf{M}f(\mathbf{q}), \tag{97}$$

$$\mathcal{C}_\mathbf{M}(\mathbf{P}f(\mathbf{q})) = -\mathbf{c}^\top \mathbf{Q}\,\mathcal{C}_\mathbf{M}f(\mathbf{q}) + \mathbf{a}^\top \mathbf{P}\,\mathcal{C}_\mathbf{M}f(\mathbf{q}). \tag{98}$$

Let now $f_\mathbf{M}(\mathbf{q}) := (\mathcal{C}_\mathbf{M}f)(\mathbf{q})$, writing $Q_j f(\mathbf{q}) = q_j f(\mathbf{q})$, $P_j f(\mathbf{q}) = -i\partial_j f(\mathbf{q})$ and $\partial_j f(\mathbf{q}) := \partial f(\mathbf{q})/\partial q_j$. Then we have

$$\mathcal{C}_\mathbf{M}(q_i f(\mathbf{q})) = \sum_j (d_{j,i} q_j + i b_{j,i} \partial_j) f_\mathbf{M}(\mathbf{q}), \tag{99}$$

$$-i\,\mathcal{C}_\mathbf{M}(\partial_i f(\mathbf{q})) = \sum_j (-c_{j,i} q_j - i a_{j,i} \partial_j) f_\mathbf{M}(\mathbf{q}). \tag{100}$$

7.2.2 Integral Transform Realization

We expect $\mathcal{C}_\mathbf{M} f(\mathbf{q})$ to be an *integral transform* of $f(\mathbf{q})$ because such are the Fourier transforms that belong to $\mathsf{Sp}(4D,\mathsf{R})$ in (90) and (91), and the Fresnel transform for (87), that is, with an integral kernel $C_\mathbf{M}(\mathbf{q},\mathbf{q}')$, and of the form

$$\mathcal{C}_\mathbf{M} f(\mathbf{q}) = \int_{\mathbb{R}^D} d^D \mathbf{q}'\, C_\mathbf{M}(\mathbf{q},\mathbf{q}') f(\mathbf{q}'). \tag{101}$$

Introducing this into (99) and (100) and integrating by parts the terms on the right-hand sides where the derivatives act on $f(\mathbf{q}')$ so that they act on the kernel, we obtain a set of differential equations that the kernel must satisfy,

$$q'_i C_\mathbf{M}(\mathbf{q},\mathbf{q}') = \sum_j (d_{j,i} q_j + \mathrm{i} b_{j,i} \partial_j) C_\mathbf{M}(\mathbf{q},\mathbf{q}'), \tag{102}$$

$$\partial'_i C_\mathbf{M}(\mathbf{q},\mathbf{q}') = \sum_j (\mathrm{i} c_{j,i} q_j - a_{j,i} \partial_j) C_\mathbf{M}(\mathbf{q},\mathbf{q}'). \tag{103}$$

Up to a constant factor, the solution is a complex Gaussian,

$$C_\mathbf{M}(\mathbf{q},\mathbf{q}') = K_\mathbf{M} \exp \mathrm{i}\left(\tfrac{1}{2} \mathbf{q}^\top \mathbf{b}^{-1} \mathbf{d} \mathbf{q} - \mathbf{q}^\top \mathbf{b}^{-1} \mathbf{q}' + \tfrac{1}{2} \mathbf{q}'^\top \mathbf{a} \mathbf{b}^{-1} \mathbf{q}'\right), \tag{104}$$

where the normalization constant $K_\mathbf{M}$ can be evaluated through asking for the Fresnel transform that corresponds to $C_{\mathbf{M}(\mathbf{b})}(\mathbf{q},\mathbf{q}') \to \delta^D(\mathbf{q}-\mathbf{q}')$ for $\mathbf{M}(\mathbf{b}) = \begin{pmatrix} 1 & \mathbf{b} \\ 0 & 1 \end{pmatrix}$ as $\mathbf{b} \to 0$. Since $\mathbf{b}$ is then a symmetric matrix due to (84) it can be diagonalized to $(\beta_1, \beta_2, \ldots, \beta_D)$, and the exponent expanded to a sum over the coordinates where for each we use the Dirac-δ convergent limit for oscillating but decreasing Gaussians in the second and fourth complex-β quadrants,

$$\lim_{\beta \to 0} \frac{1}{\sqrt{2\pi\beta}} \exp\left(\mathrm{i}\frac{(q-q')^2}{2\beta}\right) = \sigma_\beta\, e^{\mathrm{i}\pi/4}\, \delta(q-q'), \quad \sigma_\beta := \begin{cases} +1, & \arg\beta \in [-\tfrac{1}{2}\pi, 0], \\ -1, & \arg\beta \in [\tfrac{1}{2}\pi, \pi]. \end{cases} \tag{105}$$

The normalization constant in (104) is then obtained as a product of limits for each coordinate,

$$K_\mathbf{M} = \frac{1}{\sqrt{(2\pi\mathrm{i})^D \det\mathbf{b}}} = \frac{e^{-\mathrm{i}\pi D/4} \exp\left(-\mathrm{i}\tfrac{1}{2}\arg\det\mathbf{b}\right)}{\sqrt{(2\pi)^D |\det\mathbf{b}|}}. \tag{106}$$

Having reached $\mathbf{b} = 0$, the decomposition of a generic $\mathbf{M}_o = \begin{pmatrix} \mathbf{a} & 0 \\ \mathbf{c} & \mathbf{a}^{\top-1} \end{pmatrix} \in \mathsf{Sp}(2D,\mathbb{R})$ reads

$$(\mathcal{C}_{\mathbf{M}_o} f)(\mathbf{q}) = \frac{\exp\left(\mathrm{i}\tfrac{1}{2} \mathbf{q}^\top \mathbf{c}\mathbf{a}^{-1} \mathbf{q}\right)}{\sqrt{\det\mathbf{a}}} f(\mathbf{a}^{-1} \mathbf{q}). \tag{107}$$

Finally, when $\det\mathbf{b} = 0$ but $\mathbf{b} \neq 0$, we perform a similarity transformation to bring $\mathbf{b}$ to diagonal form and use (105) for its null eigenvalues.

7.2.3 Fractional Fourier and Canonical Transforms

The set of canonical integral transform kernels $C_\mathbf{M}(\mathbf{q},\mathbf{q}')$ in (102) are a *representation* of the group of 4×4 symplectic matrices $\mathbf{M} \in \mathsf{Sp}(2D, \mathsf{R})$, which we regard as a matrix of continuous, infinite rows and columns $(\mathbf{q},\mathbf{q}')$. Their basic group composition property was addressed by Moshinsky and Quesne (1971), who found that, for $\mathbf{M}_1 \mathbf{M}_2 = \mathbf{M}_3$, their kernels compose as

$$\int_{\mathsf{R}^D} d^D\mathbf{q}\, C_{\mathbf{M}_1}(\mathbf{q},\mathbf{q}')\, C_{\mathbf{M}_2}(\mathbf{q}',\mathbf{q}'') = \sigma_{1,\,2;\,3}\, C_{\mathbf{M}_2}(\mathbf{q},\mathbf{q}''), \qquad (108)$$

$$\sigma_{1,\,2;\,3} := \operatorname{sign}(\det \mathbf{b}_3 / \det \mathbf{b}_1 \det \mathbf{b}_2), \qquad (109)$$

where $\mathbf{b}_1$, $\mathbf{b}_2$, and $\mathbf{b}_3$ are the upper-right submatrices of the $\mathbf{M}$'s.

The "ambiguity sign" $\sigma_{1,\,2;\,3}$ in the group composition in (108) stems from the sign σ_β in (105) and would not go away through any redefinition of phases. Only somewhat later it was recognized that this sign is due to the multiple cover of the symplectic groups afforded by this set of kernels, whose topological features were studied by Valentin Bargmann for 1D in Bargmann (1947) and for ND in Bargmann (1970). His analysis follows the polar decomposition of complex numbers into a phase and a positive magnitude; for matrices, this is a decomposition into a unitary matrix (of the Fourier subgroup $\mathsf{U}(N)_\mathrm{F}$) and a positive definite matrix, with an appropriate choice of parameters. Unitary matrices in turn can be decomposed into the subgroup $\mathsf{SU}(N)_\mathrm{F}$ with unit determinant, and the subgroup of matrices which are the circle of isotropic fractional Fourier transforms $\mathsf{U}(1)_\mathrm{F}$ in the geometric model (90); the latter bear the onus of multivaluation because the manifolds of the other two factors are simply connected.

Fractional Fourier transforms $\mathcal{F}^\alpha$ of power α were defined in 1937 by Condon (1937) essentially on the path (103)–(105); they were rediscovered by Namias (1980), who found the kernel through the bilinear generating function of the harmonic oscillator wavefunctions, with a phase $e^{+i\alpha/2}$ which guarantees that $\mathcal{F}^{\alpha_1}\mathcal{F}^{\alpha_2} = \mathcal{F}^{\alpha_1+\alpha_2}$ and $\mathcal{F}^4 = 1$ in 1D. Comparison with the canonical transform kernel[18] in (102) for the D-dimensional version shows that for angles and powers $\phi = \tfrac{1}{2}\pi\alpha$,

$$\mathcal{C}_{\mathbf{F}(\phi)} = \exp(iD\alpha/\pi)\,\mathcal{F}^\alpha_{(D)} \quad \text{for} \quad \mathbf{F}(\phi) := \begin{pmatrix} \cos\phi\,\mathbf{1} & \sin\phi\,\mathbf{1} \\ -\sin\phi\,\mathbf{1} & \cos\phi\,\mathbf{1} \end{pmatrix}. \qquad (110)$$

---

[18] The kernel of the Fourier transform we take as $\sim e^{-iqq'}$; Namias's work uses $\sim e^{+iqq'}$ instead since he follows the harmonic oscillator evolution.

We see that while the fractional Fourier transform in Condon (1937), Namias (1980), and McBride and Kerr (1987) is a fourth root of the unit transformation, $\mathcal{F}^4 = 1$, in D dimensions we have $\mathcal{C}_F^4 = e^{-i\pi D}\mathbf{1}$, which is -1 when D is odd (including the basic 1D case), and 1 in even dimensions (including the $D = 2$ case of our concerns). The subgroup $\{\mathcal{C}_{\mathbf{F}(\phi)}\}_{\phi \in \mathsf{S}^1}$ of canonical transforms represents the quantum harmonic evolution cycle, thus distinct by a phase from the fractional Fourier transform defined in those references.

7.2.4 Canonical Transforms—Remarks and Extensions

Another special property of the wave/quantum canonical transforms is that they are generated by *second*-order differential operators, in complete algebraic correspondence with the classical counterpart of first-order Poisson operators given in (86)–(88) (Wolf, 1974). It is worth noting that the work of Lie (1888) and practically all subsequent work on Lie algebras (Gilmore, 1978) had dealt with first-order differential operators only, although the 1D time evolution generated by oscillator or waveguide Hamiltonians was used on occasion.

The triple connection between 2D×2D symplectic matrices, integral transforms, and exponentials of up-to-second-order differential operators, provides several computationally easy ways to find special function identities and Baker–Campbell–Hausdorff relations to factorize exponentials of non-commuting exponents (García-Bullé, Lassner, & Wolf, 1986).

An important special case arises when the optical setups are assumed to be axially symmetric, described by matrices $\mathbf{M} = \begin{pmatrix} a\mathbf{1} & b\mathbf{1} \\ c\mathbf{1} & d\mathbf{1} \end{pmatrix} \in \mathsf{Sp}(2,\mathsf{R})$. The integral kernel then involves Bessel functions in the radial coordinate, and phases $e^{im\theta}$ that are used to project $\sim J_m(rr'/b)$ kernels times oscillating exponentials. These *radial* canonical transform unitary kernels belong to the Bargmann *discrete* series (Bargmann, 1947) of irreducible representations of $\mathsf{Sp}(2,\mathsf{R})$, where Laguerre–Gauss beams are prominent. Another special (but less-known) case is that of *hyperbolic* canonical transforms where the $\mathbf{1}$'s in $\mathbf{M}$ are replaced by $\begin{pmatrix} 1 & 0 \\ 0 & -1 \end{pmatrix}$'s. The kernel involves Hankel functions and lies in the *continuous* irreducible representation series of this group (Healy et al., 2015, chap. 1).

Note that the parameters of $\mathsf{Sp}(2D,\mathsf{R})$ can be complexified to build the Lie algebra $\mathsf{Sp}(2D,\mathsf{C})$, with the same properties (84) and (85), but its integral transform realization is only possible for $\mathcal{L}^2(\mathsf{R}^D)$ functions when the Gaussian factor to be integrated is in the lower complex half-plane, i.e., $\mathrm{Im}\,(\mathbf{ab}^{-1})_{i,j} \leq 0$. The resulting integral transforms form a complex *semigroup* (i.e., with no guaranteed inverse), called $\mathsf{HSp}(2D,\mathsf{C})$. This allows one to treat diffusion phenomena with the same tools as for wave/quantum evolution. Thus the diffusion of harmonic, repulsive, and Airy heat distributions follows convertical ellipses, hyperbolas, and parabolas (Wolf, 1979, chap. 12), while oscillating Gaussians follow diverging lines. Moreover, with an extension of the *measure* $d^D\mathbf{q}$ to an appropriate measure $\mu(\mathbf{q}, \mathbf{q}^*)$ $d^D\text{Re}\,\mathbf{q}\,d^D\text{Im}\,\mathbf{q}$ on the complex plane, these diffusive transforms can made *unitary* (Bargmann, 1970; Wolf, 1974).

8. THE METAXIAL REGIME

Beyond the quadratic extension of the Heisenberg–Weyl algebra to the real symplectic algebra, we can propose a nested family of extensions generated by polynomials in the phase space variables or operators of degree higher than those in (80), which will generate *nonlinear* (but *canonical*) transformation of phase space. First we shall build this *covering algebra* structure for 1D images, and then consider the aberration of 2D images by aligned axis-symmetric optical setups. The resulting classification of aberrations to third order follows roughly that of Seidel (1853), who was interested in image formation rather than phase space transformations, but departs from their subsequent treatment by Buchdahl (1970) for higher-order aberrations.

8.1 Aberrations in 2D Systems

In classical 2D paraxial systems, where screens are one-dimensional lines and phase space $(q, p) \in \mathsf{R}^2$ is two-dimensional, consider the monomials

$$M_{k,m}(p,q) := p^{k+m} q^{k-m}, \quad \text{of } \begin{cases} \text{rank} & k \in \{0, \tfrac{1}{2}, 1, \tfrac{3}{2}, 2, \ldots\}, \\ \text{weight} & m \in \{k, k-1, \ldots, -k\}. \end{cases} \quad (111)$$

For $k=0$, $M_{0,0} = 1$, while for $k=\tfrac{1}{2}$, $M_{\tfrac{1}{2},\tfrac{1}{2}} = p$ and $M_{\tfrac{1}{2},-\tfrac{1}{2}} = q$ are the phase space variables. Then, for $k=1$ we have the quadratic monomials $M_{1,m} = \{p^2, pq, q^2\} \in \mathsf{sp}(2,\mathsf{R})$, $m \in \{1, 0, -1\}$, that we saw in the last section. For general k, k' their Poisson brackets are

$$\{M_{k,m}, M_{k',m'}\} = 2(km' - k'm) M_{k+k'-1, m+m'}, \quad (112)$$

so they generate a countably infinite covering algebra of the classical Heisenberg–Weyl algebra which is graded by rank k.

Under linear transformations of phase space, (86)–(88) with $ad - bc = 1$, each set of $2k + 1$ monomials $M_{k,m}$ in each rank k will transform as a *multiplet*, i.e., only among themselves as,[19]

$$\mathcal{C}\begin{pmatrix} a & b \\ c & d \end{pmatrix} : \begin{pmatrix} q \\ p \end{pmatrix} = \begin{pmatrix} a & b \\ c & d \end{pmatrix}^{-1} \begin{pmatrix} q \\ p \end{pmatrix} = \begin{pmatrix} d & -b \\ -c & a \end{pmatrix} \begin{pmatrix} q \\ p \end{pmatrix} \Rightarrow \quad (113)$$

$$\mathcal{C}\begin{pmatrix} a & b \\ c & d \end{pmatrix} : M_{k,m}(q,p) = \sum_{m'=-k}^{k} D_{m,m'}^{k}\begin{pmatrix} a & b \\ c & d \end{pmatrix}^{-1} M_{k,m'}(q,p), \quad (114)$$

$$D_{m,m'}^{k}\begin{pmatrix} a & b \\ c & d \end{pmatrix} := \sum_{m''} \binom{k+m}{m''-m'} \binom{k-m}{k-m''} \quad (115)$$
$$\times a^{k-m''} b^{m''-m} c^{m''-m'} d^{k+m+m'-m''},$$

where the matrices $D_{m,m'}^{k}(\mathbf{M})$ *represent* the linear canonical transformation $\mathcal{C}(\mathbf{M}) \equiv \mathcal{C}_{\mathbf{M}} \in \mathsf{Sp}(2,\mathsf{R})$.

Using (112) repeatedly on $\binom{q}{p}$, we find the action of the higher monomials $M_{k,m}$, $k > 1$ as generators of one-parameter groups of *nonlinear* transformations of phase space,

$$\exp(\alpha\{M_{k,m}, \circ\})\binom{q}{p} = \begin{pmatrix} q\left(1 + \sum_{n=1}^{\infty} \frac{(-\alpha)^n}{n!} c_{k,m;n}^{-} M_{n(k-1),nm}\right) \\ p\left(1 + \sum_{n=1}^{\infty} \frac{(+\alpha)^n}{n!} c_{k,m;n}^{+} M_{n(k-1),nm}\right) \end{pmatrix}, \quad (116)$$

with $c_{k,m;n}^{\sigma} := \prod_{s=0}^{n-1}(k + \sigma(2s-1)m)$.[20] In the series (116), the term linear in α ($n=1$) is of degree $2k-1$ in the phase space variables; the $\{M_{k,m}\}_{m=-k}^{k}$ are thus the generators of *aberrations* of order $A := 2k - 1$.[21]

Consider the following five aberrations of order 3, $\{M_{2,m}\}_{m=-2}^{2}$ where we now cut the series of $\exp\alpha\{M_{2,m}, \circ\}$ in (116) after the term linear in α, cubic in (q, p), and identify them by names that will be borne out through the spot diagrams of the case of 3D optics on 2D screens in the following section,[22]

---

[19] In comparing with Wolf (2004, eq. (13.5)) we note that instead of $\binom{p}{q}$ there we have here $\binom{q}{p}$. The expressions match when we exchange $a \leftrightarrow d$ and $b \leftrightarrow c$.
[20] Note that for $k=\frac{1}{2}, 1$ these coefficients become null after the first n-term.
[21] We shall here exclude half-integer ranks $\frac{1}{2}, \frac{3}{2}, \ldots$ for simplicity: they generate aberrations of even order such as due to misaligned 2D optical systems. Nevertheless they can be treated on equal footing with the axis-symmetric aberrations (Wolf, 2004, chap. 13).
[22] Perhaps I should apologize for the name of $M_{2,-2} = q^4$. It does not seem to have had any name before; its p-unfocusing property suggested the irreverent name of *pocus*.

MONOMIAL	MAP LINEAR IN α	NAME	
$M_{2,2} = p^4$,	: $\begin{pmatrix} q \\ p \end{pmatrix} \mapsto \begin{pmatrix} q - 4\alpha p^3 \\ p \end{pmatrix}$,	spherical aberration,	
$M_{2,1} = p^3 q$,	: $\begin{pmatrix} q \\ p \end{pmatrix} \mapsto \begin{pmatrix} q - 3\alpha p^2 q \\ p + \alpha p^3 \end{pmatrix}$,	coma,	
$M_{2,0} = p^2 q^2$,	: $\begin{pmatrix} q \\ p \end{pmatrix} \mapsto \begin{pmatrix} q - 2\alpha p q^2 \\ p + 2\alpha p^2 q \end{pmatrix}$,	astigmatism/ curvature of field,	(117)
$M_{2,-1} = pq^3$,	: $\begin{pmatrix} q \\ p \end{pmatrix} \mapsto \begin{pmatrix} q - \alpha q^3 \\ p + 3\alpha p q^2 \end{pmatrix}$,	distortion,	
$M_{2,-2} = q^4$,	: $\begin{pmatrix} q \\ p \end{pmatrix} \mapsto \begin{pmatrix} q \\ p + 4\alpha p^3 \end{pmatrix}$,	pocus.	

Yet except for $M_{k,\pm k}$, (117) are *not* canonical transformations of (q,p) because of terms in α^2 and higher; they will be canonical only if the full series (116) is kept. In order to cogently approximate the phase space transformations with (presumably *small*) aberrations, let us define *canonicity up to rank K* through *cutting* Poisson brackets between monomials (112), by defining

$$\{M_{k,m}, M_{k',m'}\}_{(K)} := \begin{cases} 2(km' - k'm) M_{k+k'-1, m+m'}, & k+k'-1 \leq K, \\ 0, & \text{otherwise.} \end{cases} \quad (118)$$

With the cut brackets, the transformation of (q,p) in (117) can be thus declared to be canonical—up to rank $K = 2$, or aberration order $A = 3$. Let us denote by $\mathcal{A}_k$ the linear vector space spanned by the monomials $M_{k,m}$, of elements $A_k = \sum_{m=-k}^{k} \alpha_{k,m} M_{k,m} \in \mathcal{A}_k$ with coefficients $\alpha_{k,m} \in \mathbf{R}$, which maps on itself irreducibly under linear $\mathsf{Sp}(2,\mathbf{R})$ transformations.

The next term in the series (116) is $\sim \alpha^2$; this brings in cut Poisson operator monomials $\{A_2, \{A_2, q\}\}_{(3)}$ and $\{A_2, \{A_2, p\}\}_{(3)}$, which are generally of degree 5 in the phase space variables. These terms are also brought in by the action $\{A_3, \circ\}_{(3)}$ of monomials of rank 3 (aberration order 5), which belong to the multiplet of seven generators in $\mathcal{A}_3$; the transformations will be canonical up to rank $K = 3$. With the cut Poisson brackets (118) we construct thus a Lie *group* that contains linear transformations $\mathcal{C}_M$ (generated by the quadratic polynomials in $\mathcal{A}_1$), and aberrations of orders 3 and 5, generated by $\mathcal{A}_2$ and $\mathcal{A}_3$. The structure of this group is similar to that of the Euclidean group in Section 2; it is a semidirect product whose invariant subgroup—there translations, here aberrations of ranks 2 and 3—is now *not* abelian: $\{A_2, B_2\}_{(3)} \in \mathcal{A}_3$; the factor group was there SO(3) and is here $\mathbf{M} \in \mathsf{Sp}(2,\mathsf{R})$ acting through (114) on the aberration generators that we collectively indicate by $\mathbf{A} = \{A_3, A_2\}$. We thus build the *fifth-order aberration group* with elements written in the *factored-product* parametrization (Dragt, 1982a, 1982b, 1987), as

$$\mathcal{G}_3(\mathbf{A}, \mathbf{M}) = \exp(\{A_3, \circ\}_{(3)}) \exp(\{A_2, \circ\}_{(3)}) \mathcal{C}_\mathbf{M} \in \mathcal{A}_3 \circledast \mathsf{Sp}(2,\mathsf{R}). \quad (119)$$

The product of two $\mathcal{A}_3 \mathsf{Sp}(2,\mathsf{R})$ elements, $\mathcal{G}_3(\mathbf{A}, \mathbf{M})$ and $\mathcal{G}_3(\mathbf{B}, \mathbf{N})$ is then

$$\begin{aligned}\mathcal{G}_3(\mathbf{A}, \mathbf{M}) \, \mathcal{G}_3(\mathbf{B}, \mathbf{N}) &= \mathcal{G}_3(\mathbf{A}, \mathbf{1}) \, \mathcal{C}_\mathbf{M} \, \mathcal{G}_3(\mathbf{B}, \mathbf{1}) \, \mathcal{C}_\mathbf{N} \\ &= \mathcal{G}_3(\mathbf{A}, \mathbf{1}) \, \mathcal{G}_3(\mathcal{C}_\mathbf{M} : \mathbf{B}, \mathbf{1}) \, \mathcal{C}_{\mathbf{MN}} \quad (120) \\ &= \mathcal{G}_3(\mathbf{A} \, \sharp \, (\mathcal{C}_\mathbf{M} : \mathbf{B}), \mathbf{MN}),\end{aligned}$$

where $\mathcal{C}_\mathbf{M} : \mathbf{B} = \mathcal{C}_\mathbf{M} \, \mathbf{B} \, \mathcal{C}_\mathbf{M}^{-1}$ is the linear transformation in $\mathcal{A}_3 \cup \mathcal{A}_2$, and $\mathbf{A} \, \sharp \, \mathbf{B}$ is the compounding of aberrations through the cut Poisson bracket $\{\circ, \circ\}_{(3)}$ in (112), that we called the *gato* multiplication (Wolf, 2004). The $7 + 5$ aberration polynomials indicated by $\mathbf{A}$ contain

$$A_3(p,q) = \sum_{m=-3}^{3} \alpha_{3,m} M_{3,m}(p,q), \quad A_2(p,q) = \sum_{m=-2}^{2} \alpha_{2,m} M_{2,m}(p,q), \quad (121)$$

and their *gato* product $\mathbf{C} = \mathbf{A} \, \sharp \, \mathbf{B}$ has generating polynomials

$$C_2(p,q) = A_2(p,q) + B_2(p,q), \quad (122)$$

$$C_3(p,q) = A_3(p,q) + B_3(p,q) + \tfrac{1}{2}\{A_2, B_2\}(p,q), \quad (123)$$

while $\{A_2, B_3\}_{(3)} = 0$ and $\{A_3, B_3\}_{(3)} = 0$. The group unit is $\mathcal{G}_3(\mathbf{0}, \mathbf{1})$, the inverse is $\mathcal{G}_3(\mathbf{A}, \mathbf{M})^{-1} = \mathcal{G}_3(-\mathcal{C}_\mathbf{M}^{-1} : \mathbf{A}, \mathbf{M}^{-1})$, and associativity holds. In Fig. 5 we show the maps produced by the monomial aberration functions in (116) on 2D phase space.

Application of $\mathcal{G}_3(\mathbf{A}, \mathbf{M})$ to the phase space coordinates (q, p) first performs the linear canonical transformation $\mathcal{C}_\mathbf{M}$, then acts with $\exp\{A_2, \circ\}_{(3)} = 1 + \{A_2, \circ\} + \tfrac{1}{2}\{A_2, \{A_2, \circ\}\}$ producing a polynomial of degrees up to 5 in the phase space variables, and lastly acts with $\exp\{A_3, \circ\}_{(3)} = 1 + \{A_3, \circ\}_{(3)}$, yielding the result as a polynomial with terms of degrees 1, 3, and 5. The transformation will be canonical up to rank $K = 3$ and the number of parameters $\alpha_{k,m}$, $k = 1, 2, 3$ and $m|_{-k}^{k}$, is thus $3 + 5 + 7 = 15$. The explicit

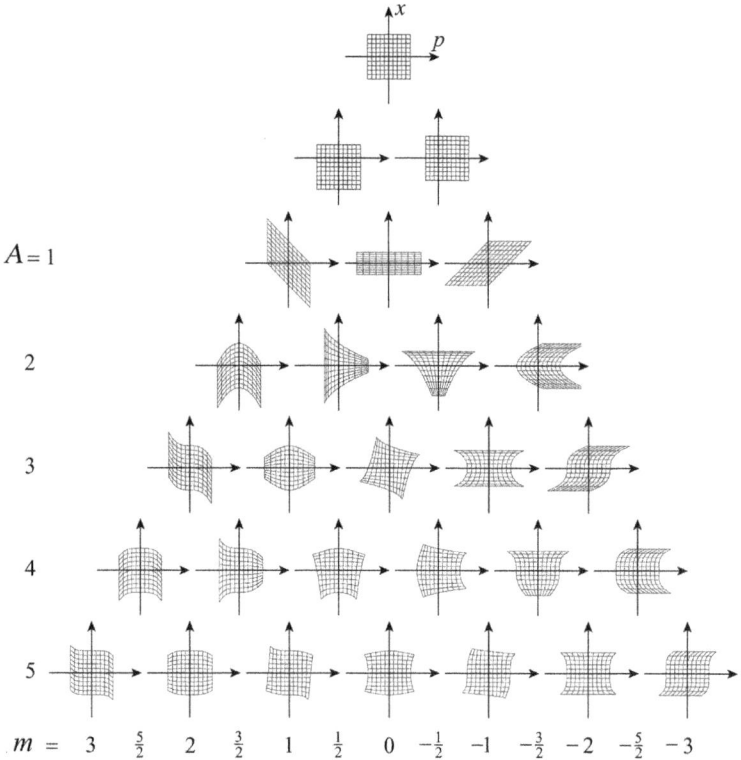

Fig. 5 Linear transformations and aberrations of classical phase space q, p, generated by the monomials (116), classified by aberration order $A = 2k - 1$ and weight m. The unit map is at the top; in the second row, Heisenberg–Weyl translations along q and p (zero aberration order). The three linear transformations ($A = 1$) are free propagation, magnification, and thin lens; higher-order aberrations of orders $A = 2, 3, 4, 5\ldots$ follow.

multiplication tables in terms of these aberration parameters for orders up to 7 are given in Wolf (2004, Part IV), which were calculated using Wolf and Krötzsch (1995).

8.2 Axis-Symmetric Aberrations in 3D Systems

Axis-symmetric linear transformations form a subgroup of the symplectic $\mathsf{Sp}(4,\mathsf{R})$ group of the paraxial model of Section 6, whose elements are block-diagonal, $\mathbf{M} = \begin{pmatrix} a\mathbf{1} & b\mathbf{1} \\ c\mathbf{1} & d\mathbf{1} \end{pmatrix}$, and whose generating Lie algebra is effectively $\mathsf{sp}(2,\mathsf{R})$. They represent 3D systems which are invariant under rotations around the z optical axis, and under reflections across planes that contain it. A basis for this algebra are the Poisson operators of the three functions[23]

$$M_{1,0,0} := |\mathbf{p}|^2, \quad M_{0,1,0} := \mathbf{p} \cdot \mathbf{q}, \quad M_{0,0,1} := |\mathbf{q}|^2. \tag{124}$$

The 3D counterpart of the monomials (111) are now

$$M_{k_+,k_0,k_-}(\mathbf{p},\mathbf{q}) := (|\mathbf{p}|^2)^{k_+} (\mathbf{p} \cdot \mathbf{q})^{k_0} (|\mathbf{q}|^2)^{k_-},$$

$$\text{of} \begin{cases} \text{rank} & k := k_+ + k_0 + k_- \in \{0,1,2,\ldots\}, \\ \text{weight} & m := k_+ - k_- \in \{k, k-1, \ldots, -k\}. \end{cases} \tag{125}$$

We can organize the monomials M_{k_+,k_0,k_-} along the axes k_σ, $\sigma \in \{+, 0, -\}$, where at a glance we see that this is the same diagram as that of the eigenstates of a 3D quantum harmonic oscillator, with k_σ energy quanta on the σ-axis. There is more than a passing analogy with boson SU(3) multiplets with (124) as quarks, of dimensions 1, 3, 6, 10, ...; it directs us to reduce each rank-k multiplet with respect to a "symplectc spin-j" into sub-multiplets that will not mix among each other under the Sp(2,R) linear canonical transformations generated by (124), which are the *same* as (115) with $D^j_{m,m'}\begin{pmatrix} a & b \\ c & d \end{pmatrix}$, where

$$\begin{aligned} \text{rank } k \text{ even} &\Rightarrow j \in \{0, 2, 4, \ldots, k\}, \\ k \text{ odd} &\Rightarrow j \in \{1, 3, 5, \ldots, k\}, \quad m, m'|_{-j}^{j}. \end{aligned} \tag{126}$$

This classification of aberrations is shown in Fig. 6, with the 'symplectic harmonics" defined as (Wolf, 2004, sect. 14.2)

$$Y_{k,j,m}(\mathbf{p},\mathbf{q}) = (\mathbf{p} \times \mathbf{q})^{k-j} Y_{j,j,m}(\mathbf{p},\mathbf{q}),$$

$$Y_{j,j,m}(\mathbf{p},\mathbf{q}) = \frac{(j+m)!\,(j-m)!}{2^j(2j-1)!!} \sum_{\nu \in N(j,m)} 2^\nu \frac{|\mathbf{p}|^{j+m-\nu}}{\left[\frac{1}{2}(j+m-\nu)\right]!} \frac{(\mathbf{p}\cdot\mathbf{q})^\nu}{\nu!} \frac{|\mathbf{q}|^{j-m-\nu}}{\left[\frac{1}{2}(j-m-\nu)\right]!}, \tag{127}$$

where $N(j, m) := \{j-|m|, j-|m|-2, \ldots, 0 \text{ or } 1\}$ and $n!! := n \cdot (n-2) \ldots 2$ or 1. We note the presence of $\mathbf{p} \times \mathbf{q}$ as factor in (127), which by itself is not symmetric under reflections, but appears with even powers and is invariant under Sp(2,R). There are *six* third-order axis-symmetric 3D aberrations; in the Cartesian basis (125), $M_{2,0,0}, M_{1,1,0}, \ldots, M_{0,0,2}$,

---

[23] We exclude the "angular momentum" function $\mathbf{p} \times \mathbf{q}$, which would be necessary for magnetic optics (Dragt, 2004).

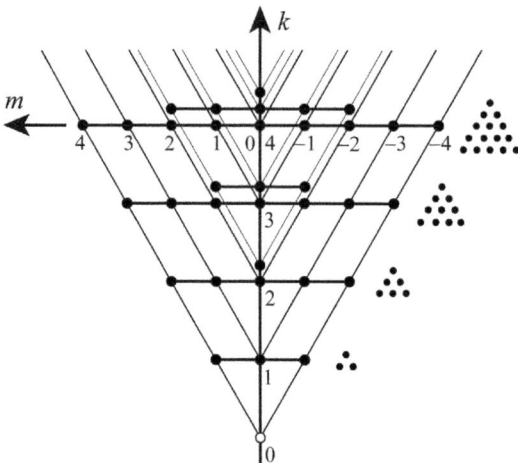

Fig. 6 The "symplectic harmonic" aberrations $Y_{k,j,m}(\mathbf{p},\mathbf{q})$ classified into multiplets j of the SO(3) rotation group (represented as *dots* joined by *horizontal lines*). For each k, on the *right*, the triangular multiplets of monomial aberrations M_{k_+,k_0,k_-} in (125) for the corresponding rank k. This is in exact analogy with the 3D quantum harmonic oscillator states.

where $M_{0,2,0}$ and $M_{1,0,1}$ have the same rank and weight k, m. In the symplectic basis (127) the last two are separated into $Y_{2,2,0} = \frac{1}{3}|\mathbf{p}|^2|\mathbf{q}|^2 + \frac{2}{3}(\mathbf{p}\cdot\mathbf{q})^2$ and $Y_{2,0,0} = (\mathbf{p}\times\mathbf{q})^2$, where the former is part of the "spin" quintuplet $Y_{2,2,m}$, and the latter is an invariant singlet. In Fig. 7 we show the spot diagrams[24] of the Cartesian and "spin" multiplets of rank $k = 2$.

The construction of the 3D axis-symmetric aberration group follows that of the 2D case in (119) and (120), except that $\mathcal{C}_\mathbf{M} : \mathbf{B}$ will now entail a 6×6 matrix in the Cartesian basis; in the symplectic spin basis this matrix is block-diagonal, with 5×5 and 1×1 submatrices. There is thus some computational advantage in using the spin basis for aberrations of higher order, but also in the geometric interpretation of the spot diagrams of these aberrations.

The use of aberration expansions may presently be obviated by fast and reliable ray-tracing computer algorithms. Still, this classification of phase space nonlinear maps clearly profits from the quantum harmonic oscillator state pattern and seems to be applicable to mechanical and other models, and extendable to a quantum/wave version. The monomials in (111) will straightforwardly "quantize" to essentially self-adjoint operators on

---

[24] A spot diagram in the optical context is the image of a nested set of cones (for various values of $|\mathbf{p}|$) that issue from a fixed point $\mathbf{q}$ away from the origin.

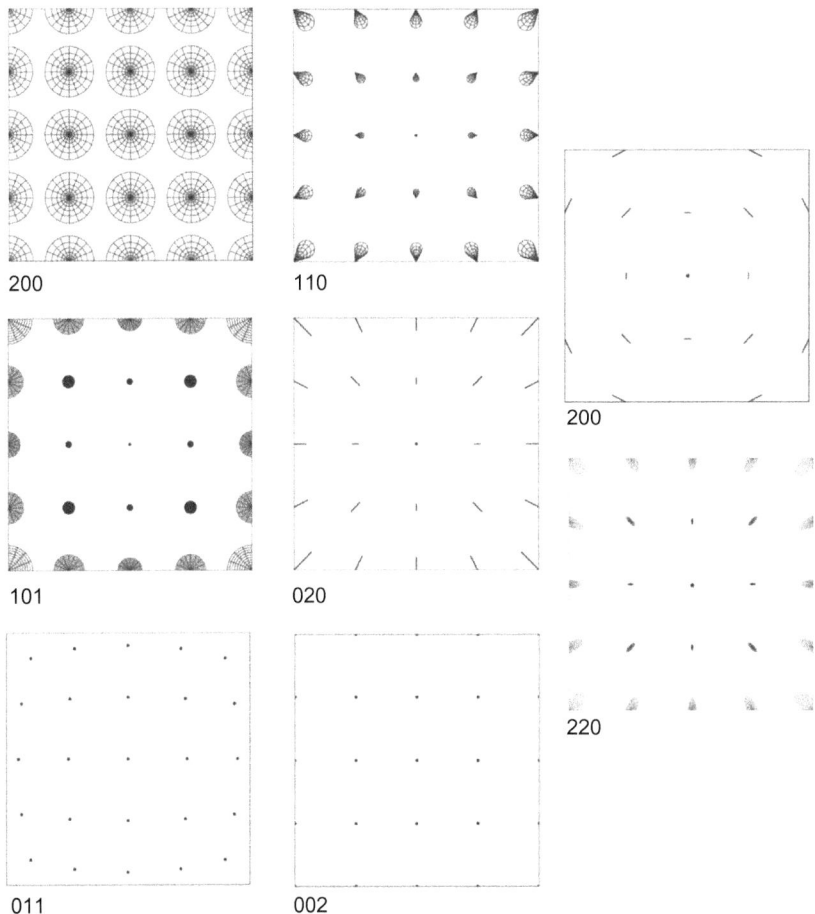

Fig. 7 *Left*: Spot diagrams of the monomial generator functions M_{k_+,k_0,k_-} in (125), indicated by the Cartesian indices $[k_+, k_0, k_-]$. *Right*: Spot diagrams of the two "symplectic harmonic" aberrations generated by $Y_{2,0,0}$ and $Y_{2,2,0}$, indicated by (k, j, m) that belong to the singlet and quintuplet irreducible representations, respectively, and are linear combinations of [1,0,1] and [0,2,0]. In [0, 1, 1] (distortion) and [0, 0, 2] (pocus) the spots are points; [0, 2, 0] and (2, 0, 0) are lines.

$\mathcal{L}^2(\mathrm{R}^2)$, only taking care to use the *Weyl symmetrization* for the noncommuting factors, which preserves their transformation properties under the linear symplectic subgroup. In Wolf (2004, chap. 15) we apply these techniques toward the correction of three fractional Fourier transform setups: a single axis-symmetric lens of polynomial surface, such a lens with a reflecting back surface, and a waveguide of polynomial refractive index profile.

9. DISCRETE OPTICAL MODELS

Many people would regard images on a finite pixelated screen as only distantly related to "continuous" geometric or wave optics; if at all, it would appear as obtained by sampling continuous images. But quite on the contrary, here we show that continuous optics is a *contraction* of the finite, pixelated, discrete model. Or conversely, we can say that the discrete model is a *deformation* of the continuous model (Boyer & Wolf, 1973; Wolf & Boyer, 1974) with a further Weyl noncompact-to-compact replacement. It is important to consider pixelated versions of optics because that is the nature of the data that is obtained from CCD sensor arrays, transformed and analyzed by computer algorithms (Pei & Ding, 2000; Pei & Yeh, 1997; Pei, Yeh, & Tseng, 1999).

With the same method we used to contract the Euclidean to the Heisenberg–Weyl Lie algebras for the paraxial model in Section 6, we shall contract the generators of rotation algebras to those of the Euclidean generators of Section 2. Then we shall show how pixelated screens contain a union of irreducible representations of the rotation group, and how they transform under the corresponding Fourier algebra (90)–(93) and group, including unitary rotations, gyrations, and preliminarily aberrations. Here, the requirement of canonicity is replaced by unitarity, i.e., reversibility and no loss of information.

9.1 The Contraction of so(4) to iso(3)

Consider the generators of rotations in a 4D space of Cartesian coordinates x_1, x_2, x_3, x_4, which span the special orthogonal Lie algebra so(4) of rotations in this space, realized as

$$\Lambda_{i,j} := i\,(x_i \partial_j - x_j \partial_i), \quad i,j|_1^4. \tag{128}$$

Now separate this set into two subsets: the generators $\{\Lambda_{1,2}, \Lambda_{1,3}, \Lambda_{2,3}\}$ of a subalgebra so(3), and the subset $\boldsymbol{\Lambda}_4 := \{\Lambda_{1,4}, \Lambda_{2,4}, \Lambda_{3,4}\}$ that transforms as a 3-vector under commutation with the former,

$$[\Lambda_{i,j}, \Lambda_{k,4}] = i\,(\delta_{j,k}\Lambda_{i,4} - \delta_{i,k}\Lambda_{j,4}). \tag{129}$$

As in (70)–(74) we introduce a change of scale on the 3-vector, defining $\boldsymbol{\Lambda}_4^{(\varepsilon)} := \varepsilon \boldsymbol{\Lambda}_4$, with a parameter ε destined to vanish. The so(3) generators $J_k := \Lambda_{i,j}$ (i, j, k cyclic) are left unscathed, as well as the transformation property (129) of $\Lambda_4^{(\varepsilon)}$ under rotation. But the commutator between any two of its components vanishes,

$$[\Lambda_{i,4}^{(\varepsilon)}, \Lambda_{j,4}^{(\varepsilon)}] = \varepsilon^2 [\Lambda_{i,4}, \Lambda_{j,4}] = i\,\varepsilon^2 \Lambda_{j,i} \xrightarrow{\varepsilon \to 0} 0. \tag{130}$$

In the limit, J_i remain as rotation generators while $T_i = \Lambda_{i,4}^{(0)}$ become translation generators of the inhomogeneous special orthogonal group $\mathsf{ISO}(3)$. Since this E_3 was referred to as the mother group of optical models, $\mathsf{SO}(4)$ could be called the *grand-mother* group. Of course, this contraction process also applies in N dimensions.

9.2 The Plane Pixelated Screen

We are interested in the *four*-dimensional Lie algebra of rotations $\mathsf{so}(4)$ because this dimension is special: it applies to the "real" case of pixelated 2D displays, and the algebra $\mathsf{so}(4)$ is—unique among all orthogonal algebras—a *direct sum*,[25]

$$\mathsf{so}(4) = \mathsf{su}(2)_x \oplus \mathsf{su}(2)_y, \tag{131}$$

This can be seen through writing the generators (128) as

$$\begin{aligned}
J_1^x &= \tfrac{1}{2}(\Lambda_{2,3} + \Lambda_{1,4}), & J_2^x &= -\tfrac{1}{2}(\Lambda_{1,3} - \Lambda_{2,4}), & J_3^x &= \tfrac{1}{2}(\Lambda_{1,2} + \Lambda_{3,4}), \\
J_1^y &= \tfrac{1}{2}(\Lambda_{2,3} - \Lambda_{1,4}), & J_2^y &= -\tfrac{1}{2}(\Lambda_{1,3} + \Lambda_{2,4}), & J_3^y &= \tfrac{1}{2}(\Lambda_{1,2} - \Lambda_{3,4}),
\end{aligned} \tag{132}$$

and verifying that for i, j, k cyclic,

$$[J_i^x, J_j^x] = i J_k^x, \quad [J_i^y, J_j^y] = i J_k^y, \quad [J_i^x, J_j^y] = 0. \tag{133}$$

The rotation algebra $\mathsf{so}(4)$ has two Casimir invariant operators, $(\vec{J}^x)^2$ and $(\vec{J}^y)^2$; their eigenvalues $j_x(j_x+1)$ and $j_y(j_y+1)$ determine that the generators will have spectra composed of equidistant points $m_x|_{-j_x}^{j_x}$ and $m_y|_{-j_y}^{j_y}$, and eigenfunction multiplets of $N_x = 2j_x+1$ and $N_y = 2j_y+1$ functions.

The gist of defining a discrete model is to associate the generators of $\mathsf{su}(2)_x \oplus \mathsf{su}(2)_y$ to *operators of position* Q_x, Q_y with eigenvalues $q_x|_{-j_x}^{j_x}$, $q_y|_{-j_y}^{j_y}$, momentum P_x, P_y, and *mode* $H_x := J_3^x + j_x \mathbf{1}$ and $H_y := J_3^y + j_y \mathbf{1}$ with eigenvalues $n_x|_0^{2j_x}$, $n_y|_0^{2j_y}$. Then we can identify the pixels of an in general $N_x \times N_y$ rectangular screen with the elements of a complex matrix with rows and columns labeled by (q_x, q_y). On these we shall determine the *mode* functions $\Psi_{n_x, n_y}^{j_x, j_y}(q_x, q_y)$ of a finite model of the harmonic oscillator. This should not

---
[25] Although as Lie algebras $\mathsf{so}(3) = \mathsf{su}(2)$, when one examines the group *manifold* one finds that indeed $\mathsf{SO}(4) = \mathsf{SU}(2)_x \otimes \mathsf{SU}(2)_y$.

be surprising, since a point on a rotating sphere projects as harmonic motion on a screen.

9.3 The Kravchuk Oscillator States

Let us work first with an $N \times 1$ pixelated screen line. The identification of phase space operators suggested above sets

$$\begin{aligned}\text{position } Q &:= J_1, & \text{momentum } P &:= -J_2, \\ K &:= J_3, & \text{mode } H &:= K + j\mathbf{1}.\end{aligned} \quad (134)$$

The su(2) commutation relations (133) then become

$$[K, Q] = -iP, \quad [K, P] = iQ, \quad [Q, P] = -iK. \quad (135)$$

The first two commutators correspond to the two Hamilton equations for quantum position and momentum operators under a harmonic oscillator Hamiltonian $H_{\text{osc}} = \frac{1}{2}(P^2 + Q^2)$. However, the third su(2) commutator differs from the usual quantum commutator $[Q, P] = i\lambdabar \mathbf{1}$, and this marks the difference between the finite and the "continuous" models of the harmonic oscillator.

The finite oscillator wavefunctions are the overlaps between the eigenfunctions of the position generator Q and the mode generator H. Since J_1 and J_3 are related by a $\frac{1}{2}\pi$ rotation around the J_2 axis, quantum angular momentum theory identifies the overlap as a Wigner *little-d* function (Biedenharn & Louck, 1981) for that angle, and given by

$$\Psi_n^j(q) := d_{n-j,q}^j\left(\tfrac{1}{2}\pi\right) \qquad n|_0^{2j}, \quad q|_{-j}^j \quad (136)$$

$$= \frac{(-1)^n}{2^j}\sqrt{\binom{2j}{n}\binom{2j}{j+q}} K_n\left(j+q; \tfrac{1}{2}, 2j\right),$$

$$K_n\left(s; \tfrac{1}{2}, 2j\right) = {}_2F_1(-n, -s; -2j; 2) = K_s\left(n; \tfrac{1}{2}, 2j\right), \quad (137)$$

for $s|_0^{2j}$, where $K_n(s; \tfrac{1}{2}, 2j)$ is a symmetric Kravchuk polynomial (Krawtchouk, 1928), $\binom{m}{n}$ are the binomial coefficients, and ${}_2F_1(a, b; c; z)$ is the Gauss hypergeometric function. These we call the *Kravchuk functions* on the discrete positions of the finite oscillator model.

The Kravchuk functions belong to su(2) multiplets and have been detailed in several papers (Atakishiyev, Pogosyan, & Wolf, 2005; Atakishiyev & Wolf, 1997). They are shown in Fig. 8, where it can be seen that the lowest n-modes $\Psi_n^j(q)$ closely resemble the continuous

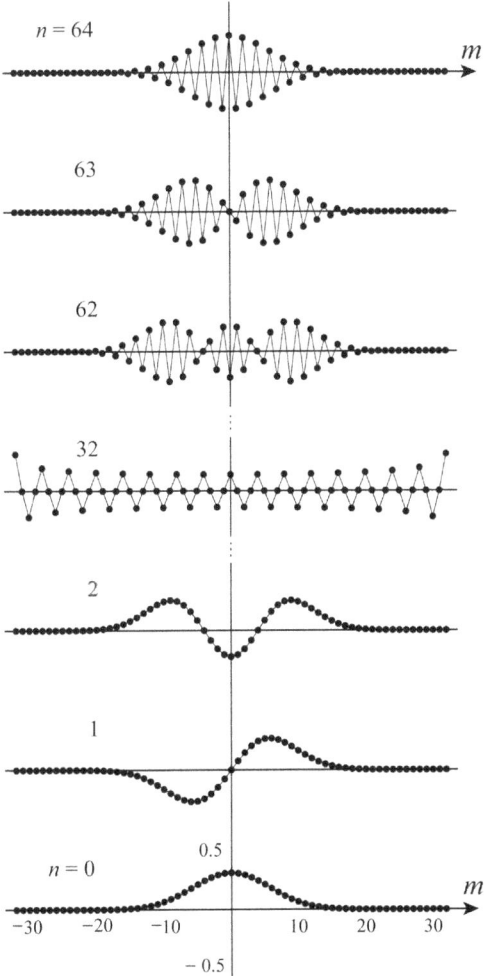

Fig. 8 The 1D Kravchuk states $\Psi_n^j(q)$ in (136), for $j = 32$, on the 65 points $m|_{-32}^{32}$ (joined by *straight lines* for visibility), and modes $n|_0^{64}$ from *bottom* to *top*. The $n = 0$ ground state of the finite oscillator is the square root of the binomial distribution, approximating the Gaussian harmonic oscillator wavefunction. The highest $n = 64$ state of the finite oscillator reproduces the ground state with a change of sign between neighboring points.

Hermite–Gauss states in continuous wave optics; they are real, have definite parity and, for higher $n > j$, they alternate in sign between each pair of neighboring points,

$$\Psi_n^j(-q) = (-1)^n \Psi_n^j(q), \quad \Psi_{2j-n}^j(q) = (-1)^q \Psi_n^j(q). \qquad (138)$$

The finite oscillator Kravchuk functions are orthonormal and complete in the N-dimensional vector space of images on the 1D pixelated screen,

$$\sum_{q=-j}^{j} \Psi_n^j(q) \Psi_{n'}^j(q) = \delta_{n,n'}, \quad \sum_{n=0}^{2j} \Psi_n^j(q) \Psi_n^j(q') = \delta_{q,q'}. \qquad (139)$$

Functions $f(q)$ of $2j+1$ points $q|_{-j}^{j}$, are expanded in the Kravchuk basis as,

$$f(q) = \sum_{n=0}^{2j} f_n \Psi_n^j(q), \quad f_n := \sum_{q=-j}^{j} f(q) \Psi_n^j(q). \qquad (140)$$

When the number and density of points grows without bound, $j \to \infty$, so that q becomes the real line, it can be shown that the Kravchuk functions (136) limit to the usual Hermite–Gaussian wavefunctions (Atakishiyev, Pogosyan, & Wolf, 2003). This is a contraction of $u(2) = u(1) \oplus su(2)$, where $u(1)$ provides the representation label j for $su(2)$, to the oscillator algebra generated by $\{1, Q, P, H\}$.

The Kravchuk eigenfunctions satisfy $Q\Psi_n^j(q) = q\Psi_n^j(q)$ and $H\Psi_n^j(q) = n\Psi_n^j(q)$; hence the finite rotation generated by $\mathcal{F}^\alpha := e^{-i\frac{1}{2}\pi\alpha H}$ (α mod 4), only multiplies them by phase,

$$\mathcal{F}^\alpha \Psi_n^j(q) = \exp\left(-i\frac{1}{2}\pi n \alpha\right) \Psi_n^j(q), \qquad (141)$$

and qualifies to be called the fractional *Fourier–Kravchuk* discrete transform.[26] This transform is realized by a matrix kernel that acts on the vector of discrete "image" values (Wolf & Krötzsch, 2007),

$$f(q) \mapsto \tilde{f}_\alpha(q) = (\mathcal{F}^\alpha f)(q) = \sum_{q'=-j}^{j} F_{q,q'}^j(\alpha) f(q'), \qquad (142)$$

$$F_{q,q'}^j(\alpha) = \sum_{n=0}^{2j} \Psi_n^j(q) e^{-i\frac{1}{2}\pi n \alpha} \Psi_n^j(q') = e^{-i\frac{1}{2}\pi(q-q')} d_{q,q'}^j(\alpha). \qquad (143)$$

Thus we have an SO(2) subgroup of rotations around the K axis with a phase built as a finite counterpart of the Namias expression (Namias, 1980) for α-fractional Fourier transforms. A rotation (141) by $\frac{1}{2}\pi$

---

[26] When the number of pixels $N = 2j + 1$ is odd, j is integer and we have a pixel at the center of the array. When N is even we are in the half-integer spin representations; the Fourier–Kravchuk transform (141) "corrects" the double spin range $\alpha \in [0, 4\pi)$ of $e^{-i\frac{1}{2}\pi K \alpha}$ with the extra phase $e^{-i\frac{1}{2}\pi j \alpha}$, as was the case in (110).

brings the Q axis of position onto the P axis of momentum, so the Fourier–Kravchuk transforms of $\Psi_n^j(q)$ are $\widetilde{\Psi}_n^j(p) = (-\mathrm{i})^n \Psi_n^j(p)$, as is familiar from quantum mechanics.[27]

9.4 2D Screens and U(2)$_F$ Transformations

Two-dimensional pixelated screens can be described basically as a Cartesian product of two one-dimensional ones. We now have two sets of generators (134) for $\mathsf{su}(2)_x \oplus \mathsf{su}(2)_y$, to name those in (132),

$$\begin{aligned} Q_x &= J_1^x, & P_x &= -J_2^x, & K_x &= J_3^x, \\ Q_y &= J_1^y, & P_y &= -J_2^y, & K_y &= J_3^y, \end{aligned} \tag{144}$$

satisfying (135) and mutually commuting—and $H_x = K_x + j_x 1$, $H_y = K_y + j_y 1$ that provide the pair of mode numbers n_x and n_y. The 2D Cartesian Kravchuk functions are (Atakishiyev, Pogosyan, Vicent, & Wolf, 2001b)

$$\begin{aligned} \Psi_{n_x, n_y}^{(j_x, j_y)}(q_x, q_y) &:= \Psi_{n_x}^{(j_x)}(q_x) \Psi_{n_y}^{(j_y)}(q_y), \\ q_x \big|_{-j_x}^{j_x}, \quad n_x \big|_0^{2j_x}, \quad q_y \big|_{-j_y}^{j_y}, \quad n_y \big|_0^{2j_y}. \end{aligned} \tag{145}$$

These $N_x N_y$ Kravchuk functions can be arranged along axes of total mode $n := n_x + n_y$ and mode difference $m := n_x - n_y$ into the rhomboid pattern shown in Fig. 9. These modes are orthonormal and complete under the natural sesquilinear inner product on $\mathbf{C}^{N_x N_y}$. We shall consider first the general case of rectangular screens, with $j_x > j_y$; the special case $j_x = j = j_y$ will deserve some extra attention in the next subsection.

9.4.1 Domestic Fourier–Kravchuk Transformations

In two dimensions we have the Fourier group $\mathsf{U}(2)_F$ generated by the Poisson operators of the classical functions $\ell_i\big|_{i=0}^3$ in (90)–(93). We evidently associate the isotropic ℓ_0 in (90), and anisotropic ℓ_1 in (91), to the fractional Fourier–Kravchuk transform seen in the last subsection. Note the factor $\frac{1}{4}$ in their expressions, and the factor $\frac{1}{2}$ in front of "physical" angular momentum in (93), which imply that we must take the *double* of the angle $\frac{1}{2}\pi\alpha$ in (141). Let $F_0 := \frac{1}{2}(H_x + H_y)$ and $F_1 := \frac{1}{2}(H_x - H_y)$; the isotropic and anisotropic fractional Fourier–Kravchuk transforms $\mathcal{F}_I(\chi) := \exp(-2\mathrm{i}\chi F_0)$ and $\mathcal{F}_A(\beta) := \exp(-2\mathrm{i}\beta F_1)$ are *domestic* to the discrete model and act on the Cartesian functions (145) as

---

[27] Note that the Fourier–Kravchuk transform is *not* "exactly" the discrete Fourier transform, but the form (143) indicates that, in the limit $j \to \infty$ referenced below, both converge to the Fourier integral transform kernel.

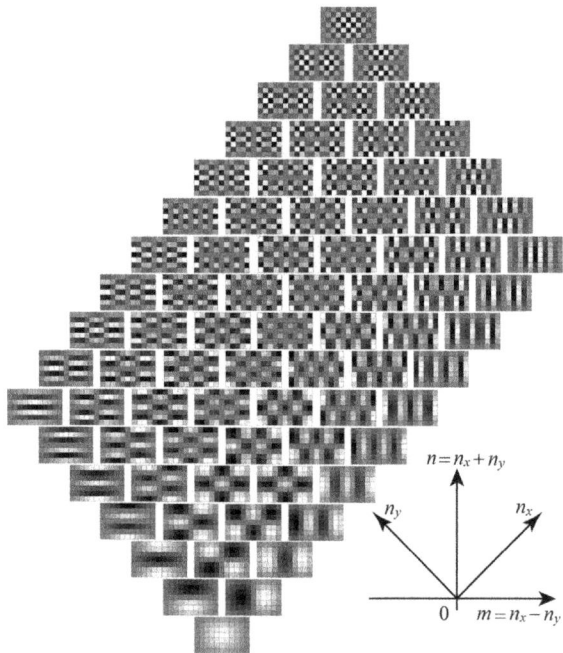

Fig. 9 The 11×7 rhomboid of Cartesian modes $\Psi_{n_x,n_y}^{(5,3)}(q_x,q_y)$ in (145), referred to axes $n_x|_0^{10}$ and $n_y|_0^6$ and also to axes n, m. In each screen, the pixels are numbered from the lower-left corner by $q_x|_{-5}^5$ and $q_y|_{-3}^3$. Gray-level densities are black and white for values from -1 to 1.

$$\mathcal{F}_I(\chi) : \Psi_{n_x,n_y}^{(j_x,j_y)}(q_x,q_y) = \exp\left[-\mathrm{i}\chi(n_x + n_y)\right] \Psi_{n_x,n_y}^{(j_x,j_y)}(q_x,q_y), \qquad (146)$$

$$\mathcal{F}_A(\beta) : \Psi_{n_x,n_y}^{(j_x,j_y)}(q_x,q_y) = \exp\left[-\mathrm{i}\beta(n_x - n_y)\right] \Psi_{n_x,n_y}^{(j_x,j_y)}(q_x,q_y). \qquad (147)$$

9.4.2 Imported Rotations

Next consider *rotations* of the pixelated image in a rectangular screen. We conjecture that if we use the well-known rotation coefficients—the Wigner $d_{\mu,\mu'}^j(2\theta)$'s with double angle—on the Cartesian Hermite–Gauss oscillator functions (Frank & van Isacker, 1994), and *import* (Barker, Çandan, Hakioğlu, Kutay, & Ozaktas, 2000) them to the discrete model, we should obtain a recognizable rotation of the pixelated image, which is real, and will be unitary (orthogonal) and thus reversible. Recall that the continuous quantum harmonic oscillator states (n_x, n_y) form an infinite pyramid with rungs $n_x + n_y = n|_0^\infty$ that are angular momentum multiplets of spin

$\lambda(n) = \frac{1}{2}n$ and z-projection $\frac{1}{2}(n_x - n_y) = \mu|_{-\lambda}^{\lambda}$. In the discrete model, however, we see in Fig. 9 that we have only the lowest part of that pyramid, in a rhombus where the spins $\lambda(n)$ and "z-projectons" μ, μ' are now constrained to the following ranges (for $j_x \geq j_y$):[28]

$$\begin{aligned}
&\text{lower triangle:}\\
&\quad 0 \leq n \leq 2j_y, \qquad \lambda(n) = \tfrac{1}{2}n, \qquad \mu = \tfrac{1}{2}(n_x - n_y);\\
&\text{mid-rhomboid:}\\
&\quad 2j_y < n < 2j_x, \qquad \lambda(n) = j_y, \qquad \mu = j_y - n_y;\\
&\text{upper triangle:}\\
&\quad 2j_x \leq n \leq 2(j_x + j_y), \quad \lambda(n) = j_x + j_y - \tfrac{1}{2}n, \quad \mu = \tfrac{1}{2}(n_x - n_y) - j_x + n_y.
\end{aligned} \qquad (148)$$

Thus we posit that rotations $\mathcal{R}(\theta)$ of the discrete modes (145) are

$$\mathcal{R}(\theta) : \Psi_{n_x, n_y}^{(j_x, j_y)}(q_x, q_y) := \sum_{n'_x + n'_y = n} d_{\mu, \mu'}^{\lambda(n)}(2\theta) \, \Psi_{n'_x, n'_y}^{(j_x, j_y)}(q_x, q_y), \qquad (149)$$

where μ, μ' are given in terms of n_x, n_y (unprimed and primed) by (148).

In Fig. 10 we rotate a 41×25 pixelated image of the letter "$\mathcal{B}$", white on black (1's on 0's), through six successive rotations by $\theta = \frac{1}{6}\pi$. We note that oscillations with small negative values appear in all intermediate positions; this type of Gibbs oscillation is common to all signal reconstruction algorithms with Fourier series when the signals have sharp edges. For $\theta = \frac{1}{2}\pi$ the image suffers expansion in x and compression in y with the concomitant oscillations, but for $\theta = \pi$ we recover the same, exact (inverted) image, as through a permutation of pixels. On square screens (Vicent & Wolf, 2008, 2011), rotations by $\theta = \frac{1}{2}\pi$ are also permutations. This would be impossible

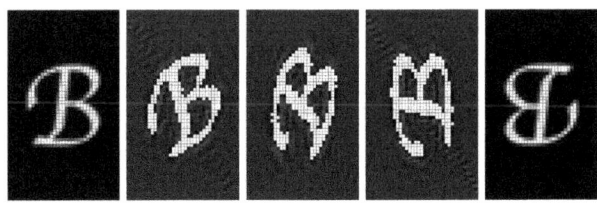

Fig. 10 Image of the letter "$\mathcal{B}$" on a 41×25 pixelated screen, $(j_x, j_y) = (20, 12)$, under successive rotations by $\frac{1}{6}\pi$ of the *left*, to $\frac{1}{3}\pi$, $\frac{1}{2}\pi$, and (extreme *right*) π. In the rotated images, the gray-level scale is rescaled so that the pixel values lie between 0 and 1.

---

[28] The ranges of n do overlap at $2j_y$ and at $2j_x$; we ascribe these to the triangles.

with any successively applied interpolation algorithm, since they inevitably loose information. Although the rotations in the present discrete model of rectangular screens are unitary, they also embody the *longest* computation algorithm, because each pixel on the transformed screen is a linear combination of *all* pixels in the original image.

9.4.3 Gyrations

Finally, consider the Fourier subgroup of *gyrations* $\mathcal{G}(\gamma)$ generated by the classical quadratic function ℓ_2 in (92) that generates rotations around the 2-axis of a mathematical "Fourier sphere." But since we already have domestic $\mathcal{F}_A(\beta)$ transformations (147) around the 1-axis, and the imported rotations $\mathcal{R}(\theta)$ around its 3-axis, remembering the double angle issue, we can write

$$\mathcal{G}(\gamma) := \mathcal{F}_A\left(\tfrac{1}{4}\pi\right) \mathcal{R}(\gamma) \mathcal{F}_A\left(-\tfrac{1}{4}\pi\right). \tag{150}$$

The action of gyrations on the basis of Cartesian Kravchuk modes is then

$$\mathcal{G}(\gamma) : \Psi_{n_x,n_y}^{(j_x,j_y)}(q_x,q_y) \\ := e^{-i\pi(n_x-n_y)/4} \sum_{n'_x+n'_y=n} d_{\mu,\mu'}^{\lambda(n)}(2\gamma)\, e^{+i\pi(n'_x-n'_y)/4}\, \Psi_{n'_x,n'_y}^{(j_x,j_y)}(q_x,q_y). \tag{151}$$

where $\lambda(n)$, μ and μ' are again determined by j_x, n_x, j_y, n_y as in (148).

9.4.4 Laguerre–Kravchuk Modes

In wave models, fractional gyrations transform continuously the Hermite–Gauss beams from $\gamma = 0$, into Laguerre–Gauss beams for $\gamma = \tfrac{1}{4}\pi$ (Alieva, Bastiaans, & Calvo, 2005; Rodrigo, Alieva, & Bastiaans, 2011; Rodrigo, Alieva, & Calvo, 2007). In the present discrete model we show in Fig. 11 gyrations of $\Psi_{n_x,n_y}^{(j_x,j_y)}(q_x,q_y)$ for the quintuplet of $\lambda = 2$ states $n = 4$, noting that after a $\tfrac{1}{4}\pi$ we indeed obtain a credible discrete analogue of Laguerre–Gauss beams which, for lack of another name, we may call (*rectangular*) *Laguerre–Kravchuk* states,

$$\Lambda_{n,m}^{(j_x,j_y)}(q_x,q_y) := e^{-i\pi(n_x-n_y)/4} \sum_{n'_x+n'_y=n} d_{\mu,\mu'}^{\lambda(n)}\left(\tfrac{1}{2}\pi\right) e^{+i\pi(n'_x-n'_y)/4}\, \Psi_{n'_x,n'_y}^{(j_x,j_y)}(q_x,q_y). \tag{152}$$

Since these are complex functions, in Fig. 11 for $\gamma = \tfrac{1}{4}\pi$ we show separately the absolute values and phases. The chosen multiplet lies in the lower triangle of (148); the upper triangle yields the same absolute values with a

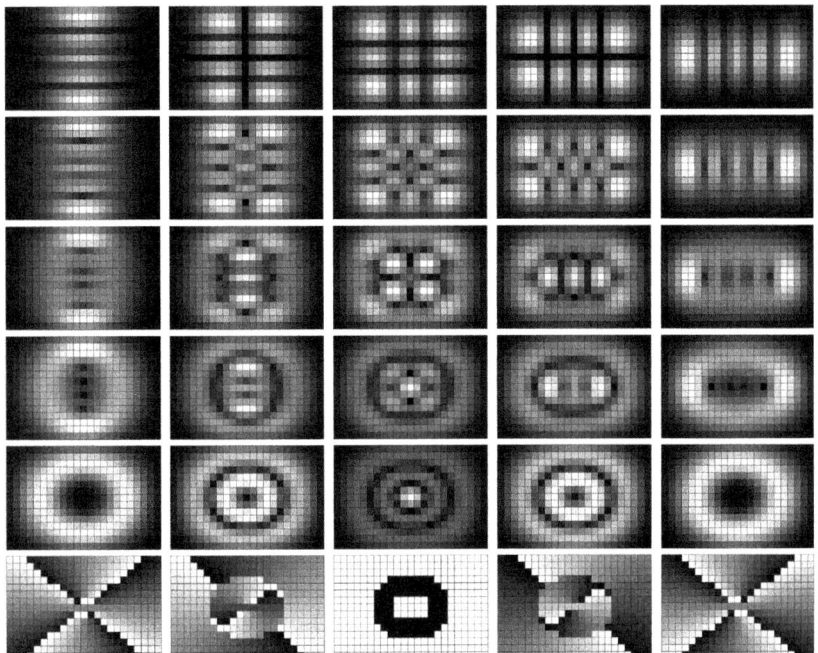

Fig. 11 Gyrations of the quintuplet $\lambda = 2$ ($m = -2, -1, 0, 1, 2$) of Cartesian Kravchuk modes $n = 4$ on 11×7 pixelated screens ($j_x = 5, j_y = 3$), through angles of $\gamma = 0, \frac{1}{16}\pi, \frac{1}{8}\pi, \frac{3}{16}\pi$, and $\frac{1}{4}\pi$ in the last two *lines*, where we show the absolute values (since the image values are complex), and the phase of the $\frac{1}{4}\pi$ gyration. The latter are the Laguerre–Kravchuk states of "rectangular angular momentum."

checkerboard of $e^{i\pi}$ differences in phase between neighbor pixels; multiplets in the mid-rhomboid follow a similar pattern. The functions (152) are also orthogonal and complete in $\mathbf{C}^{N_x N_y}$ so they can serve as an alternate basis for images; they transform under rotations by phases, with "angular momentum" number $m = 2\mu = n_x - n_y$, $|\mu| \leq \lambda(n)$, constrained by $n := n_x + n_y$ through (148).

9.5 Square and Circular Pixelated Screens

We can profit from a subalgebra chain of so(4) distinct from (131), namely the natural Gel'fand–Zetlin-type chain (Wong, 1967),

$$\mathsf{so}(4) \supset \mathsf{so}(3) \supset \mathsf{so}(2), \tag{153}$$

particularly when the screen is an $N \times N$ *square*, with $N = 2j+1$. The generators $\{\Lambda_{i,j}\}_{1=i<j}^{4}$, of so(4), with their commutation relations (129), will be

now renamed with *new* position and momentum operators identified by a circle superscript,

$$Q_x^\circ := \Lambda_{2,3} = Q_x + Q_y, \quad P_x^\circ := -\Lambda_{1,3} = P_x + P_y,$$
$$Q_y^\circ := \Lambda_{2,4} = P_x - P_y, \quad P_y^\circ := -\Lambda_{2,3} = Q_x - Q_y, \quad (154)$$
$$M^\circ := \Lambda_{3,4} = K_x - K_y, \quad K^\circ := \Lambda_{1,2} = K_x + K_y,$$

where we note that M° and K° commute and thus can be both diagonal. From (129) we can see that M° generates rotations between the x- and y-components of Q° and of P° as angular momentum operators do, while K° generates rotations between Q_i° and the corresponding P_i° as the fractional Fourier transform does. On the other hand, note that the two "components of position," Q_x° and Q_y° do *not* commute, and neither do the two P°'s.

The subalgebra so(3) in (153) that we choose, which will yield the position of pixels on a screen, contains the generators Q_x°, Q_y°, and M°; call this subalgebra so(3)$_Q$. Its Casimir invariant operator will have the usual distribution of eigenvalues

$$R_\circ^2 := (Q_x^\circ)^2 + (Q_y^\circ)^2 + (M^\circ)^2 = \rho(\rho+1)\mathbf{1}, \quad \rho|_0^{2j}, \quad (155)$$

and the basis elements are classified by the eigenvalues of $M^\circ = m\,\mathbf{1}$, $m|_{-\rho}^{\rho}$. The operator $H^\circ := K^\circ + 2j\mathbf{1}$ commutes with the generators of so(3)$_Q$ and is also diagonal with eigenvalues $n_x + n_y = n|_0^{4j}$ in the rhombus (148) with the mid-rhomboid now absent, as seen in Fig. 12.

Using the generators $\vec{J}^x$ and $\vec{J}^y$ in (132), the Cartesian modes $\Psi_{n_x,n_y}^{(j,j)} \equiv \Psi_{n,m}^j$ written now with indices n, m,[29] are

$$\text{eigenbasis of } \quad (\vec{J}^x)^2, \quad (\vec{J}^y)^2, \quad J_3^x, \quad J_3^y,$$
$$\text{with eigenvalues } \quad j(j+1), \quad j(j+1), \quad \tfrac{1}{2}(n+m)-j, \quad \tfrac{1}{2}(n-m)-j. \quad (156)$$

We define states $Q_{\rho,m}^j$ related to the "position" so(3)$_Q$ subalgebra in (155), as

$$\text{eigenbasis of } \quad (\vec{J}^x+\vec{J}^y)^2, \quad \vec{J}^x\cdot\vec{J}^y, \quad R_\circ^2, \quad M^\circ,$$
$$\text{with eigenvalues } \quad 2j(j+1), \quad 0, \quad \rho(\rho+1), \quad m. \quad (157)$$

Both $\{\Psi_{n,m}^j\}$ and $\{Q_{\rho,m}^j\}$ are orthogonal bases of $N^2 = (2j+1)^2$ states. Their *overlap* is clearly a *coupling* of two so(3) representations j to a third ρ, and thus given by Clebsch–Gordan coefficients $C_{m_1,m_2,m_3}^{j_1,\,j_2,\,j_3}$. Wigner's definition of

---
[29] Recall that $n = n_x+n_y$ and $m = n_x - n_y$, so $n_x = \tfrac{1}{2}(n+m)$ and $n_y = \tfrac{1}{2}(n-m)$.

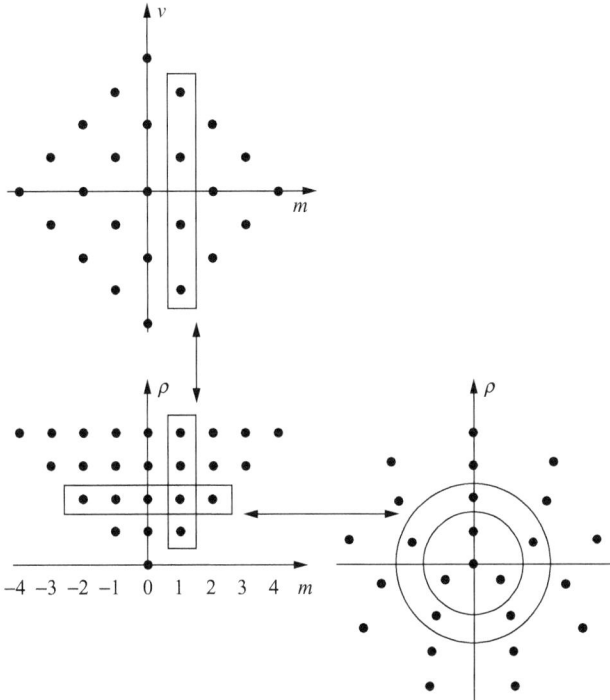

Fig. 12 *Top*: Cartesian states $\Psi^{(j,j)}_{n_x,n_y} \equiv \Psi^{j}_{n,m}$ (indicated by *dots*) in a symmetric so(4) multiplet on a square screen. *Left*: Cartesian states with the same $m = n_x - n_y$ are linearly combined (indicated by the *thin rectangles*) into "polar" states $\Phi^{j}_{\rho,m}$ with the same m, and belonging to so(3) multiplets $\rho|^{2j}_{|m|}$. *Right*: The finite Fourier transform maps the polar states $\Phi^{j}_{\rho,m}$, $m|^{\rho}_{-\rho}$ on a screen where the pixels (represented by *dots*) are on circles of radii $\rho|^{2j}_{0}$ and, on each *circle*, distributed by $2\rho+1$ angles θ_k, $k|^{\rho}_{-\rho}$.

these coefficients (Biedenharn & Louck, 1981), however, involves the subalgebra $\{\Lambda_{i,j}\}^{3}_{1 \le i < j}$, while our $\mathsf{so}(3)_Q$ is generated by $\{\Lambda_{i,j}\}^{4}_{2 \le i < j}$. A rotation is necessary, which introduces a phase and sign reversal of the J^{y}_{3} eigenvalue, yielding the overlap (Atakishiyev, Pogosyan, Vicent, & Wolf, 2001a; Vicent & Wolf, 2008),

$$R^{j}_{n,m}(\rho) := (Q^{j}_{\rho,m}, \Psi^{j}_{n,m}) = \varphi^{j,n}_{\rho,m} C^{j,\quad j,\quad \rho}_{\frac{1}{2}(m+n)-j,\,\frac{1}{2}(m-n)+j,\,m} \tag{158}$$

$$\varphi^{j,n}_{\rho,m} := (-1)^{j+\rho+\frac{1}{2}(|m|-m)} e^{i\frac{\pi}{2}n}, \tag{159}$$

which include the restrictions $0 \le \rho \le 2j$ and $|m| \le \rho$. We can regard $R^{j}_{n,m}(\rho)$ as a function of a *radius* ρ, on whose circle we have $2\rho+1$ pixels at equidistant angles ϕ_k, $k|^{\rho}_{-\rho}$, as shown in Fig. 12 (right).

Finally, we build the discrete basis of wavefields[30]

$$\Phi^j_{n,m}(\rho,\phi_k) := R^j_{n,m}(\rho)\,\frac{\exp(im\phi_k)}{\sqrt{2\rho+1}},\quad \phi_k = \frac{2\pi k}{2\rho+1}. \qquad (160)$$

These are shown in Fig. 13. By construction, they are orthonormal and complete under inner products over modes and angular momenta n, m and over *positions* of radius and angle ρ, ϕ_k on a polar-pixelated screen.[31] This pattern of pixels comes closest to contain pixels of equal size, except a bit near the origin $\rho = 0$. Another orthonormal and complete basis for modes and angular momenta n, m, are the Laguerre–Kravchuk states (152), $\Lambda^{(j,j)}_{n,m}(q_x,q_y)$ on the square screen Cartesian coordinates q_x,q_y.

We can thus transform between images f and f° on the Cartesian and polar screens through

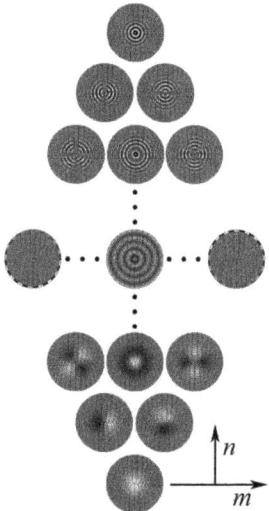

Fig. 13 The rhombus n, m of Laguerre–Kravchuk states $\Phi^j_{n,m}(\rho,\phi_k)$ in (160) for $j=32$, on the circular pixelated screen for radii $\rho|_0^{64}$ and $2\rho+1$ angles $\phi_k = 2\pi k/(2\rho+1)$. The modes are complex, $\Phi^j_{n,-m}(\rho,\phi_k) = \Phi^j_{n,m}(\rho,\phi_k)^*$, so their real parts are shown on the right-hand side $m > 0$, and their imaginary part on the left-hand side $m < 0$; the $m = 0$ modes are real.

---

[30] Please note that there is an error in Eq. (34) of Vicent and Wolf (2008).
[31] One can add fixed angles θ_ρ to the ϕ_k's on each ρ-circle in the definition (160). This will only shift the starting angle on each circle. We have found it not inconvenient to let $\theta_\rho = 0$, even if this results in one radial line of $2j$ aligned pixels.

$$f(q_x,q_y) = \sum_{n,m} U^j(\rho,\phi_k;q_x,q_y)^* f^\circ(\rho,\phi_k),$$

$$f^\circ(\rho,\phi_k) = \sum_{q_x,q_y} U^j(\rho,\phi_k;q_x,q_y) f(q_x,q_y),$$

(161)

with a kernel that is the sum over the modes and angular momenta in the two bases,

$$U^j(\rho,\phi_k;q_x,q_y) := \sum_{n,m} \Phi^j_{n,m}(\rho,\phi_k) \Lambda^{(j,j)}_{n,m}(q_x,q_y)^*.$$

(162)

Since $\Phi^j_{n,m}(\rho,\phi_k) = \Phi^j_{n,-m}(\rho,\phi_k)^*$ and $\Lambda^{(j,j)}_{n,m}(q_x,q_y)^* = \Lambda^{(j_x,j_y)}_{n,-m}(q_x,q_y)$, this kernel is real. In Fig. 14 we show an image (a letter "R") mapped between a Cartesian and a polar screen. This map can be seen as the discrete analogue of *separation of variables* between continuous Cartesian and polar coordinates.

At this point the reader may rightfully suspect that images on *rectangular* pixelated screens can also be mapped faithfully on some other screen geometry. This has been tried by simply noting that the Clebsch–Gordan coefficients in (158) would now couple $j_x > j_y$ to radii $\rho|^{j_x+j_y}_{j_x-j_y}$, forming an *annular* screen, with the same angles $\phi_k|^\rho_{-\rho}$ as in (160). This could be perhaps useful in Newtonian telescopes. The result, however, (Urzúa & Wolf, 2016) shows the images on the annular screen to be very distorted even for small j_x-j_y and hardly recognizable for larger values. We also considered possible *elliptic* screens, but the problem of distributing equally sized pixels along elliptic coordinates given by operator eigenvalues is challenging.

9.6 Aberrations of 1D Finite Discrete Signals

Transformations can be applied to finite data sets that will correspond to the geometric optical aberrations introduced in Section 8. Their study has been

Fig. 14 The image "R" on a 32 × 32 Cartesian pixelated screen (valued 0 and 1), unitarily transformed onto a circular screen of 32^2 pixels.

concentrated on 1D discrete and finite signals on their *phase space* by means of a Wigner function on sets of $N = 2j+1$ points. This Wigner function is based on the algebra $\mathsf{su}(2) = \mathsf{so}(3)$ that will be given in Appendix A. Since there are N points, the acting matrices have to be $N \times N$, and thus there cannot be more than N^2 aberrations, which may be embedded in the Lie *unitary* group $\mathsf{U}(N)$ that will contain in particular all $N!$ pixel permutations.

To build the N^2 generators of 1D aberrations in an orderly fashion, we use again the monomials of classical phase space variables $M_{k,m}(p, q) := p^{k+m} q^{k-m}$ in (111), of rank k and weight m, and replace position q and momentum p with the $N \times N$ Hermitian and traceless matrices of the $\mathsf{su}(2)$ spin j representations (Biedenharn & Louck, 1981) according to (134), with diagonal

position : $q \mapsto \mathbf{Q} = \|Q_{q,q'}\|$,
$$Q_{q,q'} = q\, \delta_{q,q'}, \quad q|_{-j}^{j}, \tag{163}$$

momentum : $p \mapsto \mathbf{P} = \|P_{q,q'}\|$,
$$P_{q,q'} = -i\tfrac{1}{2}\sqrt{(j-q)(j+q+1)}\, \delta_{q+1,q'} + i\tfrac{1}{2}\sqrt{(j+q)(j-q+1)}\, \delta_{q-1,q'}, \tag{164}$$

mode $-j$: $\mathbf{K} = \|K_{q,q'}\|$,
$$K_{q,q'} = \tfrac{1}{2}\sqrt{(j-q)(j+q+1)}\, \delta_{q+1,q'} + \tfrac{1}{2}\sqrt{(j+q)(j-q+1)}\, \delta_{q-1,q'}, \tag{165}$$

that satisfy

$$\mathbf{Q}^2 + \mathbf{P}^2 + \mathbf{K}^2 = j(j+1)\, \mathbf{1}. \tag{166}$$

Then we build a Hermitian product matrix out of the *three* matrices $\mathbf{Q}, \mathbf{P}, \mathbf{K}$ through their *Weyl-order product*. When there are n of these matrices, we sum over all $n!$ permutations of the factor operators $\mathbf{Q}, \mathbf{P}$, and $\mathbf{K}$, and divide by $n!$,

$$\{\mathbf{Q}^a, \mathbf{P}^b, \mathbf{K}^c\}_{\text{Weyl}} := \frac{1}{(a+b+c)!} \sum_{\text{permutations}} \overbrace{\mathbf{Q}\cdots\mathbf{Q}}^{a \text{ factors}} \overbrace{\mathbf{P}\cdots\mathbf{P}}^{b \text{ factors}} \overbrace{\mathbf{K}\cdots\mathbf{K}}^{c \text{ factors}}. \tag{167}$$

These will exponentiate to $N \times N$ unitary matrices, ensuring their reversibility and conservation of information. Eq. (166) is a restriction that we can choose to limit the powers of $\mathbf{K}$ to 0 or 1. In comparison with their classification in geometric optics, we thus have *two* pyramids of finite aberrations for each rank k

$$\mathbf{M}_{k,m}^0 := \{\mathbf{P}^{k+m}, \mathbf{Q}^{k-m}\}_{\text{Weyl}}, \quad m\big|_{-k}^{k}, \tag{168}$$

$$\mathbf{M}_{k,m}^1 := \{\mathbf{P}^{k-\frac{1}{2}+m}, \mathbf{Q}^{k-\frac{1}{2}-m}, \mathbf{K}\}_{\text{Weyl}}, \quad m\big|_{-k+\frac{1}{2}}^{k-\frac{1}{2}} \tag{169}$$

for integer $0 \leq 2k \leq 2j$, i.e., aberration orders $0 \leq A = 2k-1 \leq N-1$.

Through symbolic computation and numerical evaluation, these matrices have been exponentiated and applied on 1D signals, shown in Figs. 15 and 16. There we render the action single aberrations (168) and (169) on the signal and on the phase space of finite systems, determined by the Wigner function given in Appendix A, to be compared with the classical deformations of phase space in Fig. 5. The signal exposed to aberrations is a rectangle function (top of Fig. 15); overall phases are generated by $\mathbf{M}^0_{0,0} = \mathbf{1}$, followed by the SU(2)-linear transformations generated by $\mathbf{M}^0_{1/2,-1/2} = \mathbf{Q}$ and $\mathbf{M}^0_{1/2,1/2} = \mathbf{P}$ in the first pyramid. The second pyramid in Fig. 16 has on top $\mathbf{M}^1_{1/2,0} = \mathbf{K}$, corresponding to an oscillator Hamiltonian generating rotations of phase space. In the next rung, $\mathbf{M}^0_{1,m}$ are the finite counterparts of the linear canonical transformations of geometric optics, allowing for the deformation inherent in mapping the surface of the spheres on rectangles. The exponentiated aberration matrices (168) and (169) can be composed as in the geometric factored-product parametrization (119). In Rueda-Paz and Wolf (2011) we used this decomposition to simulate the aberrations of a 1D signal in a quasi-harmonic planar waveguide whose refractive index profile is

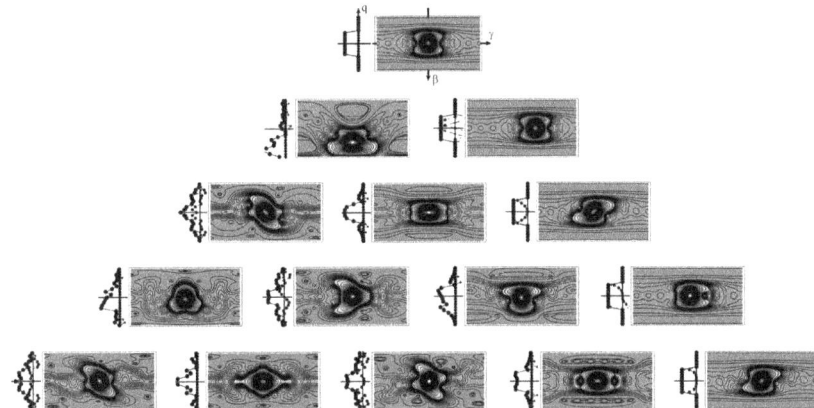

Fig. 15 Pyramid of aberrations (168) of order up to $A = 3$ on a 1D "rectangle" signal of $N = 21$ points. The signal is shown to the *left* of each subfigure, and its phase space Wigner function to the *right*. At the top is the original rectangle signal and its Wigner function on the flattened polar coordinates $0 \leq \beta \leq \pi$ and $-\pi < \gamma \leq \pi$ of the sphere. The second row has the two SO(3)-"translations" in position and momentum (rank $\frac{1}{2}$, $m = \frac{1}{2}, -\frac{1}{2}$, aberration order 0). The following row of three aberrations corresponds to the *linear* transformations in continuous signals (free flight, squeezing, and lens; rank $k = 1, m = 1, 0, -1$, aberration order $A = 1$). There follow aberrations of orders 2 and 3. To highlight the behavior of the Wigner function near zero, the contour lines are chosen at $\{0, \pm 0.0001, \pm 0.001, \pm 0.01, 0.02, 0.03, ..., 0.15, 0.2, 0.3, ..., 3.0, 3.1\}$.

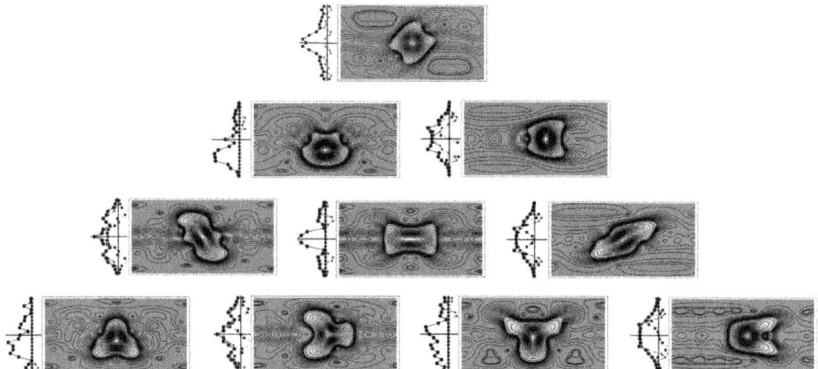

Fig. 16 Second pyramid of aberrations (169) of orders up to $A = 3$ on the same $N = 21$-point rectangle signal of the previous figure, with the same contour lines, axes, and values. At top, a 45 degree SO(3)-linear rotation generated by **K** (rank $k = \frac{1}{2}$, $m = 0$, aberration order 0), and its Wigner function. In the following rows, ranks $k = 1, \frac{3}{2}, 2$ (aberrations orders 1, 2, and 3); these are "**K**-repeaters" of aberrations of orders 0, 1, and 2 in the previous figure.

$n(q) \sim n_o q^2 + n_1 q^4$, along the z-axis of evolution. In this treatment of aberrations, we can pass directly from the geometric to the discrete model of optics.

ACKNOWLEDGMENTS

I thank the many colleagues who participated with me in the quest to find the symmetry that underlies optical models and ask leniency from those I have not mentioned. For the elaboration of the figures I acknowledge the indispensable help of Guillermo Krötzsch and of several students who worked out cases and examples. Support for the Óptica Matemática endeavors was provided by PAPIIT–DGAPA (Universidad Nacional Autónoma de México) funds along several years, the latest being project number 101115.

APPENDIX A. THE SU(2) WIGNER FUNCTION

We understand the Wigner function as the matrix elements of a Wigner *operator* that is an element of a group *ring*.[32] The Wigner operator may be defined broadly as the Fourier transform of a corresponding group. The Wigner function does not contain *more* information than the signal itself

---

[32] The group *ring* is the group G with the extra operation of *linear combination*. Its elements are of the form $\mathcal{A} = \sum_{g_i \in G} a_i g_i$ for $g_i \in G$ a discrete group and $a_i \in \mathbf{C}$ or, if the group is a continuous Lie group of elements $g(\vec{x})$, then $\mathcal{A} = \int_{\vec{x} \in G} d\mu(\vec{x}) A(\vec{x}) g(\vec{x})$, with the invariant Haar measure $d\mu(\vec{x})$ on the group manifold and $A(\vec{x}) \in \mathcal{L}^2(G)$.

(up to a total phase), but displays it in *phase space* (q, p) in a form that is friendly to the educated eye, much as a musical score is more informative than the pressure-wave register of a recorded tune (Forbes, Manko, Ozaktas, Simon, & Wolf, 2000); while the latter is meaningless to visual inspection, the former can be recognized when played on an instrument, hummed by a human voice, or simply recalled from memory.

In 1932 Wigner proposed the original quasiprobability distribution function that we can write in 1D, between two states $\phi(q)$ and $\psi(q)$ in $\mathcal{L}^2(\mathbf{R})$, with a constant λ (or $\hbar$), as

$$W(\phi,\psi|\,q,p,\lambda) = \frac{1}{2\pi\lambda} \int_{-\infty}^{\infty} dx\, \phi(q-\tfrac{1}{2}x)^* \, e^{-ixp/\lambda} \, \psi(q-\tfrac{1}{2}x). \qquad (A.1)$$

This function (Hillery, O'Connell, Scully, & Wigner, 1984; Lee, 1995) is sesquilinear, $W(\phi, \psi|\, q, p) = W(\psi, \phi|\, q, p)^*$; for $\phi = \psi$ it is real (although not quite strictly positive); it is covariant under the Heisenberg–Weyl translations and, uniquely, under linear canonical transformations (García-Calderón & Moshinsky, 1980). It also has marginals and overlaps that allow the formulation of a quantum theory of measurement. It was introduced to optical models by Adolf Lohmann in the groundbreaking articles (Bartelt, Brenner, & Lohmann, 1980; Brenner & Lohmann, 1982; Lohmann, 1980). For specific systems and phase space manifolds, several distinct "Wigner functions" have been built with these properties that will be commented on below.

A.1 The Wigner Operator

In Wolf (1996), Atakishiyev, Chumakov, and Wolf (1998), and Ali, Atakishiyev, Chumakov, and Wolf (2000) we proposed an operator belonging to the *ring* of a D-dimensional Lie group G, whose generators X_n $(n|_1^D)$ form its Lie algebra. We use the *polar* parametrization of the group,[33] which we indicate using square brackets as $g[\vec{x}] = \exp i(\sum_{n=1}^{D} x_n X_n)$ so that the group identity is $g[\vec{0}]$ and the inverse is $g[\vec{x}]^{-1} = g[-\vec{x}]$. These coordinates $\{x_n\}_{n=1}^{D}$ can be treated as a "vector" but only extend over the manifold G of the group, $\vec{x} \in G \subset \mathbf{R}^D$. Let $\vec{\xi}$ be a vector in the full real manifold $\mathbf{R}^D$ and write, with some generality,

---

[33] We assume that the group G is of *exponential* type, i.e., that all its elements can be reached with the polar parametrization. This holds for the Heisenberg–Weyl, rotation, euclidean and all compact groups, but *not* for the Sp$(2D, \mathbf{R})$ groups (Wolf, 2004, sect. 12.2).

$$\mathcal{W}(\vec{\xi}) := \int_G \mathrm{d}g[\vec{x}] \, \exp\left(-i\vec{x} \cdot \vec{\xi}\right) g[\vec{x}] = \int_G \mathrm{d}g[\vec{x}] \, \exp\left(i\vec{x} \cdot (\vec{X} - \vec{\xi})\right), \quad \text{(A.2)}$$

where $\mathrm{d}g[\vec{x}]$ is the invariant Haar measure.[34] If the generators X_n were *numbers*, the function (A.2) would be simply $(2\pi)^D \prod_{n=1}^D \delta(\xi_n - X_n)$; the fact that the X_n are *operators* that will be applied to functions of the group, of coset spaces, or finite representation multiplets of the group, is what makes this Wigner operator interesting. We shall understand the manifold $\vec{\xi} \in \mathsf{R}^D$ to be the *meta-phase space* associated to the group G, but with the ordinary Euclidean measure $\mathrm{d}^D \vec{\xi}$.

Assume we have a Hilbert space of complex functions $\phi(h), \psi(h) \in \mathcal{H}$, where $h \in H$ may be the group itself, a space of cosets, a representation multiplet, or any arena for *unitary* action $g: \phi(h)$ by $g[\vec{x}] \in \mathsf{G}$, so that $g^\dagger = g^{-1}$, and with invariant measure $\mathrm{d}h$. The matrix elements of $\mathcal{W}(\vec{\xi})$ between two such functions is their Wigner *function*,

$$W(\phi, \psi | \vec{\xi}) := \int_H \mathrm{d}h \, \phi^*(h) \, \mathcal{W}(\vec{\xi}) : \psi(h) \quad \text{(A.3)}$$

$$= \int_H \mathrm{d}h \int_G \mathrm{d}g \, \phi^*(h) \, e^{-i\vec{x} \cdot \vec{\xi}} \, (g : \psi)(h) \quad \text{(A.4)}$$

$$= \int_G \mathrm{d}g \int_H \mathrm{d}h \, (g^{-1/2} : \phi^*)(h) \, e^{-i\vec{x} \cdot \vec{\xi}} \, (g^{1/2} : \psi)(h) \quad \text{(A.5)}$$

This is the structure which, for the Heisenberg–Weyl group, yields (A.1) with its left- and right-half translations (Wolf, 1996). Note that only in the polar parametrization are the *square roots* of group elements well defined: $(g[\vec{x}])^{1/2} = g[\tfrac{1}{2}\vec{x}]$. When we are given *density matrices*[35] ρ instead of pure states ϕ, ψ, the Wigner function is defined through $W(\rho | \vec{\xi}) = \mathrm{trace}\,(\mathcal{W}(\vec{\xi}) \rho)$.

The Wigner operator (A.2) presents the following properties in $\mathcal{H}$, corresponding to those of the original Wigner function (A.1). The operator is *self-adjoint*: $\mathcal{W}(\vec{\xi})^\dagger = \mathcal{W}(\vec{\xi})$. It is *covariant* under similarity transformations by $g \in \mathsf{G}$, namely $g^{-1} \mathcal{W}(\vec{\xi}) g = \mathcal{W}(D^{\mathrm{ad}}(g)\vec{\xi})$, where $D^{\mathrm{ad}}(g)$ is the $D \times D$ adjoint matrix representation of $g \in \mathsf{G}$; this also holds even when the transformation is an *outer* automorphism of the algebra, as the Heisenberg–Weyl generators under linear canonical transformations. The product integral

---

[34] The group G is assumed to be *unimodular*, i.e., that its *right*- and *left*-invariant measures are the same; this holds for a wide class of groups, but is *not* the case for the two-parameter *affine* group of translations and dilatations relevant for wavelets. In Ali et al. (2000) the expression (A.2) is generalized for such groups.

[35] A sum of ket-bra's, or *ideal projectors* in the ring.

$\int_{\Re^D} d\vec{\xi} |\mathcal{W}(\vec{\xi})|^2 \propto 1$ is proportional to the unit operator due to the Dirac δ's produced in one of the integrations, $\int_{R^D} d^D\vec{x} \exp(i(\vec{x} - \vec{x}'))$. And finally, the Wigner operator *commutes* with all Casimir invariants of the algebra, so we may use the unitary irreducible matrix representations of the operators.

A.2 The SU(2) Wigner Matrix

Consider now the $N \times N$ representation of **SU**(2) of spin j ($N = 2j+1$), with generators $\{J_i\}_{i=1}^3$ as given in (134). The polar parametrization uses the unit axis of rotation coordinates on the sphere, $\vec{v}(\theta, \phi) = \vec{x}/|\vec{x}|$, and the length $\eta := |\vec{x}|$, which is the rotation angle. The group elements are then

$$g[\vec{x}] = \exp(i\vec{x} \cdot \vec{J}) = \exp[i\eta(v_1 J_1 + v_2 J_2 + v_3 J_3)], \quad (A.6)$$

and the Haar measure for continuous $\eta|_{-2\pi}^{2\pi}$,[36] $\theta|_0^\pi$ and $\phi|_{-\pi}^\pi$, is

$$dg[\vec{x}] = \frac{1}{2}\sin^2 \tfrac{1}{2}\eta \, d\eta \, \sin\theta \, d\theta \, d\phi. \quad (A.7)$$

Now let the Wigner operator (A.2) act on column vectors $\mathbf{f} = \{f_m\}_{m=-j}^j$ whose components are the N values of a finite signal on a 1D array of points. This will define a "Wigner *matrix*" $\mathbf{W}^j(\vec{\xi}) = \|W^j_{m,m'}(\vec{\xi})\|$ that *represents* the Wigner operator $\mathcal{W}(\vec{\xi})$, $\vec{\xi} \in \mathbf{R}^3$, for spin j,

$$\mathcal{W}(\vec{\xi}) : \mathbf{f} = \int_G dg[\vec{x}] \exp(-i\vec{x} \cdot \vec{\xi}) \, \mathbf{D}^j(g[\vec{x}]) \, \mathbf{f} =: \mathbf{W}^j(\vec{\xi}) \mathbf{f}, \quad (A.8)$$

where $\mathbf{D}^j(g[\vec{x}]) = \|D^j_{m,m'}(g[\eta, \theta, \phi])\|$ are the **SU**(2) rotation matrices (called Wigner Big-D matrices (Biedenharn & Louck, 1981)) in polar parameters. These Wigner matrices qualify to be the "Fourier transform" of the irreducible representation matrices of the group **SU**(2), because

$$\mathbf{W}^j(\vec{\xi}) = \int_{\mathsf{SU}(2)} \nu(\vec{x}) \, d^3\vec{x} \exp(-i\vec{x} \cdot \vec{\xi}) \, \mathbf{D}^j(g[\vec{x}]), \quad (A.9)$$

$$\mathbf{D}^j(g[\vec{x}]) = \frac{1}{(2\pi)^3 \nu(\vec{x})} \int_{\mathbf{R}^3} d^3\vec{\xi} \exp(i\vec{x} \cdot \vec{\xi}) \, \mathbf{W}^j(\vec{\xi}), \quad (A.10)$$

with the weight $\nu(\vec{x})$ in the Haar measure $dg[\vec{x}] = \nu(\vec{x}) d^3\vec{x}$ that we find in (A.7) as $\nu(\vec{x}) = \nu(\eta) = \frac{1}{8} \operatorname{sinc} \frac{1}{2}\eta$.

---

[36] This range ensures that the group integration will contain both an element $g(\eta, \theta, \phi)$ and its inverse with $-\eta$.

The SU(2) Wigner *function* of the N-point finite signal vectors $\mathbf{f}_1$ and $\mathbf{f}_2$, is then a vector-and-matrix product,

$$W^j(\mathbf{f}_1, \mathbf{f}_2|\vec{\xi}) := \mathbf{f}_1^\dagger \mathbf{W}^j(\vec{\xi}) \mathbf{f}_2, \quad W^j(\mathbf{f}|\vec{\xi}) := W^j(\mathbf{f}, \mathbf{f}|\vec{\xi}). \tag{A.11}$$

To complete the task of finding the Wigner matrix elements $W^j_{m,m'}(\vec{\xi})$ from (A.9) we have the Big-D matrix elements, normally written in Euler angles for spin j, $D^j_{m,m'}(\alpha,\beta,\gamma) = e^{im\alpha} d^j_{m,m'}(\beta) e^{im'\gamma}$. Here we have *polar* angles, so we can use the Wigner little-d functions to write (Biedenharn & Louck, 1981)

$$D^j_{m,m'}[\eta \, v(\theta,\phi)] = e^{i(m'-m)\phi} \sum_{m''=-j}^{j} d^j_{m,m''}(\theta) \, e^{-im''\eta} \, d^j_{m',m''}(\theta). \tag{A.12}$$

The integration over SU(2) in (A.9) can be rotated so that $\vec{v}(\theta,\phi)$ is the 3-axis unit vector $\mathbf{k}$, and where the Wigner matrix is diagonal, $W^j_{m,m'}(\eta \mathbf{k}) = \delta_{m,m'} W^j_m(\eta)$; for $W^j_m(\eta)$ we can similarly write

$$W^j_{m,m'}[\eta \, \vec{v}(\theta,\phi)] = e^{i(m'-m)\phi} \sum_{m''=-j}^{j} d^j_{m,m''}(\theta) \, W^j_{m''}(\eta) \, d^j_{m',m''}(\theta). \tag{A.13}$$

This allows us to separate the sphere manifold θ, ϕ of known functions, from the η-dependent diagonal elements of the Wigner function $W^j_m(\eta)$, $m|_{-j}^{j}$, which are the *eigenvalues* of the Wigner matrix. Calculated with some care,[37] these are

$$W^j_m(\eta) = (-1)^{2j+1} \frac{\pi}{4} \sum_{n=-j}^{j} \int_{-1}^{1} ds \, (d^j_{m,n}(\arccos s))^2 \tag{A.14}$$
$$\times \sin(2\pi\eta s) \left[\frac{1}{\eta s - n + 1} - \frac{2}{\eta s - n} + \frac{1}{\eta s - n - 1} \right].$$

This expression, replaced in (A.13), gives the matrix elements of the Wigner matrix for spin j; its dependence on the radial coordinate η is shown to be strongly peaked between j and $j+1$ (Atakishiyev et al., 1998), so we may display the Wigner function of a signal $\mathbf{f}$ on the *surface* of a *sphere* in the $\vec{\xi}$-space $\mathbf{R}^3$ at the radius $\eta = |\vec{\xi}| = j + \frac{1}{2}$.

According to (134) we may identify the three continuous coordinates $\{\xi_i\}_{i=1}^{3} \in \mathbf{R}$ of the Wigner function as position $\xi_1 = q$, momentum $\xi_2 = -p$, and $\xi_3 = \mu = n - j$ for mode $n|_0^{2j}$, each in its real line. Low modes $n \approx 0$ (see Fig. 8) register around the bottom pole of the sphere $\xi_3 \approx -j$,

---

[37] The poles from the brackets are canceled by the zeros of the sine function.

while the highest modes $n \approx 2j$ at the top pole. Plotting functions on spheres θ, ϕ is awkward in flat figures, so in Figs. 15 and 16 we resorted to show the angles $\theta|_{\frac{1}{2}\pi}^{\frac{3}{2}\pi}$ and $\phi|_{-\pi}^{\pi}$ as if they were Cartesian coordinates, with the bottom pole of the sphere at the center.[38]

A.3 Closing Remarks

This appendix mostly pertained the SU(2) Wigner function of the form (A.2), with phase space being a sphere—classically a symplectic manifold. SU(2) was also used by Agarwal et al. (Agarwal, 1981; Agarwal, Puri, & Singh, 1997; Dowling, Agarwal, & Schleich, 1994) to define a Wigner function using tensorial notation which was *prima facie* quite distinct from our presentation. Actually, the two are equivalent, as shown with some labor in Chumakov, Klimov, and Wolf (2000).

The structure of (A.2) has been used for other group rings: on the compact circle of SO(2) phase space is the set of integer points, and the Wigner function is the "sinc" interpolation between the absolute squares of the analyzed function (Nieto, Atakishiyev, Chumakov, & Wolf, 1998). The 2D Euclidean group ISO(2) phase space can be also reduced from the 3D manifold as in the present case and is a cylinder (Nieto et al., 1998). As we said before, the Heisenberg–Weyl group leads to standard Wigner function (A.1), and the 1D affine group was studied in Ali et al. (2000) to place wavelets. Phase space representations are useful when they are two-dimensional; although marginals—projections on lower-dimensional spaces—hold for all models.

Distinct "Wigner-type" functions can be obtained from (A.2) if we introduce to the integrand a function $K(\chi[\vec{x}])$ over the manifold of *equivalence classes* of the group G, so $\chi(g_c) = \chi(g\, g_c\, g^{-1})$. This plays the role of the *Cohen function* (Cohen, 1966; Lee, 1995) that defines Q-, P-, or Husimi functions, among others.

Further models that also use the basic structure of the Wigner function, with interesting properties of their own, include solutions of the Helmholtz equation (Wolf, Alonso, & Forbes, 1999), which led to in-depth studies by Gregory Forbes and Miguel Angel Alonso on electromagnetic fields (Alonso, 2009, 2011, 2015), and which settled some controversies in radiometry. Still other works have dealt where the position coordinate lies on a sphere or hyperboloid (Alonso, Pogosyan, & Wolf, 2002, 2003).

---

[38] Other maps could perhaps be better even if not as immediate.

REFERENCES

Agarwal, G. S. (1981). Relation between atomic coherent-state representation, state multipoles, and generalized phase-space distributions. *Physical Review A, 24*, 2889–2896.

Agarwal, G. S., Puri, R. R., & Singh, R. P. (1997). Atomic schrödinger cat states. *Physical Review A, 56*, 2249–2254.

Ali, S. T., Atakishiyev, N. M., Chumakov, S. M., & Wolf, K. B. (2000). The Wigner function for general Lie groups and the wavelet transform. *Annales Henri Poincaré, 1*, 685–714.

Alieva, T., Bastiaans, M. J., & Calvo, M. L. (2005). Fractional transforms in optical information processing. *EURASIP Journal in Signal Processing, 2005*, 1498–1519.

Alonso, M. A. (2009). Diffraction of paraxial partially coherent fields by planar obstacles in the Wigner representation. *Journal of the Optical Society of America A, 26*, 1588–1597.

Alonso, M. A. (2011). Wigner functions in optics: Describing beams as ray bundles and pulses as particle ensembles. *Advances in Optics and Photonics, 3*, 272–365.

Alonso, M. A. (2015). In *Mathematical optics group (Rochester) tutorial*. http://www.optics.rochester.edu/workgroups/alonso/Home.html

Alonso, M. A., Pogosyan, G. S., & Wolf, K. B. (2002). Wigner functions for curved spaces I: On hyperboloids. *Journal of Mathematical Physics, 43*, 5857–5871.

Alonso, M. A., Pogosyan, G. S., & Wolf, K. B. (2003). Wigner functions for curved spaces II: On spheres. *Journal of Mathematical Physics, 44*, 1472–1489.

Atakishiyev, N. M., Chumakov, S. M., & Wolf, K. B. (1998). Wigner distribution function for finite systems. *Journal of Mathematical Physics, 39*, 6247–6261.

Atakishiyev, N. M., Pogosyan, G. S., Vicent, L. E., & Wolf, K. B. (2001a). Finite two-dimensional oscillator II: The radial model. *Journal of Physics A, 34*, 9399–9415.

Atakishiyev, N. M., Pogosyan, G. S., Vicent, L. E., & Wolf, K. B. (2001b). Finite two-dimensional oscillator I: The Cartesian model. *Journal of Physics A, 34*, 9381–9398.

Atakishiyev, N. M., Pogosyan, G. S., & Wolf, K. B. (2003). Contraction of the finite one-dimensional oscillator. *International Journal of Modern Physics A, 18*, 317–327.

Atakishiyev, N. M., Pogosyan, G. S., & Wolf, K. B. (2005). Finite models of the oscillator. *Physics of Particles and Nuclei, 36*(Suppl. 3), 521–555.

Atakishiyev, N. M., & Wolf, K. B. (1997). Fractional Fourier-Kravchuk transform. *Journal of the Optical Society of America A, 14*, 1467–1477.

Atzema, E. J., Krötzsch, G., & Wolf, K. B. (1997). Canonical transformations to warped surfaces: Correction of aberrated optical images. *Journal of Physics A, 30*, 5793–5803.

Bargmann, V. (1947). Irreducible unitary representations of the Lorentz group. *Annals of Mathematics, 48*, 568–642.

Bargmann, V. (1970). Group representation in Hilbert spaces of analytic functions. In P. Gilbert & R. G. Newton (Eds.), *Analytical methods in mathematical physics* (pp. 27–63). New York: Gordon & Breach.

Barker, L., Çandan, Ç., Hakioğlu, T., Kutay, M. A., & Ozaktas, H. M. (2000). The discrete harmonic oscillator, Harper's equation, and the discrete fractional Fourier transform. *Journal of Physics A, 33*, 2209–2222.

Bartelt, H. O., Brenner, K.-H., & Lohmann, A. W. (1980). The Wigner distribution function and its optical production. *Optics Communication, 32*, 32–38.

Biedenharn, L. C., & Louck, J. D. (1981). Angular momentum in quantum physics. Theory and application. In G.-C. Rota (Ed.), *Encyclopædia of mathematics and its applications*. Reading, MA: Addison-Wesley.

Boyer, C. P., & Wolf, K. B. (1973). Deformation of inhomogeneous classical Lie algebras to the algebras of the linear groups. *Journal of Mathematical Physics, 14*, 1853–1859.

Brenner, K.-H., & Lohmann, A. H. (1982). Wigner distribution function display of complex 1D signals. *Optics Communication, 42*, 310–314.

Buchdahl, H. (1970). *An introduction to Hamiltonian optics*. Cambridge: Cambridge University Press.

Chumakov, S. M., Klimov, A. B., & Wolf, K. B. (2000). On the connection of two Wigner functions for spin systems. *Physical Review A, 61*(3), 034101.

Cohen, L. (1966). Generalized phase-space distribution functions. *Journal of Mathematical Physics, 7*, 781–786.

Collins, S. A., Jr. (1970). Lens-system diffraction integral written in terms of matrix optics. *Journal of the Optical Society of America, 60*, 1168–1177.

Condon, E. U. (1937). Immersion of the Fourier transform in a continuous group of functional transformations. *Proceedings of the National Academy of Sciences, 23*, 158–163.

Dowling, J. P., Agarwal, G. S., & Schleich, W. P. (1994). Wigner distribution of a general angular-momentum state: Applications to a collection of two-level atoms. *Physical Review A, 49*, 4101–4109.

Dragt, A. J. (1982a). Lectures on nonlinear orbit dynamics. In R. A. Carrigan, F. R. Huson, & M. Month (Eds.), *AIP conference proceedings* (Vol. 87, No. 1, pp. 147–313). AIP.

Dragt, A. J. (1982b). Lie algebraic theory of geometrical optics and optical aberrations. *Journal of the Optical Society of America, 72*, 372–379.

Dragt, A. J. (1987). Elementary and advanced Lie algebraic methods with applications to accelerator design, electron microscopes, and light optics. *Nuclear Instruments and Methods in Physics Research A, 258*, 339–354.

Dragt, A. J. (2004). *Lie methods for nonlinear dynamics with applications to accelerator physics*. Maryland: University of Maryland. http://www.physics.umd.edu/dsat/dsatliemethods.html.

Forbes, G. W., Manko, V. I., Ozaktas, H. M., Simon, R., & Wolf, K. B. (Eds.), (2000). Feature issue on Wigner distributions and phase space in optics. *Journal of the Optical Society of America A, 17*, 2274.

Frank, A., & van Isacker, P. (1994). *Algebraic methods in molecular and nuclear structure physics*. New York: Wiley.

García-Bullé, M., Lassner, W., & Wolf, K. B. (1986). The metaplectic group within the Heisenberg–Weyl ring. *Journal of Mathematical Physics, 27*, 29–36.

García-Calderón, G., & Moshinsky, M. (1980). Wigner distribution functions and the representation of canonical transformations. *Journal of Physics A, 13*, L185–L189.

Gilmore, R. (1978). *Lie groups, Lie algebras, and some of their applications*. New York: Wiley Interscience.

González-Casanova, P., & Wolf, K. B. (1995). Interpolation of solutions to the Helmholtz equation. *Numerical Methods of Partial Differential Equations, 11*, 77–91.

Goodman, J. W. (1968). *Introduction to Fourier optics*. New York: McGraw-Hill.

Guillemin, V., & Sternberg, S. (1984). *Symplectic techniques in physics*. Cambridge: Cambridge University Press.

Healy, J. J., Alper Kutay, M., Ozaktas, H. M., & Sheridan, J. T. (Eds.). (2015). *Linear canonical transforms, theory and applications: Vol. 198. Springer Series in Optical Sciences*. New York: Springer.

Hillery, M., O'Connell, R. F., Scully, M. O., & Wigner, E. P. (1984). Distribution functions in physics: Fundamentals. *Physics Reports, 106*, 121–167.

Kauderer, M. (1994). *Symplectic matrices. First order systems and special relativity*. Singapore: World Scientific.

Khan, S. A., & Wolf, K. B. (2002). Hamiltonian orbit structure of the set of paraxial optical systems. *Journal of the Optical Society of America A, 19*, 2436–2444.

Krawtchouk, M. (1928). Sur une généralisation des pôlynomes d'hermite. *Comptes Rendus de l'Académie des Sciences Series I - Mathématique, 189*, 620–622.

Lee, H.-W. (1995). Theory and application of the quantum phase-space distribution functions. *Physics Reports*, *259*, 147–211.
Liberman, S., & Wolf, K. B. (2015). Independent simultaneous discoveries visualized through network analysis: The case of linear canonical transforms. *Scientometrics*, *104*, 715–735.
Lie, S. (1888). *Theorie der transformationsgruppen* (Vol. 1). Leipzig: B. G. Teubner.
Lohmann, A. (1980). The Wigner function and its optical production. *Optics Communication*, *42*, 32–37.
Luneburg, R. K. (1964). *Mathematical theory of optics*. Berkeley: University of California.
McBride, A. C., & Kerr, F. H. (1987). On Namias's fractional Fourier transforms. *IMA Journal of Applied Mathematics*, *39*, 159–175.
Mendlovic, D., & Ozaktas, H. M. (1993). Fractional Fourier transforms and their optical implementation: I. *Journal of the Optical Society of America A*, *10*, 1875–1881.
Moshinsky, M., & Quesne, C. (1971). Linear canonical transformations and their unitary representation. *Journal of Mathematical Physics*, *12*, 1772–1780.
Namias, V. (1980). The fractional order Fourier transform and its applications in quantum mechanics. *IMA Journal of Applied Mathematics*, *25*, 241–265.
Navarro Saad, M., & Wolf, K. B. (1986). Factorization of the phase–space transformation produced by an arbitrary refracting surface. *Journal of the Optical Society of America A*, *3*, 340–346.
Nieto, L. M., Atakishiyev, N. M., Chumakov, S. M., & Wolf, K. B. (1998). Wigner distribution function for finite systems. *Journal of Physics A*, *31*, 3875–3895.
Ozaktas, H. M., & Mendlovic, D. (1993a). Fourier transforms of fractional order and their optical interpretation. *Optics Communication*, *101*, 163–169.
Ozaktas, H. M., & Mendlovic, D. (1993b). Fractional Fourier transforms and their optical implementation: II. *Journal of the Optical Society of America A*, *10*, 2522–2531.
Ozaktas, H. M., Zalevsky, Z., & Kutay, M. A. (2001). *The fractional Fourier transform with applications in optics and signal processing*. Chichester: Wiley.
Pei, S.-C., & Ding, J.-J. (2000). Closed-form discrete fractional and affine transforms. *IEEE Transactions on Signal Processing*, *48*, 1338–1353.
Pei, S.-C., & Yeh, M.-H. (1997). Improved discrete fractional transform. *Optics Letters*, *22*, 1047–1049.
Pei, S.-C., Yeh, M.-H., & Tseng, C.-C. (1999). Discrete fractional Fourier transform based on orthogonal projections. *IEEE Transactions on Signal Processing*, *47*, 1335–1348.
Quesne, C., & Moshinsky, M. (1971). Linear canonical transformations and matrix elements. *Journal of Mathematical Physics*, *12*, 1780–1783.
Rashed, R. (1990). A pioneer in anaclastics—Ibn Sahl on burning mirrors and lenses. *ISIS*, *81*, 464–491.
Rashed, R. (1993). *Géométrie et dioptrique au xe siècle: Ibn Sahl, al-qūhī, et Ibn al-Hayatham, Collection Sciences et Philosophie Arabes, Textes et Études*. Paris: Les Belles Lettres.
Rodrigo, J. A., Alieva, T., & Bastiaans, M. J. (2011). Phase space rotators and their applications in optics. In G. Cristóbal, P. Schelkens, & H. Thienpont (Eds.), *Optical and digital image processing: Fundamentals and applications* (pp. 251–271). Weinheim: Wiley-VCH Verlag.
Rodrigo, J. A., Alieva, T., & Calvo, M. L. (2007). Gyrator transform: Properties and applications. *Optics Express*, *15*, 2190–2203.
Rueda-Paz, J., & Wolf, K. B. (2011). Finite signals in planar waveguides. *Journal of the Optical Society of America A*, *28*, 641–650.
Seidel, L. (1853). Zur dioptrik. *Astronomische Nachrichten*, *871*, 105–120.
Simon, R., & Wolf, K. B. (2000). Fractional Fourier transforms in two dimensions. *Journal of the Optical Society of America A*, *17*, 2368–2381.

Steinberg, S. (1986). Lie series, Lie transformations, and their applications. In J. Sánchez-Mondragón & K. B. Wolf (Eds.), *Lie methods in optics: Vol. 250. Lecture notes in physics.* Heidelberg: Springer-Verlag

Steinberg, S., & Wolf, K. B. (1981). Invariant inner products on spaces of solutions of the Klein-Gordon and Helmholtz equations. *Journal of Mathematical Physics, 22*, 1660–1663.

Sudarshan, E. C. G., & Mukunda, N. (1974). *Classical dynamics: a modern perspective.* New York: Wiley.

Sudarshan, E. C. G., Mukunda, N., & Simon, R. (1985). Realization of first order optical systems using thin lenses. *Optica Acta, 32*, 855–872.

Urzúa, A. R., & Wolf, K. B. (2016). Unitary rotation and gyration of pixellated images on rectangular screens. *Journal of the Optical Society of America A, 33*, 642–647.

Vicent, L. E., & Wolf, K. B. (2008). Unitary transformation between Cartesian- and polar-pixellated screens. *Journal of the Optical Society of America A, 25*, 1875–1884.

Vicent, L. E., & Wolf, K. B. (2011). Analysis of digital images into energy-angular momentum modes. *Journal of the Optical Society of America A, 28*, 808–814.

Wigner, E. (1932). On the quantum correction for thermodynamic equilibrium. *Physics Review, 40*, 749–759.

Wolf, K. B. (1974). Canonical transforms I. Complex linear transforms. *Journal of Mathematical Physics, 15*, 1295–1301.

Wolf, K. B. (1975). The Heisenberg-Weyl ring in quantum mechanics. In E. M. Loebl (Ed.), *Group theory and its applications* (Vol. 3, pp. 190–247). New York: Academic Press.

Wolf, K. B. (1979). *Integral transforms in science and engineering.* New York: Plenum Publishing Corporation.

Wolf, K. B. (1989). Elements of euclidean optics. In K. B. Wolf (Ed.), *Lie methods in optics, II workshop: Vol. 352. Lecture notes in physics* (pp. 115–162). Heidelberg: Springer-Verlag.

Wolf, K. B. (1996). Wigner distribution function for paraxial polychromatic optics. *Optics Communication, 132*, 343–352.

Wolf, K. B. (2004). *Geometric optics on phase space.* Heidelberg: Springer-Verlag.

Wolf, K. B., Alonso, M. A., & Forbes, G. W. (1999). Wigner functions for Helmholtz wavefields. *Journal of the Optical Society of America A, 16*, 2476–2487.

Wolf, K. B., & Boyer, C. P. (1974). The algebra and group deformations $I^m[SO(n) \otimes SO(m)] \Rightarrow SO(n,m)$, $I^m[U(n) \otimes U(m)] \Rightarrow U(n,m)$, and $I^m[Sp(n) \otimes Sp(m)] \Rightarrow Sp(n,m)$, for $1 \leq m \leq n$. *Journal of Mathematical Physics, 15*, 2096–2101.

Wolf, K. B., & Krötzsch, G. (1995). `mexLIE 2`, *a set of symbolic computation functions for geometric aberration optics.* IIMAS–UNAM Manual No. 10. Mexico City.

Wolf, K. B., & Krötzsch, G. (2007). Geometry and dynamics in the fractional discrete Fourier transform. *Journal of the Optical Society of America A, 24*, 651–658.

Wong, M. K. F. (1967). Representations of the orthogonal group. I. Lowering and raising operators of the orthogonal group and matrix elements of the generator. *Journal of Mathematical Physics, 8*, 1899–1911.

Wybourne, B. G. (1974). *Classical groups for physicists.* New York: Wiley.

CHAPTER FIVE

Classical Coherence of Blackbody Radiation

Kasimir Blomstedt, Ari T. Friberg, Tero Setälä
Institute of Photonics, University of Eastern Finland, Joensuu, Finland

Contents

1. Introduction — 293
2. Blackbody Radiation Inside a Cavity — 298
  2.1 Frequency Domain — 298
  2.2 Time Domain — 309
3. Blackbody Radiation in an Aperture — 317
  3.1 Frequency Domain — 317
  3.2 Time Domain — 321
4. Blackbody Radiation in the Far Zone of an Aperture — 325
  4.1 Frequency Domain — 325
  4.2 Time Domain — 330
5. Summary — 335
Acknowledgments — 336
Appendix A. Blackbody Coherence in Cavities — 336
  Appendix A.1. Maxwell's Equations — 336
  Appendix A.2. Fluctuation-Dissipation Theorem and Blackbody Radiation — 337
Appendix B. Derivation of Aperture and Far-Field Mutual Coherence Matrices — 342
  Appendix B.1. Aperture Matrix — 342
  Appendix B.2. Far-Field Matrix — 344
References — 344

1. INTRODUCTION

A blackbody is an object that absorbs all light incident on it (Planck, 1914). A classic textbook example of a blackbody is an aperture in the wall of a large cavity. Light entering the cavity via an opening cannot, in practice, escape but is eventually absorbed by the inner walls. The opening is taken so small that it does not affect the thermal equilibrium at fixed temperature that is assumed to prevail within the cavity. Under these conditions the spectrum of the radiation inside the cavity and of that escaping from it through the aperture is given by Plank's spectrum (or Planck's law). The spectral shape is universal in the sense that it depends only on temperature and not on the composition or other details of the body itself. As the temperature increases the spectral peak shifts to shorter wavelengths, an effect known as Wien's displacement law explaining the color changes of heated bodies.

Besides the unique spectral characteristics, a blackbody such as the cavity aperture absorbs independently of the direction of the incident radiation. Likewise it emits isotropically in all directions. Radiation emanating from the aperture can be viewed as an isotropic superposition of plane waves propagating away from the aperture into a half-space, i.e., the propagation directions of the waves occupy half of the full solid angle. While all these waves are present within the aperture, only one of them contributes to the far field in a given direction, the one propagating in this direction. In contrast, the field inside the cavity is composed of waves traveling in all directions and their superposition is isotropic in the full solid angle. The waves are created in independent spontaneous emission processes that the atoms in the cavity walls undergo, rendering the waves uncorrelated and unpolarized both inside and outside of the cavity. A superposition of such mutually uncorrelated waves leads to a statistically homogeneous field whose two-point spatial coherence properties depend only on the vector separation of the points. Homogeneity applies to the field inside the cavity and within the aperture but not to the far field. In addition to statistical homogeneity the field inside the blackbody cavity is also statistically isotropic characterized by a coherence matrix of a specific general form. Inside the aperture or in its far zone the field is no longer statistically isotropic. Strictly speaking, the term blackbody radiation is devoted to the statistically homogeneous and isotropic field inside the cavity, but we broadly refer to the field in the aperture, in its far zone, or an analogous field present in other geometries as blackbody radiation.

While blackbody radiation had at the beginning of the 20th century a landmark role in the development of quantum physics and in bypassing the ultraviolet catastrophe, its coherence properties, the topic of this review, were analyzed only much later in the 1960s. The first assessment of classical space–time coherence of blackbody radiation in an enclosure was presented in Bourret (1960) where expressions were derived for the correlation functions of the (real) electric field components in special cases by making use of an analogy to isotropic turbulence of an incompressible fluid. Blackbody radiation is generally regarded as the most incoherent light field and the results by Bourret demonstrated coherence over a small space–time volume. Some of his results were reproduced also from a quantum-mechanical approach (Sarfatt, 1963). Soon after Bourret's work the temporal coherence of blackbody radiation was reanalyzed in terms of the complex analytic signals and the temporal complex degree of coherence of parallel field components was found to be expressible by the generalized Riemann zeta function (Kano & Wolf, 1962). Using only the fact that the blackbody spectrum obeys Planck's law, it was also shown that two definitions of the coherence time and the effective bandwidth of blackbody radiation have exact expressions involving the zeta function (Mehta, 1963). In addition, in both cases the coherence time was found to be inversely proportional to temperature while the bandwidth was proportional to it. Explicit expressions for the full coherence matrices of blackbody radiation within a cavity were derived by Mehta and Wolf in a series of papers. They introduced the space–time domain electric, magnetic, and mixed coherence matrices by employing both classical (Mehta & Wolf, 1964a) and quantum (Mehta & Wolf, 1964b) field theory and derived expressions for the corresponding cross-spectral density matrices of the space-frequency domain (Mehta & Wolf, 1967). In particular, the spectral coherence length was found to be on the order of wavelength independently of temperature. In the time domain the coherence time and coherence length were found to be on the order of the mean wavelength and the mean period of the radiation (Mandel & Wolf, 1965), respectively. Both of these time-domain quantities depend on temperature as posed by Wien's displacement law. The above early results on blackbody radiation concern the field inside a large cavity in thermal equilibrium, where the field can be viewed as an isotropic (in full solid angle) superposition of unpolarized and uncorrelated plane waves with Planck's spectrum. It is known that such a superposition leads to the electric cross-spectral density matrix which is proportional to the imaginary part of the free-space Green's function (Setälä, Kaivola, & Friberg, 2003), i.e., the spectral spatial coherence properties are specified by the geometry alone. This result holds more generally as shown in Agarwal (1975) where blackbody fluctuations were considered in finite geometries.

The coherence properties of blackbody radiation within an aperture made to the cavity wall or of the far field emitted from that opening were considered some 30 years after the cavity results. In the 1990s it was shown that the far field produced by a planar, finite, secondary blackbody source (the aperture field) is spectrally unpolarized in every direction and the radiant intensity obeys Lambert's cosine law (James, 1994). In 2008 the far-field angular coherence by such a source was considered and an expression for the cross-spectral density matrix was derived in the paraxial region (Lahiri & Wolf, 2008). In these two works the electric cross-spectral density matrix within the aperture was half of that inside the cavity and information only on the two transverse components were used although the aperture field contains also the longitudinal component. While the above far-field results are correct the coherence matrix within the aperture is not half of the cavity matrix but, as was shown very recently, it contains an additional term related to the longitudinal component (Blomstedt, Setälä, Tervo, Turunen, & Friberg, 2013). The refined cross-spectral density matrix reproduces all the earlier far-field results and allows to evaluate the far-field two-point coherence also in the nonparaxial directions. In Blomstedt et al. (2013) it was also rigorously demonstrated, paralleling intuitive expectations, that an unpolarized (two-component) beam which is spatially completely incoherent (δ-correlated) is a highly accurate model for a secondary, blackbody (thermal) source. Using the electromagnetic van Cittert–Zernike theorem (Tervo, Setälä, Turunen, & Friberg, 2013) such a source field leads to the same paraxial far-field cross-spectral density matrix as the accurate aperture blackbody coherence matrix.

The correction introduced in Blomstedt et al. (2013) does not affect the diagonal elements of the aperture coherence matrix and, therefore, both the complete and incomplete coherence matrix of the (homogeneous) aperture field have the same trace, explicitly given by a sinc function. This property deserves some attention. In particular, in the context of scalar fields it has been known for long that the degree of coherence of the low-frequency part of a secondary Lambertian source necessarily is of a sinc form (Carter & Wolf, 1975). In the electromagnetic context, the sinc trace of the blackbody coherence matrix and its connection to Lambertian radiation pattern was pointed out in Carter and Wolf (1975). However, a rigorous justification that the trace of the low-frequency coherence matrix of any Lambertian source has this form was published later (Blomstedt et al., 2016), along with the observation that a planar blackbody source is not the only possible electromagnetic Lambertian source. The sinc trace of the cross-spectral density matrix of the homogeneous aperture field also conforms to an electromagnetic scaling law (Hassinen, Tervo, Setälä, Turunen, & Friberg, 2013). This is consistent with the known fact that the normalized spectrum of the aperture field and that of the far field in all directions are identical and given by the normalized Planck's spectrum.

Within the last 15 years or so, the coherence properties of blackbody radiation have attracted renewed interest in part because of the progress in electromagnetic coherence theory (Friberg & Setälä, 2016). Indeed, the blackbody coherence matrices inside the cavity, in aperture, or in the aperture far zone are explicitly known, and in the case of cavity both in time and frequency domain. The related fields therefore well serve as test fields for several concepts and results pertaining to coherence of electromagnetic fields. For example, the electromagnetic theory of coherent modes was introduced in Tervo, Setälä, and Friberg (2004) and the mode construction of blackbody radiation inside a cavity was put forward in Setälä, Lindberg, Blomstedt, Tervo, and Friberg (2005). As far as we know, this is the only three-dimensional electromagnetic coherent-mode representation that has appeared in the literature so far. The recently defined electromagnetic degree of coherence (Tervo, Setälä, & Friberg, 2003; Tervo et al., 2004) has compact expressions in cavity (Setälä et al., 2005; Leppänen, Friberg, & Setälä, 2016b), aperture (Blomstedt et al., 2013), and in the far zone (Blomstedt et al., 2013) and allows to assess the coherence time, the coherence length (spectral and temporal), as well as the angular extent of the far-field blackbody coherence. Although involving different mathematical expressions, all physical conclusions are consistent with the early results derived in the 1960s. In addition, according to a recent definition of the degree of polarization proposed for fields with three electric field components (Setälä, Shevchenko, Kaivola, & Friberg, 2002), the blackbody radiation is unpolarized inside the cavity (Setälä et al., 2005; Leppänen et al., 2016b) and in the aperture (Blomstedt et al., 2013), as one would intuitively expect. A theoretical framework to treat the polarization dynamics, i.e., the time evolution of the instantaneous polarization state, its rate and magnitude, of a random paraxial and nonparaxial fields was introduced recently and applied to (unpolarized) blackbody radiation inside the cavity (Voipio, Setälä, Shevchenko, & Friberg, 2010) and in the far zone (Shevchenko, Setälä, Kaivola, & Friberg, 2009). The rate of the ultrafast polarization variations may be characterized by the polarization time over which the instantaneous polarization state does not significantly change. This quantity, which is physically different from the coherence time, was found for blackbody fields to be inversely proportional to temperature. Therefore, the higher the temperature the faster are the polarization-state fluctuations.

As is evident from the above paragraphs, various results concerning electromagnetic blackbody coherence are scattered in the literature. The aim of this review is to collect them in one place, introduce some missing pieces, and present all in a unified manner. In Section 2, the coherence of a blackbody field inside a large cavity is considered. As is customary, the large cavity corresponds to free space and we refer to it as all-space cavity. The effects of a cavity wall (that induces reflections) to the spatial coherence and especially to the polarization are briefly addressed. Some of the results are new. Appendix A contains a demonstration based on the fluctuation-dissipation theorem that in any geometry, such as free space, the electric cross-spectral density matrix of the blackbody field is determined by the imaginary part of Green's function. In Section 3, the aperture-field coherence is analyzed. The time-domain mutual coherence matrix of this field has not appeared in the literature earlier and its derivation is included in Appendix B. Section 4 discusses the coherence of the far field emitted from the aperture. The time-domain coherence matrix related to this case likewise has not been presented so far. In general, full quantification of the spectrospatial and spatiotemporal blackbody coherence in all-space cavity, aperture, and in its far zone is given. In all three geometries, the coherence time, coherence length (temporal and spectral), and polarization time are discussed in terms of proper quantities of electromagnetic coherence theory. All coherence matrices depend only on temperature, as do the characteristic time scales and the space–time coherence length. Throughout the work we consider the coherence of the electric field component only, although many of the results could straightforwardly be extended to the magnetic field and between the electric and the magnetic field. There are slight differences in the underlying definitions used to derive the results given in the literature, leading to formulas which differ by constant factors from each other. This means that although we here present the results in a self-consistent framework, our formulas necessarily also differ by constant factors from some of the formulas found in the literature. We explicitly comment on this for selected formulas discussed herein.

2. BLACKBODY RADIATION INSIDE A CAVITY

2.1 Frequency Domain

2.1.1 Spatial Coherence

The blackbody field inside a cavity contains three Cartesian electric field components. The electric field is represented, at a point $\mathbf{r}$ and at angular frequency ω, by the column vector $\mathbf{E}(\mathbf{r}, \omega) = [E_x(\mathbf{r}, \omega), E_y(\mathbf{r}, \omega), E_z(\mathbf{r}, \omega)]^{\mathrm{T}}$, where T denotes the matrix transpose. The field is zero-mean, stationary, and its spectral spatial coherence properties at two points $\mathbf{r}_1$ and $\mathbf{r}_2$ are described by the 3 × 3 cross-spectral density matrix (Mandel & Wolf, 1995; Tervo et al., 2004)

$$\mathbf{W}(\mathbf{r}_1, \mathbf{r}_2, \omega) = \langle \mathbf{E}^*(\mathbf{r}_1, \omega) \mathbf{E}^\mathrm{T}(\mathbf{r}_2, \omega) \rangle, \qquad (1)$$

where $\langle \cdots \rangle$ and * denote ensemble averaging and complex conjugation, respectively. We mention at this stage that although the vector $\mathbf{E}(\mathbf{r}, \omega)$ in Eq. (1) represents the electric field at angular frequency ω, it is not the Fourier transform at this frequency of the actual field. This is because the stationarity of the electromagnetic field renders such components infinite and thus necessitates a more subtle approach, which we briefly address later in Section 2.2.1.

The spectral degree of coherence of an electromagnetic field is defined, in squared form, as (Setälä, Tervo, & Friberg, 2004a; Tervo et al., 2004)

$$\mu^2(\mathbf{r}_1, \mathbf{r}_2, \omega) = \frac{\mathrm{tr}[\mathbf{W}^\mathrm{H}(\mathbf{r}_1, \mathbf{r}_2, \omega) \mathbf{W}(\mathbf{r}_1, \mathbf{r}_2, \omega)]}{\mathrm{tr}[\mathbf{W}(\mathbf{r}_1, \mathbf{r}_1, \omega)] \mathrm{tr}[\mathbf{W}(\mathbf{r}_2, \mathbf{r}_2, \omega)]}, \qquad (2)$$

where H stands for the matrix Hermitian transpose and tr denotes the matrix trace. This quantity is bounded between 0 and 1, with the two limits corresponding to complete spatial incoherence and complete coherence, respectively, at frequency ω. In the former case, no correlation exists between any of the orthogonal field components at $\mathbf{r}_1$ and $\mathbf{r}_2$, while in the latter case the components are fully correlated. When the field is fully coherent within a volume, the cross-spectral density matrix factors with respect to the two spatial variables (Setälä et al., 2004a). For a beam-like field, the degree of coherence describes the contrasts (visibilities) associated with the modulations of the Stokes parameters (polarization and intensity) of the field in two-pinhole interference (Setälä, Tervo, & Friberg, 2006).

We also consider the complex degree of correlation (complex correlation coefficient) between two electric field components, which is defined by

$$\mu_{ij}(\mathbf{r}_1, \mathbf{r}_2, \omega) = \frac{\hat{\mathbf{u}}_i^\mathrm{T} \mathbf{W}(\mathbf{r}_1, \mathbf{r}_2, \omega) \hat{\mathbf{u}}_j}{[\hat{\mathbf{u}}_i^\mathrm{T} \mathbf{W}(\mathbf{r}_1, \mathbf{r}_1, \omega) \hat{\mathbf{u}}_i \hat{\mathbf{u}}_j^\mathrm{T} \mathbf{W}(\mathbf{r}_2, \mathbf{r}_2, \omega) \hat{\mathbf{u}}_j]^{1/2}}, \qquad (3)$$

where $i, j \in \{x, y, z\}$ and $\hat{\mathbf{u}}_i$ denotes the unit vector pointing in the (positive) i-direction. Hence the factor $\hat{\mathbf{u}}_i^\mathrm{T} \mathbf{W}(\mathbf{r}_1, \mathbf{r}_2, \omega) \hat{\mathbf{u}}_j$ is the ij-element of the matrix $\mathbf{W}(\mathbf{r}_1, \mathbf{r}_2, \omega)$.

For blackbody radiation it can be shown, using for example the fluctuation-dissipation theorem, that the cross-spectral density matrix corresponding to blackbody radiation inside a cavity can be expressed in terms of the cavity Green function as (see Appendix A)

$$\mathbf{W}(\mathbf{r}_1,\mathbf{r}_2,\omega) = \frac{4\pi a_0(\omega)}{k_0} \text{Im}\{\mathbf{G}_\text{E}(\mathbf{r}_1,\mathbf{r}_2,\omega)\}, \tag{4}$$

where Im denotes the imaginary part and $k_0 = \omega/c$ is the free-space wave number (c is the vacuum speed of light). The function $4a_0(\omega)$ is equal to Planck's spectrum, i.e.,

$$4a_0(\omega) = \frac{\hbar\omega^3}{\pi^2 c^3} \frac{1}{\exp(\hbar\omega/k_\text{B}T) - 1}, \tag{5}$$

where $\hbar$ is the reduced Planck constant, k_B denotes the Boltzmann constant, and T is the absolute (equilibrium) temperature of the blackbody cavity. We note that in some of the cited papers this angular frequency representation of Planck's spectrum is multiplied by the factor 2π. The Green function in Eq. (4), $\mathbf{G}_\text{E}(\mathbf{r}_1, \mathbf{r}_2, \omega)$, corresponds to the vector wave equation of the electric field in the cavity geometry with zero boundary conditions (no static currents nor fields), as is shown in Appendix A. Eq. (4) shows that the spatial spectral coherence properties of the field are described by the imaginary part of the Green function, i.e., they are fixed by the cavity geometry alone. A corresponding connection holds for the distribution of a field sourced by a stationary, homogeneous, and isotropic source distribution in a low-loss region both for scalar (Nussenzveig, Foley, Kim, & Wolf, 1987; Ponomarenko & Wolf, 2001, see also Foley, Carter, & Wolf, 1986) and electromagnetic (Setälä, Blomstedt, Kaivola, & Friberg, 2003) fields.

An explicit expression for the cross-spectral density matrix of blackbody radiation can be obtained in a number of different cavity geometries (with different Green functions) including, in addition to free-space, for example vacuum regions surrounded on one (half-space geometry) or two sides by dielectrics or metals (Agarwal, 1975). In this work, we will specifically discuss the all-space (free-space) geometry and the half-space cavity geometry, which are the ones most often considered. We further note that the cross-spectral density matrix in Eq. (4) only describes the correlations between the components of the electric field. Here we will concentrate on such blackbody-field correlations, but similar coherence matrices can be computed also for the components of the magnetic field, as well as for the correlations between the electric field and the magnetic field components (Agarwal, 1975; Blomstedt, Setälä, & Friberg, 2015).

2.1.1.1 All-Space Cavity

For blackbody radiation within a cavity that encompasses all space the Green function is that of free space, and it reads as (Tai, 1971)

$$\mathbf{G}(\mathbf{r}_1,\mathbf{r}_2,\omega) = \left(\mathbf{U} + \frac{1}{k_0^2}\nabla\nabla^T\right)\frac{\exp(ik_0 R)}{R}, \qquad (6)$$

where $\mathbf{U}$ denotes the 3×3 unit matrix and $R=|\mathbf{R}|$ with $\mathbf{R}=\mathbf{r}_1-\mathbf{r}_2$. Furthermore, we put $\hat{\mathbf{R}}=\mathbf{R}/R$ in what follows and use these notations from now on without further mention. Inserting the free-space Green function into Eq. (4) gives for the cross-spectral density matrix of blackbody radiation inside the all-space cavity the expression (Mehta & Wolf, 1967; Setälä, Kaivola, et al., 2003)

$$\mathbf{W}^{(as)}(\mathbf{r}_1,\mathbf{r}_2,\omega) = 4\pi a_0(\omega)\left\{\left[j_0(k_0 R) - \frac{j_1(k_0 R)}{k_0 R}\right]\mathbf{U} + j_2(k_0 R)\hat{\mathbf{R}}\hat{\mathbf{R}}^T\right\}, \qquad (7)$$

where $j_n(x)$ denotes the spherical Bessel function of order n. The superscript (as) refers to all space. As can be seen, the all-space blackbody radiation is statistically homogeneous and isotropic, since its coherence matrix is of the form (Batchelor, 1970; Gbur, James, & Wolf, 1999; Setälä, Blomstedt, et al., 2003)

$$\mathbf{W}(\mathbf{r}_1,\mathbf{r}_2,\omega) = A(R,\omega)\mathbf{U} + B(R,\omega)\hat{\mathbf{R}}\hat{\mathbf{R}}^T, \qquad (8)$$

where $A(R,\omega)$ and $B(R,\omega)$ are scalar functions connected by a continuity equation.

A free electromagnetic field can be represented as a superposition of vectorial plane waves as (Wolf, 1976)

$$\mathbf{E}(\mathbf{r},\omega) = \int_{4\pi}\mathbf{A}(\hat{\mathbf{u}},\omega)\exp(ik_0\hat{\mathbf{u}}\cdot\mathbf{r})d\hat{\mathbf{u}}, \qquad (9)$$

where $\mathbf{A}(\hat{\mathbf{u}},\omega)$ is the amplitude of the plane wave propagating in the direction specified by the unit vector $\hat{\mathbf{u}}$ and 4π refers to integration over the full solid angle. As we discuss below, all-space blackbody radiation can be represented by an ensemble of plane waves, which propagate in different directions and are mutually uncorrelated, unpolarized, and their spectral distribution is given by Planck's spectrum. The angular correlation function corresponding to such plane waves is of the form (Setälä, Kaivola, et al., 2003; Blomstedt et al., 2013)

$$\langle \mathbf{A}^*(\hat{\mathbf{u}},\omega)\mathbf{A}^{\mathrm{T}}(\hat{\mathbf{u}}',\omega)\rangle = a_0(\omega)\Delta(\hat{\mathbf{u}}'-\hat{\mathbf{u}})(\hat{\mathbf{s}}\hat{\mathbf{s}}^{\mathrm{T}}+\hat{\mathbf{p}}\hat{\mathbf{p}}^{\mathrm{T}}) \tag{10}$$
$$= a_0(\omega)\Delta(\hat{\mathbf{u}}'-\hat{\mathbf{u}})(\mathbf{U}-\hat{\mathbf{u}}\hat{\mathbf{u}}^{\mathrm{T}}), \tag{11}$$

where $\Delta(\hat{\mathbf{u}}'-\hat{\mathbf{u}})$ is the spherical delta function (Nieto-Vesperinas, 1991) and the unit vectors $\hat{\mathbf{s}}$ and $\hat{\mathbf{p}}$ define the s and p polarization components of the plane wave propagating in the $\hat{\mathbf{u}}$ direction. From Eq. (9) we then get for the cross-spectral density matrix of the blackbody field in free space the expression

$$\mathbf{W}^{(\mathrm{pw})}(\mathbf{r}_1,\mathbf{r}_2,\omega) = a_0(\omega)\int_{4\pi}\int_{4\pi}\Delta(\hat{\mathbf{u}}_2-\hat{\mathbf{u}}_1)\left(\mathbf{U}-\hat{\mathbf{u}}_1\hat{\mathbf{u}}_2^{\mathrm{T}}\right)$$
$$\times \exp[ik_0(\hat{\mathbf{u}}_1\cdot\mathbf{r}_1-\hat{\mathbf{u}}_2\cdot\mathbf{r}_2)]\mathrm{d}\hat{\mathbf{u}}_1\mathrm{d}\hat{\mathbf{u}}_2 \tag{12}$$
$$= a_0(\omega)\int_{4\pi}\left(\mathbf{U}-\hat{\mathbf{u}}\hat{\mathbf{u}}^{\mathrm{T}}\right)\exp(ik_0\hat{\mathbf{u}}\cdot\mathbf{R})\mathrm{d}\hat{\mathbf{u}}, \tag{13}$$

where the superscript (pw) refers to the plane-wave representation. Performing the integrations in Eq. (13) results in Eq. (7) (Setälä, Kaivola, et al., 2003). That is, the all-space blackbody cavity field can be described as a uniform distribution of uncorrelated and unpolarized plane waves, which is precisely the result Planck arrives at when starting from first principles and considering the passage of light rays (plane waves) in a large blackbody cavity (Planck, 1914). An analogous plane-wave approach has also been discussed in the context of scalar fields (Gori, Ambrosini, & Bagini, 1994).

The coherent-mode decomposition of blackbody radiation within a large, but finite spherical vacuum-filled region is also known (Setälä et al., 2005). For its description we need the vector wave functions defined by (Setälä et al., 2005; Gbur, 2011; see also Tai, 1971)

$$\mathbf{M}_{mn}(\mathbf{r},\omega) = \nabla\times\left[j_n(k_0 r)Y_n^m(\hat{\mathbf{r}})\mathbf{r}\right], \tag{14}$$
$$\mathbf{N}_{mn}(\mathbf{r},\omega) = \frac{1}{k_0}\nabla\times\nabla\times\left[j_n(k_0 r)Y_n^m(\hat{\mathbf{r}})\mathbf{r}\right], \tag{15}$$

where $Y_n^m(\hat{\mathbf{r}}) = Y_n^m(\theta,\phi)$ is the spherical harmonic of order $n = 1, 2, \ldots$ and degree $m = -n, \ldots, n$, and θ and ϕ are the polar and azimuthal angle of the spherical coordinate system representation of $\mathbf{r}$. Let us denote the finite sphere where the field is considered by $\Omega \subset \mathbb{R}^3$. The all-space cross-spectral density matrix $\mathbf{W}^{(\mathrm{as})}(\mathbf{r}_1, \mathbf{r}_2, \omega)$ of Eq. (7) can be interpreted as an operator over the Hilbert space of vector-valued square-integrable functions in Ω, $\mathcal{L}_2(\Omega)$. As such, it is a compact self-adjoint operator, and thus it has a countable set of orthonormal eigenvectors (Hochstadt, 1973; Kreyszig, 1978). These eigenvectors are given by

$$\boldsymbol{\psi}_{mn}^{(K)}(\mathbf{r},\omega) = \frac{1}{\|\mathbf{K}_{mn}(\omega)\|_{\mathcal{L}_2(\Omega)}} \mathbf{K}_{mn}(\mathbf{r},\omega), \quad (16)$$

with the corresponding real $(2n+1)$-fold degenerate eigenvalues

$$\lambda_n^{(K)}(\omega) = \frac{(4\pi)^2}{n(n+1)} a_0(\omega) \|\mathbf{K}_{mn}(\omega)\|_{\mathcal{L}_2(\Omega)}^2. \quad (17)$$

In these equations $\mathbf{K}_{mn}(\mathbf{r},\omega)$ denotes either $\mathbf{M}_{mn}(\mathbf{r},\omega)$ or $\mathbf{N}_{mn}(\mathbf{r},\omega)$ so that $K \in \{M, N\}$ in the superscripts, and $\|\mathbf{K}(\omega)\|_{\mathcal{L}_2(\Omega)}$ denotes the norm of the square-integrable function $\mathbf{K}(\mathbf{r}, \omega)$ over Ω. The eigenfunctions $\boldsymbol{\psi}_{mn}^{(K)}(\mathbf{r},\omega)$ in Eq. (16) together with the corresponding eigenvalues $\lambda_n^{(K)}(\omega)$ satisfy a homogeneous Fredholm integral equation of the second kind (Hochstadt, 1973) with the cross-spectral density matrix of Eq. (7) as kernel. The behavior of the eigenvalues and two of the eigenfunctions are shown in Fig. 1 for a sphere with the radius 1000λ.

It follows that the cross-spectral density matrix can be represented in terms of its eigensystem by a Mercer series expansion (Hochstadt, 1973; Setälä et al., 2005) as

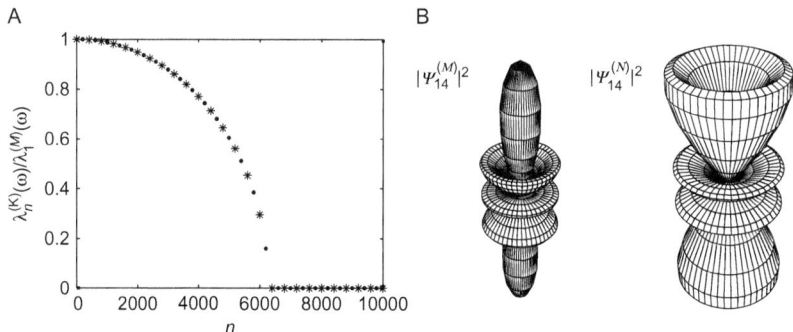

Fig. 1 (A) Behavior of the eigenvalues $\lambda_n^{(M)}(\omega)$ (dots, $n \in \{200, 600, 1000, \ldots\}$) and $\lambda_n^{(N)}(\omega)$ (asterisks, $n \in \{1, 400, 800, \ldots\}$) as a function of the index n. (B) Squared magnitudes of the eigenfunctions $\boldsymbol{\psi}_{14}^{(M)}(\mathbf{r},\omega)$ and $\boldsymbol{\psi}_{14}^{(N)}(\mathbf{r},\omega)$. The origin is in the middle of the rotationally symmetric (around the z-axis) shapes. In both (A) and (B) the radius of the spherical region equals 1000λ. *Modified from Setälä, T., Lindberg, J., Blomstedt, K., Tervo, J., & Friberg, A. T. (2005). Coherent-mode representation of a statistically homogeneous and isotropic electromagnetic field in spherical volume. Physical Review E, 71, 036618.*

$$\mathbf{W}^{(as)}(\mathbf{r}_1,\mathbf{r}_2,\omega) = \sum_{n=1}^{\infty}\sum_{m=-n}^{n}\sum_{K\in\{M,N\}} \lambda_n^{(K)}(\omega)\boldsymbol{\psi}_{mn}^{(K)*}(\mathbf{r}_1,\omega)\boldsymbol{\psi}_{mn}^{(K)\mathrm{T}}(\mathbf{r}_2,\omega),$$
(18)

where $\mathbf{r}_1, \mathbf{r}_2 \in \Omega$. Each term in the above sum is of a factored form and hence corresponds to a field, which in view of Eq. (2) is spatially fully coherent at frequency ω (Tervo et al., 2004). We may therefore refer to Eq. (18) as the coherent-mode decomposition of the free-space blackbody field within a finite vacuum-filled spherical volume. As far as we know, this is the only explicit example of a three-dimensional electromagnetic coherent-mode representation presented in the literature.

Inserting Eq. (7) into Eq. (2), we get for the degree of coherence related to the all-space blackbody field the expression (Setälä et al., 2005)

$$\mu^{(as)}(\mathbf{r}_1,\mathbf{r}_2,\omega) = \frac{1}{\sqrt{3}}\left[j_0^2(k_0R) + \frac{1}{2}j_2^2(k_0R)\right]^{1/2}.$$
(19)

This quantity as a function of R is illustrated in Fig. 2 by the solid blue graph. We observe that the quantity drops from $\mu(\mathbf{r},\mathbf{r},\omega) = 1/\sqrt{3}$, corresponding to an unpolarized three-component field (to be discussed

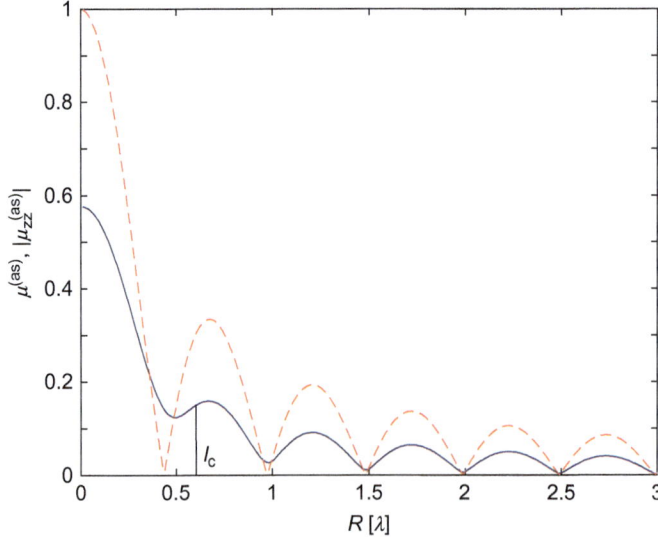

Fig. 2 Illustration of the frequency-domain spatial coherence of all-space blackbody radiation: degree of coherence (*solid blue*) and $|\mu_{zz}^{(as)}(\mathbf{R},\omega)|$ in the $z = 0$ plane (*dashed red*). The *vertical line* marks the spectral coherence length $l_c = 0.60\lambda$.

shortly), to its first local minimum at $R \approx \lambda/2$, where $\lambda = 2\pi/k_0$ is the wavelength of the field. We could take this distance as the coherence length, but following earlier works (Mehta, 1963) we define the coherence length explicitly as

$$l_c^2 = \frac{\int_0^\infty R^2 \mu^2(R,\omega) dR}{\int_0^\infty \mu^2(R,\omega) dR}, \qquad (20)$$

which in the current case yields $l_c = 0.60\lambda$. An analogous root-mean-square definition for the coherence time has been employed, e.g., in Mandel and Wolf (1995). Therefore, two samples from a blackbody field exhibit significant coherence (correlation) if they are taken from points closer to each other than about half the wavelength. This estimate is consistent with what is found in the literature (see, for example, Mehta & Wolf, 1967; Mandel & Wolf, 1965).

The correlation coefficient between a pair of field components can be obtained from Eq. (3), and is given by

$$\mu_{ij}^{(as)}(\mathbf{R},\omega) = \frac{3}{2}\left\{\left[j_0(k_0R) - \frac{j_1(k_0R)}{k_0R}\right]\delta_{ij} + \frac{R_iR_j}{R^2}j_2(k_0R)\right\}, \qquad (21)$$

where δ_{ij} is the Kronecker delta and $i,j \in \{x,y,z\}$. As an example of these correlation functions, the dashed red graph in Fig. 2 represents the behavior of $|\mu_{zz}^{(as)}(\mathbf{R},\omega)|$ in the $z = 0$ plane. Similar conclusions on the range of field correlations may be drawn as from the full electromagnetic coherence function $\mu^{(as)}(\mathbf{r}_1,\mathbf{r}_2,\omega)$. Notice that both normalized quantities are independent of the temperature and the coherence length is inversely dependent on the angular frequency of the radiation field.

2.1.1.2 Half-Space Cavity

When the half-spaces $z < 0$ and $z > 0$ are separated by an infinite, planar, perfect conductor at $z = 0$, with the $z < 0$ half space playing the role of the cavity, the blackbody field is no longer either isotropic or homogeneous due to the reflections from the cavity wall. Mathematically, the related cross-spectral density matrix is obtained from the all-space quantity $\mathbf{W}^{(as)}(\mathbf{R},\omega)$ of Eq. (7) as (Agarwal, 1975; Blomstedt et al., 2015)

$$\mathbf{W}^{(hs)}(\mathbf{r}_1,\mathbf{r}_2,\omega) = \mathbf{W}^{(as)}(\mathbf{r}_1,\mathbf{r}_2,\omega) + \Delta\mathbf{W}(\mathbf{r}_1,\mathbf{r}_2,\omega), \qquad (22)$$

where (hs) refers to half-space. The perturbation due to reflections is given by

$$\Delta \mathbf{W}(\mathbf{r}_1, \mathbf{r}_2, \omega) = -\mathbf{W}^{(\text{as})}(\mathbf{r}_1, \mathbf{Q}\mathbf{r}_2, \omega)\mathbf{Q}, \tag{23}$$

with the matrix $\mathbf{Q} = \mathbf{I} - 2\hat{\mathbf{z}}\hat{\mathbf{z}}^T$ being the Householder reflection that flips the sign of the z-component of the vector it operates on. From the plane wave representations of Eq. (12) and Eq. (13) we find for the perturbation part the plane-wave expansion (Blomstedt et al., 2015)

$$\Delta \mathbf{W}(\mathbf{r}_1, \mathbf{r}_2, \omega) = -a_0(\omega) \int_{4\pi}\int_{4\pi} \Delta(\hat{\mathbf{u}}_2 - \hat{\mathbf{u}}_1)(\mathbf{U} - \hat{\mathbf{u}}_1\hat{\mathbf{u}}_2^T)\mathbf{Q} \tag{24}$$
$$\times \exp[ik_0(\hat{\mathbf{u}}_1 \cdot \mathbf{r}_1 - \hat{\mathbf{u}}_2 \cdot \mathbf{Q}\mathbf{r}_2)] d\hat{\mathbf{u}}_1 d\hat{\mathbf{u}}_2$$
$$= -a_0(\omega) \int_{4\pi} (\mathbf{U} - \hat{\mathbf{u}}\hat{\mathbf{u}}^T)\mathbf{Q} \exp[ik_0\hat{\mathbf{u}} \cdot (\mathbf{r}_1 - \mathbf{Q}\mathbf{r}_2)] d\hat{\mathbf{u}}. \tag{25}$$

The perturbation represents the correlations that exist between fields propagating in the positive z-direction and their reflections in the wall at $z = 0$, which propagate in the negative z-direction.

Although Agarwal (1975) has derived expressions for different non-all-space blackbody cavity geometries, the simple half-space geometry already illustrates many of the important features (induced by the cavity walls) that are not present in the all-space geometry. We will explore this further at the beginning of Section 3.

2.1.2 Polarization
2.1.2.1 All-Space Cavity
The correlations that exist between the field components at a single point are described by the 3 × 3 spectral polarization matrix of the field, which can be obtained from the cross-spectral density matrix as

$$\mathbf{\Phi}_3(\mathbf{r}, \omega) = \mathbf{W}(\mathbf{r}, \mathbf{r}, \omega). \tag{26}$$

A quantity that characterizes the total amount of correlation that exists among the components of a three-component electromagnetic field is the 3D (three-dimensional) degree of polarization, defined in squared form as (Setälä, Kaivola, & Friberg, 2002; Setälä, Shevchenko, et al., 2002)

$$P_3^2(\mathbf{r}, \omega) = \frac{3}{2}\left\{ \frac{\text{tr}[\mathbf{\Phi}_3^2(\mathbf{r}, \omega)]}{\text{tr}^2[\mathbf{\Phi}_3(\mathbf{r}, \omega)]} - \frac{1}{3} \right\}. \tag{27}$$

This quantity is bounded as $0 \leq P_3(\mathbf{r}, \omega) \leq 1$, with the lower and upper limit corresponding to a fully unpolarized and a completely polarized three-component field, respectively. The former case is encountered for a field whose components at a single point are completely uncorrelated and of equal intensity, whereas the latter case occurs when the field components are fully correlated regardless of their intensities. In other cases $P_3(\mathbf{r}, \omega)$ represents the intensity weighted average of the correlations and reduces to the square root of the mean of correlations squared when the three intensities are equal. The equal-intensity situation can be realized for any field by a suitable rotation of the coordinate system.

For all-space cavity blackbody fields it follows from Eq. (7) that the spectral polarization matrix is given by

$$\mathbf{\Phi}_3^{(as)}(\mathbf{r}, \omega) = \frac{8\pi}{3} a_0(\omega) \mathbf{U}. \tag{28}$$

It immediately follows from the definition in Eq. (27) that

$$P_3^{(as)}(\mathbf{r}, \omega) = 0, \tag{29}$$

for the radiation everywhere inside an all-space blackbody cavity. The blackbody radiation in such a cavity is therefore a fully unpolarized three-component field, consistent with what one might intuitively expect. Setting $\mathbf{r}_1 = \mathbf{r}_2 = \mathbf{r}$ in Eq. (2) results in $\mu^2(\mathbf{r}, \mathbf{r}, \omega) = (2/3) P_3^2(\mathbf{r}, \omega) + 1/3$. For unpolarized light we then find that $\mu(\mathbf{r}, \mathbf{r}, \omega) = 1/\sqrt{3}$, which explains the single-point value of the degree of coherence in Fig. 2.

2.1.2.2 Half-Space Cavity

In the half-space cavity the spectral polarization matrix can be computed from Eq. (22) with Eqs. (23) and (7), and it attains the form

$$\begin{aligned}\mathbf{\Phi}_3^{(hs)}(\mathbf{r}, \omega) = \frac{4\pi}{3} a_0(\omega) \{&[2 - 2j_0(2k_0 z) + j_2(2k_0 z)] \mathbf{U} \\&+ [4j_0(2k_0 z) + j_2(2k_0 z)] \hat{\mathbf{u}}_z \hat{\mathbf{u}}_z^T\},\end{aligned} \tag{30}$$

which is seen to approach the all-space polarization matrix given in Eq. (28) as $O(z^{-1})$, when $z \to -\infty$. The xx and zz components of this matrix as functions of the distance from the wall, $-z/\lambda$, are plotted relative to their asymptotic values at $z \to -\infty$ in Fig. 3A. It can be seen that the convergence is fairly rapid and the asymptotic form is approximately attained already at

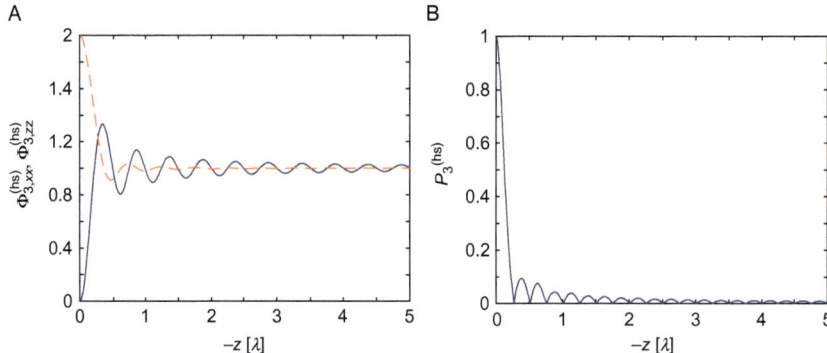

Fig. 3 Illustration of the effect of the cavity wall on polarization in the immediate vicinity of the wall. (A) The *xx* (*solid blue*) and *zz* (*dashed red*) elements of the polarization matrix $\mathbf{\Phi}_3^{(hs)}(\mathbf{r},\omega)$ normalized by their values at $z \to -\infty$. (B) 3D degree of polarization inside the half-space cavity. All quantities are plotted as functions of the distance from the wall, $-z/\lambda$ (z is negative).

about three wavelengths away from the wall. At the cavity wall we have from Eq. (30) the result

$$\mathbf{\Phi}_3^{(\text{hswall})}(\mathbf{r},\omega) = \frac{16\pi}{3} a_0(\omega) \hat{\mathbf{u}}_z \hat{\mathbf{u}}_z^T, \quad (31)$$

where $\mathbf{r}$ is in the plane $z = 0$. When we introduce this result into the definition in Eq. (27), we get for the 3D polarization at the wall the result

$$P_3^{(\text{hswall})}(\mathbf{r},\omega) = 1, \quad (32)$$

that is, the half-space cavity field is fully polarized at the wall, and specifically so in the $\hat{\mathbf{u}}_z$ direction, as can be seen from Eq. (31). Hence the polarization state of the half-space cavity field goes from fully polarized at the wall to unpolarized infinitely far from it, with a smooth transition in between. This is illustrated in Fig. 3B, where the convergence is seen to be extremely rapid, with the field becoming almost unpolarized at about half a wavelength from the wall. We can thereby conclude that the presence of the cavity wall is insignificant for the polarization properties of the field inside the cavity, apart from the immediate vicinity of the wall.

2.1.3 Spectrum

The spectral density of an electromagnetic field is given by the sum of the intensities of each component separately, whereby the spectral density can be written in terms of the 3×3 spectral polarization matrix as

$$S(\mathbf{r},\omega) = \text{tr}[\mathbf{\Phi}_3(\mathbf{r},\omega)]. \tag{33}$$

For the all-space blackbody radiation, Eq. (28) implies that

$$S^{(as)}(\mathbf{r},\omega) = 8\pi a_0(\omega), \tag{34}$$

which shows that the spectral density is constant throughout space and follows the Planck's spectrum.

2.2 Time Domain
2.2.1 Spatiotemporal Coherence

The space–time coherence of the electric component of a vectorial light field is described by the electric mutual coherence matrix. For an ensemble $\{\mathbf{E}(\mathbf{r},t)\}$ of stationary, zero-mean electric fields expressed as complex analytic signals corresponding to the actual real-valued fields, the mutual coherence matrix is defined as (Mandel & Wolf, 1995)

$$\mathbf{\Gamma}(\mathbf{r}_1,\mathbf{r}_2,\tau) = \langle \mathbf{E}^*(\mathbf{r}_1,t)\mathbf{E}^T(\mathbf{r}_2,t+\tau)\rangle, \tag{35}$$

where τ denotes the temporal difference between the two observation points and $\langle \cdots \rangle$ is again the ensemble average. For ergodic fields this average is equal to the temporal average. Furthermore, the mutual coherence matrix and the cross-spectral density matrix of the same ensemble of electric fields are related by the Wiener–Khintchine theorem through a Fourier transform pair as (Mandel & Wolf, 1995)

$$\mathbf{W}(\mathbf{r}_1,\mathbf{r}_2,\omega) = \frac{1}{2\pi}\int_{-\infty}^{\infty} \mathbf{\Gamma}(\mathbf{r}_1,\mathbf{r}_2,\tau)\exp(i\omega\tau)d\tau, \tag{36}$$

$$\mathbf{\Gamma}(\mathbf{r}_1,\mathbf{r}_2,\tau) = \int_0^{\infty} \mathbf{W}(\mathbf{r}_1,\mathbf{r}_2,\omega)\exp(-i\omega\tau)d\omega, \tag{37}$$

where the lower integration limit in the latter formula is zero due to the symmetry imposed by the analytic signal representation. The time-domain polarization matrix is readily obtained as

$$\mathbf{J}(\mathbf{r}) = \mathbf{\Gamma}(\mathbf{r},\mathbf{r},0). \tag{38}$$

We remark that despite the connection established by the Wiener–Khintchine theorem, the time and frequency domain polarization properties may be very different in general, as has recently been exemplified in the case of beam fields (Setälä, Nunziata, & Friberg, 2009; Lahiri, 2009; Réfrégier, Setälä, & Friberg, 2012).

The Wiener–Khintchine theorem presents the solution to the conundrum of how to represent the spectra of stationary fields (see, for example, Mandel & Wolf, 1995). Indeed, if $\tilde{\mathbf{E}}(\mathbf{r}, \omega)$ denotes the temporal Fourier transform of a stationary field $\mathbf{E}(\mathbf{r}, t)$, it follows that

$$\left\langle \tilde{\mathbf{E}}^*(\mathbf{r}_1, \omega_1) \tilde{\mathbf{E}}^T(\mathbf{r}_2, \omega_2) \right\rangle = \mathbf{W}(\mathbf{r}_1, \mathbf{r}_2, \omega_1) \delta(\omega_2 - \omega_1), \tag{39}$$

whereby the naive definition of spectral power diverges. This expression also illustrates the renormalization that the Wiener–Khintchine theorem implies by effectively dividing out the troublesome δ-function. It should, however, be noted that the δ-function in Eq. (39) is not unitless, but carries the unit of time. Thereby the cross-spectral density matrix $\mathbf{W}(\mathbf{r}_1, \mathbf{r}_2, \omega)$ equals in units the correlation function of electric fields, as given by the left-hand side of Eq. (39), divided by the unit of time. Consequently the fields $\mathbf{E}(\mathbf{r}, \omega)$ defined previously and used in the definition in Eq. (1) equal in units the fields $\tilde{\mathbf{E}}(\mathbf{r}, \omega)$ divided by the square root of the unit of time. Therefore, and in view of Eq. (39), it follows that the fields $\mathbf{E}(\mathbf{r}, \omega)$ are not strictly the angular frequency representations of electric fields. In practice, however, these fields can be used as such representations since, apart from the difference in units, they satisfy the same equations as do the actual fields (Mandel & Wolf, 1995).

Similarly to the spectral coherence functions defined in Eqs. (2) and (3), we may introduce the analogous time-domain quantities. Indeed, the squared degree of coherence (Tervo et al., 2003) and the correlation coefficient of two field components are, respectively, given by

$$\gamma^2(\mathbf{r}_1, \mathbf{r}_2, \tau) = \frac{\text{tr}[\boldsymbol{\Gamma}^H(\mathbf{r}_1, \mathbf{r}_2, \tau) \boldsymbol{\Gamma}(\mathbf{r}_1, \mathbf{r}_2, \tau)]}{\text{tr}[\mathbf{J}(\mathbf{r}_1)] \text{tr}[\mathbf{J}(\mathbf{r}_2)]}, \tag{40}$$

and

$$\gamma_{ij}(\mathbf{r}_1, \mathbf{r}_2, \tau) = \frac{\hat{\mathbf{u}}_i^T \boldsymbol{\Gamma}(\mathbf{r}_1, \mathbf{r}_2, \tau) \hat{\mathbf{u}}_j}{\left[\hat{\mathbf{u}}_i^T \mathbf{J}(\mathbf{r}_1) \hat{\mathbf{u}}_i \hat{\mathbf{u}}_j^T \mathbf{J}(\mathbf{r}_2) \hat{\mathbf{u}}_j\right]^{1/2}}, \tag{41}$$

where $i, j \in \{x, y, z\}$ in the last expression. The degree of coherence is bounded as $0 \leq \gamma(\mathbf{r}_1, \mathbf{r}_2, \tau) \leq 1$ with the two limits corresponding to incoherence (noncorrelation) and complete coherence (correlation), respectively, at two space–time points. In the case of full coherence at all time separations for all pairs of points within a volume, the mutual coherence matrix is of a factored form (Setälä, Tervo, & Friberg, 2004b). For beam fields, the space–time degree of coherence finds applications, for example, in the analysis of the interferometers of Young (Leppänen, Saastamoinen, Friberg, & Setälä, 2014) and Michelson (Leppänen, Friberg, & Setälä, 2016a). There it describes the interference-induced modulation contrasts of the Stokes parameters which represent both the polarization state and the intensity of the field. When studying polarization dynamics, one encounters the quantity

$$\gamma_1(\mathbf{r}_1,\mathbf{r}_2,\tau) = \frac{\text{tr}[\mathbf{\Gamma}(\mathbf{r}_1,\mathbf{r}_2,\tau)]}{\{\text{tr}[\mathbf{J}(\mathbf{r}_1)]\text{tr}[\mathbf{J}(\mathbf{r}_2)]\}^{1/2}}, \qquad (42)$$

which for beam fields describes the visibility of the intensity fringes in Young's interferometer (Karczewski, 1963).

For blackbody radiation in an all-space cavity the electric mutual coherence matrix is known to be (Mehta & Wolf, 1964a; Leppänen et al., 2016b, see also Kano & Wolf, 1962; Agarwal, 1975)

$$\mathbf{\Gamma}^{(\text{as})}(\mathbf{r}_1,\mathbf{r}_2,\tau) = \frac{4\hbar c}{\pi}\{[f_1(R,\tau) - f_2(R,\tau)]\mathbf{U} + f_2(R,\tau)\hat{\mathbf{R}}\hat{\mathbf{R}}^{\text{T}}\}, \qquad (43)$$

where $R = |\mathbf{R}|$ and $\hat{\mathbf{R}} = \mathbf{R}/R$ with $\mathbf{R} = \mathbf{r}_1 - \mathbf{r}_2$, as previously. The auxiliary functions are defined by

$$f_1(R,\tau) = \sum_{n=1}^{\infty} \frac{1}{\left[(n\alpha + ic\tau)^2 + R^2\right]^2}, \qquad (44)$$

$$f_2(R,\tau) = \sum_{n=1}^{\infty} \frac{2R^2}{\left[(n\alpha + ic\tau)^2 + R^2\right]^3}, \qquad (45)$$

where $\alpha = \hbar c/k_\text{B}T$ (with a unit of distance). We observe from Eq. (43) that the time-domain all-space blackbody field, besides being stationary in time, is (spatially) statistically homogeneous and isotropic for all τ, i.e., of the form of Eq. (8) (with $\tau \to \omega$). With the above representations in place, the spatio-temporal degree of coherence in Eq. (40), and the normalized correlation coefficient in Eq. (41), assume the forms

$$\gamma^{(\text{as})}(\mathbf{r}_1,\mathbf{r}_2,\tau) = \frac{30\alpha^4}{\pi^4}\left[|f_1(R,\tau)|^2 + 2|f_1(R,\tau) - f_2(R,\tau)|^2\right]^{1/2}, \qquad (46)$$

$$\gamma_{ij}^{(\text{as})}(\mathbf{r}_1,\mathbf{r}_2,\tau) = \frac{90\alpha^4}{\pi^4}\left\{[f_1(R,\tau) - f_2(R,\tau)]\delta_{ij} + f_2(R,\tau)\frac{R_iR_j}{R^2}\right\}, \qquad (47)$$

which were originally introduced in Leppänen et al. (2016b) and Mehta and Wolf (1964a), respectively.

The temporal coherence alone, associated with all-space blackbody radiation, is characterized by the single-spatial-point quantities

$$\gamma^{(as)}(\mathbf{r},\mathbf{r},\tau) = \frac{90}{\sqrt{3}\pi^4}|\zeta(4,1+ic\tau/\alpha)|, \tag{48}$$

$$\gamma_{ij}^{(as)}(\mathbf{r},\mathbf{r},\tau) = \frac{90}{\pi^4}\zeta(4,1+ic\tau/\alpha)\delta_{ij}, \tag{49}$$

where $\zeta(q,a)$ is the Hurwitz (generalized Riemann) zeta function $\zeta(q,a) = \sum_{n=0}^{\infty}(a+n)^{-q}$ (Olver, Lozier, Boisvert, & Clark, 2010). On the other hand, with a zero time difference we get for the spatial coherence the expressions

$$\gamma^{(as)}(\mathbf{r}_1,\mathbf{r}_2,0) = \frac{15}{4s^4}\left[3A^2(s) + 2A(s)B(s) + B^2(s)\right]^{1/2}, \tag{50}$$

$$\gamma_{ij}^{(as)}(\mathbf{r}_1,\mathbf{r}_2,0) = \frac{45}{4s^4}\left[A(s)\delta_{ij} + B(s)\frac{R_iR_j}{R^2}\right], \tag{51}$$

where $s = (\pi/\alpha)R$ is a dimensionless variable and

$$A(s) = -s\coth s - s^2\operatorname{csch}^2 s - 2s^3\operatorname{csch}^2 s\coth s + 4, \tag{52}$$

$$B(s) = 3s\coth s + 3s^2\operatorname{csch}^2 s + 2s^3\operatorname{csch}^2 s\coth s - 8. \tag{53}$$

Eqs. (47), (49), and (51) were originally derived in Mehta and Wolf (1964a). The spatiotemporal coherence of blackbody radiation inside a large cavity has later been briefly addressed, e.g., in Mandel and Wolf (1965), Peřina (1985), and Brosseau (1998).

Fig. 4 illustrates the temporal and spatial coherence of all-space blackbody radiation in terms of the degree of coherence (solid blue) and the magnitude of the zz correlation coefficient in the $z=0$ plane (dashed red). The coherence time τ_c, computed from its definition

$$\tau_c^2 = \frac{\int_{-\infty}^{\infty}\tau^2\gamma^2(\mathbf{r},\mathbf{r},\tau)d\tau}{\int_{-\infty}^{\infty}\gamma^2(\mathbf{r},\mathbf{r},\tau)d\tau}, \tag{54}$$

is found to be $\tau_c = 0.44\alpha/c$ (vertical line in Fig. 4A). The peak wavelength λ_p of Planck's spectrum obeys Wien's displacement law $\lambda_p T = b$, where $b \approx 2.8978 \times 10^{-3}$ mK is Wien's displacement constant. Consequently, we may express the coherence time as $\tau_c \approx 0.35 T_p$, where $T_p = \lambda_p/c$ is the

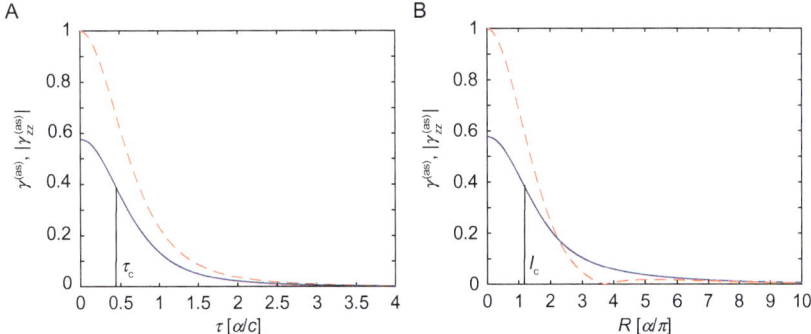

Fig. 4 Illustration of the time-domain coherence of all-space blackbody radiation: (A) Temporal degree of coherence $\gamma^{(as)}(\mathbf{r}, \mathbf{r}, \tau)$ (*solid blue*) and the temporal correlation coefficient $|\gamma_{zz}^{(as)}(\mathbf{r},\mathbf{r},\tau)|$ at the $z = 0$ plane (*dashed red*) as functions of τ. (B) Spatial degree of coherence $\gamma^{(as)}(\mathbf{r}_1, \mathbf{r}_2, 0)$ (*solid blue*) and the spatial correlation coefficient $|\gamma_{zz}^{(as)}(\mathbf{r}_1,\mathbf{r}_2,0)|$ in the $z = 0$ plane (*dashed red*) as functions of $R = |\mathbf{r}_1 - \mathbf{r}_2|$. The *vertical lines* indicate the coherence time $\tau_c = 0.44 \alpha/c$ and the coherence length $l_c = 1.23 \alpha/\pi$, respectively.

oscillation period corresponding to the peak wavelength. The space–time coherence length, given by

$$l_c^2 = \frac{\int_0^\infty R^2 \gamma^2(R,0) \mathrm{d}R}{\int_0^\infty \gamma^2(R,0) \mathrm{d}R}, \tag{55}$$

with $R = |\mathbf{r}_1 - \mathbf{r}_2|$, has the value $l_c = 1.23\alpha/\pi$ (vertical line in Fig. 4B) which can be expressed also as $l_c \approx 0.31 \lambda_p$. We, therefore, conclude that in the space–time domain the coherence length and the coherence time are, respectively, about half of the peak wavelength and half of the cycle time related to that wavelength. Both quantities are inversely proportional to temperature, since $\alpha \propto 1/T$. As an example, at $T = 300$ K we have $\tau_c \approx 11$ fs and $l_c \approx 3.0$ µm. Notice that the equal-time equal-point degree of coherence represented in Figs. 4A and B is $1/\sqrt{3} \approx 0.577$, corresponding to an unpolarized three-component field in the time domain. In contrast, $\gamma_{zz}^{(as)}(\mathbf{r},\mathbf{r},0) = 1$, as required for a normalized autocorrelation function.

2.2.2 Polarization

We define the time-domain 3D degree of polarization in analogy to Eq. (27) of the frequency domain as

$$P_3^2(\mathbf{r}) = \frac{3}{2}\left\{\frac{\text{tr}[\mathbf{J}^2(\mathbf{r})]}{\text{tr}^2[\mathbf{J}(\mathbf{r})]} - \frac{1}{3}\right\}, \tag{56}$$

with the polarization matrix $\mathbf{J}(\mathbf{r})$ given in Eq. (38). The properties of $P_3(\mathbf{r})$ are analogous to those of its spectral counterpart, listed below Eq. (27).

For all-space blackbody radiation we find from Eq. (43) that

$$\mathbf{J}^{(\text{as})}(\mathbf{r}) = \frac{2\pi^3 \hbar c}{45\alpha^4}\mathbf{U}, \tag{57}$$

whereby it immediately follows from Eq. (56) that

$$P_3^{(\text{as})}(\mathbf{r}) = 0, \tag{58}$$

for radiation everywhere inside an all-space blackbody cavity. As in the frequency domain, the all-space cavity blackbody field is fully unpolarized also in the time domain.

2.2.3 Polarization Dynamics

As observed earlier, blackbody radiation within an all-space cavity is represented by an unpolarized three-component field both in the time and frequency domains. In the time domain, therefore, the instantaneous polarization ellipse and the plane in which it lies evolve randomly with time and the field exhibits (ultra fast) polarization dynamics. However, the polarization state varies in a continuous manner and within a sufficiently short-time interval the instantaneous state may be regarded as constant and the field highly or even fully polarized. We refer to this characteristic temporal time scale as the polarization time and denote it by τ_P. A theoretical formalism for the analysis of polarization dynamics and the assessment of the polarization time was introduced for three-component light fields and applied to free-space blackbody radiation in Voipio et al. (2010). Two approaches were demonstrated: a geometric one based on the Poincaré vectors in an eight-dimensional Stokes-parameter space and one based on the energy distribution in Jones vectors. Since the methods lead to equivalent results, we employ here only the latter, as it has a more intuitive physical description.

The Jones-vector approach leads to the polarization correlation function put forward in Voipio et al. (2010), which reads as

$$\gamma_P(\tau) = \frac{\left\langle|\mathbf{E}^H(t)\mathbf{E}(t+\tau)|^2\right\rangle}{\langle I(t)I(t+\tau)\rangle}, \tag{59}$$

where $I(t) = \mathbf{E}^H(t)\mathbf{E}(t)$ denotes the instantaneous intensity of the random electric field $\mathbf{E}(t)$. The position variable has been dropped as the field is considered at a single spatial point. Physically $\gamma_P(\tau)$ describes how much energy at time $t + \tau$ is, on average, in the polarization state in which the field was at time t. The intensity correlation function in the denominator is the maximum value of the numerator, obtained when the polarization states with τ time difference are the same. Consequently, $0 \leq \gamma_P(\tau) \leq 1$ and $\gamma_P(0) = 1$. Furthermore, $\gamma_P(\tau) = 1$ for all τ if the instantaneous polarization state does not change with time and no polarization dynamics takes place. When $\gamma_P(\tau) = 0$ the polarization states separated by the time τ are orthogonal on average.

The blackbody field within a cavity is generated by a large number of independent radiators, whereby the central limit theorem implies that the corresponding field fluctuations obey Gaussian statistics. Thus the polarization correlation function of Eq. (59) involving fourth-order field correlations can be expressed in terms of the second-order correlation functions. When the resulting form is simplified, we arrive at the expression (Voipio et al., 2010)

$$\gamma_P(\tau) = \frac{1 + 2P_3^2 + 3|\gamma_I(\tau)|^2}{3 + 3\gamma^2(\tau)}, \tag{60}$$

where P_3 is the degree of polarization of Eq. (56), $\gamma_I(\tau)$ is given in Eq. (42), and $\gamma(\tau)$ is the degree of time coherence in Eq. (40). Since the field is ergodic, we find for large separations τ in time the limit $\lim_{\tau \to \infty} \gamma_P(\tau) = (2P_3^2 + 1)/3$. For fully polarized light the instantaneous polarization state does not evolve and $\lim_{\tau \to \infty} \gamma_P(\tau) = 1$, as required. On the other hand, for unpolarized light we find that $\lim_{\tau \to \infty} \gamma_P(\tau) = 1/3$, demonstrating the fact that regardless of the initial polarization state, on average one third of the energy remains indefinitely in that state. This is consistent with the notion that for unpolarized three-component light fields, such as blackbody radiation, the energy is on average equally distributed between the three orthogonal field components.

The polarization parameter $\gamma_P^{(as)}(\tau)$ for a free-space blackbody field is obtained with the help of Eq. (48), calculating $\gamma_I^{(as)}(\tau)$ from Eqs. (42) and (43), and setting $P_3^{(as)} = 0$. This results in the expression

$$\gamma_P^{(as)}(\tau) = \frac{\pi^8 + 3|90\zeta(4, 1 + ic\tau/\alpha)|^2}{3\pi^8 + |90\zeta(4, 1 + ic\tau/\alpha)|^2}. \tag{61}$$

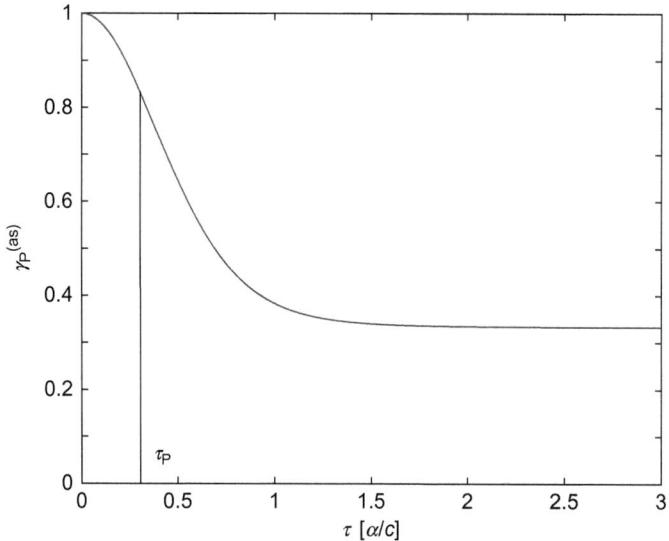

Fig. 5 Illustration of the all-space polarization correlation function $\gamma_P^{(as)}(\tau)$ as a function of τ expressed in units of α/c. The *vertical line* indicates the polarization time of $\tau_P = 0.30\alpha/c$.

The behavior of this quantity as a function of τ is shown in Fig. 5. As seen there, $\gamma_P^{(as)}(\tau)$ drops from $\gamma_P^{(as)}(0) = 1$ to $1/3$ within a time difference of about α/c. In analogy to the coherence time we may explicitly define the polarization time as

$$\tau_P^2 = \frac{\int_0^\infty \tau^2 [\gamma_P(\tau) - 1/3]^2 \, d\tau}{\int_0^\infty [\gamma_P(\tau) - 1/3]^2 \, d\tau}, \tag{62}$$

and find that for all-space blackbody radiation $\tau_P = 0.30\alpha/c$. At the temperature $T = 300$ K, the polarization time equals $\tau_P = 7.6$ fs. In general, the polarization time is inversely proportional to temperature (since $\alpha \propto 1/T$). Thus the higher the temperature, the faster is the polarization dynamics.

We emphasize that the polarization time may physically be very different from the coherence time, although their values are approximatively the same for the free-space blackbody field. For a polarized field the polarization time is infinite, but the coherence time can have any value. In addition, Eq. (60), which is valid for a field that obeys Gaussian statistics, demonstrates the relationship between the degree of temporal coherence and the polarization correlation function. The physical meaning of the connection is not particularly transparent, but we see that also from a mathematical point of view the polarization time and the coherence time, i.e., the widths of $\gamma_P(\tau)$ and $\gamma(\tau)$, respectively, are not simply related.

Note that the coherence matrices in both the frequency and the time domain, as well as the characteristic time scales, depend only on the temperature, as is expected for a blackbody field. However, the spectral coherence length is about half the wavelength independently of temperature, whereas the time-domain coherence length is again a function of temperature. Similar properties will be found in later sections for the blackbody field within an aperture in the cavity wall and for the far field of the radiation emanating from it.

3. BLACKBODY RADIATION IN AN APERTURE

3.1 Frequency Domain

3.1.1 Spatial Coherence

In order to obtain an expression for blackbody radiation outside the blackbody cavity we must first ensure that there is an outside to the cavity, which we can do by requiring the cavity to have a wall that is not everywhere asymptotically far away. For the radiation to escape the cavity we must also introduce an aperture through the wall connecting the inside of the cavity with the outside, where the radiation will be observed. However, in order to study the radiation outside the cavity we must first determine the field inside the aperture, which we will do in this section.

To begin with we note that a cavity with an aperture, through which radiation can propagate, cannot be a blackbody cavity in the strict sense since it is not an isolated system and hence it does not reach thermal equilibrium independently from the outside. Hence for the concept of blackbody radiation outside the cavity to make sense, the aperture must be such that its influence on the thermal equilibrium inside the cavity is negligible. Paradoxically this then means that we do not include the aperture itself to the cavity geometry when we determine the blackbody field inside it. Instead we first consider the field inside the cavity with no aperture and then we take this field as a given when we study how the field behaves inside the aperture. Furthermore, to avoid getting tangled up in details concerning the structure of the cavity wall, we will here assume that the cavity wall is a perfect conductor, whereby it can additionally be assumed to be infinitely thin. This ensures that the aperture has no thickness and we may, to a good degree of accuracy, assume that the field inside the aperture is described by those components of the cavity field at the aperture that propagate through it. Note that this last assumption is precisely the assumption that lies at the heart of the Rayleigh–Sommerfeld diffraction integral, which we later employ to determine the far-zone radiation field.

As discussed in Section 2.1.1, the cross-spectral density matrix of blackbody radiation of an arbitrary cavity geometry can be expressed in terms of the imaginary part of the Green function corresponding to that geometry, but for arbitrary (finite) cavity geometries this Green function cannot usually be expressed in closed form. Therefore we will here study the simplest cavity geometry that contains a wall, that is, the half-space cavity geometry considered in the previous section. Since the results presented there suggest that the influence of the wall becomes negligible already at a few wavelengths distance from the wall, we may assume that for sufficiently large blackbody cavities, the effect of one wall on the field at another wall can essentially be ignored, unless the walls are so placed that they induce resonance effects. However, such properties, which depend on the absolute size of the cavity, cannot hold universally for all angular frequencies. In fact, the Planck spectrum in Eq. (5) itself is valid only for infinite cavities with walls that are perfect conductors. For finite cavities and in cavities with partially absorbing walls the spectral distribution of the radiation field at thermal equilibrium will deviate from Planck's spectrum (Baltes & Hilf, 1976). Hence caution is required if the half-space cavity results are applied to finite cavity geometries.

Following the above discussion we now set out to determine the field inside the aperture by considering the half-space cavity field and by removing from it all waves that do not propagate into the $z > 0$ half-space. The half-space cavity cross-spectral density matrix is given by Eq. (22) and we point out that it is not equal to the all-space cavity matrix in Eq. (7). Serendipitously it turns out, however, that the correlations between the plane waves propagating into the $z > 0$ half-space are the same for both matrices (Blomstedt et al., 2015) and thus the properties of aperture coherence can be derived from either matrix, with identical results. This then validates after-the-fact the use of the all-space cavity matrix to describe the field at the aperture, which is what has typically been done in the literature. The way in which the field components that propagate into the $z < 0$ half-space are removed from the aperture field, however, comprises an additional pitfall, which was observed only recently (Blomstedt et al., 2013). Specifically, when the contribution of these components is removed from the cross-spectral density matrix, the result is not simply half of that matrix, as might be expected based on energy considerations and used in previous works (James, 1994; Lahiri & Wolf, 2008), since an additional term appears as we see below. The full all-space cavity matrix has also been used to represent the aperture field (Carter & Wolf, 1975). Although the aperture coherence matrices in these approaches are incorrect, they still lead to correct results for the far-field polarization and the far-field paraxial coherence properties of the field radiated from the aperture.

The correct removal of the contribution of the plane waves traveling in the $z < 0$ direction yields for the 3×3 cross-spectral density matrix inside the aperture of a half-space blackbody cavity the form (Blomstedt et al., 2013)

$$\mathbf{W}^{(\mathrm{ap})}(\boldsymbol{\rho}_1,\boldsymbol{\rho}_2,\omega) = 2\pi a_0(\omega) \left\{ \left[j_0(k_0\rho) - \frac{j_1(k_0\rho)}{k_0\rho} \right] \mathbf{U} \right. \\ \left. + j_2(k_0\rho)\hat{\boldsymbol{\rho}}\hat{\boldsymbol{\rho}}^{\mathrm{T}} - i\frac{J_2(k_0\rho)}{k_0\rho}(\hat{\boldsymbol{\rho}}\hat{\mathbf{u}}_z^{\mathrm{T}} + \hat{\mathbf{u}}_z\hat{\boldsymbol{\rho}}^{\mathrm{T}}) \right\}, \quad (63)$$

where (ap) denotes an aperture quantity, $\boldsymbol{\rho}_1$ and $\boldsymbol{\rho}_2$ represent two points in the aperture (in the plane $z = 0$), and $\rho = |\boldsymbol{\rho}|$ and $\hat{\boldsymbol{\rho}} = \boldsymbol{\rho}/\rho$, with $\boldsymbol{\rho} = \boldsymbol{\rho}_2 - \boldsymbol{\rho}_1$. The function $J_2(x)$ is the Bessel function of order 2. As discussed above, the matrix in Eq. (63) differs from one half of the all-space coherence matrix of Eq. (7), evaluated at the $z = 0$ plane, by the additional last term. We observe that the aperture cross-spectral density matrix depends only on the difference in the positions of the observation points, whereby it is of a statistically homogeneous form, but unlike the all-space cross-spectral density matrix evaluated at $z = 0$, it is not isotropic (cf., Eq. 8). Furthermore, it can be noted that the aperture cross-spectral density matrix is not purely real and hence not the imaginary part of any Green's function. This is in no contradiction to the general result of Eq. (4) since, as discussed previously, the aperture is not included in the cavity geometry and hence the considerations leading up to the general result are not valid for the field inside the aperture.

From the cross-spectral density matrix in Eq. (63), we obtain for the degree of coherence and for the two-component correlation coefficient of blackbody radiation inside an aperture the expressions

$$\mu^{(\mathrm{ap})}(\boldsymbol{\rho}_1,\boldsymbol{\rho}_2,\omega) = \frac{1}{\sqrt{3}} \left[j_0^2(k_0\rho) + \frac{1}{2}j_2^2(k_0\rho) + \frac{3J_2^2(k_0\rho)}{2(k_0\rho)^2} \right]^{1/2}, \quad (64)$$

and

$$\mu_{ij}^{(\text{ap})}(\boldsymbol{\rho}_1,\boldsymbol{\rho}_2,\omega) = \frac{3}{2}\left\{\left[j_0(k_0\rho) - \frac{j_1(k_0\rho)}{k_0\rho}\right]\delta_{ij}\right.$$
$$\left. + j_2(k_0\rho)\frac{\rho_i\rho_j}{\rho^2} - i\frac{j_2(k_0\rho)}{k_0\rho}(\rho_i\delta_{jz} + \delta_{iz}\rho_j)\right\}, \quad (65)$$

where $i,j \in \{x, y, z\}$, $\rho_i = \hat{\mathbf{u}}_i \cdot \boldsymbol{\rho}$, and $\rho_z = 0$ by definition. The behavior of the degree of coherence $\mu^{(\text{ap})}(\boldsymbol{\rho}_1, \boldsymbol{\rho}_2, \omega)$ in the aperture is shown in Fig. 6 (solid blue). Even though the degrees of coherence inside the cavity and inside the aperture, as given by Eqs. (19) and (64), respectively, are different, their behaviors are very similar as can be seen by comparing Figs. 2 and 6. For both functions the equal-point value is $1/\sqrt{3}$, corresponding to an unpolarized three-component field. The spectral coherence length determined from Eqs. (20) and (64) is $l_c \approx 0.60\lambda$ (the vertical line in Fig. 6), which closely agrees with the spectral coherence length inside the cavity. In Fig. 6 we also depict the behavior of the magnitude of the correlation coefficient $\mu_{xx}^{(\text{ap})}(\boldsymbol{\rho}_1,\boldsymbol{\rho}_2,\omega)$ (dashed red) in the aperture ($z = 0$) as a function of the point-separation along the x axis.

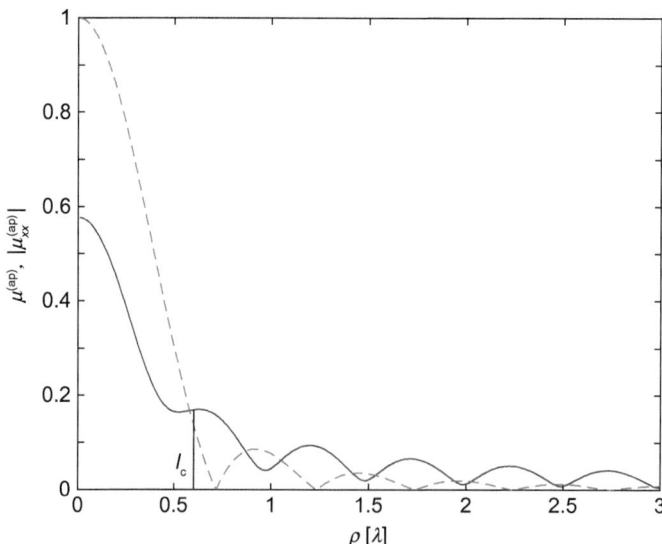

Fig. 6 Illustration of the spectrospatial coherence inside an aperture of a blackbody cavity: degree of coherence (*solid blue*) and $|\mu_{xx}^{(\text{ap})}(\rho,\omega)|$ along the x axis (*dashed red*). The *vertical line* indicates the spectral coherence length $l_c = 0.60\lambda$.

3.1.2 Polarization

From Eq. (63) of the cross-spectral density matrix inside the aperture it follows that the spectral polarization matrix therein is given by

$$\mathbf{\Phi}_3^{(\mathrm{ap})}(\boldsymbol{\rho}_0,\omega) = \frac{4\pi}{3}a_0(\omega)\mathbf{U}, \tag{66}$$

where $\boldsymbol{\rho}_0 = \boldsymbol{\rho}_1 = \boldsymbol{\rho}_2$. Therefore, the degree of polarization defined in Eq. (27) becomes

$$P_3^{(\mathrm{ap})}(\boldsymbol{\rho}_0,\omega) = 0, \tag{67}$$

that is, the blackbody radiation at every point inside the aperture and at all frequencies is a fully unpolarized three-component field.

3.1.3 Spectrum

When we introduce the polarization matrix given in Eq. (66) into Eq. (33), we get for the spectral density of the aperture field the result

$$S^{(\mathrm{ap})}(\boldsymbol{\rho}_0,\omega) = 4\pi a_0(\omega), \tag{68}$$

which shows that the aperture field has the same Planckian spectral distribution as the field inside the cavity. However, the energy of the field in the aperture is half of that in the cavity, as given by Eq. (34), since only the plane waves propagating toward the $z > 0$ half-space are present in the aperture.

3.2 Time Domain
3.2.1 Spatiotemporal Coherence

While the spectrospatial coherence properties within the aperture of a blackbody cavity have been studied earlier in the literature, the results presented in this section concerning spatiotemporal coherence are to our knowledge new. Substituting the aperture cross-spectral density matrix, as given by Eq. (63), into the Wiener–Khintchine relation in Eq. (37), we obtain for the 3×3 mutual coherence matrix of the aperture field the expression (a detailed derivation can be found in Appendix B)

$$\mathbf{\Gamma}^{(\mathrm{ap})}(\boldsymbol{\rho}_1,\boldsymbol{\rho}_2,\tau) = \frac{2\hbar c}{\pi}\{[f_1(\rho,\tau) - f_2(\rho,\tau)]\mathbf{U} + f_2(\rho,\tau)\hat{\boldsymbol{\rho}}\hat{\boldsymbol{\rho}}^\mathrm{T} + f_3(\rho,\tau)(\hat{\boldsymbol{\rho}}\hat{\mathbf{u}}_z^\mathrm{T} + \hat{\mathbf{u}}_z\hat{\boldsymbol{\rho}}^\mathrm{T})\}. \tag{69}$$

The functions $f_1(\rho, \tau)$ and $f_2(\rho, \tau)$ are given by Eqs. (44) and (45), respectively, and the new function $f_3(\rho, \tau)$ is defined as

$$f_3(\rho,\tau) = -i\frac{3}{4}\sum_{n=1}^{\infty} \frac{\rho}{\left[(n\alpha + ic\tau)^2 + \rho^2\right]^{5/2}}, \quad (70)$$

where $\alpha = \hbar c/k_B T$ as before, and the sign of the fractional power is determined from the definition $(\cdot)^{5/2} = (\sqrt{\cdot})^5$ with $\mathrm{Re}\{\sqrt{\cdot}\} \geq 0$.

When we use Eq. (69) to compute the time-domain degree of coherence from Eq. (40) and the normalized correlation coefficient from Eq. (41), we get the expressions

$$\gamma^{(\mathrm{ap})}(\boldsymbol{\rho}_1,\boldsymbol{\rho}_2,\tau) = \frac{30\alpha^4}{\pi^4}\left[|f_1(\rho,\tau)|^2 + 2|f_1(\rho,\tau) - f_2(\rho,\tau)|^2 + 2|f_3(\rho,\tau)|^2\right]^{1/2}, \quad (71)$$

and

$$\gamma_{ij}^{(\mathrm{ap})}(\boldsymbol{\rho}_1,\boldsymbol{\rho}_2,\tau) = \frac{90\alpha^4}{\pi^4}\left\{[f_1(\rho,\tau) - f_2(\rho,\tau)]\delta_{ij} + f_2(\rho,\tau)\frac{\rho_i\rho_j}{\rho^2} + f_3(\rho,\tau)\frac{\rho_i\delta_{j3} + \delta_{i3}\rho_j}{\rho}\right\}, \quad (72)$$

respectively.

As with the field inside the cavity, we consider the temporal and the spatial coherence of the aperture field separately. To study the temporal coherence, we put $\boldsymbol{\rho}_1 = \boldsymbol{\rho}_2 = \boldsymbol{\rho}_0$ in Eqs. (71) and (72), to get

$$\gamma^{(\mathrm{ap})}(\boldsymbol{\rho}_0,\boldsymbol{\rho}_0,\tau) = \frac{90}{\sqrt{3}\pi^4}|\zeta(4, 1 + ic\tau/\alpha)|, \quad (73)$$

$$\gamma_{ij}^{(\mathrm{ap})}(\boldsymbol{\rho}_0,\boldsymbol{\rho}_0,\tau) = \frac{90}{\pi^4}\zeta(4, 1 + ic\tau/\alpha)\delta_{ij}, \quad (74)$$

which are seen to be equal to the corresponding quantities inside the cavity, as given by Eqs. (48) and (49), respectively. Therefore, the conclusions on temporal coherence inside the cavity are valid in the aperture as well. The spatial coherence is described by Eqs. (71) and (72) with $\tau = 0$, that is, by

$$\gamma^{(\mathrm{ap})}(\boldsymbol{\rho}_1,\boldsymbol{\rho}_2,0) = \frac{15}{4\rho'^4}\left[3A^2(\rho') + 2A(\rho')B(\rho') + B^2(\rho') + |C(\rho')|^2\right]^{1/2}, \quad (75)$$

$$\gamma_{ij}^{(\mathrm{ap})}(\boldsymbol{\rho}_1,\boldsymbol{\rho}_2,0) = \frac{45}{4\rho'^4}\left[A(\rho')\delta_{ij} + B(\rho')\frac{\rho_i\rho_j}{\rho} + C(\rho')\frac{\rho_i\delta_{j3} + \delta_{i3}\rho_j}{\rho}\right], \quad (76)$$

where $\rho' = (\pi/a)\rho$ is a dimensionless variable and the functions $A(\rho')$ and $B(\rho')$ are given by Eqs. (52) and (53), respectively. The function $C(\rho')$ is defined as

$$C(\rho') = -i\frac{8\alpha^4 \rho'^4}{\pi^4} f_3(\alpha\rho'/\pi, 0), \tag{77}$$

where $f_3(\rho, \tau)$ is the function given in Eq. (70). With respect to the cavity quantities in Eqs. (50) and (51), the aperture quantities in Eqs. (75) and (76) have an additional term, containing the function $C(\rho')$. Therefore the spatial coherence properties inside the cavity and within the aperture are different, in contrast to the temporal coherence properties, which are the same in both regions. However, the difference is small as can be seen from Fig. 7, where the spatial degrees of coherence within the aperture (solid blue) and inside the cavity (dashed red, shown also in Fig. 4B) are illustrated. Consequently, we can draw the same conclusions about the aperture-field degree of coherence as we did in Section 2.2.1 for the field inside the cavity. For example, for the coherence length we get from Eqs. (55) and (75) the result $l_c = 1.29\alpha/\pi$ or $l_c = 0.33\lambda_p$, where

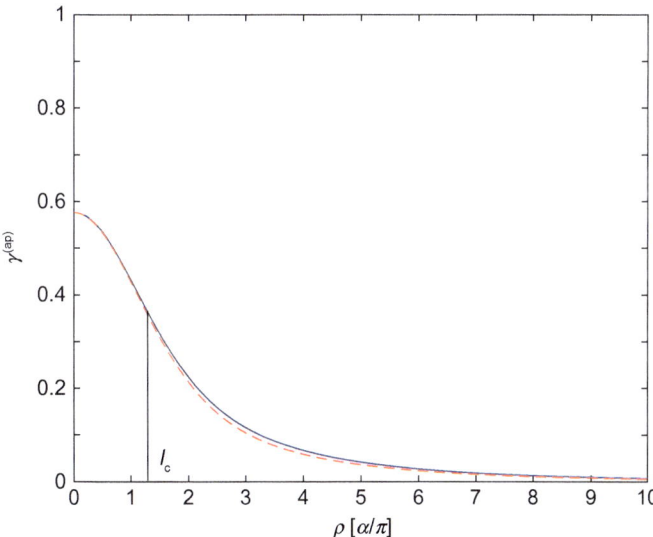

Fig. 7 Illustration of the spatial degree of coherence $\gamma^{(ap)}(\rho_1, \rho_2, 0)$ in the aperture (*solid blue*). The degree of coherence inside the cavity (*dashed red*) is shown in Fig. 4B and reproduced here for comparison. The *vertical line* depicts the coherence length $l_c = 1.29\alpha/\pi$.

λ_p is the peak wavelength of Planck's spectrum. The coherence length, which is approximately the same as the coherence length inside the cavity, is illustrated by the vertical line in Fig. 7. We note that, although not shown here explicitly, the behavior of the function $\gamma_{zz}^{(ap)}(\boldsymbol{\rho}_1,\boldsymbol{\rho}_2,0)$ in the aperture plane ($z=0$) is identical to the behavior of the corresponding cavity function shown (as dashed red) in Fig. 4B. In addition, in analogy to the cavity field, the normalized time-domain quantities of the aperture field depend on temperature, while the corresponding spectral quantities do not.

3.2.2 Polarization
The polarization matrix of the field in the aperture is obtained by setting $\boldsymbol{\rho}_1 = \boldsymbol{\rho}_2 = \boldsymbol{\rho}_0$ and $\tau = 0$ in Eq. (69), so that

$$\mathbf{J}^{(ap)}(\boldsymbol{\rho}_0) = \frac{\pi^3 \hbar c}{45\alpha^4} \mathbf{U}, \tag{78}$$

which when introduced into Eq. (56) yields for the degree of polarization the result

$$P_3^{(ap)}(\boldsymbol{\rho}_0) = 0. \tag{79}$$

Hence the field in the space–time domain is unpolarized throughout the aperture. This can also be seen in Fig. 7, where the (equal-time) single-point degree of coherence value $1/\sqrt{3}$ is unique to unpolarized light.

3.2.3 Polarization Dynamics
The polarization dynamics of the aperture field can be obtained by inserting Eqs. (42) (with Eq. 69), (73), and (79) into Eq. (60). This leads to exactly the same polarization correlation function as inside the cavity, that is, to Eq. (61), which actually was to be expected since, as we observed above, the temporal coherence properties of blackbody radiation inside the aperture are identical to those inside the cavity. Therefore, all the conclusions presented in connection with Fig. 5 concerning the polarization dynamics and polarization time of blackbody radiation in the cavity also apply to the field in the aperture.

4. BLACKBODY RADIATION IN THE FAR ZONE OF AN APERTURE

4.1 Frequency Domain

4.1.1 Far-Field Coherence

We will now study the far field of the radiation emanating from a half-space blackbody cavity through a circular aperture $\mathcal{A}$ of radius ϵ. The geometry and notation are shown in Fig. 8. The mapping of the three-component aperture field, $\mathbf{E}(\boldsymbol{\rho}, \omega)$, to the far-zone field can be described by the Rayleigh diffraction formula (Mandel & Wolf, 1995; Blomstedt et al., 2013), which reads as

$$\mathbf{E}^{(\infty)}(\mathbf{r},\omega) = \int_{\mathcal{A}} G(\mathbf{r}-\boldsymbol{\rho},\omega)\mathbf{E}(\boldsymbol{\rho},\omega)\mathrm{d}\boldsymbol{\rho}, \tag{80}$$

where the superscript (∞) refers to a far-field quantity (we use this notation in what follows). The integration in Eq. (80) is over the aperture $\mathcal{A}$ in the plane $z=0$ and the Green function is given by

$$G(\mathbf{r},\omega) = i\frac{k_0^2}{2\pi}(\hat{\mathbf{u}}_z \cdot \hat{\mathbf{r}})h_1^{(1)}(k_0 r), \tag{81}$$

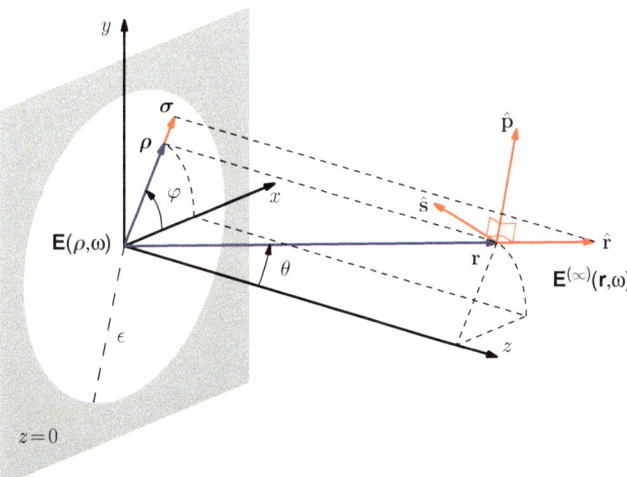

Fig. 8 Illustration of the notation for calculating the far field produced by blackbody radiation emanating from a circular aperture of radius ϵ. *Modified from Blomstedt, K., Setälä, T., Tervo, J., Turunen, J., & Friberg, A. T. (2013). Partial polarization and electromagnetic spatial coherence of blackbody radiation emanating from an aperture.* Physical Review A, 88, 013824.

where $h_1^{(1)}(x)$ is the spherical Hankel function of the first kind of order 1 and $\mathbf{r} = r\hat{\mathbf{r}}$. When $\mathbf{r}$ is in the far zone, the Green function has the asymptotic form

$$G(\mathbf{r}-\boldsymbol{\rho},\omega) \sim -i\frac{k_0}{2\pi}(\hat{\mathbf{u}}_z \cdot \hat{\mathbf{r}})\frac{\exp(ik_0 r)}{r}\exp(-ik_0\hat{\mathbf{r}}\cdot\boldsymbol{\rho}), \qquad (82)$$

whereby we get from the representation in Eq. (80) for the far field the expression

$$\mathbf{E}^{(\infty)}(\mathbf{r},\omega) = -\frac{ik_0}{2\pi}(\hat{\mathbf{u}}_z \cdot \hat{\mathbf{r}})\frac{\exp(ik_0 r)}{r}\int_{\mathcal{A}} \mathbf{E}(\boldsymbol{\rho},\omega)\exp(-ik_0\hat{\mathbf{r}}\cdot\boldsymbol{\rho})d\boldsymbol{\rho}. \qquad (83)$$

It follows from the definition in Eq. (1) that the far-field cross-spectral density matrix related to the asymptotic expression in Eq. (83) contains a four-dimensional Fourier transform of the aperture matrix in Eq. (63). It turns out that the integrations can be performed analytically resulting in the far-field cross-spectral density matrix of the form (Blomstedt et al., 2013)

$$\begin{aligned}\mathbf{W}^{(\infty)}&(r_1\hat{\mathbf{r}}_1, r_2\hat{\mathbf{r}}_2, \omega)\\ &= 2\mathcal{A}_0 a_0(\omega)\frac{J_1(k_0\epsilon\sigma)}{k_0\epsilon\sigma}(\hat{\mathbf{u}}_z \cdot \hat{\mathbf{r}}_1)(\hat{\mathbf{u}}_z \cdot \hat{\mathbf{r}}_2)\frac{\exp[ik_0(r_2 - r_1)]}{r_1 r_2}(1-\bar{\sigma}^2)^{-1/2}\\ &\quad \times \left[\mathbf{U} - (1-\bar{\sigma}^2)\hat{\mathbf{u}}_z\hat{\mathbf{u}}_z^{\mathrm{T}} - \bar{\boldsymbol{\sigma}}\bar{\boldsymbol{\sigma}}^{\mathrm{T}} - (1-\bar{\sigma}^2)^{1/2}(\hat{\mathbf{u}}_z\bar{\boldsymbol{\sigma}}^{\mathrm{T}} + \bar{\boldsymbol{\sigma}}\hat{\mathbf{u}}_z^{\mathrm{T}})\right],\end{aligned} \qquad (84)$$

where $\mathcal{A}_0 = \pi\epsilon^2$ is the area of the circular aperture, $J_1(x)$ is the Bessel function of order 1, and $a_0(\omega)$ is the Planck spectrum given by Eq. (5). Here $\sigma = |\boldsymbol{\sigma}|$ with $\boldsymbol{\sigma} = \boldsymbol{\sigma}_2 - \boldsymbol{\sigma}_1$, and $\bar{\sigma} = |\bar{\boldsymbol{\sigma}}|$ with $\bar{\boldsymbol{\sigma}} = (\boldsymbol{\sigma}_1 + \boldsymbol{\sigma}_2)/2$, where $\boldsymbol{\sigma}_i$ is the projection of $\hat{\mathbf{r}}_i$ onto the aperture plane ($z=0$) for $i \in \{1,2\}$ (see Fig. 8). The matrix $\mathbf{W}^{(\infty)}(r_1\hat{\mathbf{r}}_1, r_2\hat{\mathbf{r}}_2, \omega)$ is valid throughout the far zone, covering both paraxial and nonparaxial directions.

For paraxial directions, i.e., when $\bar{\sigma} \approx 0$, $\hat{\mathbf{u}}_z \cdot \hat{\mathbf{r}}_i \approx 1$, $i \in \{1,2\}$, and assuming that $r_1 = r_2 = r$, Eq. (84) simplifies into

$$\mathbf{W}^{(\infty)}(r\hat{\mathbf{r}}_1, r\hat{\mathbf{r}}_2, \omega) = \frac{2\mathcal{A}_0 a_0(\omega) J_1(k_0\epsilon\sigma)}{r^2}\frac{}{k_0\epsilon\sigma}(\mathbf{U} - \hat{\mathbf{u}}_z\hat{\mathbf{u}}_z^{\mathrm{T}}). \qquad (85)$$

This paraxial far-field coherence matrix of blackbody radiation was found in Lahiri and Wolf (2008), where the analysis is based on the incomplete aperture-field cross-spectral density matrix (the point discussed in Section 3.1.1). However, the far-field results there are correct since they are computed from the transverse (x and y) electric field components, which are the same as for the complete cross-spectral density matrix, and which can also alone be used to determine the far field in every direction (James, 1994).

Inserting Eq. (84) into Eq. (2) gives for the far-field spectrospatial degree of coherence the expression

$$\mu^{(\infty)}(r_1\hat{\mathbf{r}}_1, r_2\hat{\mathbf{r}}_2, \omega) = \sqrt{2}\left|\frac{J_1(k_0\epsilon\sigma)}{k_0\epsilon\sigma}\right|\left[\frac{(\hat{\mathbf{u}}_z \cdot \hat{\mathbf{r}}_1)(\hat{\mathbf{u}}_z \cdot \hat{\mathbf{r}}_2)}{1-\overline{\sigma}^2}\right]^{1/2}, \quad (86)$$

which is plotted in Fig. 9 in terms of the spherical polar coordinates (φ_1, θ_1) and (φ_2, θ_2) corresponding to the observation directions $\hat{\mathbf{r}}_1$ and $\hat{\mathbf{r}}_2$, as defined in Fig. 8. Fig. 9A and B corresponds to a paraxial direction $(\varphi_2, \theta_2) = (0, 0)$ and a nonparaxial direction $(\varphi_2, \theta_2) = (0, 3\pi/8)$, respectively, and illustrate the degree of coherence as a function of θ_1 ($\varphi_1 = 0$). The solid blue ($k_0\epsilon = 100$) and dashed red ($k_0\epsilon = 400$) curves show the effect of the aperture size ϵ. In both examples the aperture radius is chosen to be much larger than the wavelength in order to conform to the assumptions underlying the Kirchhoff boundary conditions in aperture diffraction. We observe from the figure that the angular extent of the degree of coherence is larger in the nonparaxial direction than in the paraxial direction, and this extent decreases when the aperture radius increases. In these cases, the angular coverage is of the order of milliradians and the peak (equal-angle) value is $1/\sqrt{2}$ corresponding to an unpolarized plane wave (this point will be discussed shortly).

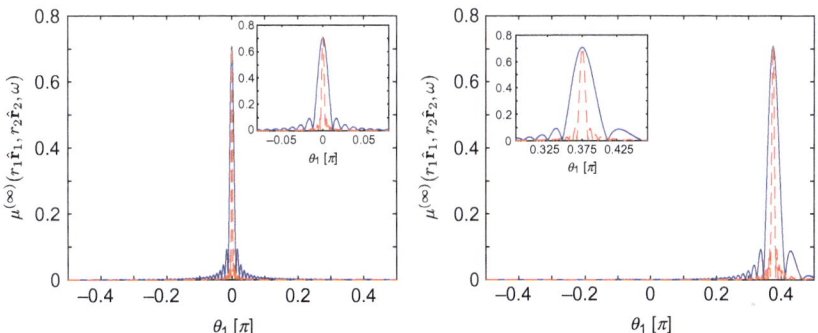

Fig. 9 Far-field degree of coherence for blackbody radiation escaping from a cavity through a circular aperture: (A) Around a paraxial direction $(\varphi_2, \theta_2) = (0, 0)$; (B) Around a nonparaxial direction $(\varphi_2, \theta_2) = (0, 3\pi/8)$. The *solid blue* and *dashed red* curves refer to the aperture sizes $k_0\epsilon = 100$ and $k_0\epsilon = 400$, respectively. The *insets* show magnifications around θ_2.

In the paraxial regime the degree of coherence in Eq. (86) reduces to

$$\mu^{(\infty)}(r_1\hat{\mathbf{r}}_1, r_2\hat{\mathbf{r}}_2, \omega) = \sqrt{2}\left|\frac{J_1(k_0\epsilon\sigma)}{k_0\epsilon\sigma}\right|, \qquad (87)$$

which exactly equals the result obtained by applying the electromagnetic van Cittert–Zernike theorem (Tervo et al., 2013) to determine the far field generated by a uniform, spatially δ-correlated (fully incoherent) and unpolarized planar circular beam-like secondary source. Hence we note that with respect to paraxial observation directions the radiation fields produced by δ-correlated sources model thermal (blackbody) radiation to a good degree of accuracy.

4.1.2 Polarization

The full 3×3 polarization matrix of the far field can be obtained from Eq. (84) as

$$\boldsymbol{\Phi}_3^{(\infty)}(r\hat{\mathbf{r}}, \omega) \equiv \mathbf{W}^{(\infty)}(r\hat{\mathbf{r}}, r\hat{\mathbf{r}}, \omega) = \frac{\mathcal{A}_0 a_0(\omega)(\hat{\mathbf{u}}_z \cdot \hat{\mathbf{r}})}{r^2}(\hat{\mathbf{s}}\hat{\mathbf{s}}^T + \hat{\mathbf{p}}\hat{\mathbf{p}}^T), \qquad (88)$$

where $\hat{\mathbf{s}}$ and $\hat{\mathbf{p}}$ are two mutually orthogonal unit vectors in the plane orthogonal to the observation direction $\hat{\mathbf{r}}$, as illustrated in Fig. 8. These unit vectors represent a complete orthonormal basis, so that, in particular, $\hat{\mathbf{s}}\hat{\mathbf{s}}^T + \hat{\mathbf{p}}\hat{\mathbf{p}}^T + \hat{\mathbf{r}}\hat{\mathbf{r}}^T = \mathbf{U}$. From Eq. (88) we note that the 3×3 matrix $\boldsymbol{\Phi}_3^{(\infty)}(r\hat{\mathbf{r}}, \omega)$ is of rank 2 and it can be represented in the local $\hat{\mathbf{s}}\hat{\mathbf{p}}$-space by the 2×2 matrix

$$\boldsymbol{\Phi}_2^{(\infty)}(r\hat{\mathbf{r}}, \omega) = \frac{\mathcal{A}_0 a_0(\omega)(\hat{\mathbf{u}}_z \cdot \hat{\mathbf{r}})}{r^2}\mathbf{U}_2, \qquad (89)$$

where $\mathbf{U}_2$ is the 2×2 unit matrix.

For far fields, which are locally planar, we may employ the conventional definition of the degree of polarization. It can be expressed in terms of the 2×2 spectral polarization matrix $\boldsymbol{\Phi}_2^{(\infty)}(r\hat{\mathbf{r}}, \omega)$ as (Wolf, 2007)

$$P_2^{(\infty)}(r\hat{\mathbf{r}}, \omega) = \left\{1 - \frac{4\det\left[\boldsymbol{\Phi}_2^{(\infty)}(r\hat{\mathbf{r}}, \omega)\right]}{\mathrm{tr}^2\left[\boldsymbol{\Phi}_2^{(\infty)}(r\hat{\mathbf{r}}, \omega)\right]}\right\}^{1/2}, \qquad (90)$$

where det denotes the matrix determinant. Using Eq. (89) we get from this definition the result

$$P_2^{(\infty)}(r\hat{\mathbf{r}}, \omega) = 0, \qquad (91)$$

which shows that the far field is fully unpolarized in every direction, a result originally found in James (1994). For the degree of polarization of three-component fields, as defined by Eq. (27), we get from the 3×3 matrix in Eq. (88) the result

$$P_3^{(\infty)}(r\hat{\mathbf{r}},\omega) = \frac{1}{2}. \tag{92}$$

The differences between the degrees of polarization of two-component and three-component electromagnetic fields are discussed in Setälä, Shevchenko, et al. (2002) and Setälä, Lindfors, and Friberg (2009).

4.1.3 Radiant Intensity

The spectral radiant intensity of the blackbody far field is obtained by applying Eq. (89) as (see James, 1994 for the original derivation)

$$\begin{aligned} J(\hat{\mathbf{r}},\omega) &= \lim_{r \to \infty} \{r^2 \operatorname{tr}[\mathbf{\Phi}_2^{(\infty)}(r\hat{\mathbf{r}},\omega)]\}, \\ &= 2\mathcal{A}_0 a_0(\omega)(\hat{\mathbf{u}}_z \cdot \hat{\mathbf{r}}), \end{aligned} \tag{93}$$

where $\hat{\mathbf{u}}_z \cdot \hat{\mathbf{r}} = \cos\theta$. This result implies that the blackbody far field obeys Lambert's cosine law, so that the aperture field thereby corresponds to a vector-valued Lambertian (secondary) source. Whereas the scalar blackbody aperture field represents, apart from evanescent contributions and a multiplicative constant, the unique scalar Lambertian source (Carter & Wolf, 1975; Starikov & Friberg, 1984), there are several kinds of vector-valued Lambertian sources, each with different polarization properties (Blomstedt et al., 2016). Of these the electromagnetic aperture blackbody field represents the kind that has an unpolarized far-field distribution.

We observe that the dependence on angular frequency ω in both the aperture and the cavity spectral density (Eqs. 68 and 34, respectively), as well as in the spectral radiant intensity (Eq. 93), is completely contained in the factor $a_0(\omega)$. This means that when these representations are normalized with respect to their integrals over all angular frequencies (total intensity), the results will agree and be equal to a function of ω alone. Specifically, the normalized far-field spectrum of blackbody radiation is thus independent of the observation direction and will equal the normalized spectrum of the aperture (cavity) field. This result can also be seen as the consequence of the scaling law for three-component electromagnetic fields (Hassinen et al., 2013), which states that such a correspondence holds between the far field and the aperture field if the trace of the normalized aperture cross-spectral density matrix, given by

$$\mu_1^{(\mathrm{ap})}(\boldsymbol{\rho}_1,\boldsymbol{\rho}_2,\omega) = \frac{\mathrm{tr}\mathbf{W}^{(\mathrm{ap})}(\boldsymbol{\rho}_1,\boldsymbol{\rho}_2,\omega)}{\left\{\mathrm{tr}\left[\mathbf{W}^{(\mathrm{ap})}(\boldsymbol{\rho}_1,\boldsymbol{\rho}_1,\omega)\right]\mathrm{tr}\left[\mathbf{W}^{(\mathrm{ap})}(\boldsymbol{\rho}_2,\boldsymbol{\rho}_2,\omega)\right]\right\}^{1/2}}, \qquad (94)$$

only depends on the separation $\rho = |\boldsymbol{\rho}_1 - \boldsymbol{\rho}_2|$. Indeed, when the blackbody aperture cross-spectral density matrix in Eq. (63) is inserted into Eq. (94), we get the result

$$\mu_1^{(\mathrm{ap})}(\boldsymbol{\rho}_1,\boldsymbol{\rho}_2,\omega) = j_0(k_0\rho) = \frac{\sin(k_0\rho)}{k_0\rho}, \qquad (95)$$

which is precisely of the required (sinc) form.

4.2 Time Domain

4.2.1 Spatiotemporal Coherence

From the far-zone blackbody cross-spectral density matrix of Eq. (84), together with the Wiener–Khintchine theorem in Eq. (37), we get for the related mutual coherence matrix the expression (see Appendix B)

$$\begin{aligned}\boldsymbol{\Gamma}^{(\infty)}(r\hat{\mathbf{r}}_1, r\hat{\mathbf{r}}_2, \tau) = &\frac{3\hbar c A_0}{2\pi^2 r^2} f_4(\epsilon\sigma,\tau)(\hat{\mathbf{u}}_z\cdot\hat{\mathbf{r}}_1)(\hat{\mathbf{u}}_z\cdot\hat{\mathbf{r}}_2)(1-\overline{\sigma}^2)^{-1/2}\\ &\times\left[\mathbf{U}-(1-\overline{\sigma}^2)\hat{\mathbf{u}}_z\hat{\mathbf{u}}_z^{\mathrm{T}}-\overline{\boldsymbol{\sigma}}\overline{\boldsymbol{\sigma}}^{\mathrm{T}}-(1-\overline{\sigma}^2)^{1/2}(\hat{\mathbf{u}}_z\overline{\boldsymbol{\sigma}}^{\mathrm{T}}+\overline{\boldsymbol{\sigma}}\hat{\mathbf{u}}_z^{\mathrm{T}})\right],\end{aligned} \qquad (96)$$

where we have taken both observation points to be at the same distance r. In addition, the function $f_4(s,\tau)$ is defined by

$$f_4(s,\tau) = \sum_{n=1}^{\infty}\frac{n\alpha + ic\tau}{\left[(n\alpha + ic\tau)^2 + s^2\right]^{5/2}}, \qquad (97)$$

with the same sign-convention for the fractional power as in Eq. (70).

We obtain for the equal-distance degree of coherence, by substituting Eq. (96) into Eq. (40), the expression

$$\gamma^{(\infty)}(r\hat{\mathbf{r}}_1, r\hat{\mathbf{r}}_2, \tau) = \frac{90\alpha^4}{\sqrt{2}\pi^4}|f_4(\epsilon\sigma,\tau)|\left[\frac{(\hat{\mathbf{u}}_z\cdot\hat{\mathbf{r}}_1)(\hat{\mathbf{u}}_z\cdot\hat{\mathbf{r}}_2)}{(1-\overline{\sigma}^2)}\right]^{1/2}. \qquad (98)$$

Notice that the geometrical factor in this expression is the same as in the corresponding spectral quantity given in Eq. (86). In addition, we observe that this equation does not depend on temperature, whereas Eq. (98) does. From Eq. (41) we, on the other hand, have

$$\gamma_{ij}^{(\infty)}(r\hat{\mathbf{r}}_1, r\hat{\mathbf{r}}_2, \tau) = \frac{90\alpha^4}{\pi^4} f_4(\epsilon\sigma, \tau) \left[\frac{(\hat{\mathbf{u}}_z \cdot \hat{\mathbf{r}}_1)(\hat{\mathbf{u}}_z \cdot \hat{\mathbf{r}}_2)}{(1-\overline{\sigma}^2)} \right]^{1/2}$$
$$\times \frac{\delta_{ij} - \overline{\sigma}_i \overline{\sigma}_j}{\sqrt{(1-\hat{r}_{1i}^2)(1-\hat{r}_{2j}^2)}}, \quad (99)$$

for $i, j \in \{x, y\}$, while

$$\gamma_{iz}^{(\infty)}(r\hat{\mathbf{r}}_1, r\hat{\mathbf{r}}_2, \tau) = \gamma_{zi}^{(\infty)}(r\hat{\mathbf{r}}_1, r\hat{\mathbf{r}}_2, \tau) = 0, \quad i \in \{x, y\}, \quad (100)$$

and

$$\gamma_{zz}^{(\infty)}(r\hat{\mathbf{r}}_1, r\hat{\mathbf{r}}_2, \tau) = \frac{90\alpha^4}{\pi^4} f_4(\epsilon\sigma, \tau) \left[\frac{(\hat{\mathbf{u}}_z \cdot \hat{\mathbf{r}}_1)(\hat{\mathbf{u}}_z \cdot \hat{\mathbf{r}}_2)}{(1-\overline{\sigma}^2)} \right]^{1/2}$$
$$\times \frac{\overline{\sigma}^2}{\sqrt{(1-\hat{r}_{1z}^2)(1-\hat{r}_{2z}^2)}}, \quad (101)$$

where $\hat{r}_{ki} = \hat{\mathbf{u}}_i \cdot \hat{\mathbf{r}}_k$, with $k \in \{1, 2\}$ and $i \in \{x, y, z\}$.

Fig. 10 shows the equal-distance degree of coherence $\gamma^{(\infty)}(r\hat{\mathbf{r}}_1, r\hat{\mathbf{r}}_2, \tau)$ evaluated from Eq. (98) around a paraxial direction $(\varphi_2, \theta_2) = (0, 0)$ and

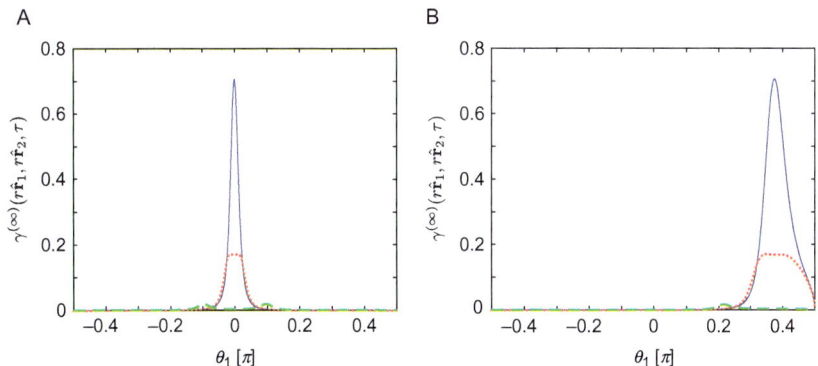

Fig. 10 Far-field equal-distance degree of coherence for blackbody radiation emanating from a circular aperture of radius $\epsilon = 100$ µm: (A) Around a paraxial direction $(\varphi_2, \theta_2) = (0, 0)$; (B) Around a nonparaxial direction $(\varphi_2, \theta_2) = (0, 3\pi/8)$. The *solid blue, dotted red,* and *dashed green* curves correspond to $\tau = 0$, $\tau = \alpha/c$, and $\tau = 4\alpha/c$, respectively. Here $T = 300$ K for which $\alpha/c \approx 25$ fs.

a nonparaxial direction $(\varphi_2, \theta_2) = (0, 3\pi/8)$ at several time differences τ. We observe that the angular extent of coherence is very narrow (a few mrad) and it decreases with increasing τ, as expected.

The far-field temporal coherence is quantified by the mutual coherence matrix of Eq. (96) with $r\hat{\mathbf{r}}_1 = r\hat{\mathbf{r}}_2 = r\hat{\mathbf{r}}$, i.e., by

$$\boldsymbol{\Gamma}^{(\infty)}(r\hat{\mathbf{r}}, r\hat{\mathbf{r}}, \tau) = \frac{3\hbar c \mathcal{A}_0}{2\pi^2 \alpha^4 r^2}(\hat{\mathbf{u}}_z \cdot \hat{\mathbf{r}})\zeta(4, 1 + ic\tau/\alpha)(\hat{\mathbf{s}}\hat{\mathbf{s}}^T + \hat{\mathbf{p}}\hat{\mathbf{p}}^T), \quad (102)$$

whereby the related single-point degree of coherence and the normalized correlation function, respectively, become

$$\gamma^{(\infty)}(r\hat{\mathbf{r}}, r\hat{\mathbf{r}}, \tau) = \frac{90}{\sqrt{2}\pi^4}|\zeta(4, 1 + ic\tau/\alpha)|, \quad (103)$$

$$\gamma_{ij}^{(\infty)}(r\hat{\mathbf{r}}, r\hat{\mathbf{r}}, \tau) = \frac{90}{\pi^4}\zeta(4, 1 + ic\tau/\alpha)\left(\delta_{ij} - \frac{r_i r_j}{r^2}\right), \quad (104)$$

where $i, j \in \{x, y, z\}$ in the latter expression. We note that the equal-point degree of coherence $\gamma^{(\infty)}(r\hat{\mathbf{r}}, r\hat{\mathbf{r}}, \tau)$ characterizing the far-field temporal coherence is independent of the observation direction. In addition, apart from the numerical factor ($\sqrt{2}$ instead of $\sqrt{3}$ related to the 2D/3D nature of the field), the temporal degree of coherence and the associated coherence time in the far zone coincide with those in the aperture (Eq. 73) and inside the cavity (Eq. 48).

4.2.2 Polarization
From Eq. (102) we get for the far-field polarization matrix the expression

$$\mathbf{J}^{(\infty)}(r\hat{\mathbf{r}}) = \frac{\pi^2 \hbar c \mathcal{A}_0}{60\alpha^4 r^2}(\hat{\mathbf{u}}_z \cdot \hat{\mathbf{r}})(\hat{\mathbf{s}}\hat{\mathbf{s}}^T + \hat{\mathbf{p}}\hat{\mathbf{p}}^T). \quad (105)$$

It immediately follows that the far-field degree of polarization, given by

$$P_2^{(\infty)}(r\hat{\mathbf{r}}) = \left\{1 - \frac{4\det\left[\mathbf{J}^{(\infty)}(r\hat{\mathbf{r}})\right]}{\text{tr}^2\left[\mathbf{J}^{(\infty)}(r\hat{\mathbf{r}})\right]}\right\}^{1/2}, \quad (106)$$

vanishes, that is,

$$P_2^{(\infty)}(r\hat{\mathbf{r}}) = 0, \quad (107)$$

in every direction.

4.2.3 Radiant Intensity

With the help of Eq. (105) the far-field radiant intensity becomes

$$J^{(\infty)}(\hat{\mathbf{r}}) = \lim_{r\to\infty}\left\{r^2\,\mathrm{tr}\left[\mathbf{J}^{(\infty)}(r\hat{\mathbf{r}})\right]\right\} = \frac{\pi^2\hbar c\mathcal{A}_0}{30\alpha^4}(\hat{\mathbf{u}}_z\cdot\hat{\mathbf{r}}). \qquad (108)$$

The factor $(\hat{\mathbf{u}}_z\cdot\hat{\mathbf{r}}) = \cos\theta$ shows that the radiant intensity follows Lambert's cosine law.

4.2.4 Polarization Dynamics

The far-field polarization dynamics of blackbody radiation can be quantified by Eq. (59) that we employed earlier for the three-component field inside the cavity and in the aperture. However, instead of a three-component representation we express the far-zone electric field in the local $\hat{\mathbf{s}}\hat{\mathbf{p}}$-space. Since blackbody radiation obeys Gaussian statistics the polarization correlation function takes for a two-component electric field the form (Shevchenko et al., 2009)

$$\gamma_P^{(\infty)}(r\hat{\mathbf{r}},\tau) = \frac{1 + P_2^2(r\hat{\mathbf{r}}) + 2|\gamma_1(r\hat{\mathbf{r}},r\hat{\mathbf{r}},\tau)|^2}{2[1 + \gamma^2(r\hat{\mathbf{r}},r\hat{\mathbf{r}},\tau)]}, \qquad (109)$$

with $P_2(r\hat{\mathbf{r}})$, $\gamma_1(r\hat{\mathbf{r}},r\hat{\mathbf{r}},\tau)$, and $\gamma(r\hat{\mathbf{r}},r\hat{\mathbf{r}},\tau)$ given by Eqs. (91), (42), and (40), respectively. The above polarization correlation function is based on considering the time evolution of the instantaneous two-component Jones vector in the local $\hat{\mathbf{s}}\hat{\mathbf{p}}$-space. Hence it describes the time evolution of the local polarization state, i.e., the variations in the shape of the polarization ellipse lying in a fixed plane orthogonal to the propagation direction or, which is the same, it describes the rate and amount of energy transfer between the orthogonal polarization states. An equivalent formalism can be established in terms of the instantaneous Poincaré vector (Shevchenko et al., 2009). Unlike with three-component fields inside the cavity and in the aperture, for which the Poincaré vector consists of eight components, the far-field Poincaré vector in the local $\hat{\mathbf{s}}\hat{\mathbf{p}}$-space only has three components, making it easier to analyze (Setälä, Shevchenko, Kaivola, & Friberg, 2008).

When we introduce Eqs. (107) and (103) into Eq. (109), and use Eq. (42), we get for the far-field polarization correlation function the expression

$$\gamma_P^{(\infty)}(r\hat{\mathbf{r}},\tau) = \frac{\pi^8 + 2|90\zeta(4,1+ic\tau/\alpha)|^2}{2\pi^8 + |90\zeta(4,1+ic\tau/\alpha)|^2}, \qquad (110)$$

which is independent of direction, that is, the rate and amplitude of the polarization changes are identical in every direction in the far zone. However, the instantaneous polarization state along a surface of constant distance from the aperture is generally different in different directions. The behavior of $\gamma_P^{(\infty)}(r\hat{\mathbf{r}},\tau)$ as a function of τ is shown in Fig. 11, where we note that its value drops from 1 to 1/2 when the time difference τ increases from 0 to about a/c. We may define the polarization time τ_P for two-component fields similarly as was done in Eq. (62) for three-component fields, to get

$$\tau_P^2 = \frac{\int_0^\infty \tau^2 [\gamma_P^{(\infty)}(r\hat{\mathbf{r}},\tau) - 1/2]^2 d\tau}{\int_0^\infty [\gamma_P^{(\infty)}(r\hat{\mathbf{r}},\tau) - 1/2]^2 d\tau}, \tag{111}$$

which implies that $\tau_P = 0.31 a/c$. When $T = 300$ K we thus have $\tau_P = 7.8$ fs. In general $a \propto 1/T$, so that the polarization time is inversely proportional to the absolute temperature. In addition, the polarization time, as well as the other quantities related to blackbody radiation, only has the

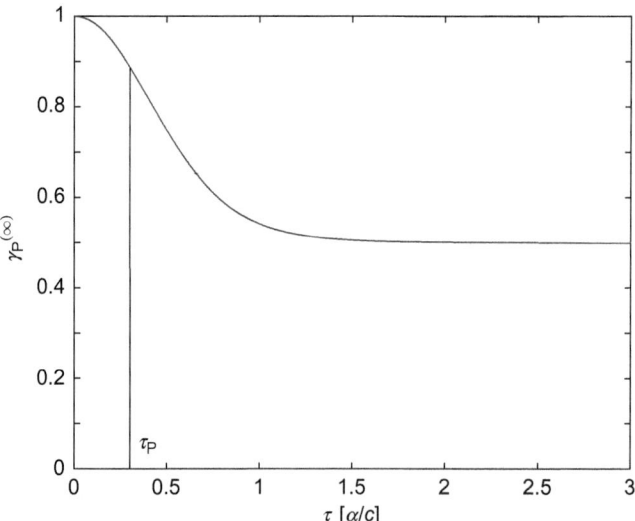

Fig. 11 Presentation of the far-field blackbody polarization correlation function $\gamma_P^{(\infty)}(\tau)$ as a function of τ expressed in units of a/c. The *vertical line* shows the polarization time at $\tau_P = 0.31 a/c$.

absolute temperature as a parameter. We also note that the polarization dynamics practically evolves at the same rate in the far field as inside the cavity or within the aperture. A significant difference, however, is that in the far field the polarization changes happen in a fixed plane while in the other cases the orientation of this polarization plane also evolves with time.

5. SUMMARY

In this review, we have given, both in the time and in the frequency domains, a unified treatment of classical electromagnetic coherence of blackbody radiation inside an all-space cavity, within an aperture in the cavity wall, and in the far zone of the aperture. Some of the results, namely, the effect of cavity wall on polarization inside the cavity and the derivations of the two-point mutual coherence matrices in the aperture and in the far zone, have not been presented in the literature before. The full temporal and spectral coherence matrices in the cavity, in the aperture, and in the far zone are therefore now known and they make it possible to assess various classical two-point (spatial coherence) or one-point (polarization, spectrum) spectral or spatiotemporal coherence properties of blackbody radiation. In all three geometries we determined the spectral coherence length as well as the coherence time, the temporal coherence length, and the polarization time; their values are summarized in Table 1.

Table 1 Collection of Quantities Characterizing Both the Spatiotemporal and the Spectrospatial Coherence of Blackbody Radiation Inside the All-Space Cavity, Within an Aperture in the Cavity Wall, and in the Far Zone of the Aperture

		Time Domain			Frequency Domain
	Coherence Time τ_c	Spatial Coherence ($\tau = 0$)	Polarization Time τ_P		Spatial Coherence
Cavity	$0.44\alpha/c = 0.35 T_p$	$l_c = 1.23\alpha/\pi = 0.31\lambda_p$	$0.30\alpha/c = 0.24 T_p$		$l_c = 0.60\lambda$
Aperture	As in cavity	$l_c = 1.29\alpha/\pi = 0.33\lambda_p$	As in cavity		$l_c = 0.60\lambda$
Far zone	As in cavity	A few mrad	$0.31\alpha/c = 0.25 T_p$		A few mrad

The quantity l_c refers to the coherence length, while λ_p and T_p are the peak wavelength and the corresponding oscillation period, respectively. When $T = 300$ K, for example, the values are $\lambda_p \approx 9.7$ μm, $T_p \approx 32$ fs, and $\alpha/c \approx 25$ fs. The time-domain characteristic scales are inversely proportional to temperature as $\alpha = \hbar c/k_B T$.

The coherence time as well as the temporal and spectral coherence lengths are consistent with the early results on blackbody radiation, although they are here evaluated with the recently introduced quantities of electromagnetic coherence theory. In the frequency domain, the spatial coherence length at a specific frequency in the cavity and in the aperture is on the order of half of the wavelength related to that frequency, independently of temperature. In the time domain, the two coherence lengths are of the order of half of the peak wavelength of Planck's spectrum and hence they depend on temperature via Wien's displacement law. The coherence time is approximately equal to half of the oscillation period corresponding to the peak frequency of the spectrum inside the cavity, within the aperture, and in the far zone alike. The polarization times of the 3D blackbody fields in the cavity and in the aperture are the same while for the far field, due to its 2D nature, the value is slightly different. In practice, however, the polarization time is the same in all three regions. All time scales are inversely proportional to temperature. Regarding the far-field spatial coherence, we considered the two-point coherence on an equidistance surface and found that the angular extent of the coherence is a few milliradians in both the time and the frequency domain. The results summarized in this work should find use in a wide variety of situations where natural light sources and propagation of thermal radiation are considered.

ACKNOWLEDGMENTS

The authors thank the Academy of Finland for funding (projects 268480, 268705). Also Dean's special support for coherence research at the University of Eastern Finland (projects 930350 and 931726) is acknowledged. Collaboration with Andriy Shevchenko, Jani Tervo, Jari Turunen, Bernhard Hoenders, Jari Lindberg, Timo Hassinen, Timo Voipio, and Lasse-Petteri Leppänen on the topics of this review is gratefully acknowledged.

APPENDIX A. BLACKBODY COHERENCE IN CAVITIES
Appendix A.1. Maxwell's Equations

In the absence of primary sources Maxwell's equations are represented in the Gaussian system, at a specific angular frequency ω, by the reduced equations

$$\nabla \times \mathbf{E}(\mathbf{r},\omega) = ik_0[\mathbf{H}(\mathbf{r},\omega) + 4\pi\mathbf{M}(\mathbf{r},\omega)], \quad (A.1)$$

and

$$\nabla \times \mathbf{H}(\mathbf{r},\omega) = -ik_0[\mathbf{E}(\mathbf{r},\omega) + 4\pi\mathbf{P}(\mathbf{r},\omega)], \quad (A.2)$$

where $k_0 = \omega/c$ is the vacuum wave number, $\mathbf{E}(\mathbf{r}, \omega)$ denotes the electric field, $\mathbf{H}(\mathbf{r}, \omega)$ denotes the magnetic field, $\mathbf{P}(\mathbf{r}, \omega)$ is the polarization vector, and $\mathbf{M}(\mathbf{r}, \omega)$ is the magnetization vector. By solving for the magnetic field in the first and for the electric field in the second of these equations, we can express one field in terms of the other as

$$\mathbf{H}(\mathbf{r},\omega) = -\frac{i}{k_0}\nabla \times \mathbf{E}(\mathbf{r},\omega) - 4\pi \mathbf{M}(\mathbf{r},\omega), \quad (A.3)$$

and

$$\mathbf{E}(\mathbf{r},\omega) = \frac{i}{k_0}\nabla \times \mathbf{H}(\mathbf{r},\omega) - 4\pi \mathbf{P}(\mathbf{r},\omega). \quad (A.4)$$

For Maxwell's equations to possess a unique solution they must be accompanied by boundary conditions for the electric and magnetic fields. In view of the above equations, we note that such boundary conditions can also readily be expressed in terms of either field alone.

When we introduce the representations given in Eqs. (A.3) and (A.4) into Eqs. (A.2) and (A.1), we obtain the electric field and magnetic field vector wave equations

$$(\nabla \times \nabla \times -k_0^2)\mathbf{E}(\mathbf{r},\omega) = -i4\pi k_0 \mathbf{P}(\mathbf{r},\omega) + 4\pi k_0 \nabla \times \mathbf{M}(\mathbf{r},\omega), \quad (A.5)$$

and

$$(\nabla \times \nabla \times -k_0^2)\mathbf{H}(\mathbf{r},\omega) = i4\pi k_0 \mathbf{M}(\mathbf{r},\omega) - 4\pi k_0 \nabla \times \mathbf{P}(\mathbf{r},\omega), \quad (A.6)$$

respectively. Either of these equations together with the electromagnetic boundary conditions constitutes a complete description of the behavior of the whole electromagnetic field.

Appendix A.2. Fluctuation-Dissipation Theorem and Blackbody Radiation

The fluctuation-dissipation theorem (Agarwal, 1975; Carminati & Greffet, 1999; Joulain, Mulet, Marquier, Carminati, & Greffet, 2005; Shchegrov, Joulain, Carminati, & Greffet, 2000) connects the response of an electromagnetic field to local disturbances (expressed in terms of the polarization and magnetization) in the correlation properties of a stationary field. Since the electromagnetic boundary conditions represent nonlocal sources, we must begin by removing their influence. To do this we observe that when the boundary conditions together with Maxwell's equations constitute a well-posed problem, there exists a unique electromagnetic field that satisfies the boundary conditions and Maxwell's equations with zero polarization and zero magnetization. If we denote the electric and magnetic components of this field by $\mathbf{E}'(\mathbf{r},\omega)$ and $\mathbf{H}'(\mathbf{r},\omega)$, it then immediately follows that the difference fields $\Delta \mathbf{E}(\mathbf{r},\omega) = \mathbf{E}(\mathbf{r},\omega) - \mathbf{E}'(\mathbf{r},\omega)$ and $\Delta \mathbf{H}(\mathbf{r},\omega) = \mathbf{H}(\mathbf{r},\omega) - \mathbf{H}'(\mathbf{r},\omega)$ satisfy the Maxwell equations (Eqs. A.1 and A.2) with the electromagnetic boundary conditions replaced by zero boundary conditions. Since the mapping from the zero-boundary-condition fields to the full fields is straightforward and since we are not here concerned with field properties related to external sources, we will for simplicity assume that the fields considered here correspond to zero-boundary conditions from the onset and hence we effectively drop the Δ in front of the field quantities in what follows. Since we are here only interested in electric field correlations, we will henceforth also assume that there is no magnetization, that is, $\mathbf{M}(\mathbf{r}, \omega) = \mathbf{0}$. For the general case see the derivations in Blomstedt et al. (2015), which we follow closely below, correcting a small error at the last step of those derivations. Before continuing we note that since the boundary conditions are linear in the fields, any multiple of a field that satisfies the zero boundary conditions will also satisfy the zero boundary conditions, albeit possibly with a different unit for the zero. In what follows we use the term zero boundary conditions to label all such variations.

With the assumptions put forward above it immediately follows from the magnetic vector wave equation (Eq. A.6) that

$$\nabla \cdot \mathbf{H}(\mathbf{r},\omega) = 0, \tag{A.7}$$

whereby we can use the vector equality

$$\nabla \times \nabla \times \mathbf{F} = \nabla(\nabla \cdot \mathbf{F}) - \nabla^2 \mathbf{F} \tag{A.8}$$

to develop Eq. (A.6) into

$$(\nabla^2 + k_0^2)\mathbf{H}(\mathbf{r},\omega) = 4\pi i k_0 \nabla \times \mathbf{P}(\mathbf{r},\omega), \tag{A.9}$$

which is the magnetic vector Helmholtz equation. Together with the divergence condition of Eq. (A.7) this equation is equivalent to the magnetic vector wave equation (Eq. A.6), when the magnetization is zero. Thereby the solution to the magnetic vector Helmholtz equation also provides the solution to the magnetic vector wave equation. When the magnetic zero boundary conditions are taken into account this solution is given by

$$\mathbf{H}(\mathbf{r},\omega) = -ik_0 \int \mathbf{g}_H(\mathbf{r},\mathbf{r}',\omega) \nabla' \times \mathbf{P}(\mathbf{r}',\omega) d\mathbf{r}', \qquad (A.10)$$

where the magnetic vector Green function $\mathbf{g}_H(\mathbf{r},\mathbf{r}',\omega)$ is the unique function of $\mathbf{r}$ that satisfies the Helmholtz equation

$$(\nabla^2 + k_0^2)\mathbf{g}_H(\mathbf{r},\mathbf{r}',\omega) = -4\pi\delta(\mathbf{r}-\mathbf{r}')\mathbf{U}, \qquad (A.11)$$

and the magnetic zero boundary conditions augmented by the divergence condition of Eq. (A.7) for all $\mathbf{r}'$. The magnetic Green function can be written as

$$\mathbf{g}_H(\mathbf{r},\mathbf{r}',\omega) = G(\mathbf{r},\mathbf{r}',\omega)\mathbf{U} + \Delta\mathbf{g}_H(\mathbf{r},\mathbf{r}',\omega), \qquad (A.12)$$

where

$$G(\mathbf{r},\mathbf{r}',\omega) = \frac{\exp(ik_0|\mathbf{r}-\mathbf{r}'|)}{|\mathbf{r}-\mathbf{r}'|} \qquad (A.13)$$

is the Green function of the scalar Helmholtz equation that satisfies the so-called radiation condition (Nieto-Vesperinas, 1991), and $\Delta\mathbf{g}_H(\mathbf{r},\mathbf{r}',\omega)$ satisfies the homogeneous counterpart to Eq. (A.11). By introducing the magnetic field representation of Eq. (A.10) into Eq. (A.4), we get the representation

$$\mathbf{E}(\mathbf{r},\omega) = \nabla \times \int \mathbf{g}_H(\mathbf{r},\mathbf{r}',\omega) \nabla' \times \mathbf{P}(\mathbf{r}',\omega) d\mathbf{r}' - 4\pi \mathbf{P}(\mathbf{r},\omega) \qquad (A.14)$$

for the corresponding electric field.

The fluctuation-dissipation theorem connects the electric field correlation function to the susceptibility matrix $\chi(\mathbf{r}_1, \mathbf{r}_2, \omega)$ between the source polarization at $\mathbf{r}_2$ and the electric field at a point $\mathbf{r}_1$ as (see Agarwal, 1975, but note that there the expression is multiplied by the factor 2π due to differences in where this factor enters into the Wiener–Khintchine theorem)

$$\mathbf{W}(\mathbf{r}_1,\mathbf{r}_2,\omega) = \langle \mathbf{E}^*(\mathbf{r}_1,\omega)\mathbf{E}^T(\mathbf{r}_2,\omega)\rangle = \frac{4\pi a_0(\omega)}{k_0^3}\text{Im}\{\chi(\mathbf{r}_1,\mathbf{r}_2,\omega)\}, \qquad (A.15)$$

where $a_0(\omega)$ is the Planck spectrum defined in Eq. (5). The susceptibility matrix is given by linear response theory as the perturbation of the average electric field at $\mathbf{r}_1$ with respect to changes in the polarization at $\mathbf{r}_2$ (Agarwal,

1975). To make this explicit, we consider a polarization vector of the form $\mathbf{P}_n(\mathbf{r}, \omega) = \mathbf{P}(\omega)\delta_n(\mathbf{r} - \mathbf{r}_2)$, where $\{\delta_n(\mathbf{r})\}$ is a sequence of smooth, real-valued functions that converge toward the δ-function, when $n \to \infty$. This definition reflects the connection that implicitly exists between physical and mathematical point sources and by taking the limit only at the end of our computations, all integrands below will be well behaved. We can then rewrite Eq. (A.14) as

$$\mathbf{E}_n(\mathbf{r}_1,\omega) = \left\{ \nabla_1 \times \int \mathbf{g}_\mathrm{H}(\mathbf{r}_1,\mathbf{r}',\omega) \times \nabla'\delta_n(\mathbf{r}' - \mathbf{r}_2)\mathrm{d}\mathbf{r}' - 4\pi\delta_n(\mathbf{r}_1 - \mathbf{r}_2) \right\} \cdot \mathbf{P}(\omega),$$
(A.16)

so that

$$\begin{aligned}
\mathrm{Im}\{\chi(\mathbf{r}_1,\mathbf{r}_2,\omega)\} &= \lim_{n \to \infty} \mathrm{Im}\left\{\frac{\delta\langle \mathbf{E}_n(\mathbf{r}_1,\omega)\rangle}{\delta \mathbf{P}(\omega)}\right\}, \\
&= \lim_{n \to \infty} \nabla_1 \times \int \mathrm{Im}\{\mathbf{g}_\mathrm{H}(\mathbf{r}_1,\mathbf{r}',\omega)\} \times \nabla'\delta_n(\mathbf{r}' - \mathbf{r}_2)\mathrm{d}\mathbf{r}' \\
&= \nabla_1 \times \left[\nabla_2 \times \mathrm{Im}\{\mathbf{g}_\mathrm{H}^\mathrm{T}(\mathbf{r}_1,\mathbf{r}_2,\omega)\}\right]^\mathrm{T},
\end{aligned}$$
(A.17)

where we have used Green's theorem and the fact that the support of the δ-function sequence shrinks to a single point when $n \to \infty$, to transfer the differentiation from $\delta_n(\mathbf{r}' - \mathbf{r}_2)$ to the imaginary part of the Green function, and where the last step follows from the definition of the δ-function sequence. We observe that the δ-function sequence outside the integral vanishes at the first step since the functions are real-valued. Crucially, however, as can be seen by taking the imaginary part of the Helmholtz equation (Eq. A.11), the function $\mathrm{Im}\{\mathbf{g}_\mathrm{H}(\mathbf{r},\mathbf{r}',\omega)\}$ is regular. This is noteworthy, since the singularity associated with the real part of the Green function would make the use of Green's theorem more involved and, in particular, prohibits taking the limit explicitly (see Yaghjian, 1980 for details).

In the last expression in Eq. (A.17) we can exchange the order of taking the imaginary part and computing the differentials, and we obtain that

$$\nabla_1 \times \left[\nabla_2 \times \mathbf{g}_\mathrm{H}^\mathrm{T}(\mathbf{r}_1,\mathbf{r}_2,\omega)\right]^\mathrm{T} = k_0^2 \mathbf{G}_\mathrm{E}(\mathbf{r}_1,\mathbf{r}_2,\omega),$$
(A.18)

where $\mathbf{G}_\mathrm{E}(\mathbf{r},\mathbf{r}',\omega)$ is the Green function, corresponding to the electric zero boundary conditions, of the electric vector wave equation (Eq. A.5), where $\mathbf{M}(\mathbf{r},\omega) = \mathbf{0}$, that is

$$(\nabla \times \nabla \times -k_0^2)\mathbf{G}_{\mathrm{E}}(\mathbf{r},\mathbf{r}',\omega) = -4\pi\delta(\mathbf{r}-\mathbf{r}')\mathbf{U}. \qquad (A.19)$$

If we use the partitioning of Eq. (A.12), we can compute

$$\nabla_1 \times \left[\nabla_2 \times \frac{1}{k_0^2}\mathbf{g}_{\mathrm{H}}^{\mathrm{T}}(\mathbf{r}_1,\mathbf{r}_2,\omega)\right]^{\mathrm{T}} = \frac{1}{k_0^2}\nabla_1 \times \left\{\nabla_2 \times [G(\mathbf{r}_1,\mathbf{r}_2)\mathbf{U}]^{\mathrm{T}}\right\}^{\mathrm{T}}$$
$$+ \frac{1}{k_0^2}\nabla_1 \times \left[\nabla_2 \times \Delta\mathbf{g}_{\mathrm{H}}^{\mathrm{T}}(\mathbf{r}_1,\mathbf{r}_2,\omega)\right]^{\mathrm{T}}. \qquad (A.20)$$

Here the second term satisfies the divergence condition and hence we can use the vector equality in Eq. (A.8) to obtain

$$(\nabla_1 \times \nabla_1 \times -k_0^2)\frac{1}{k_0^2}\nabla_1 \times \left[\nabla_2 \times \Delta\mathbf{g}_{\mathrm{H}}^{\mathrm{T}}(\mathbf{r}_1,\mathbf{r}_2,\omega)\right]^{\mathrm{T}}$$
$$= -(\nabla_1^2 + k_0^2)\frac{1}{k_0^2}\nabla_1 \times \left[\nabla_2 \times \Delta\mathbf{g}_{\mathrm{H}}^{\mathrm{T}}(\mathbf{r}_1,\mathbf{r}_2,\omega)\right]^{\mathrm{T}} = 0, \qquad (A.21)$$

where the last step follows when we exchange the order of the differentiations and use the fact that $\Delta\mathbf{g}_{\mathrm{H}}^{\mathrm{T}}(\mathbf{r}_1,\mathbf{r}_2,\omega)$ satisfies the homogeneous magnetic vector Helmholtz equation by assumption. Starting from the definition in Eq. (A.13), it is straightforward to show that

$$\frac{1}{k_0^2}\nabla_1 \times \left\{\nabla_2 \times [G(\mathbf{r}_1,\mathbf{r}_2,\omega)\mathbf{U}]^{\mathrm{T}}\right\}^{\mathrm{T}} = \mathbf{G}(\mathbf{r}_1,\mathbf{r}_2,\omega), \qquad (A.22)$$

where

$$\mathbf{G}(\mathbf{r}_1,\mathbf{r}_2,\omega) = \left(\mathbf{U} + \frac{1}{k_0^2}\nabla_1\nabla_1^{\mathrm{T}}\right)G(\mathbf{r}_1,\mathbf{r}_2,\omega) \qquad (A.23)$$

is the vector Green function, which satisfies the vector wave equation given in Eq. (A.19) together with the Silver–Müller radiation condition (Nieto-Vesperinas, 1991). Thereby, it follows from Eqs. (A.18) and (A.21) that $\nabla_1 \times \left[\nabla_2 \times \mathbf{g}_{\mathrm{H}}^{\mathrm{T}}(\mathbf{r}_1,\mathbf{r}_2,\omega)\right]^{\mathrm{T}}$ satisfies the same wave equation. On the other hand, since $\mathbf{g}_{\mathrm{H}}^{\mathrm{T}}(\mathbf{r}_1, \mathbf{r}_2, \omega)$ satisfies the magnetic zero boundary conditions for all $\mathbf{r}_2$, the same is true for any linear combination of these functions with different $\mathbf{r}_2$, and thus specifically for the function $\left[\nabla_2 \times \mathbf{g}_{\mathrm{H}}^{\mathrm{T}}(\mathbf{r}_1,\mathbf{r}_2,\omega)\right]^{\mathrm{T}}$. The definition in Eq. (A.4) then suggests that $\nabla_1 \times \left[\nabla_2 \times \mathbf{g}_{\mathrm{H}}^{\mathrm{T}}(\mathbf{r}_1,\mathbf{r}_2,\omega)\right]^{\mathrm{T}}$ satisfies the electric zero boundary conditions, whereby the uniqueness of Green functions implies that the equality in Eq. (A.18) holds. We use that equality in Eq. (A.17) and introduce the result into Eq. (A.15) to obtain the expression in Eq. (4), which connects the cross-spectral correlation matrix of a blackbody field in a cavity to the electric vector Green function of the corresponding cavity geometry.

APPENDIX B. DERIVATION OF APERTURE AND FAR-FIELD MUTUAL COHERENCE MATRICES

Here we present the derivations of the aperture and the far-field mutual coherence matrices, which as far as we know have not previously appeared in the literature.

Appendix B.1. Aperture Matrix

We begin by noting that the aperture cross-spectral density matrix in Eq. (63) can be written in terms of the all-space cross-spectral density matrix of Eq. (7) as

$$\mathbf{W}^{(\mathrm{ap})}(\boldsymbol{\rho}_1,\boldsymbol{\rho}_2,\omega) = \frac{1}{2}\mathbf{W}^{(\mathrm{as})}(\boldsymbol{\rho}_1,\boldsymbol{\rho}_2,\omega) - i2\pi a_0(\omega)\frac{J_2(k_0\rho)}{k_0\rho}(\hat{\boldsymbol{\rho}}\hat{\mathbf{u}}_z^{\mathrm{T}} + \hat{\mathbf{u}}_z\hat{\boldsymbol{\rho}}^{\mathrm{T}}).$$
(B.1)

When this is inserted into the Wiener–Khintchine transform (Eq. 37), we obtain for the mutual coherence matrix of the aperture the expression

$$\mathbf{\Gamma}^{(\mathrm{ap})}(\boldsymbol{\rho}_1,\boldsymbol{\rho}_2,\tau) = \frac{1}{2}\mathbf{\Gamma}^{(\mathrm{as})}(\boldsymbol{\rho}_1,\boldsymbol{\rho}_2,\tau) + \frac{2\hbar c}{\pi}f_3(\rho,\tau)(\hat{\boldsymbol{\rho}}\hat{\mathbf{u}}_z^{\mathrm{T}} + \hat{\mathbf{u}}_z\hat{\boldsymbol{\rho}}^{\mathrm{T}}),$$
(B.2)

which, in view of Eq. (43), is seen to reproduce Eq. (69), provided that the function

$$\begin{aligned}f_3(\rho,\tau) &= -i\frac{\pi^2}{\hbar c}\int_0^\infty a_0(\omega)\frac{J_2(k_0\rho)}{k_0\rho}\exp(-i\omega\tau)d\omega \\ &= -\frac{i}{4\rho}\int_0^\infty \frac{k_0^2}{\exp(\alpha k_0)-1}J_2(\rho k_0)\exp(-ic\tau k_0)dk_0\end{aligned}$$
(B.3)

matches the definition in Eq. (70). Here the latter expression follows, when we use Eq. (5) and change the integration variable from ω to $k_0 = \omega/c$.

To evaluate the integral in Eq. (B.3), we use the power series representation

$$\frac{1}{\exp(\alpha k_0)-1}=\sum_{n=1}^{\infty}\exp(-n\alpha k_0), \tag{B.4}$$

valid for all $\alpha k_0 > 0$, and the Bessel function representation (Olver et al., 2010)

$$J_m(z)=\frac{i^{-m}}{\pi}\int_0^{\pi}\cos(m\theta)\exp(iz\cos\theta)d\theta, \tag{B.5}$$

whereby we can develop Eq. (B.3) as

$$f_3(\rho,\tau)=\frac{i}{4\pi\rho}\sum_{n=1}^{\infty}\int_0^{\pi}\cos(2\theta)$$
$$\times\int_0^{\infty}k_0^2\exp\{[-n\alpha+i(\rho\cos\theta-c\tau)]k_0\}dk_0 d\theta \tag{B.6}$$
$$=\frac{i}{2\pi\rho}\sum_{n=1}^{\infty}\int_0^{\pi}\frac{\cos(2\theta)}{[(n\alpha+ic\tau)-i\rho\cos\theta]^3}d\theta,$$

where $n\alpha > 0$ ensures that the inner integral converges. Next we make the change of variables $\theta = 2\arctan x$ and put $\chi_{n,\pm} = n\alpha + ic\tau \pm i\rho$, which transforms the integral in Eq. (B.6) (after straightforward rearrangements) into

$$f_3(\rho,\tau)=\frac{i}{\pi\rho}\sum_{n=1}^{\infty}\chi_{n,-}^{-3}\int_0^{\infty}\frac{1+x^4-6x^2}{[1+(\chi_{n,+}/\chi_{n,-})x^2]^3}dx, \tag{B.7}$$

where we in turn set $\xi=\sqrt{\chi_{n,+}/\chi_{n,-}}\,x$, with $\mathrm{Re}\{\sqrt{\cdot}\}\geq 0$, to get

$$f_3(\rho,\tau)=\frac{i}{\pi\rho}\sum_{n=1}^{\infty}\chi_{n,-}^{-3}\sqrt{\chi_{n,-}/\chi_{n,+}}\int_0^{\sqrt{\chi_{n,+}/\chi_{n,-}}\infty}\frac{1}{(1+\xi^2)^3} \tag{B.8}$$
$$\times[1+(\chi_{n,-}/\chi_{n,+})^2\xi^4-6(\chi_{n,-}/\chi_{n,+})\xi^2]d\xi,$$
$$=i\frac{3}{16\rho}\sum_{n=1}^{\infty}\chi_{n,-}^{-3}\sqrt{\chi_{n,-}/\chi_{n,+}}\,[1+(\chi_{n,-}/\chi_{n,+})^2-2(\chi_{n,-}/\chi_{n,+})] \tag{B.9}$$
$$=-i\frac{3}{4}\sum_{n=1}^{\infty}\frac{\rho}{[(n\alpha+ic\tau)^2+\rho^2]^{5/2}},$$

which finally verifies Eq. (70), and therefore also the expression in Eq. (69). We note that here the integration limits of the integral can be changed into $[0,\infty)$ by applying Cauchy's theorem, since the integrand vanishes as R^{-2}

on the origin-centered circular arc from R to $\sqrt{\chi_-/\chi_+}R$ when $R \to \infty$, and since the poles of the integrand lie at $\xi = \pm i$.

Appendix B.2. Far-Field Matrix

Suppose then that the far-field cross-spectral density matrix of Eq. (84) is introduced into the Wiener–Khintchine transform (Eq. 37). The result equals the mutual coherence matrix in Eq. (96) with $r_1 = r_2$, if we put

$$f_4(s,\tau) = \frac{4\pi^2}{3\hbar c} \int_0^\infty a_0(\omega) \frac{J_1(k_0 s)}{k_0 s} \exp(-i\omega\tau) d\omega. \tag{B.10}$$

Analogously to how we above showed that the integral representation in Eq. (B.3) equals the series representation of Eq. (70), it is possible to show that the integral representation in Eq. (B.10) equals the series representation of Eq. (97). This correspondence then proves the expression in Eq. (96).

REFERENCES

Agarwal, G. S. (1975). Quantum electrodynamics in the presence of dielectrics and conductors. I. Electromagnetic-field response functions and black-body fluctuations in finite geometries. *Physical Review A, 11*, 230–242.

Baltes, H. P., & Hilf, E. R. (1976). *Spectra of finite systems*. Mannheim: Bibliographisches Institut.

Batchelor, G. K. (1970). *The theory of homogeneous turbulence*. Cambridge: Cambridge University Press.

Blomstedt, K., Setälä, T., & Friberg, A. T. (2015). Blackbody aperture radiation: Effect of cavity wall. *Physical Review A, 91*, 063805.

Blomstedt, K., Setälä, T., Tervo, J., Hoenders, B. J., Turunen, J., & Friberg, A. T. (2016). Vector-valued Lambertian fields and their sources. *Physical Review A, 93*, 053813.

Blomstedt, K., Setälä, T., Tervo, J., Turunen, J., & Friberg, A. T. (2013). Partial polarization and electromagnetic spatial coherence of blackbody radiation emanating from an aperture. *Physical Review A, 88*, 013824.

Bourret, R. C. (1960). Coherence properties of blackbody radiation. *Il Nuovo Cimento, 18*, 347–356.

Brosseau, C. (1998). *Fundamentals of polarized light: A statistical optics approach*. New York: Wiley.

Carminati, R., & Greffet, J.-J. (1999). Near-field effects in spatial coherence of thermal sources. *Physical Review Letters, 82*, 1660–1663.

Carter, W. H., & Wolf, E. (1975). Coherence properties of Lambertian and non-Lambertian sources. *Journal of the Optical Society of America, 65*, 1067–1071.

Foley, J. T., Carter, W. H., & Wolf, E. (1986). Field correlations within a completely incoherent primary spherical source. *Journal of the Optical Society of America, 3*, 1090–1096.

Friberg, A. T., & Setälä, T. (2016). Electromagnetic theory of optical coherence. *Journal of the Optical Society of America A, 33*, 2431–2442.

Gbur, G. J. (2011). *Mathematical methods for optical physics and engineering*. Cambridge: Cambridge University Press.

Gbur, G., James, D., & Wolf, E. (1999). Energy conservation law for randomly fluctuating electromagnetic fields. *Physical Review E, 59*, 4594–4599.

Gori, F., Ambrosini, D., & Bagini, V. (1994). Field correlations within a homogeneous and isotropic source. *Optics Communications, 107*, 331–334.

Hassinen, T., Tervo, J., Setälä, T., Turunen, J., & Friberg, A. T. (2013). Spectral invariance and the scaling law with random electromagnetic fields. *Physical Review A, 88*, 043804.

Hochstadt, H. (1973). *Integral equations*. New York: Wiley.

James, D. F. V. (1994). Polarization of light radiated by black-body sources. *Optics Communications, 109*, 209–214.

Joulain, K., Mulet, J.-P., Marquier, F., Carminati, R., & Greffet, J.-J. (2005). Surface electromagnetic waves thermally excited: Radiative heat transfer, coherence properties and Casimir forces revisited in the near field. *Surface Science Reports, 57*, 59–112.

Kano, Y., & Wolf, E. (1962). Temporal coherence of black body radiation. *Proceedings of the Physical Society, 80*, 1273–1276.

Karczewski, B. (1963). Degree of coherence of the electromagnetic field. *Physics Letters, 5*, 191–192.

Kreyszig, E. (1978). *Introductory functional analysis with applications*. New York: Wiley.

Lahiri, M. (2009). Polarization properties of stochastic light beams in the space-time and space-frequency domains. *Optics Letters, 34*, 2936–2938.

Lahiri, M., & Wolf, E. (2008). Cross-spectral density matrix of the far field generated by a blackbody source. *Optics Communications, 281*, 3241–3244.

Leppänen, L.-P., Friberg, A. T., & Setälä, T. (2016a). Temporal electromagnetic degree of coherence and Stokes-parameter modulations in Michelson's interferometer. *Applied Physics B, 122*, 32.

Leppänen, L.-P., Friberg, A. T., & Setälä, T. (2016b). Connection of electromagnetic degrees of coherence in space-time and space-frequency domains. *Optics Letters, 41*, 1821–1824. (The factor $128\pi^2 \hbar c$ of Equation (22) in this paper should be replaced with $4\hbar c/\pi$).

Leppänen, L.-P., Saastamoinen, K., Friberg, A. T., & Setälä, T. (2014). Interferometric interpretation for the degree of polarization of classical optical beams. *New Journal of Physics, 16*, 113059.

Mandel, L., & Wolf, E. (1965). Coherence properties of optical fields. *Reviews of Modern Physics, 37*, 231–287.

Mandel, L., & Wolf, E. (1995). *Coherence and quantum optics*. Cambridge: Cambridge University Press.

Mehta, C. L. (1963). Coherence-time and effective bandwidth of blackbody radiation. *Il Nuovo Cimento, 28*, 401–408.

Mehta, C. L., & Wolf, E. (1964a). Coherence properties of blackbody radiation. I. Tensors of the classical field. *Physical Review, 134*, A1143–A1149.

Mehta, C. L., & Wolf, E. (1964b). Coherence properties of blackbody radiation. II. Correlation tensors of the quantized field. *Physical Review, 134*, A1149–A1153.

Mehta, C. L., & Wolf, E. (1967). Coherence properties of blackbody radiation. III. Cross-spectral tensors. *Physical Review, 161*, 1328–1334.

Nieto-Vesperinas, M. (1991). *Scattering and diffraction in physical optics*. New York: Wiley.

Nussenzveig, H. M., Foley, J. T., Kim, K., & Wolf, E. (1987). Field correlations within a fluctuating homogeneous medium. *Physical Review Letters, 58*, 218–221.

Olver, F. W. J., Lozier, D. W., Boisvert, R. F., & Clark, C. W. (2010). *NIST handbook of mathematical functions*. Cambridge: Cambridge University Press.

Peřina, J. (1985). *Coherence of light*. Dordrecht: Reidel.

Planck, M. (1914). *The theory of heat radiation*. Philadelphia: P. Blakiston's Son & Co.

Ponomarenko, S. A., & Wolf, E. (2001). Universal structure of field correlations within a fluctuating medium. *Physical Review E, 65*, 016602.

Réfrégier, P., Setälä, T., & Friberg, A. T. (2012). Maximal polarization order of random optical beams: Reversible and irreversible polarization variations. *Optics Letters*, *37*, 3750–3752.

Sarfatt, J. (1963). Quantum-mechanical correlation theory of electromagnetic fields. *Il Nuovo Cimento*, *27*, 1119–1129.

Setälä, T., Blomstedt, K., Kaivola, M., & Friberg, A. T. (2003). Universality of electromagnetic-field correlations within homogeneous and isotropic sources. *Physical Review E*, *67*, 026613.

Setälä, T., Kaivola, M., & Friberg, A. T. (2002). Degree of polarization in near fields of thermal sources: Effects of surface waves. *Physical Review Letters*, *88*, 123902.

Setälä, T., Kaivola, M., & Friberg, A. T. (2003). Spatial correlations and degree of polarization in homogeneous electromagnetic fields. *Optics Letters*, *28*, 1069–1071.

Setälä, T., Lindberg, J., Blomstedt, K., Tervo, J., & Friberg, A. T. (2005). Coherent-mode representation of a statistically homogeneous and isotropic electromagnetic field in spherical volume. *Physical Review E*, *71*, 036618.

Setälä, T., Lindfors, K., & Friberg, A. T. (2009). Degree of polarization in 3D optical fields generated from a partially polarized plane wave. *Optics Letters*, *34*, 3394–3396.

Setälä, T., Nunziata, F., & Friberg, A. T. (2009). Differences between partial polarizations in the space-time and space-frequency domains. *Optics Letters*, *34*, 2924–2926.

Setälä, T., Shevchenko, A., Kaivola, M., & Friberg, A. T. (2002). Degree of polarization for optical near fields. *Physical Review E*, *66*, 016615.

Setälä, T., Shevchenko, A., Kaivola, M., & Friberg, A. T. (2008). Polarization time and length for random optical beams. *Physical Review A*, *78*, 033817.

Setälä, T., Tervo, J., & Friberg, A. T. (2004a). Complete electromagnetic coherence in the space-frequency domain. *Optics Letters*, *29*, 328–340.

Setälä, T., Tervo, J., & Friberg, A. T. (2004b). Theorems on complete electromagnetic coherence in the space-time domain. *Optics Communications*, *238*, 229–236.

Setälä, T., Tervo, J., & Friberg, A. T. (2006). Contrasts of Stokes parameters in Young's interference experiment and electromagnetic degree of coherence. *Optics Letters*, *31*, 2669–2671.

Shchegrov, A. V., Joulain, K., Carminati, R., & Greffet, J.-J. (2000). Near-field spectral effects due to electromagnetic surface excitations. *Physical Review Letters*, *85*, 1548–1551.

Shevchenko, A., Setälä, T., Kaivola, M., & Friberg, A. T. (2009). Characterization of polarization fluctuations in random electromagnetic beams. *New Journal of Physics*, *11*, 073004.

Starikov, A., & Friberg, A. T. (1984). One-dimensional Lambertian sources and the associated coherent-mode representation. *Applied Optics*, *23*, 4261–4268.

Tai, C.-T. (1971). *Dyadic Green's functions in electromagnetic theory*. Scranton: Intext.

Tervo, J., Setälä, T., & Friberg, A. T. (2003). Degree of coherence for electromagnetic fields. *Optics Express*, *11*, 1137–1143.

Tervo, J., Setälä, T., & Friberg, A. T. (2004). Theory of partially coherent electromagnetic fields in the space–frequency domain. *Journal of the Optical Society of America A*, *21*, 2205–2215.

Tervo, J., Setälä, T., Turunen, J., & Friberg, A. T. (2013). Van Cittert-Zernike theorem with Stokes parameters. *Optics Letters*, *38*, 2301–2303.

Voipio, T., Setälä, T., Shevchenko, A., & Friberg, A. T. (2010). Polarization dynamics and polarization time of random three-dimensional electromagnetic fields. *Physical Review A*, *82*, 063807.

Wolf, E. (1976). New theory of radiative energy transfer in free electromagnetic fields. *Physical Review D*, *13*, 869–886.

Wolf, E. (2007). *Introduction to the theory of coherence and polarization of light*. Cambridge: Cambridge University Press.

Yaghjian, A. D. (1980). Electric dyadic Green's function in the source region. *Proceedings of the IEEE*, *68*, 248–263.

AUTHOR INDEX

Note: Page numbers followed by "*f*" indicate figures.

A

Abouraddy, A.F., 24, 44, 53, 62–63
Abrams, D.S., 62–63
Ádám, A., 2
Agarwal, G.S., 287, 294–295, 300, 305–306, 311–312, 337–340
Agrawal, G.P., 164–166
Ahmed, N., 162–163
Akhmediev, N., 157–158
Alarify, Y.S., 160–162
Ali, S.T., 283–284, 287
Alieva, T., 274–275
Alkelly, A.A., 160–162
Allen, L., 162–163
Allen, L.J., 38–39
Alonso, M.A., 287
Alper Kutay, M., 253–254, 257
Al-Qasimi, A., 13–14, 18, 20–21
Amarande, S., 158–160
Ambrosini, D., 160–162, 178, 301–302
Andrews, J.C., 38–39
Andrews, L.C., 37, 157–160
Ansari, N.A., 157–160
Anzolin, G., 162–163
Apostol, A., 157–158
Aragon, J.L., 113
Arce, G.R., 157–158
Arinaga, S., 157–158
Artal, P., 113
Aspect, A., 2–3
Atakishiyev, N.M., 268–271, 276–277, 283–284, 286–287
Atzema, E.J., 241
Avramov-Zamurovic, S., 164–166, 168–170, 198–199, 207–208

B

Bache, M., 24, 30–32
Bagini, V., 160–162, 178, 301–302
Baldwin, K.G.H., 38
Baltes, H.P., 318
Barach, G., 168–170, 208–211, 211–212*f*

Baraniuk, R.G., 24–25
Barbier, M., 39, 42
Barbieri, C., 162–163
Bargmann, V., 256–258
Barker, L., 272–273
Barnett, S.M., 162–163
Bartelt, H.O., 283
Basano, L., 25
Bastiaans, M.J., 160–162, 178, 274–275
Basu, S., 164–166, 168–170, 188–189, 207–208, 207*f*
Batchelor, G.K., 301
Baykal, Y., 160–162, 164–166
Baym, G., 2–3
Beijersbergen, M.W., 162–163
Bennett, V.P., 149–150, 150*f*
Bennink, R.S., 24
Bentley, S.J., 24, 62–63
Berardi, V., 24–25
Berman, G.P., 158–160
Berne, B.J., 63
Beyerlein, M., 149
Bhaduri, B., 32
Bianchini, A., 162–163
Biedenharn, L.C., 268, 276–277, 280, 285–286
Biggerstaff, D.N., 44–45
Bizheva, K., 44–45, 51–52, 61–62
Blomstedt, K., 293–344
Boggatyryova, V.G., 162–163
Boiron, D., 2–3
Boisvert, R.F., 312, 342–344
Boitier, F., 50–51
Bondani, M., 24
Bonnet, G., 63
Borghi, R., 4–5, 14–15, 32, 158–168, 184–188, 187–188*f*
Born, M., 5–6, 32, 46–47
Borra, E.F., 2–3
Boto, A.N., 62–63
Bourret, R.C., 294–295
Bowman, A., 24–25

347

Bowman, R., 24–25
Boyd, R.W., 24–25, 36–37, 44–45, 50–51, 62–63
Boyer, C.P., 266
Brambilla, E., 2–3, 24, 30–32
Brangaccio, D.J., 132
Brannen, E., 2
Braunstein, S.L., 62–63
Brenner, K.-H., 283
Brock, N., 135
Bromberg, Y., 44–45, 50
Brosseau, C., 312
Brown, D.P., 157–158, 168–170
Brown, T.G., 157–158, 168–170
Bruning, J.H., 132, 142
Bruns, D.G., 148
Buccini, C.J., 99–100
Buchdahl, H., 258
Burch, J.M., 101
Burow, R., 149

C

Caetano, D.P., 158–160
Cai, Y., 14, 24, 31–32, 157–213, 175f, 180f, 190f, 193f, 199f, 209f
Calvo, M.L., 274–275
Campos-García, M., 114–115
Çandan, Ç., 272–273
Canovas, C., 113
Cantelli, V., 38
Cao, H., 168–170, 208–211
Carminati, R., 337–338
Carrigan, R.A., 260–261
Carter, W.H., 296, 299–300, 318–319, 329
Caulfield, H.J., 149
Chan, K.W.C., 37
Chang, W., 43–44
Chapman, H.N., 38–39
Chekhova, M.V., 62–63
Chen, D., 162–163, 181–184
Chen, J., 162–163, 183–184, 184–185f
Chen, M., 162–163, 183–184
Chen, X.H., 25
Chen, Y., 14, 157–213, 199f, 209f
Chen, Z., 39
Cheng, J., 24, 32, 37–38
Cheng, Y.-Y., 151
Cho, Y.W., 157–158
Choma, M.A., 45

Chow, W.W., 150
Chriki, R., 168–170, 208–211, 211–212f
Chu, B., 63
Chumak, A.A., 158–160
Chumakov, S.M., 283–284, 286–287
Ciddor, P.E., 59
Cincotti, G., 167–168
Clark, C.W., 312, 342–344
Cohen, L., 287
Collett, E., 158–160, 173–174
Collins, S.A., 253–254
Condon, E.U., 256–257
Cornejo, A., 109
Cornejo-Rodriguez, A., 107–108
Creath, K., 142, 151
Cristóbal, G., 274–275

D

D'Angelo, M., 24, 62–63
Dall, H.E., 147
Dall, R.G., 38
D'Angelo, M., 158–160
Darudi, A., 123
Davenport, M.A., 24–25
Davidson, F.M., 157–160
Davidson, N., 168–170, 208–211
Dayan, B., 44–45, 50
De Martini, F., 162–163
De Santis, P., 158–160, 174–175, 174f
Deacon, K.S., 24, 36–37
Delaye, P., 50–51
Delisle, C., 157–158
Denis, S., 39
Devaux, F., 39
Dholakia, K., 162–163, 183–184
Díaz-Uribe, R., 114–115
Diels, J.C., 40, 46
Ding, C., 168–170
Ding, J.-J., 266
Dixon, P.B., 37
Doblado, M., 114
Dogariu, A., 157–160, 164–166, 186
Dong, Y., 160, 162–167, 181–184, 182f, 191–195, 193f
Dörband, B., 149
Dowling, J.P., 62–63, 287
Dragt, A.J., 260–263
Drexler, W., 43–44, 56–57
Drouin, M., 157–158

Du, G., 38–39
Duan, K., 158–160
Duarte, M.F., 24–25
Dubreuil, N., 50–51
Dudley, J.M., 39, 42
Duker, J.S., 44
Dvore, D., 149

E

Edgar, M.P., 24–25
Elson, E.L., 63
Elzaiat, S.Y., 45
Ensslin, K., 2–3
Epstein, A., 149
Erkmen, B.I., 2–3, 23–25, 30–31, 44–45
Escalante, A.Y., 162–163, 183–184
Esslinger, T., 2–3
Eyyuboğlu, H.T., 160–162, 164–166

F

Fabre, C., 24, 50–51
Fainman, Y., 63
Fante, R.L., 37
Fazal, I.M., 162–163
Felde, V.C., 162–163
Feng, F., 166–167
Fercher, A.F., 43–45, 56–57
Ferguson, H.I.S., 2
Ferri, F., 24, 30–32
Fienup, J.R., 38–39
Firth, W.J., 157–158
Fischer, D.G., 157–160
Fleischer, J.W., 168–170
Foellmi, C., 2–3
Foley, J.T., 158–160, 173–174, 299–300
Fonseca, E.J.S., 158–160, 164–166, 188–189, 189f
Forbes, A., 162–163, 181–182, 182f
Forbes, G.W., 88, 282–283, 287
Foreman, M.R., 164–166, 168–170
Forsythe, G.E., 80–81
Foucault, L.M., 104
Frank, A., 272–273
Freischlad, K.R., 135–136
Friberg, A.T., 2–4, 16–17, 22, 30–32, 36, 38–39, 41–45, 48, 51–53, 57–58, 62–63, 158–162, 164–166, 172, 174–175, 178, 178–179f, 293–344

Friesem, A.A., 44–45, 50, 168–170, 208–211, 211–212f
Fujimoto, J.G., 43–44

G

Gallagher, J.E., 132
Gamiz, V.L., 8–9
Gan, C.H., 157–158, 168–170
Gao, H., 162–163, 181–184
Garbusi, E., 151
García-Bullé, M., 257
García-Calderón, G., 283
Gatti, A., 2–3, 24, 30–32
Gbur, G., 13–14, 157–160, 162–163, 167–170, 195–196, 301
Gbur, G.J., 302–303
Ge, D., 160–162
Genty, G., 39, 42, 44–45, 58
Ghozeil, I., 111, 114
Gilbert, P., 256–258
Gilmore, R., 229, 257
Glauber, R.J., 1–2
Godard, A., 50–51
Goelz, S., 113
Gomes, J.V., 2–3
Gong, W., 32
González-Casanova, P., 244–245
Goodman, J.W., 36, 232, 249
Gori, F., 4–5, 14–15, 32, 157–172, 174–175, 174f, 178, 184–188, 187f, 195–196, 301–302
Gorshkov, V.N., 158–160
Goudail, F., 4, 22
Gouët, J.L., 44–45
Greffet, J.-J., 337–338
Greivenkamp, J.E., 129, 132
Grier, D.G., 162–163
Gross, H., 83
Gu, J., 168–170, 198–199, 200–201f
Gu, Y., 167–170
Guattari, G., 4–5, 158–162, 167–168, 174–175, 174f, 195–196
Guillemin, V., 249–250
Guo, H., 167–168
Guo, L., 14, 164–170, 195
Gureyev, T.E., 157–158
Guth, S., 168–170, 198–199, 207–208

H

Hakioğlu, T., 272–273
Han, S., 24, 32, 38–39
Han, Y., 162–163, 183–184, 183f
Hanbury Brown, R., 1–2
Hannonen, A., 62–63
Hardy, N.D., 37
Hariharan, P., 59, 151
Hasselbach, F., 2–3
Hassinen, T., 2–3, 16–17, 22, 296, 329–330
Häusler, G., 45
Hayes, J., 135
He, Q.S., 158–160
He, S., 158–162, 175–176
Healy, J.J., 253–254, 257
Heinzel, T., 2–3
Helen, S.S., 59
Henny, M., 2–3
Henson, B.M., 38
Hernandez-Aranda, R.I., 162–163, 181–184, 182f
Hernández-Gómez, G., 120
Herriot, D.R., 132
Hickmann, J.M., 158–160, 164–166, 188–189, 189f
Hilf, E.R., 318
Hillery, M., 283
Hitzenberger, C.K., 43–45, 56–57
Hochstadt, H., 302–304
Hodgman, S.S., 38
Hoenders, B.J., 296, 329
Hoffer, H., 113
Hogervorst, W., 2–3
Holland, M., 2–3
Holzner, C., 38–39
Hong, C.K., 52–53, 58
Hoover, B.G., 8–9
Hopkins, H.H., 75
Hoppeler, R., 2–3
Houston, J.B., 99–100
Howell, J.C., 37
Howland, G.A., 37
Hradil, Z., 162–163
Hu, L., 160–162
Huang, D., 43–44
Huang, H., 162–163
Huang, Y., 14, 164–166
Hudgin, R.H., 125

Huson, F.R., 260–261
Hyde, M.W., 164–166, 168–170, 188–189, 207–208, 207f

I

Ichikawa, K., 126
Ichioka, Y., 131
Ina, H., 130
Inuya, M., 131
Izatt, J.A., 45

J

Jain, P., 2–3
James, D., 301
James, D.F.V., 2–4, 7–8, 11, 13–14, 18, 20–21, 163–164, 295–296, 318–319, 326–329
Jánossy, L., 2
Jeltes, T., 2–3
Jensen, S.C., 150
Joulain, K., 337–338

K

Kaivola, M., 294–295, 297, 299–302, 306–307, 328–329, 333
Kaltenbaek, R., 44–45
Kamp, G., 45
Kandpal, H.C., 164–166
Kano, Y., 294–295, 311–312
Kanseri, B., 164–166
Karamata, B., 44
Karczewski, B., 310–311
Karimi, E., 162–163
Kato, Y., 157–158
Kauderer, M., 249–250
Kellock, H., 30–31, 38, 164–166
Kelly, K.F., 24–25
Kerr, F.H., 256–257
Khakimov, R.I., 38
Khan, S.A., 253
Kiesel, H., 2–3
Kim, C.-J., 82
Kim, J., 2–3
Kim, K., 299–300
Kim, Y.H., 157–158
Kimbrough, B., 135
Kingslake, R., 75
Kinzly, R.E., 157–158

Kitagawa, Y., 157–158
Klimov, A.B., 287
Klyshko, D.N., 23, 25–27
Ko, T.H., 44
Kobayashi, S., 130
Köhl, M., 2–3
Kok, P., 62–63
Koliopoulos, C.L., 135–136, 150
Kolner, B.H., 39–40
Korotkova, O., 8–9, 11–12, 14–15, 157–170, 172–173, 180–182, 182f, 184–186, 188–189, 191–195, 193f, 197–199, 201–208, 204f
Kothiyal, M.P., 59
Kowalczyk, A., 44
Krachmalnicoff, V., 2–3
Krawtchouk, M., 268
Kreyszig, E., 302–303
Krichevsky, O., 63
Krolikowski, W., 157–158
Krötzsch, G., 240–241, 261–262, 270–271
Kuebel, D., 10, 13–14, 18, 20–21
Kuittinen, M., 168–170, 207–208, 208–209f
Kujawinska, M., 130
Kurokawa, T., 63
Kutay, M.A., 251–252, 272–273

L

Labeyrie, A., 2
Lahiri, M., 10, 13–14, 18, 20–21, 295–296, 309, 318–319, 326–327
Lajunen, H., 39, 41–42, 44–45, 48, 168–170, 197
Lancis, J., 39, 41–42, 44–45, 48
Lantz, E., 39
Laska, J.N., 24–25
Lasser, T., 43–44, 56–57
Lassner, W., 257
Lavery, M.P.J., 162–163
Lavoie, J., 44–45
Lawrence, G.N., 150
Lee, H.-W., 283, 287
Lee, W.H., 149
Lehtolahti, J., 168–170, 207–208, 208–209f
Leitgeb, R., 45
Leppänen, L.-P., 297, 310–312
Li, C., 37
Li, H., 39

Li, Y., 2–3, 8–9, 13–14, 39
Li, Z., 166–170, 195, 201–204
Liang, C., 168–170, 197–199
Liberman, S., 253–254
Lie, S., 257
Lim, R., 168–170
Lin, C.P., 43–44
Lin, Q., 160–162, 178
Lindberg, J., 297, 302–305
Lindfors, K., 328–329
Lindlein, N., 149
Lindner, M.W., 45
Lipson, S.G., 2
Liu, H., 32
Liu, L., 14, 157–213, 202–206f
Liu, Q., 25
Liu, R., 162–163, 181–184
Liu, R.C., 2–3
Liu, X., 14, 157–213, 184f, 190f
Liu, Y., 32, 38–39, 162–163, 183–184
Liu, Y.-M., 150
Loebl, E.M., 248
Lohmann, A., 283
Lohmann, A.H., 283
Lohmann, A.W., 126, 283
Loomis, J.S., 149
López-Ramírez, D., 114
Lou, Q., 167–168
Louck, J.D., 268, 276–277, 280, 285–286
Loudon, R., 53
Lozier, D.W., 312, 342–344
Lü, B., 158–160
Lu, R., 38–39
Lu, X., 158–160
Lugiato, L.A., 2–3, 24, 30–32
Luis, A., 4, 22
Lundeen, J.S., 44–45, 51–52, 61–62
Luneburg, R.K., 230
Luo, K.H., 25

M

Ma, L., 168–170
Ma, X., 157–158
MacGovern, A.J., 149
Macías-Romero, C., 168–170
Magatti, D., 24, 30–32
Magde, D., 63

Mahajan, V.N., 82–83
Malacara, D., 82–83, 88–89, 104, 107–109, 111, 114, 132, 142
Malacara-Doblado, D., 74–152
Malacara-Hernández, D., 74–152
Malacara-Hernández, Z., 120, 129–130, 132, 135
Maleev, I.D., 162–163, 180–184
Malek-Madani, R., 164–166, 168–170
Malvimat, V., 2–3
Mandel, L., 2–6, 9–12, 46–47, 52–53, 58, 157–158, 170–171, 175–176, 294–295, 298–299, 304–305, 309–310, 312, 325–326
Man̄ko, V.I., 282–283
Marathay, A.S., 162–163, 180–184
Marquier, F., 337–338
Marrucci, L., 162–163
Martínez-Herrero, R., 168–170
Mayo, S.C., 157–158
Mazilu, M., 162–163, 183–184
Mazurek, M.D., 44–45
Mazzoni, A., 150
McBride, A.C., 256–257
McBride, W., 38–39
McNamara, J.M., 2–3
Mehta, C.L., 294–295, 301, 304–305, 311–312
Mei, Z., 168–170, 197–198
Meinel, A.B., 147–148
Meinel, M.P., 147–148
Mejia-Barbosa, Y., 114–115
Mejías, P.M., 168–170
Melozzi, M., 150
Menchaca, C., 88
Mendlovic, D., 251–252
Meneses-Fabián, C., 164–166, 188–189
Merano, M., 158–160
Meyers, R.E., 24, 36–37
Miao, J., 38–39
Millerd, J., 135
Mima, K., 157–158
Mir, M., 32
Mistura, G., 158–160
Mitchell, M.W., 44–45, 51–52, 61–62
Miyanaga, N., 157–158
Momose, A., 38–39

Mondello, A., 14, 164–166, 184–188, 187–188f
Month, M., 260–261
Moreau, P.A., 39
Moshinsky, M., 253–254, 256, 283
Motka, L., 162–163
Mourka, A., 162–163, 183–184
Movilla, J.M., 160–162, 178
Mueller, P., 149
Mukunda, N., 160–162, 177–178, 229, 251–252
Mulet, J.-P., 337–338
Murphy, T.E., 50–51
Murty, M.V.R.K., 102, 118

N

Nagali, E., 162–163
Nakatsuka, M., 157–158
Namias, V., 256–257, 270–271
Nasiri, S., 123
Nasr, M.B., 44, 53
Navarro Saad, M., 240
Negro, J.E., 150–151
Nelson, C., 164–166, 168–170, 198–199, 207–208
Nelson, J., 38–39
Newton, R.G., 256–258
Nieto, L.M., 287
Nieto-Vesperinas, M., 301–302, 338–342
Nisenson, P., 2
Nixon, M., 168–170, 208–211, 211–212f
Noll, R.J., 125
North-Morris, M., 135
Novak, M., 135
Nugent, K.A., 157–158
Nunziata, F., 309
Nussenzveig, H.M., 299–300
Nyyssonen, D., 157–158

O

O'Connell, R.F., 283
O'Leary, N.L., 38–39
O'Neill, P.K., 99–100
O'Sullivan-Hale, C., 37
Oberholzer, S., 2–3
Offner, A., 147
Ofir, A., 2–3
Oh, J.E., 157–158

Ojeda-Castañeda, J., 104
Oliver, W.D., 2–3
Olver, F.W.J., 312, 342–344
Olvera-Santamaría, M.A., 164–166, 188–189
O'Neill, P.K., 150, 150*f*
Ono, A., 149
Oreb, B.F., 150–151
Osten, W., 149, 151
Ostrovsky, A.S., 164–166, 188–189
Öttl, A., 2–3
Ottonello, P., 25
Ou, Z.Y., 52–53, 58
Ozaktas, H.M., 251–254, 257, 272–273, 282–283

P

Padgett, M.J., 24–25, 162–163
Padilla, J.M., 132
Paganin, D.M., 38, 157–158
Pal, V., 168–170, 208–211, 211–212*f*
Palacios, D.M., 162–163, 180–184
Palma, C., 158–160, 167–168, 174–175, 174*f*
Pan, L., 168–170
Panasenko, D., 63
Papoulis, A., 125–126
Paris, M.G.A., 24
Parks, R.E., 147–148
Pastor, J., 149
Pe'er, A., 44–45, 50
Pecora, R., 63
Pei, S.-C., 266
Pelliccia, D., 38
Perez-Garcia, B., 162–163, 181–184, 182*f*
Peřina, J., 312
Perrin, A., 2–3
Perry, J.W., 50–51
Peschel, U., 160–162
Peterman, E.J.G., 167–168, 195–196, 196*f*
Pezzati, L., 150
Phillips, R.L., 37, 157–160
Pianetta, P., 38–39
Piccirillo, B., 162–163
Pickles, A., 126–127
Piquero, G., 14, 158–160, 164–166, 184–188, 187–188*f*
Piredda, G., 44–45, 50–51
Pittman, T.B., 23, 25–27
Planck, M., 293–294, 301–302
Platt, B.C., 112–113
Plonus, M.A., 158–160
Pogosyan, G.S., 268–271, 276–277, 287
Polyanskii, P.V., 162–163
Ponomarenko, S.A., 160–166, 168–170, 186, 198–199, 207–208, 209–210*f*, 299–300
Popescu, G., 32
Prevedel, R., 44–45
Prieto, P.M., 113
Priss, C., 151
Pruss, C., 149
Pu, J., 37
Purcell, E.M., 2
Puri, R.R., 287
Puvanathasan, P., 44–45, 51–52, 61–62

Q

Qamar, S., 24–25
Qu, J., 168–170
Quesne, C., 253–254, 256
Quiroga, J.A., 132

R

Rack, A., 38
Raghunathan, S.B., 164–168, 195–196, 196*f*
Ralston, J.P., 2–3
Ramírez-Sánchez, V., 14–15, 164–166
Rao, R., 37
Rashed, R., 235
Rath, S., 164–166
Rayces, J.L., 78–79
Redding, B., 168–170, 208–211
Réfrégier, P., 4, 22, 309
Rehacek, J., 162–163
Reichelt, S., 149
Reid, G.T., 130
Renz, A., 2–3
Resch, K.J., 44–45, 51–52, 61–62
Ribak, E.N., 2–3, 113
Rickenstorff-Parrao, C., 164–166, 188–189
Ricklin, J.C., 157–160
Rimmer, M.P., 120
Ritter, S., 2–3
Robb, G.R.M., 157–158

Robinson, D.W., 130
Roddier, C., 116, 123–127
Roddier, F., 116, 123–127
Roddier, N., 124–125
Rodenburg, B., 37
Rodrigo, J.A., 274–275
Rodríguez-Zurita, G., 164–166, 188–189
Romanini, P., 186–188, 187–188f
Ronchi, V., 107–108
Rosencher, E., 50–51
Rosenfeld, D.P., 132
Rota, G.-C., 268, 276–277, 280, 285–286
Roth, J.M., 50–51
Roussey, M., 63
Roychowdhury, H., 11, 13–15, 164–166, 186
Rudolf, W., 40, 46
Rueda-Paz, J., 281–282
Rumi, M., 50–51
Ryczkowski, P., 39, 42, 44–45, 58

S

Saastamoinen, K., 310–311
Saastamoinen, T., 164–166, 168–170, 197
Saha, P., 2–3
Sahin, S., 164–166, 168–170, 197
Sako, N., 63
Saleh, B.E.A., 2–3, 24, 32, 44–45, 53, 62–63
Salem, M., 14–15, 164–166, 184–186
Sanchez, V.R., 168–172
Sánchez-Mondragón, J., 237–238
Sánchez-Soto, L.L., 162–163
Santarsiero, M., 4–5, 14–15, 32, 158–172, 178, 184–188, 187–188f
Sarfatt, J., 294–295
Sarunic, M.V., 45
Sasian, J.M., 147
Saunders, J.B., 116–117, 119
Sayre, D., 38–39
Scarcelli, G., 24–25, 157–160
Scheel, M., 38
Schelkens, P., 274–275
Schellekens, M., 2–3
Schleich, W.P., 287
Schmidt, J., 142
Schönenberger, C., 2–3
Schouten, H.F., 164–166

Schreibert, H., 142
Schreiter, K.M., 44–45
Schulz, G., 85
Schuman, J.S., 43–44
Schwider, J., 142, 149
Sciarrino, F., 162–163
Scully, M.O., 283
Seidel, L., 258
Senthilkumaran, P., 168–170
Sergienko, A.V., 23–27, 37, 44, 53, 62–63
Serna, J., 160–162, 178
Servín, M., 129–130, 132, 135
Setälä, T., 2–4, 16–17, 22, 30–32, 36, 38–39, 41–43, 62–63, 164–166, 172, 293–344
Shack, R.V., 112–113
Shannon, R., 82
Shao, L.Z., 147–148
Shapiro, J.H., 2–3, 23–25, 30–31, 36–37, 44–45, 50–51
Sharma, M.K., 168–170
Shchegrov, A.V., 337–338
Shchepakina, E., 168–170
Shen, X., 32
Shen, Y., 162–163
Sheridan, J.T., 253–254, 257
Shevchenko, A., 63, 297, 306–307, 314–315, 328–329, 333
Shi, J., 39
Shih, Y.H., 23–27, 36–37, 62–63, 158–160
Shimizu, F., 2–3
Shin, D.K., 38
Shirai, T., 1–63, 158–160, 164–166, 168–172, 188–189
Shomali, R., 123
Shukri, M.A., 160–162
Siegman, A.E., 40
Silberberg, Y., 44–45, 50
Simon, D.S., 37
Simon, R., 14, 160–162, 164–166, 177–178, 184–186, 251–252, 282–283
Singer, B., 113
Singer, W., 83
Singh, R.K., 168–170
Singh, R.P., 287
Sirohi, R.S., 59
Situ, G., 168–170
Sjödahl, M., 150–151

Smart, R.N., 103–104
Snyder, A.W., 157–158
Som, S.C., 157–158
Soskin, M.S., 162–163
Speirits, F.C., 162–163
Spreeuw, R., 162–163
Srinivasan, V.J., 44
Starikov, A., 329
Starikov, S., 158–160, 174–175
Steel, W.H., 103–104
Steibl, N., 125–126
Steinberg, S., 237–238, 244–245
Sternberg, S., 249–250
Stevenson, A.W., 157–158
Sticker, M., 44
Stinson, W.G., 43–44
Stockton, A., 126–127
Stoklasa, B., 162–163
Strekalov, D.V., 23, 25–27
Strunk, C., 2–3
Sudarshan, E.C.G., 229, 251–252
Sun, B., 24–25
Sun, T., 24–25
Sundar, K., 160–162
Swanson, E.A., 43–44
Swartzlander, G.A., 162–163, 180–184, 182f

T

Tai, C.-T., 301–303
Takeda, M., 63, 126, 130
Takhar, D., 24–25
Tamburini, F., 162–163
Tanaka, Y., 63
Teague, M.R., 125–126
Teich, M.C., 2–3, 24, 32, 44–45, 53, 62–63
Tervo, J., 2–4, 16–17, 22, 164–166, 168–170, 172, 207–208, 208–209f, 295–299, 301–305, 310–311, 318–319, 325–326, 328–330
Tervonen, E., 158–162, 174–175, 178, 178–179f
Thienpont, H., 274–275
Thompson, B.J., 157–158
Tiziani, H.J., 149
Tong, Z., 160–162, 164–166, 168–170
Török, P., 164–166, 168–170
Torres-Company, V., 39, 41–42, 44–45, 48

Totzek, M., 83
Toussaint, K.C., 62–63
Tradonsky, C., 168–170, 208–211, 211–212f
Truscott, A.G., 38
Tseng, C.-C., 266
Tsuda, H., 63
Turunen, J., 44–45, 58, 158–162, 164–166, 168–170, 174–175, 178, 178f, 207–208, 208f, 295–296, 328–330
Turunen, T., 295–297, 301–302, 318–319, 325–326
Twiss, R.Q., 1–2
Twyman, F., 96

U

Umbriaco, G., 158–160, 162–163
Urzúa, A.R., 279

V

Valencia, A., 24, 158–160
van Dijk, T., 157–160, 162–163, 167–168, 195–196, 196f
van Isacker, P., 272–273
Varga, P., 2
Vargas-Martín, F., 113
Vassen, W., 2–3
Venkatraman, D., 44–45
Vicalvi, S., 4–5
Vicent, L.E., 271, 273–274, 276–278
Vidal, I., 158–160, 164–166, 188–189, 189f
Vinu, R.V., 168–170
Visser, T.D., 2–3, 10, 13–14, 16–18, 157–160, 162–170, 195–196, 196f
Vittert, L.E., 24–25
Voelz, D.G., 164–166, 168–170, 188–189, 207–208, 207f
Voipio, T., 297, 314–315
Volkov, S.N., 2–4, 7–8, 11

W

Waller, L., 168–170
Wang, F., 14, 158–170, 175–177, 175f, 180–184, 180f, 188–189, 190f, 191–195, 193f, 197–199, 199f, 201–206
Wang, H., 168–170, 201–204
Wang, J., 162–163, 168–170, 201–204
Wang, L.G., 24–25

Wang, R., 32
Wang, S.C.H., 158–160
Wang, T., 37, 164–166
Wang, W., 158–160
Wang, Y., 158–160, 162–166, 181–184, 182f
Watson, E., 164–166
Webb, W.W., 63
Wei, C., 158–160
Wei, Q., 32
Welsh, S., 24–25
Westbrook, C.I., 2–3
White, A.D., 132
Wigner, E.P., 283
Wilkins, S.W., 157–158
Williams, C.P., 62–63
Williams, D., 113
Woerdman, J.P., 162–163
Wojtkowski, M., 44
Wolf, E., 2–16, 18, 20–23, 32, 37, 46–47, 52–53, 85, 157–160, 162–166, 170–176, 184–186, 188–189, 294–296, 298–302, 304–305, 309–312, 318–319, 325–329
Wolf, K.B., 226–287
Wong, F.N.C., 2–3, 44–45
Wong, M.K.F., 275–276
Wu, G., 2–3, 13–14, 16–18, 157–160, 163–168, 181–182, 182f, 188–189, 190f, 191–193, 204–206
Wu, J., 45
Wu, L.A., 25
Wucknitz, O., 2–3
Wyant, J.C., 82, 120, 142, 149–152, 150f
Wybourne, B.G., 229

X

Xiao, T., 38–39
Xiao, X., 164–166, 168–170, 188–189, 207–208, 207f
Xie, H., 38–39
Xu, C., 50–51

Y

Yaghjian, A.D., 339–340
Yamamoto, Y., 2–3
Yan, Y., 162–163
Yang, C., 45
Yang, J., 162–163
Yang, Y., 162–163, 183–184
Yao, M., 164–170
Yaqoob, Z., 45
Yasuda, M., 2–3
Yeh, M.-H., 266
Yepiz, A., 162–163, 181–182, 182f
Young, T., 157–158
Yu, F., 157–158
Yu, H., 38–39
Yu, J., 157–213, 184–185f
Yuan, Y., 166–170, 195, 197–199
Yuan, Z., 167–168

Z

Zahid, M., 162–163
Zalevsky, Z., 251–252
Zawadzki, R., 44
Zeng, G., 39
Zernike, F., 157–158
Zerom, P., 44–45, 50–51
Zhan, Q., 160, 163–164, 166–167
Zhang, L., 158–160
Zhang, M., 14, 32, 164–166, 191
Zhang, Y., 14, 162–166, 168–170, 181–182, 182f
Zhao, C., 157–170, 176–177, 180–184, 182f, 191–195, 197–199, 201–206, 202f
Zhao, D., 14, 164–166, 168–170
Zhao, H., 167–168
Zhao, Z., 168–170
Zhou, J., 167–168
Zhou, Y., 162–163, 181–184
Zhu, D., 38–39
Zhu, R., 32
Zhu, S.Y., 24–25, 31–32, 157–170, 180, 180f, 188–189, 190f, 195, 201–204
Zhu, W., 37
Zhu, X., 166–167, 195
Zhuang, S., 157–158
Zubairy, M.S., 24–25, 157–160, 162–163, 173–174

SUBJECT INDEX

Note: Page numbers followed by "*f*" indicate figures, "*t*" indicate tables, and "*np*" indicate footnotes.

A

Aberrations, 76–77, 76*t*. *See also* Transverse aberrations
   axis-symmetric, in 3D system, 262–265, 264–265*f*
   interferogram for primary, 100*f*
   linear transformation and, 262*f*
   of 1D finite discrete signals, 279–282
   pyramid of, 281–282*f*
   Seidel, 75
   symplectic harmonic, 264*f*
   in 2D system, 258, 262*f*
   wavefront transverse, 78–79, 78*f*
Acousto-optic coherence control technique, 178
Aperture matrix, 342–344
Aspherical surfaces testing, 142–152
   autocompensating configurations, 143–146, 143–146*f*
   compensators, 146–150
      synthetic hologram, 149–150
   testing hyperboloids with autocollimating configurations, 147–148
   two wavelengths measurement, 151–152
   wavefront stitching, 150–151
Aspheric optical surface, 85–88, 86*f*
Aspheric wavefront, 142–143
Auxiliary optical system, 146
Axis-symmetric aberration, in 3D system, 262–265
Axis-symmetric linear transformation, 262–263

B

Beam. *See also specific types of beam*
   cylindrical vector, 166–167
   He–Ne laser, 204–206
   Hermite–Gaussian correlated, 168–170
   Laguerre–Gaussian, 162–163
   Laguerre–Gaussian correlated Schell-model, 168–170, 197–198, 202–203*f*
   laser, 157–158
   light, 157–158
   linearly polarized, 9
   multi-Gaussian correlated Schell-model, 168–170, 197, 207–208
   nonuniform multi-Gaussian correlated, 198
   parameters, 190–191, 191*f*
   pseudo-Bessel correlated, 167–168
   Schell-model, 195–196, 196*f*, 210*f*
   SCRP, 201–204
   specially correlated radially polarized, 201–206, 204*f*, 206*f*
   stochastic, propagation of, 13–14
   TGSM, 177–178
   unpolarized, 10
   vector, 157–158, 166–167
      cylindrical, 166–167
   vector Hermite–Gaussian correlated, 168–170
Beam splitter (BS), 25, 43*f*, 48*f*, 175–176
Bessel–Gauss beam, 160–163, 167–168
Blackbody coherence, in cavity
   Fluctuation-dissipation theorem, 337–342
   Maxwell's equations, 336–337
Blackbody radiation, 294–295, 297, 337–342
   coherence properties of, 295–296
   coherent-mode decomposition of, 302–303
   cross-spectral density matrix of, 300
Blackbody radiation, in aperture
   frequency domain
      polarization, 321
      spatial coherence, 317–320, 320*f*
      spectrum, 321
   time domain

Blackbody radiation, in aperture (*Continued*)
  polarization, 324
  polarization dynamics, 324
  spatiotemporal coherence, 321–324, 323*f*
Blackbody radiation, in cavity
  polarization
    all-space cavity, 306–307
    half-space cavity, 307–308, 308*f*
  spatial coherence, 298–306
    all-space cavity, 301–305, 303–304*f*
    half-space cavity, 305–306
  spectrum, 308–309
  time domain
    polarization, 313–314
    polarization dynamics, 314–317, 316*f*
    spatiotemporal coherence, 309–313, 313*f*
Blackbody radiation, in far zone of aperture
  frequency domain
    far-field coherence, 325–328, 325*f*, 327*f*, 331*f*
    polarization, 328–329
    radiant intensity, 329–330
  time domain
    polarization, 332
    polarization dynamics, 333–335, 334*f*
    radiant intensity, 333
    spatiotemporal coherence, 330–332
BS. *See* Beam splitter (BS)

C

Canonical integral transform, wave model, 253–258
Canonical transform, 237–238, 257–258
  Fractional Fourier and, 256–257
Cartesian Kravchuk modes, 274
Casimir invariants, 243–244
Cavity laser, 208–211, 211*f*
CGH. *See* Computer generated/synthetic hologram (CGH)
Charge-coupled device (CCD), 175–176
Classical light, ghost imaging
  and diffraction with, 23–43, 26*f*
  of pure phase objects with, 32–36, 33*f*, 35*f*
Classical optics, 1–2, 24–25
Coherence matrices, 297–298

Coherent beam, partially, 157–158
  characterization and generation of various, 170–211
  coherence lattices in, 209*f*
  generation of, 181–182, 182*f*, 193*f*, 199, 199*f*, 209*f*
  intensity distribution and state of polarization, 192–193, 193*f*
  normalized distribution of log, 185*f*
  with prescribed degrees of coherence, 167–170, 195–211
  with prescribed phases, 160–163, 177–184
  with prescribed states of polarization, 163–167, 184–195
  radially polarized, 193–195, 194*f*
Coherent beam, polarization, 163–167
Coherent modes, electromagnetic theory of, 297
Common path interferometers, 101–104
Computer generated/synthetic hologram (CGH), 149, 150*f*
Conic surface, 86*t*
Coset parameter
  Euclidean group and, 232–234, 233*f*
  for wave model, 242–243, 262*f*
Cross-spectral density (CSD), 157–158, 170–171, 196–197
  GSM beam, 173–174
  matrix, 171–173
  SCRP beam, 201–204
  TGSM beam, 177–178
Curvature measurement test
  Hartmann test, 120–121
  irradiance transport equation, 121–127, 122–124*f*
Cylindrical vector beam, 166–167

D

Dall compensator, 147, 147*f*
Deformation, in optical system, 74
Diffraction, ghost imaging
  with classical light, 23–43, 26*f*, 40*f*
  in time domain, 39–43
  with X-rays, 38–39
Diffuser, 207–208, 208*f*
  fabricated, 209*f*
Digital interferometry, 129–130, 129*f*

Discrete optical model, 266–282
  contraction of so (4) to iso (3), 266–267
  imported rotations, 272–274, 273f
  Kravchuk oscillator states, 268–271, 269f
  plane pixelated screen, 267–268
  2D screens and $U(2)_F$ transformations, 271–275, 272f
Dispersion cancelation, quantum-mimetic intensity-interferometric OCT with, 51–62
Domestic Fourier–Kravchuk transformations, 271–272
Dynamic postulate, 234–235

E

EGSM beam. *See* Electromagnetic Gaussian Schell-model (EGSM) beam
Electric cross-spectral density matrix, 295–296
Electromagnetic beam
  HBT effect with classical, 3–4
  stochastic, correlations between intensity fluctuations in, 4–11, 5f, 18f
Electromagnetic field, 297, 299, 301–302, 306–309
Electromagnetic Gaussian Schell-model (EGSM) beam, 14–15, 164–166, 184–186, 187–188f
  experimental setup for generating, 186–189, 187f, 190f
  sources, 18–20
Electromagnetic theory, of coherent modes, 297
Electromagnetic van Cittert–Zernike theorem, 295–296, 328
Euclidean generators, 243–244
Euclidean group, 228–230
  and coset parameters, 232–234, 233f

F

Fabricated diffuser, 209f
Far-field matrix, 344
Fizeau interferometer, 93–95, 93–97f, 129–130, 151–152
Fluctuation-dissipation theorem, 337–342
Foucault knife-edge test, 104–107, 104–107f
Fourier transform, 52f, 252f, 254–255, 277f, 309
  in interferogram with linear carrier, 131f
Fractional Fourier and canonical transforms, 256–257
Fundamental object of model, 230–232

G

Gaussian amplitude filter (GAF), 158–160
Gaussian intensity filter (GF), 186–188
Gaussian moment theorem, 49
Gaussian Schell-model (GSM), 14, 31–32
Gaussian Schell-model (GSM) beam, 158–160, 173–177, 174f
  CSD of, 173–174
  experimental setup
    for generating beam, 174–175, 174f, 176f
    for measuring the degree of coherence, 175–176, 175f
  fourth-order correlation function, 158–160
  interaction, 158–160
  rotation of the twisted, 179f
  into TGSM beam, conversion, 178, 178f
Gaussian Schell-model vortex (GSMV) beam, 180
  experimental setup for generating, 180, 180–181f
Geometric model
  Heisenberg–Weyl algebra, 247–248
  of light, 230, 232
    canonical and optical transformations, 237–238
    Euclidean group and coset parameters, 232–234, 233f
    geometric and dynamic postulates, 234–235
    Hamiltonian structure and phase space, 236–237
    refracting surfaces and root transformation, 239–241, 239f
    linear transformations of phase space, 249–253
Geometric postulate, 234–235
Ghost diffraction, 23–24, 29–30

Ghost imaging
  and diffraction
    with classical light, 23–43, 26f
    in time domain, 39–43
    with X-rays, 38–39
  of pure phase objects with classical light, 32–36, 33f, 35f
  through turbulence, 36–38
Gibbs oscillation, 273–274
Gram–Schmidt orthogonalization, 81–82
Green function, 299–301, 318, 325–326
Group delay dispersion (GDD) parameter, 40, 40f
Group velocity dispersion (GVD), 40, 44
GSM. See Gaussian Schell-model (GSM)
GSMV beam. See Gaussian Schell-model vortex (GSMV) beam
GVD. See Group velocity dispersion (GVD)
Gyrations, 274, 275f

H

Hamiltonian system, 227, 236–237
Hanbury Brown–Twiss (HBT) effect, 2–3, 24–25, 164–166
  alternative approach to basic problem, 21–23
  with classical electromagnetic beam, 3–4
  correlations between intensity fluctuations, 13–21
  degree of cross-polarization, 11–13
Hanbury Brown–Twiss interferometry. See Intensity interferometry
Harmonic aberration, symplectic, 264f
Hartmann test
  curvature measurement test, 120–121
  transverse aberrations measurement, 111–112, 111f
HBT effect. See Hanbury Brown–Twiss (HBT) effect
Heisenberg–Weyl algebra, 245, 249, 258
  contraction of Euclidean to, 246–247
  geometric model, 247–248
  linear maps of, 254
  quadratic extension of, 258
  wave model, 248
Helmholtz equation, 338–339
Helmholtz fields, 244–245
Helmholtz model, 241–245

Helmholtz wavefields, Hilbert space for, 244–245
He–Ne laser, 199
  beam, 204–206
Hermite–Gaussian correlated beam, 168–170
Hermite-Gaussian correlated Schell-model (HGCSM) beam, 198–204, 200–201f
Hilbert space, for Helmholtz wavefields, 244–245
Homomorphism, 253np
Hong-Ou-Mandel (HOM) interferometer, 52–53
Huygens–Fresnel principle, 37
Huygens wavelet, 125–126

I

Integral transform realization, wave model, 254–255
Intensity correlation, 23–27, 30–31, 38–39, 50, 54, 57, 61–62
Intensity interferometry, 1–3, 23
  into OCT, 46–51
Interferogram, 89
  analysis with spatial carrier, 130–132, 130–131f
  for primary aberrations, 100f
  sparse sampling of fringes, 128–129, 128f
Interferometer, 89
  common path, 101–104
  digital, 129–130, 129f
  Fizeau, 93–95, 93–97f
  Hong-Ou-Mandel, 52–53
  intensity, into OCT, 46–51
  lateral shearing, 116–120, 117f, 119f
  Michelson, 96
  Newton, 90–93, 90–92f
  phase shifting, 132–142
  point diffraction, 103–104, 103f
  scatter plate, 101–103, 101f, 103f
  stellar intensity, 43
  Twyman–Green, 96–100, 97f, 133f
Irradiance transport equation, 121–127, 122–124f

J

Jones-vector approach, 314–315

K

Kravchuk oscillator states, 268–271, 269f

L

Laguerre–Gaussian (LG) beam, 162–163
Laguerre–Gaussian (LG) correlated Schell-model (LGCSM) beam, 168–170, 197–198, 202–203f
Laguerre–Kravchuk modes, 274–275, 278f
Lambertian radiation pattern, 296
Lambert's cosine law, 295–296
Laser beam, 157–158
Lateral shearing interferometer, 116–120, 117f, 119f
Laue diffraction, 38
Least squares fitting, 79–81
Light beam, 157–158
Linearly polarized beam, 9
Linear transformation
 and aberrations, 262f
 axis-symmetric, 262–263
 of phase space, 249
  geometric model, 249–253

M

Mach–Zehnder interferometer, 186–189
Magnetic vector wave equation, 338–339
Maxwell's equation, 336–337
Metaxial regime, 258
 aberrations in 2D system, 258
Michelson interferometer, 96
Michelson stellar interferometry, 43
Monochromatic light source, 91
Multi-Gaussian correlated Schell-model (MGCSM) beam, 168–170, 197, 207–208

N

Newton interferometer, 90–93, 90–92f
Nonlinear crystal (NLC), 43f
Nonuniform multi-Gaussian correlated beam, 198
Nyquist condition, 146, 150

O

OAM. See Orbital angular momentum (OAM)
OCT. See Optical coherence tomography (OCT)
1D finite discrete signal, aberrations of, 279–282
Optical coherence lattices, 210f
Optical coherence tomography (OCT), 43–45. See also specific types of optical coherence tomography
 intensity interferometry into, 46–51
 signals, 57f
Optical models, discrete, 266–282
Optical path difference (OPD), 98–99
Optical system
 components in modern, 74
 wavefront deformations in, 74
Optical test, 75, 89
Optical transformation, 237–238
Orbital angular momentum (OAM), 160–162

P

Paraxial model, 245–248
Partially coherent sources, numerical modeling of, 168–170
Phase shifting interferometry, 129–130
 algorithms, 135–142
  five steps separated 90 degree, 140–142, 141–142f
  four steps separated 90 degree, 138–139, 138–139f
  four steps separated 120 degree, 140, 140–141f
  three steps separated 120 degree, 137, 137–138f
 instrumentation, 132–135, 133f
Photodetector, semiconductor, 50–51
Photomultiplier tube (PMT), 50–51
π phase shift, 59np
Pixelated images, 272–274
Pixelated screens, square and circular, 275–279
Planck's law, 294–295
Plank's spectrum, 293–296, 299–302
Point diffraction interferometer, 103–104, 103f
Pseudo-Bessel correlated beam, 167–168

Q

Q-OCT. *See* Quantum optical coherence tomography (Q-OCT)
Quantum-mechanical approach, 294–295
Quantum-mimetic intensity-interferometric OCT, 51–62
Quantum-mimetic technology, 62–63
Quantum optical coherence tomography (Q-OCT), 43f, 44–45, 51–54

R

Rayleigh diffraction, 325–326
Resonator theory, for stochastic electromagnetic field, 164–166
RGGPs. *See* Rotating ground glass plates (RGGPs)
Riemann zeta function, 294–295
Ronchi test, 107–110, 108–110f
Rotating ground glass disk (RGGD), 158–160
Rotating ground glass plates (RGGPs), 186–188

S

Scatter plate interferometer, 101–103, 101f, 103f
Schell-model beam, 195–196, 196f, 210f
SD-OCT. *See* Spectral domain optical coherence tomography (SD-OCT)
Seidel aberrations, 75
Semiconductor photodetector, 50–51
Shack–Hartmann test, 112–116, 113–116f
Shack–Hartmann wavefront sensor, 162–163
Silver–Müller radiation condition, 340–342
Single-spatial-point quantities, 312
SLM. *See* Spatial light modulator (SLM)
Space–time coherence length, 312–313
Spatial light modulator (SLM), 162–163, 181–182, 199
Specially correlated radially polarized (SCRP) beam, 201–206, 204f, 206f
Spectral coherence length, 294–295
Spectral domain optical coherence tomography (SD-OCT), 45, 52f
practical method of realization, 58–60, 60–61f
theory in, 52–58, 52f, 57f
Spherical convex, 91–92
Spherical wavefront, 146–147
Standoff sensing, 36–37
Stellar intensity interferometry, 43
Stochastic beam, propagation of, 13–14
Stochastic electromagnetic beam
correlations between intensity fluctuations in, 4–11, 5f
correlation singularities in, 164–166
Stochastic electromagnetic field, resonator theory for, 164–166
Stokes parameter, 172–173
Symplectic harmonic aberrations, 264f
Synthetic hologram compensator, 149–150

T

TD-OCT. *See* Time domain optical coherence tomography (TD-OCT)
TGSM beam. *See* Twisted Gaussian Schell-model (TGSM) beam
3D system, axis-symmetric aberrations in, 262–265, 264–265f
Time domain optical coherence tomography (TD-OCT), 43f, 47–50, 48f
Time domain, theory in, 61–62
TPA. *See* Two-photon absorption (TPA)
Transverse aberrations, 78–79, 78f
measurement, 104–120
Foucault knife-edge test, 104–107, 104–107f
Hartmann test, 111–112, 111f
lateral shearing interferometers, 116–120, 117f, 119f
Ronchi test, 107–110, 108–110f
Shack–Hartmann and modified Hartmann test, 112–116, 113–116f
wavefront deformations and, 78–79, 78f
Turbulence, ghost imaging through, 36–38
Twisted Gaussian Schell-model (TGSM) beam, 160–162
CSD of, 177–178
GSM beam into, conversion, 178, 178f
in physical-optics, 178
2D system, aberrations in, 258, 262f
Two-photon absorption (TPA), 50–51
Two-wavelengths holography, 151–152

Twyman–Green interferometer, 96–100, 97f, 129–130, 133f, 151–152

U
Unpolarized beam, 10

V
van Cittert–Zernike theorem, 208–211, 295–296, 328
Vector beam, 157–158, 166–167
  cylindrical, 166–167
Vector Hermite–Gaussian correlated beam, 168–170

W
Wavefront
  aspheric, 142–143
  aspheric optical surface representation, 85–88, 86f, 86t
  characteristics, 74–88
  distortions, 89–104
  Gram–Schmidt orthogonalization, 81–82
  least squares fitting, 79–81
  mathematical, 74–77
  stitching, 150–151
  transverse aberrations, 78–79, 78f
  two-wave interferogram, 89, 89f
  zernike polynomials, 82–84, 84t
Wavefront deformations
  measurement, 89–104
    common path interferometers, 101–104
    Fizeau interferometer, 93–95, 93–97f
    Newton interferometer, 90–93, 90–92f
    Point diffraction interferometer, 103–104, 103f
    scatter plate interferometer, 101–103, 101f, 103f
    Twyman–Green interferometer, 96–100, 97–99f
  in optical system, 74
  and transverse aberrations, 78–79, 78f
Wave model, 241–245
  canonical integral transforms, 253–258
  coset parameters for, 242–243, 262f
  Euclidean generators and Casimir invariants, 243–244
  Heisenberg–Weyl algebra, 248
  integral transform realization, 254–255
Wave transform, 244
Wiener–Khintchine theorem, 309–310, 330, 342
Wien's displacement law, 293–294, 312–313
Wigner function, 227, 279–280
  SU(2), 282–287
Wigner matrix, SU (2), 285–287
Wigner operator, 283–285

X
X-rays, ghost imaging and diffraction with, 38–39

Z
Zernike polynomials, 82–84, 84t, 85f

CUMULATIVE INDEX – VOLUMES 1–62[*]

Abdullaev, F. and J. Garnier: Optical solitons in random media	**48**, 35
Abdullaev, F.Kh., S.A. Darmanyan and J. Garnier: Modulational instability of electromagnetic waves in inhomogeneous and in discrete media	**44**, 303
Abelès, F.: Methods for determining optical parameters of thin films	**2**, 249
Abella, I.D.: Echoes at optical frequencies	**7**, 139
Abitbol, C.I., *see* Clair, J.J.	**16**, 71
Abraham, N.B., P. Mandel and L.M. Narducci: Dynamical instabilities and pulsations in lasers	**25**, 1
Aegerter, C.M. and G. Maret: Coherent backscattering and Anderson localization of light	**52**, 1
Agarwal, G.S.: Master equation methods in quantum optics	**11**, 1
Agranovich, V.M. and V.L. Ginzburg: Crystal optics with spatial dispersion	**9**, 235
Agrawal, G.P.: Single-longitudinal-mode semiconductor lasers	**26**, 163
Agrawal, G.P., *see* Essiambre, R.-J.	**37**, 185
Allen, L. and D.G.C. Jones: Mode locking in gas lasers	**9**, 179
Allen, L., M.J. Padgett and M. Babiker: The orbital angular momentum of light	**39**, 291
Alfalou, A. and C. Brosseau: Recent advances in optical image processing	**60**, 119
Ammann, E.O.: Synthesis of optical birefringent networks	**9**, 123
Andersen, U.L. Filip, R.: Quantum feed-forward control of light	**53**, 365
Anderson, R., *see* Carriere, J.	**41**, 97
Apresyan, L.A., *see* Kravtsov, Yu.A.	**36**, 179
Arimondo, E.: Coherent population trapping in laser spectroscopy	**35**, 257
Armstrong, J.A. and A.W. Smith: Experimental studies of intensity fluctuations in lasers	**6**, 211
Arnaud, J.A.: Hamiltonian theory of beam mode propagation	**11**, 247
Asakura, T., *see* Okamoto, T.	**34**, 183
Asakura, T., *see* Peiponen, K.-E.	**37**, 57
Asatryan, A.A., *see* Kravtsov, Yu.A.	**39**, 1
Babiker, M., *see* Allen, L.	**39**, 291
Baby, V., *see* Glesk, I.	**45**, 53
Backman, V., *see* Çapoğlu, İ.R.	**57**, 1
Baltes, H.P.: On the validity of Kirchhoff's law of heat radiation for a body in a nonequilibrium environment	**13**, 1
Banaszek, K., *see* Juan P. Torres	**56**, 227
Barabanenkov, Yu.N., Yu.A. Kravtsov, V.D. Ozrin and A.I. Saichev: Enhanced backscattering in optics	**29**, 65
Barakat, R.: The intensity distribution and total illumination of aberration-free diffraction images	**1**, 67
Barrett, H.H.: The Radon transform and its applications	**21**, 217
Bashkin, S.: Beam-foil spectroscopy	**12**, 287

---

[*] Volumes I–XL were previously distinguished by roman rather than by arabic numerals.

Bassett, I.M., W.T. Welford and R. Winston: Nonimaging optics for
    flux concentration   **27**, 161
Beckmann, P.: Scattering of light by rough surfaces   **6**, 53
Bellini, M. and Zavatta, A.: Manipulating light states by single-photon addition
    and subtraction   **55**, 41
Benisty, H. and C. Weisbuch: Photonic crystals   **49**, 177
Beran, M.J. and J. Oz-Vogt: Imaging through turbulence in the atmosphere   **33**, 319
Bernard, J., *see* Orrit, M.   **35**, 61
Berry, M.V. and C. Upstill: Catastrophe optics: morphologies of caustics and
    their diffraction patterns   **18**, 257
Bertero, M. and C. De Mol: Super-resolution by data inversion   **36**, 129
Bertolotti, M., *see* Chumash, V.   **36**, 1
Bertolotti, M., *see* Mihalache, D.   **27**, 227
Bertolotti, M., F. Bovino and C. Sibilia: Quantum state engineering:
    generation of single and pairs of photons   **60**, 1
Beverly III, R.E.: Light emission from high-current surface-spark discharges   **16**, 357
Bhaduri, B., *see* Mir, M.   **57**, 133
Bialynicki-Birula, I.: Photon wave function   **36**, 245
Biener, G., *see* Hasman, E.   **47**, 215
Björk, G., A.B. Klimov and L.L. Sánchez-Soto: The discrete
    Wigner function   **51**, 469
Bloembergen, N.: From millisecond to attosecond laser pulses   **50**, 1
Blomstedt, K., A.T. Friberg and T. Setälä: Classical coherence of blackbody
    radiation   **62**, 293
Bloom, A.L.: Gas lasers and their application to precise length measurements   **9**, 1
Bokor, N. and N. Davidson: Curved diffractive optical elements: Design and
    applications   **48**, 107
Bokor, N., *see* Davidson, N.   **45**, 1
Borghi, R.: Computational optics through sequence transformations   **61**, 1
Bouman, M.A., W.A. Van De Grind and P. Zuidema: Quantum fluctuations in
    vision   **22**, 77
Bousquet, P., *see* Rouard, P.   **4**, 145
Bovino, F., *see* Bertolotti, M.   **60**, 1
Boyd, R.W. and D.J. Gauthier: "Slow" and "fast" light   **43**, 497
Braat, J.J.M., S. van Haver, A.J.E.M. Janssen and P. Dirksen: Assessment
    of optical systems by means of point-spread functions   **51**, 349
Brambilla, E., *see* Gatti, A.   **51**, 251
Brosseau, C., *see* Alfalou, A.   **60**, 119
Brosseau, C. and A. Dogariu: Symmetry properties and polarization descriptors
    for an arbitrary electromagnetic wavefield   **49**, 315
Brosseau, C.: Polarization and coherence optics: Historical perspective, status,
    and future directions   **54**, 149
Brown, G.S., *see* DeSanto, J.A.   **23**, 1
Brown, R., *see* Orrit, M.   **35**, 61
Brown, T. G.: Unconventional Polarization States: Beam Propagation,
    Focusing, and Imaging   **56**, 81
Brunner, W. and H. Paul: Theory of optical parametric amplification and
    oscillation   **15**, 1

Bryngdahl, O.: Applications of shearing interferometry   **4**, 37
Bryngdahl, O.: Evanescent waves in optical imaging   **11**, 167
Bryngdahl, O., T. Scheermesser and F. Wyrowski: Digital halftoning: synthesis of binary images   **33**, 389
Bryngdahl, O. and F. Wyrowski: Digital holography - computer-generated holograms   **28**, 1
Bunning, T., see De Sio L.   **58**, 1
Burch, J.M.: The metrological applications of diffraction gratings   **2**, 73
Butterweck, H.J.: Principles of optical data-processing   **19**, 211
Bužek, V. and P.L. Knight: Quantum interference, superposition states of light, and nonclassical effects   **34**, 1

Cagnac, B., see Giacobino, E.   **17**, 85
Cai, Y., Y. Chen, J. Yu, X. Liu and L. Liu: Generation of partially coherent beams   **62**, 157
Calvo, M.L., see Velasco, A.V.   **59**, 159
Cao, H.: Lasing in disordered media   **45**, 317
Çapoğlu, İ.R., J.D. Rogers, A. Taflove and V. Backman: The Microscope in a Computer: Image Synthesis from Three-Dimensional Full-Vector Solutions of Maxwell's Equations at the Nanometer Scale   **57**, 1
Carmichael, H.J., G.T. Foster, L.A. Orozco, J.E. Reiner and P.R. Rice: Intensity-field correlations of non-classical light   **46**, 355
Carriere, J., R. Narayan, W.-H. Yeh, C. Peng, P. Khulbe, L. Li, R. Anderson, J. Choi and M. Mansuripur: Principles of optical disk data storage   **41**, 97
Cartwright, N.A. and K.E. Oughstun: Precursors and dispersive pulse dynamics, a century after the Sommerfeld-Brillouin Theory: Part I. The Original Theory   **59**, 209
Cartwright, N.A. and K.E. Oughstun: Precursors and dispersive pulse dynamics: a century after the Sommerfeld–Brillouin theory: Part II. the modern asymptotic theory   **60**, 263
Casasent, D. and D. Psaltis: Deformation invariant, space-variant optical pattern recognition   **16**, 289
Cattaneo, S., see Kauranen, M.   **51**, 69
Ceglio, N.M. and D.W. Sweeney: Zone plate coded imaging: theory and applications   **21**, 287
Cerf, N.J. and J. Fiurášek: Optical quantum cloning   **49**, 455
Chang, R.K., see Fields, M.H.   **41**, 1
Charnotskii, M.I., J. Gozani, V.I. Tatarskii and V.U. Zavorotny: Wave propagation theories in random media based on the path-integral approach   **32**, 203
Cheben, P., see Velasco, A.V.   **59**, 159
Chen, J., see Cai, Y.   **62**, 157
Chen, R.T. and Z. Fu: Optical true-time delay control systems for wideband phased array antennas   **41**, 283
Chen, Z., L. Hua and J. Pu: Tight Focusing of Light Beams: Effect of Polarization, Phase, and Coherence   **57**, 219
Chiao, R.Y. and A.M. Steinberg: Tunneling times and superluminality   **37**, 345
Chmelik, R., M. Slaba, V. Kollarova, T. Slaby, M. Lostak, J. Collakova, and Z. Dostal: The role of coherence in image formation in holographic microscopy   **59**, 267

Choi, J., *see* Carriere, J.	41, 97
Christensen, J.L., *see* Rosenblum, W.M.	13, 69
Christov, I.P.: Generation and propagation of ultrashort optical pulses	29, 199
Chumash, V., I. Cojocaru, E. Fazio, F. Michelotti and M. Bertolotti: Nonlinear propagation of strong laser pulses in chalcogenide glass films	36, 1
Clair, J.J. and C.I. Abitbol: Recent advances in phase profiles generation	16, 71
Clarricoats, P.J.B.: Optical fibre waveguides - a review	14, 327
Cohen-Tannoudji, C. and A. Kastler: Optical pumping	5, 1
Cojocaru, I., *see* Chumash, V.	36, 1
Cole, T.W.: Quasi-optical techniques of radio astronomy	15, 187
Collakova, J., *see* Chmelik, R.	59, 267
Colombeau, B., *see* Froehly, C.	20, 63
Cook, R.J.: Quantum jumps	28, 361
Courtès, G., P. Cruvellier and M. Detaille: Some new optical designs for ultraviolet bidimensional detection of astronomical objects	20, 1
Creath, K.: Phase-measurement interferometry techniques	26, 349
Crewe, A.V.: Production of electron probes using a field emission source	11, 223
Crosignani, B., *see* DelRe, E.	53, 153
Crowe, I.F., *see* Roschuck, T.	58, 251
Cruvellier, P., *see* Courtès, G.	20, 1
Cummins, H.Z. and H.L. Swinney: Light beating spectroscopy	8, 133
Dainty, J.C.: The statistics of speckle patterns	14, 1
Dändliker, R.: Heterodyne holographic interferometry	17, 1
Danilishin, S.L., *see* Khalili, F.Ya.	61, 113
Darmanyan, S.A., *see* Abdullaev, F.Kh.	44, 303
Dattoli, G., L. Giannessi, A. Renieri and A. Torre: Theory of Compton free electron lasers	31, 321
Davidson, N. and N. Bokor: Anamorphic beam shaping for laser and diffuse light	45, 1
Davidson, N., *see* Bokor, N.	48, 107
Davidson, N., *see* Oron, R.	42, 325
De Mol, C., *see* Bertero, M.	36, 129
De Sterke, C.M. and J.E. Sipe: Gap solitons	33, 203
De Sio, L., N. Tabiryan, T. Bunning, B.R. Kimball, and C. Umeton: Dynamic Photonic Materials Based on Liquid crystals	58, 1
Decker Jr, J.A., *see* Harwit, M.	12, 101
Delano, E. and R.J. Pegis: Methods of synthesis for dielectric multilayer filters	7, 67
DelRe, E., Crosignani, B. and Di Porto, P.: Photorefractive solitons and their underlying nonlocal physics	53, 153
Demaria, A.J.: Picosecond laser pulses	9, 31
Demkowicz-Dobrzański, R., M. Jarzyna and J. Kołodyński: Quantum limits in optical interferometry	60, 345
Derevyanko, S., *see* Levy, U.	61, 237
DeSanto, J.A. and G.S. Brown: Analytical techniques for multiple scattering from rough surfaces	23, 1
Dennis, M.R., O'Holleran, K. and Padgett, M.J.: Singular optics: Optical vortices and polarization singularities	53, 293

Desyatnikov, A.S., Y.S. Kivshar and L.L. Torner: Optical vortices and vortex
 solitons   **47**, 291
Detaille, M., *see* Courtès, G.   **20**,   1
Dexter, D.L., *see* Smith, D.Y.   **10**, 165
Di Porto, P., *see* DelRe, E.   **53**, 153
Dickey, F.M., *see* Romero, L.A.   **54**, 319
Dirksen, P., *see* Braat, J.J.M.   **51**, 349
Dogariu, A., *see* Brosseau, C.   **49**, 315
Domachuk, P., *see* Eggleton, B.J.   **48**,   1
Dostal, Z., *see* Chmelik, R.   **59**, 267
Dragoman, D.: The Wigner distribution function in optics and optoelectronics   **37**,   1
Dragoman, D.: Phase space correspondence between classical optics and
 quantum mechanics   **43**, 433
Drexhage, K.H.: Interaction of light with monomolecular dye layers   **12**, 163
Duguay, M.A.: The ultrafast optical Kerr shutter   **14**, 161
Dušek M., N. Lütkenhaus and M. Hendrych: Quantum cryptography   **49**, 381
Dutta, N.K. and J.R. Simpson: Optical amplifiers   **31**, 189
Dutta Gupta, S.: Nonlinear optics of stratified media   **38**,   1

Eberly, J.H.: Interaction of very intense light with free electrons   **7**, 359
Eggleton, B.J., P. Domachuk, C. Grillet, E.C. Mägi, H.C. Nguyen, P.
 Steinvurzel and M.J. Steel: Laboratory post-engineering of microstructured
 optical fibers   **48**,   1
Englund, J.C., R.R. Snapp and W.C. Schieve: Fluctuations, instabilities and
 chaos in the laser-driven nonlinear ring cavity   **21**, 355
Ennos, A.E.: Speckle interferometry   **16**, 233
Erez, N., *see* Greenberger, D.M.   **50**, 275
Essiambre, R.-J. and G.P. Agrawal: Soliton communication systems   **37**, 185
Etrich, C., F. Lederer, B.A. Malomed, T. Peschel and U. Peschel: Optical
 solitons in media with a quadratic nonlinearity   **41**, 483
Evers, J., *see* Kiffner, M.   **55**, 85

Fabelinskii, I.L.: Spectra of molecular scattering of light   **37**, 95
Fabre, C., *see* Reynaud, S.   **30**,   1
Facchi, P. and S. Pascazio: Quantum Zeno and inverse quantum Zeno effects   **42**, 147
Fante, R.L.: Wave propagation in random media: a systems approach   **22**, 341
Fazio, E., *see* Chumash, V.   **36**,   1
Fercher, A.F. and C.K. Hitzenberger: Optical coherence tomography   **44**, 215
Ficek, Z. and H.S. Freedhoff: Spectroscopy in polychromatic fields   **40**, 389
Fields, M.H., J. Popp and R.K. Chang: Nonlinear optics in microspheres   **41**,   1
Filip, R.: *see* Andersen, U.L.   **53**, 365
Fiorentini, A.: Dynamic characteristics of visual processes   **1**, 253
Fiurášek, J., *see* Cerf, N.J.   **49**, 455
Florjańczyk, M., *see* Velasco, A.V.   **59**, 159
Flytzanis, C., F. Hache, M.C. Klein, D. Ricard and Ph. Roussignol: Nonlinear
 optics in composite materials. 1. Semiconductor and metal crystallites in
 dielectrics   **29**, 321
Focke, J.: Higher order aberration theory   **4**,   1

Forbes, G.W., see Kravtsov, Yu.A.	**39**, 1
Foster, G.T., see Carmichael, H.J.	**46**, 355
Françon, M. and S. Mallick: Measurement of the second order degree of coherence	**6**, 71
Franta, D., see Ohlídal, I.	**41**, 181
Freedhoff, H.S., see Ficek, Z.	**40**, 389
Freilikher, V.D. and S.A. Gredeskul: Localization of waves in media with one-dimensional disorder	**30**, 137
Friberg, A.T., see Blomstedt, K.	**62**, 293
Friberg, A.T., see Genty, G.	**61**, 71
Friberg, A.T., see Turunen, J.	**54**, 1
Frieden, B.R.: Evaluation, design and extrapolation methods for optical signals, based on use of the prolate functions	**9**, 311
Friesem, A.A., see Oron, R.	**42**, 325
Froehly, C., B. Colombeau and M. Vampouille: Shaping and analysis of picosecond light pulses	**20**, 63
Fry, G.A.: The optical performance of the human eye	**8**, 51
Fu, Z., see Chen, R.T.	**41**, 283
Gabitov, I.R., see Litchinitser, N.M.	**51**, 1
Gabor, D.: Light and information	**1**, 109
Gallion, P., F. Mendieta and S. Jiang: Signal and quantum noise in optical communications and cryptography	**52**, 149
Gamo, H.: Matrix treatment of partial coherence	**3**, 187
Gandjbakhche, A.H. and G.H. Weiss: Random walk and diffusion-like models of photon migration in turbid media	**34**, 333
Gantsog, Ts., see Tanaś, R.	**35**, 355
Gao, W., see Yin,J.	**45**, 119
Garcia-Ojalvo, J., see Uchida, A.	**48**, 203
Garnier, J., see Abdullaev, F.	**48**, 35
Garnier, J., see Abdullaev, F.Kh.	**44**, 303
Gatti, A., E. Brambilla and L. Lugiato: Quantum imaging	**51**, 251
Gauthier, D.J.: Two-photon lasers	**45**, 205
Gauthier, D.J., see Boyd, R.W.	**43**, 497
Gbur, G.: Nonradiating sources and other "invisible" objects	**45**, 273
Gbur, G. and Visser, T.D.: The structure of partially coherent fields	**55**, 285
Gbur, G.: Invisibility Physics: Past, Present, and Future	**58**, 65
Gea-Banacloche, J.: Optical realizations of quantum teleportation	**46**, 311
Genty, G., A.T. Friberg and J. Turunen: Coherence of supercontinuum light	**61**, 71
Ghatak, A. and K. Thyagarajan: Graded index optical waveguides: a review	**18**, 1
Ghatak, A.K., see Sodha, M.S.	**13**, 169
Giacobino, E. and B. Cagnac: Doppler-free multiphoton spectroscopy 17, 85	
Giacobino, E., see Reynaud, S.	**30**, 1
Giannessi, L., see Dattoli, G.	**31**, 321
Ginzburg, V.L.: Radiation by uniformly moving sources. Vavilov-Cherenkov effect, Doppler effect in a medium, transition radiation and associated phenomena	**32**, 267
Ginzburg,V.L., see Agranovich, V.M.	**9**, 235

Giovanelli, R.G.: Diffusion through non-uniform media   2, 109
Glaser, I.: Information processing with spatially incoherent light   24, 389
Glesk, I., B.C. Wang, L. Xu, V. Baby and P.R. Prucnal: Ultra-fast all-optical switching in optical networks   45, 53
Gniadek, K. and J. Petykiewicz: Applications of optical methods in the diffraction theory of elastic waves   9, 281
Goodman, J.W.: Synthetic-aperture optics   8, 1
Gozani, J., see Charnotskii, M.I.   32, 203
Graham, R.: The phase transition concept and coherence in atomic emission   12, 233
Gredeskul, S.A., see Freilikher, V.D.   30, 137
Greenberger, D.M., N. Erez, M.O. Scully, A.A. Svidzinsky and M.S. Zubairy: Planck, photon statistics, and Bose-Einstein condensation   50, 275
Grillet, C., see Eggleton, B.J.   48, 1

Hache, F., see Flytzanis, C.   29, 321
Hall, D.G.: Optical waveguide diffraction gratings: coupling between guided modes   29, 1
Halsall, M.P., see Roschuk T.   58, 251
Hariharan, P.: Colour holography   20, 263
Hariharan, P.: Interferometry with lasers   24, 103
Hariharan, P.: The geometric phase   48, 149
Hariharan, P. and B.C. Sanders: Quantum phenomena in optical interferometry   36, 49
Harwit, M. and J.A. Decker Jr: Modulation techniques in spectrometry   12, 101
Hasegawa, A., see Kodama, Y.   30, 205
Hasman, E., G. Biener, A. Niv and V. Kleiner: Space-variant polarization manipulation   47, 215
Hasman, E., see Oron, R.   42, 325
Haus, J.W., see Sakoda, K.   54, 271
He, G.S., Stimulated scattering effects of intense coherent light   53, 201
Heidmann, A., see Reynaud, S.   30, 1
Hello, P.: Optical aspects of interferometric gravitational-wave detectors   38, 85
Helstrom, C.W.: Quantum detection theory   10, 289
Hendrych, M., see Dušek, M.   49, 381
Herriot, D.R.: Some applications of lasers to interferometry   6, 171
Herzig, H.P., see Kim, M.S.   58, 115
Hess, O., see Wuestner, S.   59, 1
Hitzenberger, C.K., see Fercher, A.F.   44, 215
Horner, J.L., see Javidi, B.   38, 343
Hua, L., see Chen, Z.   57, 219
Huang, T.S.: Bandwidth compression of optical images   10, 1

Ichioka, Y., see Tanida, J.   40, 77
Imoto, N., see Yamamoto, Y.   28, 87
Ishii, Y.: Laser-diode interferometry   46, 243
Itoh, K.: Interferometric multispectral imaging   35, 145
Iwata, K.: Phase imaging and refractive index tomography for X-rays and visible rays   47, 393

Jacobsson, R.: Light reflection from films of continuously varying refractive
  index **5**, 247
Jacquinot, P. and B. Roizen-Dossier: Apodisation **3**, 29
Jacquod, Ph., *see* Türeci, H.E. **47**, 75
Jaeger, G. and A.V. Sergienko: Multi-photon quantum interferometry **42**, 277
Jahns, J.: Free-space optical digital computing and interconnection **38**, 419
Jamroz, W. and B.P. Stoicheff: Generation of tunable coherent vacuum-
  ultraviolet radiation **20**, 325
Janssen, A.J.E.M., *see* Braat, J.J.M. **51**, 349
Jarzyna, M., *see* Demkowicz-Dobrzański, R. **60**, 345
Javidi, B. and J.L. Horner: Pattern recognition with nonlinear techniques in the
  Fourier domain **38**, 343
Jesús Lancis, *see* Victor Torres-Company **56**, 1
Jiang, S., *see* Gallion, P. **52**, 149
Jones, D.G.C., *see* Allen, L. **9**, 179
Joshi, A. and M. Xiao: Controlling nonlinear optical processes in multi-level
  atomic systems **49**, 97
Juan P. Torres, K. Banaszek, and I. A. Walmsley: Engineering nonlinear optic
  sources of photonic entanglement **56**, 227

Kapale, K.T.: Subwavelength Atom Localization **58**, 199
Kartashov, Y.V., V.A. Vysloukh and L. Torner: Soliton shape and mobility
  control in optical lattices **52**, 63
Kastler, A., *see* Cohen-Tannoudji, C. **5**, 1
Kauranen, M. and S. Cattaneo: Polarization techniques for surface nonlinear
  optics **51**, 69
Keitel, C.H., *see* Kiffner, M. **55**, 85
Keller, O.: Local fields in linear and nonlinear optics of mesoscopic systems **37**, 257
Keller, O.: Optical works of L.V. Lorenz **43**, 195
Keller, O.: Historical papers on the particle concept of light **50**, 51
Keller, U.: Ultrafast solid-state lasers **46**, 1
Kemp, B.A.: Macroscopic theory of optical momentum **60**, 437
Khalili, F.Ya. and S.L. Danilishin: Quantum optomechanics **61**, 113
Khoo, I.C.: Nonlinear optics of liquid crystals **26**, 105
Khulbe, P., *see* Carriere, J. **41**, 97
Kielich, S.: Multi-photon scattering molecular spectroscopy **20**, 155
Kiffner, M., Macovei, M., Evers, J. and Keitel, C.H.: Vacuum-induced
  processes in multilevel atoms **55**, 85
Kilin, S., *see* Mogilevtsev, D. **54**, 89
Kilin, S.Ya.: Quanta and information **42**, 1
Kimball, B.R., *see* De Sio L. **58**, 1
Kinosita, K.: Surface deterioration of optical glasses **4**, 85
Kim, M.S., T. Scharf, C. Rockstuhl, H.P. Herzig: Phase Anomalies in Micro-
  Optics **58**, 115
Kitagawa, M., *see* Yamamoto, Y. **28**, 87
Kivshar, Y.S., *see* Desyatnikov, A.S. **47**, 291
Kivshar, Y.S., *see* Saltiel, S.M. **47**, 1
Klein, M.C., *see* Flytzanis, C. **29**, 321

Kleiner, V., see Hasman, E.	47, 215
Klimov, A.B., see Björk, G.	51, 469
Klyatskin, V.I.: The imbedding method in statistical boundary-value wave problems	33, 1
Knight, P.L., see Bužek, V.	34, 1
Knights, A.P., see Roschuk, T.	58, 251
Kodama, Y. and A. Hasegawa: Theoretical foundation of optical-soliton concept in fibers	30, 205
Kollarova, V., see Chmelik, R.	59, 267
Kołodyński, J., see Demkowicz-Dobrzański, R.	60, 345
Koppelman, G.: Multiple-beam interference and natural modes in open resonators	7, 1
Kottler, F.: The elements of radiative transfer	3, 1
Kottler, F.: Diffraction at a black screen, Part I: Kirchhoff's theory	4, 281
Kottler, F.: Diffraction at a black screen, Part II: electromagnetic theory	6, 331
Kowalewska-Kudlaszyk, A. see W. Leoński	56, 131
Kozhekin, A.E., see Kurizki, G.	42, 93
Kravtsov, Yu.A.: Rays and caustics as physical objects	26, 227
Kravtsov, Yu.A. and L.A. Apresyan: Radiative transfer: new aspects of the old theory	36, 179
Kravtsov, Yu.A., G.W. Forbes and A.A. Asatryan: Theory and applications of complex rays	39, 1
Kravtsov, Yu.A., see Barabanenkov, Yu.N.	29, 65
Kroisova, D.: Microstructures and Nanostructures in Nature	57, 93
Kubota, H.: Interference color	1, 211
Kuittinen, M., see Turunen, J.	40, 343
Kurizki, G., A.E. Kozhekin, T. Opatrný and B.A. Malomed: Optical solitons in periodic media with resonant and off-resonant nonlinearities	42, 93
Labeyrie, A.: High-resolution techniques in optical astronomy	14, 47
Lakhtakia, A., see Mackay, T.G.	51, 121
Lean, E.G.: Interaction of light and acoustic surface waves	11, 123
Lederer, F., see Etrich, C.	41, 483
Lee, W.-H.: Computer-generated holograms: techniques and applications	16, 119
Leith, E.N. and Upatnieks, J.: Recent advances in holography	6, 1
Leonhardt, U. and Philbin, T.G.: Transformation optics and the geometry of light	53, 69
Leoński, W., and A. Kowalewska-Kudlaszyk: Quantum scissors finite-dimensional states engineering	56, 131
Letokhov, V.S.: Laser selective photophysics and photochemistry	16, 1
Leuchs, G., see Sizmann, A.	39, 373
Levi, L.: Vision in communication	8, 343
Levy, U., S. Derevyanko and Y. Silberberg: Light modes of free space	61, 237
Li, G.: Adaptive lens	55, 199
Li, L., see Carriere, J.	41, 97
Lipson, H. and C.A. Taylor: X-ray crystal-structure determination as a branch of physical optics	5, 287

Litchinitser, N.M., I.R. Gabitov, A.I. Maimistov and V.M. Shalaev: Negative refractive index metamaterials in optics   **51**, 1
Liu, L., see Cai, Y.   **62**, 157
Liu, X., see Cai, Y.   **62**, 157
Lohmann, A.W., D. Mendlovic and Z. Zalevsky: Fractional transformations in optics   **38**, 263
Lohmann, A.W., see Zalevsky, Z.   **40**, 271
Lostak, M., see Chmelik, R.   **59**, 267
Lounis, B., see Orrit, M.   **35**, 61
Lugiato, L., see Gatti, A.   **51**, 251
Lugiato, L.A.: Theory of optical bistability   **21**, 69
Luis, A.: Polarization in quantum optics   **61**, 283
Luis, A. and L.L. Sánchez-Soto: Quantum phase difference, phase measurements and Stokes operators   **41**, 419
Lukš, A. and V. Peřinová: Canonical quantum description of light propagation in dielectric media   **43**, 295
Lukrš, A., see Peřinová, V.   **33**, 129
Lukrš, A., see Peřinová, V.   **40**, 115
Lütkenhaus, N., see Dušek, M.   **49**, 381

Machida, S., see Yamamoto, Y.   **28**, 87
Mackay, T.G. and A. Lakhtakia: Electromagnetic fields in linear bianisotropic mediums   **51**, 121
Macovei, M., see Kiffner, M.   **55**, 85
Mägi, E.C., see Eggleton, B.J.   **48**, 1
Mahajan, V.N.: Gaussian apodization and beam propagation   **49**, 1
Maimistov, A.I., see Litchinitser, N.M.   **51**, 1
Mainfray, G. and C. Manus: Nonlinear processes in atoms and in weakly relativistic plasmas   **32**, 313
Malacara, D.: Optical and electronic processing of medical images   **22**, 1
Malacara, D., see Vlad, V.I.   **33**, 261
Malacara-Doblado, D., see Malacara-Hernández, D.   **62**, 73
Malacara-Hernández, D. and D. Malacara-Doblado: Optical testing and interferometry   **62**, 73
Mallick, S., see Francon, M.   **6**, 71
Malomed, B.A.: Variational methods in nonlinear fiber optics and related fields   **43**, 71
Malomed, B.A., see Etrich, C.   **41**, 483
Malomed, B.A., see Kurizki, G.   **42**, 93
Mandel, L.: Fluctuations of light beams   **2**, 181
Mandel, L.: The case for and against semiclassical radiation theory   **13**, 27
Mandel, P., see Abraham, N.B.   **25**, 1
Mansuripur, M., see Carriere, J.   **41**, 97
Manus, C., see Mainfray, G.   **32**, 313
Maradudin, A.A., see Shchegrov, A.V.   **46**, 117
Marchand, E.W.: Gradient index lenses   **11**, 305
Maret, G., see Aegerter, C.M.   **52**, 1
Maria Chekhova: Polarization and Spectral Properties of Biphotons   **56**, 187

Martin, P.J. and R.P. Netterfield: Optical films produced by ion-based techniques	**23**, 113
Martínez-Corral, M. and Saavedra, G.: The resolution challenge in 3D optical microscopy	**53**, 1
Masalov, A.V.: Spectral and temporal fluctuations of broad-band laser radiation	**22**, 145
Maystre, D.: Rigorous vector theories of diffraction gratings	**21**, 1
Meessen, A., *see* Rouard, P.	**15**, 77
Mehta, C.L.: Theory of photoelectron counting	**8**, 373
Méndez, E.R., *see* Shchegrov, A.V.	**46**, 117
Mendieta, F., *see* Gallion, P.	**52**, 149
Mendlovic, D., *see* Lohmann, A.W.	**38**, 263
Mendlovic, D., *see* Zalevsky, Z.	**40**, 271
Meystre, P.: Cavity quantum optics and the quantum measurement process	**30**, 261
Meystre, P., *see* Search, C.P.	**47**, 139
Michelotti, F., *see* Chumash, V.	**36**, 1
Mihalache, D., M. Bertolotti and C. Sibilia: Nonlinear wave propagation in planar structures	**27**, 227
Mikaelian, A.L.: Self-focusing media with variable index of refraction	**17**, 279
Mikaelian, A.L. and M.L. Ter-Mikaelian: Quasi-classical theory of laser radiation	**7**, 231
Mills, D.L. and K.R. Subbaswamy: Surface and size effects on the light scattering spectra of solids	**19**, 45
Milonni, P.W.: Field quantization in optics	**50**, 97
Milonni, P.W. and B. Sundaram: Atoms in strong fields: photoionization and chaos	**31**, 1
Mir, M.: Quantitative Phase Imaging	**57**, 133
Miranowicz, A., *see* Tanaś, R.	**35**, 355
Miyamoto, K.: Wave optics and geometrical optics in optical design	**1**, 31
Mogilevtsev, D. and Kilin, S.: Theoretical tools for quantum optics in structured media	**54**, 89
Mollow, B.R.: Theory of intensity dependent resonance light scattering and resonance fluorescence	**19**, 1
Murata, K.: Instruments for the measuring of optical transfer functions	**5**, 199
Musset, A. and A. Thelen: Multilayer antireflection coatings	**8**, 201
Nakwaski, W. and M. Osiński: Thermal properties of vertical-cavity surface-emitting semiconductor lasers	**38**, 165
Narayan, R., *see* Carriere, J.	**41**, 97
Narducci, L.M., *see* Abraham, N.B.	**25**, 1
Navrátil, K., *see* Ohlídal, I.	**34**, 249
Netterfield, R.P., *see* Martin, P.J.	**23**, 113
Nguyen, H.C., *see* Eggleton, B.J.	**48**, 1
Nishihara, H. and T. Suhara: Micro Fresnel lenses	**24**, 1
Niv, A., *see* Hasman, E.	**47**, 215
Noethe, L.: Active optics in modern large optical telescopes	**43**, 1
Novotny, L.: The history of near-field optics	**50**, 137
Nussenzveig, H.M.: Light tunneling	**50**, 185

Obod, Yu.A., *see* Shvartsburg, A.B.	**60**, 489
Ohlídal, I. and D. Franta: Ellipsometry of thin film systems	**41**, 181
Ohlídal, I., K. Navrátil and M. Ohlídal: Scattering of light from multilayer systems with rough boundaries	**34**, 249
Ohlídal, M., *see* Ohlídal, I.	**34**, 249
O'Holleran, K., *see* Dennis, M.R.	**53**, 293
Ohtsu, M. and T. Tako: Coherence in semiconductor lasers	**25**, 191
Ohtsubo, J.: Chaotic dynamics in semiconductor lasers with optical feedback	**44**, 1
Okamoto, T. and T. Asakura: The statistics of dynamic speckles	**34**, 183
Okoshi, T.: Projection-type holography	**15**, 139
Omenetto, F.G.: Femtosecond pulses in optical fibers	**44**, 85
Ooue, S.: The photographic image	**7**, 299
Opatrný, T., *see* Kurizki, G.	**42**, 93
Opatrný, T., *see* Welsch, D.-G.	**39**, 63
Oron, R., N. Davidson, A.A. Friesem and E. Hasman: Transverse mode shaping and selection in laser resonators	**42**, 325
Orozco, L.A., *see* Carmichael, H.J.	**46**, 355
Orrit, M., J. Bernard, R. Brown and B. Lounis: Optical spectroscopy of single molecules in solids	**35**, 61
Osiński, M., *see* Nakwaski, W.	**38**, 165
Ostrovskaya, G.V. and Yu.I. Ostrovsky: Holographic methods of plasma diagnostics	**22**, 197
Ostrovsky, Yu.I. and V.P. Shchepinov: Correlation holographic and speckle interferometry	**30**, 87
Ostrovsky,Yu.I., *see* Ostrovskaya, G.V.	**22**, 197
Oughstun, K.E.: Unstable resonator modes	**24**, 165
Oughstun, K.E., *see* Cartwright, N.A.	**59**, 209
Oughstun, K.E., *see* Cartwright, N.A.	**60**, 263
Oz-Vogt, J., *see* Beran, M.J.	**33**, 319
Ozrin,V.D., *see* Barabanenkov, Yu.N.	**29**, 65
Padgett, M.J., *see* Allen, L.	**39**, 291
Padgett, M.J., *see* Dennis, M.R.	**53**, 293
Pal, B.P.: Guided-wave optics on silicon: physics, technology and status	**32**, 1
Paoletti, D. and G. Schirripa Spagnolo: Interferometric methods for artwork diagnostics	**35**, 197
Pascazio, S., *see* Facchi, P.	**42**, 147
Patorski, K.: The self-imaging phenomenon and its applications	**27**, 1
Paul, H., *see* Brunner, W.	**15**, 1
Pedro Andrés, *see* Víctor Torres-Company	**56**, 1
Pegis, R.J.: The modern development of Hamiltonian optics	**1**, 1
Pegis, R.J., *see* Delano, E.	**7**, 67
Peiponen, K.-E., E.M. Vartiainen and T. Asakura: Dispersion relations and phase retrieval in optical spectroscopy	**37**, 57
Peng, C., *see* Carriere, J.	**41**, 97
Peřina Jr, J.: Spontaneous parametric down-conversion in nonlinear layered structures	**59**, 89
Peřina Jr, J. and J. Peřina: Quantum statistics of nonlinear optical couplers	**41**, 359

Peřina, J.: Photocount statistics of radiation propagating through random and
  nonlinear media   **18**, 127
Peřina, J., *see* Peřina Jr, J.   **41**, 359
Peřinová, V. and A. Lukš: Quantum statistics of dissipative nonlinear oscillators   **33**, 129
Peřinová, V. and A. Lukš: Continuous measurements in quantum optics   **40**, 115
Peřinová, V., *see* Lukš, A.   **43**, 295
Pershan, P.S.: Non-linear optics   **5**, 83
Peschel, T., *see* Etrich, C.   **41**, 483
Peschel, U., *see* Etrich, C.   **41**, 483
Petite, G., *see* Shvartsburg, A.B.   **44**, 143
Petykiewicz, J., *see* Gniadek, K.   **9**, 281
Philbin, T.G., *see* Leonhardt, U.   **53**, 69
Picht, J.: The wave of a moving classical electron   **5**, 351
Pollock, C.R.: Ultrafast optical pulses   **51**, 211
Popescu, G., *see* Mir, M.   **57**, 133
Popov, E.: Light diffraction by relief gratings: a macroscopic and microscopic view **31**, 139
Popp, J., *see* Fields, M.H.   **41**, 1
Porter, R.P.: Generalized holography with application to inverse scattering and
  inverse source problems   **27**, 315
Premaratne, M.: Optical pulse propagation in biological media: theory and
  numerical methods   **55**, 1
Presnyakov, L.P.: Wave propagation in inhomogeneous media: phase-shift
  approach   **34**, 159
Prucnal, P.R., *see* Glesk, I.   **45**, 53
Pryde, G.J., *see* Ralph, T.C.   **54**, 209
Psaltis, D. and Y. Qiao: Adaptive multilayer optical networks   **31**, 227
Psaltis, D., *see* Casasent, D.   **16**, 289
Pu, J., *see* Chen, Z.   **57**, 219

Qiao, Y., *see* Psaltis, D.   **31**, 227
Qiu, M., *see* Yan, M.   **52**, 261

Ralph, T.C. and Pryde, G.J.: Optical quantum computation   **54**, 209
Raymer, M.G. and I.A. Walmsley: The quantum coherence properties of
  stimulated Raman scattering   **28**, 181
Reiner, J.E., *see* Carmichael, H.J.   **46**, 355
Renieri, A., *see* Dattoli, G.   **31**, 321
Reynaud, S., A. Heidmann, E. Giacobino and C. Fabre: Quantum fluctuations
  in optical systems   **30**, 1
Ricard, D., *see* Flytzanis, C.   **29**, 321
Rice, P.R., *see* Carmichael, H.J.   **46**, 355
Riseberg, L.A. and M.J. Weber: Relaxation phenomena in rare-earth
  luminescence   **14**, 89
Risken, H.: Statistical properties of laser light   **8**, 239
Rockstuhl, C., *see* Kim, M.S.   **58**, 115
Roddier, F.: The effects of atmospheric turbulence in optical astronomy   **19**, 281
Rogers, J.D., *see* Çapoğlu, İ.R.   **57**, 1
Rogister, F., *see* Uchida, A.   **48**, 203

Roizen-Dossier, B., *see* Jacquinot, P.	**3**, 29
Romero, L.A. and Dickey, F.M.: The mathematical theory of laser beam-splitting gratings	**54**, 319
Ronchi, L., *see* Wang Shaomin	**25**, 279
Rosanov, N.N.: Transverse patterns in wide-aperture nonlinear optical systems	**35**, 1
Roschuk, T., I.F. Crowe, A.P. Knights, M.P. Halsall: Low-Dimensional Silicon Structures for Use in Photonic Circuits	**58**, 251
Rosenblum, W.M. and J.L. Christensen: Objective and subjective spherical aberration measurements of the human eye	**13**, 69
Rothberg, L.: Dephasing-induced coherent phenomena	**24**, 39
Rouard, P. and P. Bousquet: Optical constants of thin films	**4**, 145
Rouard, P. and A. Meessen: Optical properties of thin metal films	**15**, 77
Roussignol, Ph., *see* Flytzanis, C.	**29**, 321
Roy, R., *see* Uchida, A.	**48**, 203
Rubinowicz, A.: The Miyamoto-Wolf diffraction wave	**4**, 199
Rudolph, D., *see* Schmahl, G.	**14**, 195
Saavedra, G., *see* Martínez-Corral, M.	**53**, 1
Saichev, A.I., *see* Barabanenkov, Yu.N.	**29**, 65
Saito, S., *see* Yamamoto, Y.	**28**, 87
Sakai, H., *see* Vanasse, G.A.	**6**, 259
Sakoda, K. and Haus, J.W.: Science and engineering of photonic crystals	**54**, 271
Saleh, B.E.A., *see* Teich, M.C.	**26**, 1
Saltiel, S.M., A.A. Sukhorukov and Y.S. Kivshar: Multistep parametric processes in nonlinear optics	**47**, 1
Sánchez-Soto, L.L., *see* Björk, G.	**51**, 469
Sánchez-Soto, L.L., *see* Luis, A.	**41**, 419
Sanders, B.C., *see* Hariharan, P.	**36**, 49
Scharf, T., *see* Kim, M.S.,	**58**, 115
Scheermesser, T., *see* Bryngdahl, O.	**33**, 389
Schieve, W.C., *see* Englund, J.C.	**21**, 355
Schirripa Spagnolo, G., *see* Paoletti, D.	**35**, 197
Schmahl, G. and D. Rudolph: Holographic diffraction gratings	**14**, 195
Schubert, M. and B. Wilhelmi: The mutual dependence between coherence properties of light and nonlinear optical processes	**17**, 163
Schulz, G.: Aspheric surfaces	**25**, 349
Schulz, G. and J. Schwider: Interferometric testing of smooth surfaces	**13**, 93
Schwefel, H.G.L., *see* Türeci, H.E.	**47**, 75
Schwider, J.: Advanced evaluation techniques in interferometry	**28**, 271
Schwider, J., *see* Schulz, G.	**13**, 93
Scully, M.O. and K.G. Whitney: Tools of theoretical quantum optics	**10**, 89
Scully, M.O., *see* Greenberger, D.M.	**50**, 275
Search, C.P. and P. Meystre: Nonlinear and quantum optics of atomic and molecular fields	**47**, 139
Senitzky, I.R.: Semiclassical radiation theory within a quantum-mechanical framework	**16**, 413
Sergienko, A.V., *see* Jaeger, G.	**42**, 277
Setälä, A.T., *see* Blomstedt, K.	**62**, 293

Shalaev, V.M., *see* Litchinitser, N.M.	**51**, 1
Sharma, S.K. and D.J. Somerford: Scattering of light in the eikonal approximation	**39**, 213
Shchegrov, A.V., A.A. Maradudin and E.R. Méndez: Multiple scattering of light from randomly rough surfaces	**46**, 117
Shchepinov, V.P., *see* Ostrovsky, Yu.I.	**30**, 87
Shirai, T.: Modern aspects of intensity interferometry with classical light	**62**, 1
Shvartsburg, A.B. and G. Petite: Instantaneous optics of ultrashort broadband pulses and rapidly varying media	**44**, 143
Shvartsburg, A.B., Yu.A. Obod and O.D. Volpian: Tunneling of electromagnetic waves in all-dielectric gradient metamaterials	**60**, 489
Sibilia, C., *see* Bertolotti, M.	**60**, 1
Sibilia, C., *see* Mihalache, D.	**27**, 227
Silberberg, Y., *see* Levy, U.	**61**, 237
Simpson, J.R., *see* Dutta, N.K.	**31**, 189
Sipe, J.E., *see* De Sterke, C.M.	**33**, 203
Sipe, J.E., *see* Van Kranendonk, J.	**15**, 245
Sittig, E.K.: Elastooptic light modulation and deflection	**10**, 229
Sizmann, A. and G. Leuchs: The optical Kerr effect and quantum optics in fibers	**39**, 373
Slaba, M., *see* Chmelik, R.	**59**, 267
Slaby, T., *see* Chmelik, R.	**59**, 267
Slusher, R.E.: Self-induced transparency	**12**, 53
Smith, A.W., *see* Armstrong, J.A.	**6**, 211
Smith, D.Y. and D.L. Dexter: Optical absorption strength of defects in insulators	**10**, 165
Smith, R.W.: The use of image tubes as shutters	**10**, 45
Snapp, R.R., *see* Englund, J.C.	**21**, 355
Sodha, M.S., A.K. Ghatak and V.K. Tripathi: Self-focusing of laser beams in plasmas and semiconductors	**13**, 169
Somerford, D.J., *see* Sharma, S.K.	**39**, 213
Soroko, L.M.: Axicons and meso-optical imaging devices	**27**, 109
Soskin, M.S. and M.V. Vasnetsov: Singular optics	**42**, 219
Spreeuw, R.J.C. and J.P. Woerdman: Optical atoms	**31**, 263
Steel, M.J., *see* Eggleton, B.J.	**48**, 1
Steel, W.H.: Two-beam interferometry	**5**, 145
Steinberg, A.M., *see* Chiao, R.Y.	**37**, 345
Steinvurzel, P., *see* Eggleton, B.J.	**48**, 1
Stoicheff, B.P., *see* Jamroz, W.	**20**, 325
Stone, A.D., *see* Türeci, H.E.	**47**, 75
Strohbehn, J.W.: Optical propagation through the turbulent atmosphere	**9**, 73
Stroke, G.W.: Ruling, testing and use of optical gratings for high-resolution spectroscopy	**2**, 1
Subbaswamy, K.R., *see* Mills, D.L.	**19**, 45
Suhara, T., *see* Nishihara, H.	**24**, 1
Sukhorukov, A.A., *see* Saltiel, S.M.	**47**, 1
Sundaram, B., *see* Milonni, P.W.	**31**, 1
Svelto, O.: Self-focusing, self-trapping, and self-phase modulation of laser beams	**12**, 1
Svidzinsky, A.A., *see* Greenberger, D.M.	**50**, 275
Sweeney, D.W., *see* Ceglio, N.M.	**21**, 287
Swinney, H.L., *see* Cummins, H.Z.	**8**, 133

Tabiryan, N., *see* De Sio L	
Taflove, A., *see* Çapoğlu, İ.R.	**57**, 1
Tako, T., *see* Ohtsu, M.	**25**, 191
Tanaka, K.: Paraxial theory in optical design in terms of Gaussian brackets	**23**, 63
Tanaś, R., A. Miranowicz and Ts. Gantsog: Quantum phase properties of nonlinear optical phenomena	**35**, 355
Tango, W.J. and R.Q. Twiss: Michelson stellar interferometry	**17**, 239
Tanida, J. and Y. Ichioka: Digital optical computing	**40**, 77
Tatarskii, V.I. and V.U. Zavorotnyi: Strong fluctuations in light propagation in a randomly inhomogeneous medium	**18**, 204
Tatarskii,V.I., *see* Charnotskii, M.I.	**32**, 203
Taylor, C.A., *see* Lipson, H.	**5**, 287
Teich, M.C. and B.E.A. Saleh: Photon bunching and antibunching	**26**, 1
Ter-Mikaelian, M.L., *see* Mikaelian, A.L.	**7**, 231
Thelen, A., *see* Musset, A.	**8**, 201
Thompson, B.J.: Image formation with partially coherent light	**7**, 169
Thyagarajan, K., *see* Ghatak, A.	**18**, 1
Tonomura, A.: Electron holography	**23**, 183
Torner, L., *see* Kartashov, Y.V.	**52**, 63
Torner, L.L., *see* Desyatnikov, A.S.	**47**, 291
Torre, A.: The fractional Fourier transform and some of its applications to optics	**43**, 531
Torre, A., *see* Dattoli, G.	**31**, 321
Tripathi,V.K., see Sodha, M.S.	**13**, 169
Tsujiuchi, J.: Correction of optical images by compensation of aberrations and by spatial frequency filtering	**2**, 131
Türeci, H.E., H.G.L. Schwefel, Ph. Jacquod and A.D. Stone: Modes of wave-chaotic dielectric resonators	**47**, 75
Turunen, J., *see* Genty, G.	**61**, 71
Turunen, J., M. Kuittinen and F. Wyrowski: Diffractive optics: electromagnetic approach	**40**, 343
Turunen, J. and Friberg, A.T.: Propagation-invariant optical fields	**54**, 1
Twiss, R.Q., *see* Tango, W.J.	**17**, 239
Uchida, A., F. Rogister, J. García-Ojalvo and R. Roy: Synchronization and communication with chaotic laser systems	**48**, 203
Umeton, C., *see* De Sio L.	**58**, 1
Upatnieks, J., *see* Leith, E.N.	**6**, 1
Upstill, C., see Berry, M.V.	**18**, 257
Ushioda, S.: Light scattering spectroscopy of surface electromagnetic waves in solids	**19**, 139
Vampouille, M., *see* Froehly, C.	**20**, 63
Van De Grind, W.A., *see* Bouman, M.A.	**22**, 77
van Haver, S., *see* Braat, J.J.M.	**51**, 349
Van Heel, A.C.S.: Modern alignment devices	**1**, 289
Van Kranendonk, J. and J.E. Sipe: Foundations of the macroscopic electromagnetic theory of dielectric media	**15**, 245

Vanasse, G.A. and H. Sakai: Fourier spectroscopy — **6**, 259
Vartiainen, E.M., *see* Peiponen, K.-E. — **37**, 57
Vasnetsov, M.V., *see* Soskin, M.S. — **42**, 219
Velasco, A.V., P. Cheben, M. Florjańczyk, and M.L. Calvo: Spatial heterodyne fourier-transform waveguide spectrometers — **59**, 159
Vernier, P.J.: Photoemission — **14**, 245
Víctor Torres-Company, Jesús Lancis, and Pedro Andrés: Space-time analogies in optics — **56**, 1
Visser, T.D., *see* Gbur, G. — **55**, 285
Vlad, V.I. and D. Malacara: Direct spatial reconstruction of optical phase from phase-modulated images — **33**, 261
Vogel, W., see Welsch, D.-G. — **39**, 63
Volpian, O.D., *see* Shvartsburg, A.B. — **60**, 489
Vysloukh,V.A., *see* Kartashov, Y.V. — **52**, 63

Walmsley, I.A., *see* Raymer, M.G. — **28**, 181
Walmsley, I. A., *see* Juan P. Torres — **56**, 227
Wang Shaomin, and L. Ronchi: Principles and design of optical arrays — **25**, 279
Wang, B.C., *see* Glesk, I. — **45**, 53
Wang, T., *see* Zhao, D. — **57**, 261
Weber, M.J., *see* Riseberg, L.A. — **14**, 89
Weigelt, G.: Triple-correlation imaging in optical astronomy — **29**, 293
Weisbuch, C., *see* Benisty, H. — **49**, 177
Weiss, G.H., *see* Gandjbakhche, A.H. — **34**, 333
Welford, W.T.: Aberration theory of gratings and grating mountings — **4**, 241
Welford, W.T.: Aplanatism and isoplanatism — **13**, 267
Welford, W.T., see Bassett, I.M. — **27**, 161
Welsch, D.-G., W. Vogel and T. Opatrný: Homodyne detection and quantum-state reconstruction — **39**, 63
Whitney, K.G., see Scully, M.O. — **10**, 89
Wilhelmi, B., *see* Schubert, M. — **17**, 163
Winston, R., *see* Bassett, I.M. — **27**, 161
Woerdman, J.P., *see* Spreeuw, R.J.C. — **31**, 263
Wolf, E.: The influence of Young's interference experiment on the development of statistical optics — **50**, 251
Wolf, K.B.: Optical models and symmetries — **62**, 225
Woliński, T.R.: Polarimetric optical fibers and sensors — **40**, 1
Wolter, H.: On basic analogies and principal differences between optical and electronic information — **1**, 155
Wuestner, S. and O. Hess: Active optical metamaterials — **59**, 1
Wynne, C.G.: Field correctors for astronomical telescopes — **10**, 137
Wyrowski, F., *see* Bryngdahl, O. — **28**, 1
Wyrowski, F., *see* Bryngdahl, O. — **33**, 389
Wyrowski, F., *see* Turunen, J. — **40**, 343

Xiao, M., *see* Joshi, A. — **49**, 97
Xu, L., *see* Glesk, I. — **45**, 53

Yan, M., W. Yan and M. Qiu: Invisibility cloaking by coordinate transformation **52**, 261
Yan, W., see Yan, M. **52**, 261
Yamaguchi, I.: Fringe formations in deformation and vibration measurements using laser light **22**, 271
Yamaji, K.: Design of zoom lenses **6**, 105
Yamamoto, T.: Coherence theory of source-size compensation in interference microscopy **8**, 295
Yamamoto, Y., S. Machida, S. Saito, N. Imoto, T. Yanagawa and M. Kitagawa: Quantum mechanical limit in optical precision measurement and communication **28**, 87
Yanagawa, T., see Yamamoto, Y. **28**, 87
Yaroslavsky, L.P.: The theory of optimal methods for localization of objects in pictures **32**, 145
Yeh, W.-H., see Carriere, J. **41**, 97
Yin, J., W. Gao and Y. Zhu: Generation of dark hollow beams and their applications **45**, 119
Yoshinaga, H.: Recent developments in far infrared spectroscopic techniques **11**, 77
Yu, F.T.S.: Principles of optical processing with partially coherent light **23**, 221
Yu, F.T.S.: Optical neural networks: architecture, design and models **32**, 61
Yu, J., see Cai, Y. **62**, 157

Zalevsky, Z., D. Mendlovic and A.W. Lohmann: Optical systems with improved resolving power **40**, 271
Zalevsky, Z., see Lohmann, A.W. **38**, 263
Zavatta, A., see Bellini, M. **55**, 41
Zavorotny, V.U., see Charnotskii, M.I. **32**, 203
Zavorotnyi, V.U., see Tatarskii, V.I. **18**, 204
Zhao, D. and T. Wang: Direct and Inverse Problems in the Theory of Light Scattering **57**, 261
Zhu, R., see Mir, M. **57**, 133
Zhu, Y., see Yin, J. **45**, 119
Zubairy, M.S., see Greenberger, D.M. **50**, 275
Zuidema, P., see Bouman, M.A. **22**, 77

CPI Antony Rowe
Chippenham, UK
2017-04-07 21:04